Recent Advances in
the Aerospace Sciences

Professor Luigi Crocco

Recent Advances in the Aerospace Sciences

In Honor of Luigi Crocco
on His Seventy-fifth Birthday

Edited by

Corrado Casci

Chairman, Department of Energetics
Polytechnic of Milan
Milan, Italy

Assistant Editor

Claudio Bruno

National Research Council, CNPM
Milan, Italy

PLENUM PRESS • NEW YORK AND LONDON

Library of Congress Cataloging in Publication Data

Main entry under title:

Recent advances in the aerospace sciences.
  Includes bibliographical references and index.
  1. Aerothermodynamics—Addresses, essays, lectures. 2. Combustion—Addresses, essays,
lectures. 3. Crocco, Luigi. I. Crocco, Luigi. II. Casci, Corrado. III. Bruno, Claudio.
TL574.A45R43   1985   629.1′2   84-24833
ISBN-13: 978-1-4684-4300-4   e-ISBN-13: 978-1-4684-4298-4
DOI: 10.1007/978-1-4684-4298-4

© 1985 Plenum Press, New York
Softcover reprint of the hardcover 1st edition 1985
A Division of Plenum Publishing Corporation
233 Spring Street, New York, N.Y. 10013

Contributors

M. Y. Bahadori, University of California, San Diego, La Jolla, CA, U.S.A.

R. Borghi, Laboratoire de Thermodynamique, Faculté des Sciences et des Techniques de Rouen, Mont-Saint-Aignan, France.

F. V. Bracco, Department of Mechanical and Aerospace Engineering, Princeton University, Princeton, NJ, U.S.A.

M. R. Brambley, University of California, San Diego, La Jolla, CA, U.S.A.

K. Brezinsky, Department of Mechanical and Aerospace Engineering, Princeton University, Princeton, NJ, U.S.A.

C. Casci, Dipartimento di Energetica — Politecnico di Milano, Milano, Italy.

C. Cercignani, Dipartimento di Matematica, Politecnico di Milano, Milano, Italy.

S. I. Cheng, Princeton University, Princeton, NJ, U.S.A. and (Consultant) Princeton Combustion Research Laboratories, Inc., Princeton, NJ, U.S.A.

J. F. Clarke, Aerodynamics, Cranfield Institute of Technology, Bedford, England.

I. Glassman, Department of Mechanical and Aerospace Engineering, Princeton University, Princeton, NJ, U.S.A.

A. Gomez, Department of Mechanical and Aerospace Engineering, Princeton University, Princeton, NJ, U.S.A.

S. H. Lam, Department of Mechanical and Aerospace Engineering, Princeton University, Princeton, NJ, U.S.A.

P. A. Libby, Department of Applied Mechanics and Engineering Sciences, University of California, San Diego, La Jolla, CA, U.S.A. and Systems Science and Software, La Jolla, CA, U.S.A.

A. S. C. Ma, Computational Fluid Dynamics Unit, Imperial College of Science and Technology, London, U.K.

F. E. Marble, California Institute of Technology, Pasadena, CA, U.S.A.

A. Miele, Rice University, Department of Mechanical Engineering and Materials Science, P.O. Box 1892, Houston, TX, U.S.A.

P. Orlandi, Scuola di Ingegneria Aerospaziale, Istituto di Aerodinamica, Università degli Studi di Roma, Rome, Italy.

S. S. Penner, University of California, San Diego, La Jolla, CA, U.S.A.

R. Peyret, Département de Mathématiques, Université de Nice, Nice, France, and ONERA, Chatillon, France.

A. Reggiori, Dipartimento di Energetica, Politecnico di Milano, Milano, Italy.

W. A. Sirignano, Department of Mechanical Engineering, Carnegie-Mellon University, Pittsburgh, PA, U.S.A.

D. B. Spalding, Computational Fluid Dynamics Unit, Imperial College of Science and Technology, London, U.K.

L. C. Squire, Cambridge University Engineering Department, Trumpington Street, Cambridge, England.

W. C. Strahle, School of Aerospace Engineering, Georgia Institute of Technology, Atlanta, GA, U.S.A.

R. L. T. Sun, Computational Fluid Dynamics Unit, Imperial College of Science and Technology, London, U.K.

F. Takahashi, Department of Mechanical and Aerospace Engineering, Princeton University, Princeton, NJ, U.S.A.

P. Venkataraman, Rice University, Department of Mechanical Engineering and Materials Science, Houston, TX, U.S.A.

H. Viviand, ONERA, Chatillon, France.

F. A. Williams, Department of Mechanical and Aerospace Engineering, Princeton University, Princeton, NJ, U.S.A.

Preface

This volume, published in honor of Prof. Luigi Crocco, appears when Luigi Crocco celebrates his 75th birthday of a life devoted to study, research, and teaching.

The events in his life and World War II forced Luigi Crocco, as well as other Italian scientists, to look to foreign countries for the calm haven so vital to study. This notwithstanding, his scientific activity was never interrupted, and this volume is an acknowledgment of scientists and researchers to his work and life.

Prefazione

Questo volume in onore del prof. ing. Luigi Crocco vede la luce quando Luigi Crocco compie i 75 anni di una vita dedicata allo studio, alla ricerca e all'insegnamento.

Le vicende della vita, ed anche della 2^a guerra mondiale, hanno costretto Luigi Crocco, come altri scienziati italiani, a dover cercare in altri Paesi quella serenità necessaria per dedicarsi allo studio. Ma la sua attività scientifica non ha avuto interruzioni e questo volume essere la testimonianza di studiosi e di ricercatori alla sua opera e alla sua vita.

Corrado Casci
Editor

Claudio Bruno
Assistant Editor

Contents

Curriculum Vitae

Luigi Crocco was born in Palermo, Italy, on 2 February 1909 to G. A. Crocco, a great scientist, and Bice Patti, an artist. From his parents he inherited a rigorous mind, scientific liveliness, and a love for the arts.

Luigi (Gino) Crocco began life in a hut at Vigna di Valle on the Lake of Bracciano, where his parents lived. There his father designed and built the first dirigible airship of the Italian Air Force which, on 31 October 1908, flew over Rome during one of those evenings that can be enjoyed only in Rome. Gino's parents lived happily and were full of love both for science and their children.

From Bice Crocco's diary*:

Gino has a cube, a cylinder and a sphere at hand.
"How many sides does the cube have?" his mother asks.
"Six," he answers promptly.
"And the cylinder?"
"Three," he says, pointing to the circular one.
"And the sphere?"
He thinks for a while, then without hesitation:
"One."
"Well done, for a 3-year-old geometer!"

The father to Alfredo (Gino's elder brother):
"How much is five minus four?"
"One."
"And four minus five?"
"It can't be done."
Gino intervenes:
" −1."

---

* B. Crocco, *Questa terra non ci basta*, Cappelli Editore, Bologna (1957).

"Why?"
"Because it takes one to make zero."
Gino was not yet five!

I report these episodes to show the mathematical precocity of Gino Crocco, who was not an infant prodigy. On the contrary, he educated and sharpened his natural intelligence through deep and serious studies. In fact, after High School he entered the School of Engineering of Rome and graduated as a Mechanical Engineer in 1931. During this time, simultaneously with his studies, he was involved for four years in research on the field of rocketry, in association with his father, the well-known pioneer of aeronautics and astronautics G. Arturo Crocco, who was Professor at the newly created School of Aeronautical Engineering of the University of Rome.

The research on rockets was initiated in 1927 and sponsored by the Italian Army General Staff. After 1928, it was entrusted entirely to Luigi Crocco because his father had been named to important governmental positions, preventing him from devoting to this research more than a supervisory role. The first few years were devoted to study of the suitability of double-base powders as rocket propellants. Many of the presently known combustion laws of these powders were determined in those years, both experimentally and theoretically, by Luigi Crocco. However, they remained unpublished, since the principle of restricted technical publications had not yet matured at that time, at least in Italy. Small solid-propellant rockets were fired in those years with some success.

In 1929, the research became oriented toward liquid propellants. As a result, in 1930, a small regeneratively cooled rocket chamber was operated successfully for ten minutes with the storable bipropellant combination of gasoline and nitrogen tetroxide.

In 1931, Luigi Crocco left the University of Rome for his military duties. His research came to a temporary halt, which lasted until 1933. Free from military duties, he became involved not only in rocket research, but also in nonatomospheric propulsion. For the latter research, the sponsorship switched from the Army General Staff to the Ministry of Aeronautics.

Around 1933, the emphasis of the research changed and Crocco concentrated on the study of liquid monopropellants. During this study, the exceptional properties of mononitromethane were discovered from theoretical considerations. This compound was already known for its excellent properties as a solvent and was produced commercially and utilized in the U.S. However, the fact that it was one of the most powerful monopropellants was still unknown.

Experimental work started immediately on the possibility of using mononitromethane not only in rockets, but also in other types of motors, not requiring the use of air as an oxidizer. In the course of the experiments a bad explosion took place in which Crocco, who was personally conducting certain injection tests, was severely wounded (1934). The accident revealed that all was not well with the properties of nitromethane and that, under certain circumstances, contrary to common knowledge, it could become explosive. This dictated more caution during subsequent developments, directed primarily at underwater and stratospheric propulsion.

The above experimental research came to a stop in 1937, when Crocco became Assistant Professor of Aviation Engines at the School of Aeronautical Engineering of the University of Rome. One of the reasons was that Crocco, whose mind was theoretically inclined, found it frustrating that he was being prevented from publishing the results of his activity.

During all those years of engineering research, Crocco was becoming increasingly attracted to a more fundamental field: aerodynamics of high speed. His first publication, in 1931, was on high-speed boundary layers; it introduced what became known later as the "Crocco energy integral."

It is interesting to observe that, with the exception of only four papers dealing with piston engines and jet engines, all of Crocco's publications dealt with problems of aerodynamics and gasdynamics. This statement refers to the period from 1937 to 1949, the year in which he left Italy for the United States. It is also worth mentioning that Crocco published in 1935 a fundamental theoretical study on the relative merits of different types of supersonic wind tunnels. Theodore von Karman used to refer to this study as the "bible" of supersonic wind tunnels.

In 1939, a nationwide competition was called to fill the vacant chair of Aviation Engines at the School of Aeronautical Engineering of the University of Rome. Crocco took part in the competition and won first place, thus becoming Full Professor. He held this position until 1949, when he left the University of Rome for Princeton University.

During the war years, Crocco served in the Aeronautical Technical Corps of Italy, first as a Captain and then as a Major. His activity in the Research Center of Guidonia was confined to scientific and technical fields. He continued with his teaching at the University of Rome and organized a section of the School of Aeronautical Engineering dealing with aviation motors. Probably, this section was the first of its kind in the world, its principal aim being to approach engine design and operational problems on a scientific basis, of course with due consideration paid to important practical and engineering problems.

Toward the end of the war, Crocco's attention was attracted strongly to the field of jet engines. The scientific approach helped considerably. This led to a unitary and general view of the field, where all kinds of engines were studied within the same unifying scheme. An interesting and voluminous publication was the result.

In 1947, while continuing his activities at the University of Rome, Crocco initiated an intense collaboration with the French War Ministry in Paris on the use of nitromethane in rocket motors. The research lasted two years and, after solving a substantial number of problems, culminated with the successful operation of a monopropellant combustion chamber. It was interrupted by the departure of Crocco for the U.S. in 1949.

In 1949, Harry Guggenheim, faithful to his family traditions, created two Jet Propulsion Centers, one at California Institute of Technology in Pasadena and one at Princeton University. Crocco was invited to join the staff of the Princeton Center, first as Visiting Professor, later as the holder of the Robert H. Goddard Chair of Jet Propulsion. In 1957, Crocco became a U.S. citizen.

Crocco spent some 20 years at the newly created Department of Aeronautical Engineering of Princeton University. It was a fortunate coincidence that these years were the crucial years of intense development of the aeronautical and space sciences. The presence and prestige of Crocco helped in attracting a considerable number of excellent graduate students to the MS and PhD programs at Princeton University. In turn, the new PhDs became the leaders of other PhD programs developing elsewhere in the U.S.

During the initial years of his activity in Princeton, Crocco became deeply involved in a theoretical and experimental program on combustion instability in liquid-propellant rocket motors. Indeed, this was becoming an essential problem, since rockets of increasingly larger thrust and size were required. Although the mechanism of low-frequency instability was already rather well understood, high-frequency instability was appearing increasingly often, but no theoretical explanation had as yet been arrived at.

Crocco supplied the first theoretical explanation of high-frequency instability. For the next twenty years, Crocco's results provided the only available rules for the prediction and control of this phenomenon. A theoretical and experimental program ensued for the next two decades, with the sponsorship of the Bureau of Aeronautics and NASA.

Crocco's program on combustion instability gained interest because of the advent of ICBM vehicles and also because of the space program, which culminated with the moon mission. It must be noted that problems of combustion instability appeared in the development of the F-1 rocket of the Saturn program, and Crocco was called by NASA to help in their elimination. Possibly, this was the most important contribution of knowledge acquired theoretically and experimentally to a very advanced and important problem of engineering.

In the years following the Sputnik, the space aspects of aerodynamics and rocketry grew considerably and became overwhelming with respect to the aeronautical aspects. Everywhere, the emphasis of departments of aeronautical engineering switched to the more advanced and demanding field, and the word "aeronautical" was replaced by the word "aerospace." Courses were adapted to the new emphasis.

At Princeton University, for instance, the course on jet propulsion was replaced by a course on aerospace propulsion, including not only rockets but also the more exotic types of space propulsion. A new course on space flight was born, devoted not only to boosting and reentry problems but also to the flight mechanics of thrusted space vehicles, developed on the basis of celestial mechanics. These new courses were created and taught by Crocco; they were chosen preferentially by large student audiences.

Of course, these new trends were reflected in Crocco's publications. During those twenty years at Princeton University the majority of Crocco's publications dealt with combustion of liquid propellants and combustion in general. But there were publications in the field of space flight and trajectory optimization. Also, numerous publications dealt with gas-dynamics and compressible viscous flows, which were Crocco's preferred fields of theoretical speculation.

The Princeton years were also marked by a substantial amount of consultation and scientific advisory activity to a variety of firms in the fields of rocketry, space flight, gasdynamics, viscous flows, and other areas.

The Princeton period came to an end in 1970. In view of the health conditions of his wife, Crocco took the difficult decision of resigning his professorship and returning to Europe. He spent one year at the University of Rome as a Fulbright Professor; then, he became a member of the teaching staff of the École Centrale Polytechnique, where he taught for eight years courses on boundary-layer theory, singular-perturbation methods, rockets, waves and tides, and biomechanics. At the same time, he became a consultant to ONERA on laminar-flow separation and turbulence.

In France, Luigi Crocco reached the retirement age in 1978. He remained in Paris until 1981, returning to Rome after the death of his beloved wife. His latest interest is the field of turbulence, and he continues working in this area in cooperation with members of the staffs of the School of Aeronautical Engineering of the University of Rome and the Department of Energetics of the Polytechnic of Milan.

When Crocco left Princeton University he interrupted his studies on combustion and devoted himself to the theoretical studies he preferred; but the studies done in that period (mechanics of the formation of drops in sprays and their combustion) are still valid.

This short biography has focused on the scientific activity of Luigi Crocco, but he must also be remembered as a teacher. In fact many of his pupils used his teachings as a starting point for their research activities.

Those who had the chance of attending his classes and seminars at the University still remember the clarity and precision of his speech which, together with his conciseness, showed the total command he had (and has) over the several scientific subjects he used to teach.

When we speak of Luigi Crocco, we cannot forget Simone, "his Simone." Gino met that "pretty and lively Parisian" at Mondello, Sicily. He loved her from that moment throughout her life and devoted his life to her. I think it was the match of two noble and deeply religious souls: their wedding was celebrated in 1940 in the church of Porziuncola, Assisi, where St. Francis lived.

The health of Simone began to deteriorate already while they were living in Princeton; perhaps she was already feeling the effects of the illness which would eventually cause her death.

Gino was always near her. For her he turned down all the successes that his life as scientist and professor could offer him. Once, when Simone was still alive, he told me: "Since you know Simone, you can understand me. I had to choose, and the soul is more important to me than the mind." Gino ministered to her with love and devotion, and except for a short period of time following her operation, he kept her at home where she died in his arms.

Gino's love for Simone is still alive. To honor and remember her to their friends, he dedicated to her a reproduction of thirty water colors — the most difficult kind of painting — that Simone, a very valued and sensitive artist, had painted to fix in time moments and sensations of her life.

As this book is dedicated to Gino Crocco, I want to dedicate it to him with the same words of a poem ("Appel") that Simone wrote for him:

À mon Gino

"......
Ton beau regard d'enfant
pour m'aider à mourir."

Prof. Corrado Casci

Curriculum Vitae

Luigi Crocco è nato a Palermo il 2 febbraio 1909 da G. Arturo Crocco, mente di scienziato, e da Bice Patti, anima di artista, e dai genitori ha preso il rigore e la vivacità scientifica della mente e l'amore per il culto artistico dell'anima.

Luigi (Gino) Crocco iniziò la sua carriera di essere umano a Vigna di Valle, in una baracca — sul lago di Bracciano, dove il padre risiedeva perchè qui progettò e realizzò il primo dirigibile dell'Aviazione Italiana, dirigibile che il 31 ottobre 1908 volò su Roma in una di quelle sere di ottobre che solo Roma può offrire. E la vita dei genitori di Gino trascorreva felice e quella vita era piena di amore: per la scienza e la tecnica e per i figli e la loro educazione.

Dal diario di Bice Crocco*:

Gino ha sottomano il cubo, il cilindro e la sfera.

"Quante facce ha il cubon?" chiede la mamma.
"Sei," risponde pronto.
"Quante il cilindro?"
"Tre," dice accennando a quella circolare.
"E la sfera?"
Si ferma un po' pensoso, poi afferma senza esitazione:
"Una."
"Bravo il piccolo geometra di tre anni!"

Il babbo ad Alfredo**
"Quanto fa cinque meno quattro?"
"Uno."
"E quattro meno cinque?"

---

* Bice Crocco, *Questa terra non ci basta*, Cappelli Editore, Bologna (1957).
** Fratello maggiore di Gino.

"Non si può!"
Gino interviene:
"Meno uno!"
"Perchè?"
"Perchè ... ce ne manca uno a fare zero."

Ho voluto riportare questo episodio per attestare la precocità "matematica" di Gino Crocco, il quale non è stato un bambino prodigio, ma ha educato ed affinato la sua innata intelligenza con studi seri e profondi. Infatti compiuti gli studi liceali a Roma si iscrisse alla Scuola di Ingegneria di Roma nel 1926 e si laureò in Ingegneria Meccanica nel 1931. In questo periodo, contemporaneamente agli studi, fu coinvolto per 4 anni nella ricerca nel campo della missilistica insieme a suo padre, che era passato dagli studi "sul più leggero dell'aria" a quelli "sul più pesante dell'aria" e con visione veramente profetica iniziò in quel periodo gli studi di aeronautica ed astronautica e fondò e diresse la Scuola di Ingegneria Aeronautica, oggi Scuola di Ingegneria Aerospaziale, di Roma.

La ricerca sui missili ebbe inizio nel 1927 e fu sponsorizzata dallo Stato Maggiore dell'Esercito Italiano. Dopo il 1928, rimase interamente affidata a Luigi Crocco: suo padre era stato chiamato a importanti cariche governative che gli consentivano di dedicare a questa ricerca solo un controllo dall'alto. I primi anni furono dedicati allo studio della convenienza di polveri a doppia base come propellenti per missili. Molte leggi sulla combustione conosciute oggi su queste polveri furono determinate in quegli anni, sia sperimentalmente che teoricamente, da Luigi Crocco. Non vennero comunque pubblicate, poichè il concetto di ristrette pubblicazioni techniche non era ancora maturato a quel tempo, almeno in Italia. Piccoli missili a propellente solido furono lanciati con successo in quegli anni.

Nel 1929, la ricerca cominciò ad orientarsi verso i propellenti liquidi. Ne risultò che nel 1930 una piccola camera di missile raffreddata a recupero venne fatta funzionare con successo per 10 minuti con la combinazione immagazzinabile di bipropellente composto da benzina e tetrossido di azoto.

Nel 1931, Luigi Crocco lasciò l'Università di Roma per assolvere al servizio militare. La sua ricerca ebbe un arresto temporaneo che durò fino al 1933. Libero dagli obblighi militari, fu coinvolto non solo nella ricerca missilistica, ma anche nella propulsione nonatmosferica. Per quest'ultima ricerca la sponsorizzazione venne non più dallo Stato Maggiore dell'Esercito ma dal Ministero dell'Aeronautica.

Verso il 1933, l'enfasi della ricerca cambiò e Crocco concentrò i suoi studi sui monopropellenti liquidi. Nel corso di questo studio vennero scoperte da considerazioni teoriche le eccezionali proprietà del mononitrometano. Questo composto era già conosciuto per le sue eccellenti

proprietà di solvente ed era prodotto commercialmente e utilizzato negli
Stati Uniti, ma il fatto che fosse uno dei più potenti monopropellenti era
ancora sconosciuto.

Il lavoro sperimentale iniziò immediatamente sulla possibilità di usare
il mononitrometano non solo nei missili, ma anche in altri tipi di motori che
non richiedevano l'uso di aria come ossidante. Nel corso degli esperimenti
si verificò una brutta esplosione in cui Crocco, che stava conducendo
personalmente alcuni tests di iniezione, fu seriamente ferito (1934). L'in-
cidente rivelò che non tutto andava bene con le proprietà del mononitrometano e che in certe circostanze, al contrario di ciò che si conosceva,
poteva diventare esplosivo. Ciò dettò una maggiore cautela negli sviluppi
successivi, diretti principalmente alla propulsione sottomarina e alla propulsione stratosferica. La menzionata ricerca sperimentale si interruppe
nel 1937 quando Gino Crocco divenne assistente di Motori per Aeromobili
presso la Scuola di Ingegneria Aeronautica dell'Università di Roma. Una
delle ragioni dell'interruzione fu che Gino Crocco, più incline alla ricerca
teorica, trovò frustrante il fatto che gli impedissero di pubblicare i risultati
della sua attività. In tutti quegli anni di ricerca ingegneristica, Crocco era
sempre più attratto da un campo fondamentale: l'aerodinamica delle alte
velocità. La sua prima pubblicazione, nel 1931, trattava degli strati limite
delle alte velocità e introduceva quello che più tardi venne conosciuto col
nome di "integrale dell'energia di Crocco".

E' interessante osservare che con l'unica eccezione di 4 saggi che
trattavano di motori a stantuffo e motori a reazione, tutte le pubblicazioni
di Crocco trattarono problemi di aerodinamica e gasdinamica. Questi studi
si riferiscono al periodo dal 1937 al 1949, anno in cui lasciò l'Italia per gli
Stati Uniti. Vale anche la pena menzionare che Crocco pubblicò nel 1935
uno studio teorico fondamentale sui meriti relativi di diversi tipi di gallerie
del vento supersoniche. Theodore Von Karman si riferiva a questo studio
definendolo la "bibbia" delle gallerie del vento supersoniche.

Nel 1939 venne bandito un concorso nazionale per coprire la cattedra
vacante di Motori per Aeromobili alla Scuola di Ingegneria Aeronautica
dell'Università di Roma. Gino Crocco partecipò al concorso e risultò
primo, diventando professore di ruolo e mantenne tale posizione fino al
1949 quando lasciò l'Università di Roma per l'Università di Princeton.

Negli anni della guerra, Crocco prestò servizio nel Genio Aeronautico
Italiano, prima come Capitano e quindi come Maggiore. La sua attività nel
Centro di Ricerca di Guidonia fu limitata a campi tecnici e scientifici.
Proseguì la sua attività didattica all'Università di Roma e organizzò una
sezione della Scuola di Ingegneria Aeronautica che trattava di "Motori per
Aeromobili". Probabilmente questa sezione fu la prima del mondo nel suo
genere: lo scopo principale era quello di avvicinarsi alla progettazione dei
motori e ai problemi operazionali su basi scientifiche, naturalmente

tenendo in dovuta considerazione importanti problemi practici e ingegneristici.

Verso la fine della guerra, l'attenzione di Crocco fu fortemente attratta dal campo dei motori a reazione. L'approccio scientifico diede un impulso considervole a questo settore. Questo studio condusse infatti ad una visione unitaria e generale del campo, dove tutti i tipi di motori erano studiati secondo lo stesso schema unificante. Il risultato fu una pubblicazione voluminosa e interessante.

Nel 1947, pur continuando le sue attività all'Università di Roma, Crocco iniziò un'intensa colaborazione con il Ministero della Guerra francese a Parigi sull'uso del nitrometano nei motori per missili riprendendo gli studi della sua prima giovinezza. La ricerca durò 2 anni e, dopo aver risolto un notevole numero di problemi, culminò con il buon esito di una camera di combustione per monopropellente. Queste ricerche furono interrotte dalla partenza di Crocco per gli Stati Uniti nel 1949.

Nel 1949 Harry Guggenheim, fedele alle tradizioni familiari, creava due centri per la Propulsione a Reazione, uno a Pasadena al California Institute of Technology, e uno all'Università di Princeton. Crocco fu invitato ad unirsi allo staff della Università di Princeton prima come Visiting Professor poi come detentore della cattedra Robert H. Goddard di Propulsione a Reazione. Nel 1957 Crocco divenne cittadino americano.

Gino Crocco trascorse quasi 20 anni all'appena creato Dipartimento di Ingegneria Aeronautica dell'Università di Princeton. Fu una fortunata coincidenza che quegli anni siano stati gli anni cruciali di un intenso sviluppo delle scienze aeronautiche e spaziali. La presenza ed il prestigio di Crocco aiutarono ad attrarre un considerevole numero di eccellenti studenti laureati al programma di Master of Science e a quello di Doctor of Philosophy dell'università di Princeton. A loro volta coloro che avevano seguito il PhD diventarono i leaders di altri programmi PhD che si stavano sviluppando in altre parti degli Stati Uniti.

Durante gli anni iniziali della sua attività a Princeton, Crocco fu profondamente coinvolto in programmi teorici e sperimentali sull'instabilità della combustione nei motori per missili a propellente liquido. In realtà questo stava diventando un problema essenziale, dal momento che erano richiesti missili di dimensioni e spinta maggiori. Sebbene il meccanismo dell'instabilità della bassa frequenza era già stato abbastanza ben compreso, l'instabilità dell'alta frequenza appariva sempre più spesso, ma non si era ancora arrivati ad una spiegazione teorica.

Crocco fornì la prima spiegazione teorica dell'instabilità dell'alta frequenza. Nei successivi 20 anni, i risultati di Crocco fornirono le uniche regole disponibili per la predizione e il controllo di questo fenomeno. Un programma teorico e sperimentale seguì nelle successive due decadi, con la sponsorizzazione del Bureau of Aeronautics e della NASA.

Il programma di Crocco sull'instabilità della combustione acquistò interesse sia con l'avvento dei veicoli ICBM, sia con il programma spaziale che culminò con la missione sulla Luna. Si deve notare che problemi di instabilità di combustione apparvero nello sviluppo del razzo F-1 del programma Saturno, e Crocco fu chiamato dalla NASA per aiutare ad eliminarlo. Probabilmente questo fu il più importante contributo della conoscenza acquisita teoricamente e sperimentalmente a un problema molto avanzato ed importante di ingegneria.

Negli anni che seguirono lo Sputnik, gli aspetti spaziali dell'aerodinamica e della missilistica crebbero considerevolmente e divennero schiaccianti rispetto agli aspetti aeronautici. Dappertutto, l'enfasi dei dipartimenti di ingegneria aeronautica si spostò al campo più avanzato e la parola "aeronautica" venne sostituita dalla parola "aerospaziale". Anche i corsi di insegnamento vennero adattati alla nuvova necessità.

All'Università di Princeton, ad esempio, il corso di propulsione a reazione venne rimpiazzato da un corso sulla propulsione aerospaziale, che comprendeva non solo i missili ma anche tipi più esotici di propulsione spaziale. Nacque un nuovo corso di volo spaziale, dedicato non solo ai problemi di lancio e di rientro, ma anche alla meccanica del volo dei veicoli spaziali a spinta, sviluppata sulla base della meccanica celeste. Questi nuovi corsi furono creati e tenuti da Crocco e vennero seguiti da un gran numero di studenti.

Naturalmente queste nuove tendenze si rifletterono nelle pubblicazioni di Crocco. Durante i vent'anni trascorsi all'Università di Princeton la maggior parte delle pubblicazioni di Crocco trattò della combustione di propellenti liquidi e della combustione in genere; ma ci furono anche pubblicazioni nel campo del volo spaziale e dell'ottimizzazione della traiettoria. Inoltre numerose pubblicazioni trattarono di gasdinamica e di fluidi viscosi comprimibili che erano i campi di speculazione teorica preferiti da Crocco.

Gli anni di Princeton furono anche segnati da una notevole attività di consulenza e da una attività consultiva scientifica per numerose industrie nel campo della missilistica, del volo spaziale, della gasdinamica, dei fluidi viscosi e in altri campi.

Il periodo di Princeton terminò nel 1970. A causa delle condizioni di salute della moglie, Crocco prese la dura decisione di rassegnare le sue dimissioni da professore e ritornare in Europa. Trascorse un anno all'Università di Parigi come Fulbright Professor; quindi divenne membro del collegio docenti dell'Ecole Centrale Polytechnique, dove tenne per otto anni corsi sulla teoria degli strati limite, metodi di perturbazione singolare, missili, onde e maree, e biomeccanica. Nello stesso tempo divenne consulente dell'ONERA sulla separazione e la turbolenza dei flussi laminari.

In Francia, Luigi Crocco si ritirò in pensione nel 1978. Rimase a Parigi fino al 1981, tornando a Roma dopo la morte della sua amata consorte. Il suo interesse più recente è nel campo della turbolenza e continua a lavorare in quest'area in collaborazione con gli staffs del Dipartimento di Energetica di Milano e della Scuola di Ingegneria Aeronautica dell'Università di Roma.

Come si vede, lasciando l'Università di Princeton Gino Crocco interruppe i suoi studi relativi alla combustione per dedicarsi agli studi teorici da lui preferiti, ma gli studi fatti da lui in quel periodo (mi riferisco alla meccanica della formazione delle goccioline negli spray e la loro combustione) sono ancora validi.

Si è ricordato in questa breva biografia di Luigi Crocco la sua attività scientifica, ma è giusto anche ricordare i suoi meriti di Maestro e di docente. Infatti molti sono gli allievi da lui formati e che hanno continuato le sue ricerche e i i suoi insegnamenti, insegnamenti che vivono ancora per merito di quelli che un tempo furono suoi allievi. Coloro che hanno avuto la ventura di ascoltare le sue lezioni, sia nei corsi universitari sia nei seminari da lui impartiti, ricordano ancora la chiarezza dell'impostazione, la sintetica esposizione unita ad una eleganza di linguaggio che dimostrano la completa padronanza e la chiarezza di idee dei molteplici argomenti scientifici oggetto di insegnamento.

Parlando di Luigi Crocco non si può tacere di Simone, la "sua Simone". Gino incontrò "quella graziosa e vivace parigina" a Mondello in Sicilia: la amò fino da allora e l'amò per tutta la vita e a Lei dedicò la sua vita. Credo che quello fu l'incontro di due anime elette di profonda fede cristiana; e questo incontro fu suggellato nella chiesa della Porziuncola di Assisi nel 1940. Forti disturbi minarono la salute di Simone quando ancora vivevano a Princeton; forse già covava in Lei il terribile male che la avrebbe portata alla tomba. Gino le è stato sempre vicino e per Lei eveva rinunciato a tanti allori che potevano dargli la vita di scienziato ed accademico. Un giorno parlandomi, e Simone era ancora in vita, mi disse: "Tu che hai conosciuto Simone puoi comprendermi: io ho fatto una scelta e i beni dell'anima sono più forti di quelli della mente". Gino l'ha assistita con amore e con dedizione ed escluso il periodo dell'intervento chirurgico l'ha tenuta in casa e l'ha assistita solo lui ed è morta fra le sue braccia.

E l'Amore che Gino ha avuto per Simone non si è esaurito: infatti a Lei ha dedicato, per onorarla e per ricordarla agli amici, una riproduzione di XXX acquerelli — la più difficile delle pitture — che Simone, pittrice fine e molto apprezzata, aveva dipinto fissando momenti ed impressioni della sua vita.

E poichè questo volume è dedicato a Luigi Crocco voglio dedicarlo con le stesse parole di una poesia — Appel — di Simone dedicata proprio a

Gino:

À mon Gino

"......
Ton beau regard d'enfant
pour m'aider à mourir."

Prof. Ing. Corrado Casci

Honors

Accademia Nazionale dei Lincei, Rome (Member)
Accademia della Scienze, Turin (Member)
International Academy of Astronautics, Paris (Member)
National Academy of Engineering, Washington (Member)
A.I.A.A., New York (Honorary fellow)
A.I.D.A., Rome (Member)
Academie Nationale de l'Air et de l'Espace, Toulose (Member)

Awards

Pendray Award 1965
Wilde Award 1969
Columbus International Prize and Gold Medal 1973

Consultant for: FIAT, Italy • French Ministry of Defense, France • Curtiss-Wright, U.S.A. • Bendix, U.S.A. • Aerojet General, U.S.A. • General Dynamics, U.S.A. • Reaction Motors, U.S.A. • Arthur D. Little, U.S.A. • AGARD, France • G. E. Steam Turbines, U.S.A • G. E. Rocket Station, U.S.A. • G. E. Jet Engines, U.S.A. • G. E. Space Technology, U.S.A. • R.C.A., U.S.A. • Thomson-Ramo-Woolridge, U.S.A. • N.A.S.A., U.S.A. • O.N.E.R.A., France • European Space Agency, France

Luigi Crocco — Publications

1. Su di un Valore Massimo del Coefficiente di Trasmissione del Calore da una Lamina Piana a un Fluido Scorrente, *Rendiconti della R. Accademia Nazionale dei Lincei, Classe di Scienze Fisiche, Matematiche e Naturali*, Vol. XIV, Serie 6ª, 2° sem., Fasc. II, Dicembre 1931.
2. **Sulla Trasmissione del Calore de una Lamina Piana a un Fluido Scorrente ad Alta Velocità,** *L'Aerotecnica*, **Vol. XII, No. 2, pp. 181–197, Febbraio 1932.**
3. Gallerie Aerodinamiche per Alte Velocità, *L'Aerotecnica*, Vol. XV, No. 3, 7–8, Marzo, Luglio-Agosto 1935. Also published as: *Tunnels Aérodynamiques pour Grandes Vitesses*, Extrait du Mémorial de L'Artillerie, Imprimerie Nationale, Paris, France, pp. 358–442, 1938.
4. Una Nuova Funzione di Corrente per lo Studio del Moto Rotazionale dei Gas, *Rendiconti della R. Accademia Nazionale dei Lincei, Classe di Scienze Fisiche, Matematiche e Naturali*, Vol. XXIII, Serie 6ª, 1° sem., Fasc. II, Roma, Febbraio 1936. Also published as: Eine Neue Stromfunktion fur die Erforschung der Bewegung der Gase Mit Rotation, *Zeitschrift fur Angewandte Mathematik und Mechanik*, Band 17, Heft 1, Februar 1937.
5. Singolarità della Corrente Gassosa Iperacustica nell'Intorno di una Prora a Diedro, *L'Aerotecnica*, Vol. XVII, No. 6, Giugno 1937.
6. Una Proprietà Approssimata delle Eliche e Sua Applicazione al Calcolo delle Caratteristiche di un Motovelivolo, *L'Aerotecnica*, Vol. XVII, No. 7, pp. 1–19, Luglio 1937.
7. Moderni Problemi sul Raffreddamento dei Motori D'Aviazione, *L'Aerotecnica*, Istituto Poligrafico dello Stato, Libreria, 1937 (monograph).
8. Una Proprietà del Meccanismo Manovella-Biella-Stantuffo e sue Applicazioni al Meccanismo a Biella Madre e Bielletta, *Atti di Guidonia*, pp. 61–68, Vol. XVII, Aprile 1939.
9. Una Caratteristica Trasformazione delle Equazioni dello Strato Limite nei Gas, *Atti di Guidonia*, Vol. XVII, No. 7, pp. 105–120, Maggio 1939.
10. **Sullo Strato Limite Laminare nei Gas lungo una Parete Piana,** *Rendiconti di Matematica e delle sue Applicazioni*, Serie V, Vol. II, Fasc. II, pp. 138–152, Giugno 1941.
11. Lo Strato Limite Laminare nei Gas, *Monografie Scientifiche di Aeronautica*, No. 3, Dicembre 1946.
12. On a Kind of Stress-Function for the Study of Non Isentropic Two-Dimensional Motion of Gases, *Proc. of the Int'l. Meet. of Appl. Mech.*, London, 1946. Also published as: Una Nuova Funzione Potenziale per lo Studio del

Moto Bidimensionale Non Isentropico dei Gas, *L'Aerotecnica*, Vol. XXIX, No. 6; pp. 347–355, Dicembre 1949.

13. Diagrammi Termodinamici dei Gas di Combustione, *Monografie Scientifiche di Aeronautica*, No. 2, Marzo 1947.

14. Il Turboreattore a Due Flussi, *Monografie Scientifiche di Aeronautica*, No. 6, Guigno 1947.

15. Prese d'Aria e Diffusori Supersonici, *L'Aerotecnica*, Vol. XXIX, Fasc. 3°, pp. 131–138, 1949.

16. Instruction and Research in Jet Propulsion, *Journal of the American Rocket Society*, March 1950.

17. Transformations of the Hodograph Flow Equation and the Introduction of Two Generalized Potential Functions, NACA Technical Note 2432, August 1951.

18. Aspects of Combustion Instability in Liquid Propellant Rocket Motors. Part I: Fundamentals. Low Frequency Instability with Monopropellants. Part II: Low Frequency Instability with Bipropellants. High Frequency Instability, *Journal of the American Rocket Society*, Vols. 21 and 22, November 1951 and January 1952.

19. A Mixing Theory for the Interaction Between Dissipative Flows and Nearly Isentropic Streams, *Journal of the Aeronautical Sciences*, Vol. 19, No. 10, October 1952. With L. Lees.

20. An Approximate Theory of Porous, Sweat or Film Cooling with Reactive Fluids, *Journal of the American Rocket Society*, pp. 331–338, November–December 1952.

21. High Frequency Combustion Instability in Rockets with Concentrated Combustion, *Proceedings of the 8th International Congress on Theoretical and Applied Mechanics*, 1953. With S. I. Cheng.

22. Comments on the paper "Investigation of Annular Liquid Flow with Concurrent Air Flow in Horizontal Tubes," by A. E. Abramson, *Transactions ASME*, Vol. 74, pp. 267–274. Comments found in *Transactions ASME*, Vol. 75, p. 311, 1953.

23. High Frequency Combustion Instability in Rockets with Distributed Combustion, *Proceedings of the 4th International Symposium on Combustion*, Pittsburgh, 1953. With S. I. Cheng.

24. Supercritical Gaseous Discharge with High Frequency Oscillations, *L'Aerotecnica*, No. 1, 1953.

25. Compressible Laminar Boundary Layer with Heat Transfer and Pressure Gradient, *50 Jahre Grenzschichtforschung*, Verlag Friedr. Vieweg and Sohn, Braunschweig, pp. 280–293, 1954. With C. B. Cohen.

26. Comments on a paper of C. C. Ross and P. P. Datner. *Proceedings of and AGARD Combustion Colloquium, "Selected Combustion Problems,"* Cambridge University, 1953, Butterworths Scientific Publications, pp. 397–400, 1954.

27. Combustion Instability in Liquid Propellant Rocket Motors, American Rocket Society Paper 205–55, June 1955. With J. Grey.

28. Measurements of the Combustion Time-Lag in a Liquid Bipropellant Rocket

Motor, American Rocket Society, 205–55, June 1955. With J. Grey and G. B. Matthews.

29. Theory of Combustion Instability in Liquid Propellant Rocket Motors, *AGARD-ograph No. 8*, Butterworth, 1955. With S. I. Cheng (monograph).

30. Considerations on the Shock-Boundary Layer Interaction, *Proceedings of the Conference on High-Speed Aeronautics*, Brooklyn Polytechnic Institute, pp. 75–112, 1955.

31. Considerations on the Problem of Scaling Rocket Motors, *Selected Combustion Problems, II*, Butterworths Scientific Publications, January 1956.

32. A Flow Reactor for High Temperature Reaction Kinetics, *Jet Propulsion*, pp. 1266–1267, December 1957. With I. Glassman and I. E. Smith.

33. Measurement of High-Frequency Limits of Stability in a Liquid Bipropellant Rocket Motor, *Aerotecnica* No. 38, p. 135, April 1958.

34. Comments on the paper "An Experimental Study of High-Frequency Combustion Pressure Oscillations," by M. J. Zucrow and J. R. Osborn. *Jet Propulsion*, Vol. 28, December 1958.

35. On the Importance of the Sensitive Time Lag in Longitudinal High-Frequency Rocket Combustion Instability, *Jet Propulsion*, Vol. 28, pp. 841–843, December 1958. With J. Grey and D. T. Harrje.

36. One Dimensional Treatment of Steady Gas Dynamics, *High Speed Aerodynamics and Jet Propulsion, Vol. III Fundamentals of Gas Dynamics*, pp. 64–348, Princeton University Press, 1958 (monograph).

37. Note on the Shock-Induced Unsteady Laminar Boundary Layer on a Semi-Infinite Flat Plate, *Journal of the Aerospace Sciences*, Vol. 26, No. 1, January 1959. With S. H. Lam.

38. Kinetics and Mechanism of Ethylene Oxide Decomposition at High Temperatures, *The Journal of Chemical Physics*, Vol. 31, pp. 506–510, August 1959.

39. Theory of Liquid Propellant Rocket Combustion Instability and its Experimental Verification, *Journal of the American Rocket Society*, Vol. 30, No. 2, pp. 159–168, February 1960.

40. Verification of Nozzle Admittance Theory by Direct Measurement of the Admittance Parameter, *Journal of the American Rocket Society*, Vol. 31, No. 6, pp. 771–775, June 1961. With J. Grey and R. Monti.

41. Calculation of Nozzle Admittance Coefficients, Report of the G. E. Company, Flight Propulsion Laboratory, Cincinnati, June 1961.

42. Dynamics of Space Propulsion, *Proceedings of Symposium on Space Propulsion in Varenna, Italy*, August 1961. Also published as: Mechanics of Space Propulsion, *Seminar on Astronautical Propulsion, Milan, 1960, Proceedings: Advances in Astronautical Propulsion*, New York, Pergamon Press, pp. 11–46, 1963.

43. Transverse Combustion Instability in Liquid Propellant Rocket Motors, *Journal of the American Rocket Society*, March 1962. With D. T. Harrje and F. Reardon.

44. Measurement of Mean Particle Sizes of Sprays from Diffractively Scattered Light, *AIAA Journal*, Vol. 1, No. 8, pp. 1882–1886, August 1963. With R. A. Dobbins and I. Glassman.

45. Wide-Range Photographic Photometry, *Review of Scientific Instruments*, Vol. 34, No. 2, pp. 162–167, February 1963. With I. Glassman and R. A. Dobbins.
46. Analytical Investigation of Several Mechanisms of Combustion Instability, *Bulletin of the Fifth Liquid Propulsion Symposium*, Chem. Prop. Inf. Agency, 13–15 November 1963. With W. C. Strahle.
47. Transformations of the Compressible Turbulent Boundary Layer with Heat Exchange, *AIAA Journal*, Vol. 1, No. 12, pp. 2723–2731, December 1963.
48. A Shock Wave Model of Unstable Rocket Combustions, *AIAA Journal*, Vol. 2, No. 7, pp. 1285–1296, July 1964. With W. A. Sirignano.
49. Combustion Instability in Liquid Propellant Rocket Motors, Experimental Methods in Combustion Research, *AGARD Manual*, New York, Pergamon Press, 1964. With D. T. Harrje.
50. Problems in Liquid Propellant Instability, *Proceedings of the Fourth Meeting of the Technical Panel on Solid Propellant Combustion Instability*, A. Ph. L. TH 371-7, April 1964. With D. T. Harrje.
51. Theoretical Studies on Liquid Propellant Rocket Instability, *Proceedings of the Tenth International Symposium on Combustion*, Cambridge, England, May 1964.
52. Periodic Solutions to a Convective Droplet Burning Problem: The Stagnation Point, *Proceedings of the Tenth International Symposium on Combustion*, Cambridge, England, August 1964. With W. C. Strahle.
53. Velocity Effects in Transverse Mode Liquid Propellant Rocket Combustion Instability, *AIAA Journal*, Vol. II, No. 9, p. 1631, September 1964. With D. T. Harrje and F. H. Reardon.
54. Nonlinear Aspects of Combustion Instability in Liquid Propellant Rocket Motors, *Proceedings of the CPIA Combustion Instability Conference*, Orlando, Florida, 16–20 November 1964. With D. T. Harrje.
55. Turbulent Boundary Layer and Mixing Coefficient, *Journal of Society for Industrial and Applied Mechanics*, Vol. 13, pp. 206–215, March 1965. (Presented at the 1964 Spring Meeting of the Society at the session in memory of Dr. T. von Karman.)
56. A Suggestion for the Numerical Solution of the Steady Navier–Stokes Equation, *AIAA Journal*, Vol. 3, No. 10, pp. 1824–1832, October 1965.
57. The Relevance of a Characteristic Time in Combustion Instability, *Proceedings of CPIA Conference*, El Segundo, California, November 1965.
58. Nonlinear Aspects of Combustion Instability in Liquid Propellant Rocket Motors, *Proceedings of CPIA Conference*, El Segundo, California, November 1965. With D. T. Harrje and W. A. Sirignano.
59. Theoretical Studies on Liquid-Propellant Rocket Instability, *10th Symposium on Combustion*, pp. 1011–1128, 1965.
60. Linearized Treatment of the Optimal Transfer of a Thrust-Limited Vehicle Between Coplanar Circular Orbits, *Astronautica Acta*, June–July 1966. With J. E. McIntyre.
61. Effect of the Transverse Velocity Component on the Non-Linear Behavior of Short Nozzles, *AIAA Journal*, Vol. 4, No. 8, August 1966, p. 1428. With W. A. Sirignano.

62. Higher Order Treatment of the Optimal Transfer of a Thrust-Limited Vehicle Between Coplanar Circular Orbits, *Astronautica Acta*, January–February 1967. With J. E. McIntyre.

63. Acoustic Liner Studies, *ICRPG Third Combustion Conference*, CPIA Publication No. 138, Vol. 1, February 1967, p. 581. With W. A. Sirignano and D. T. Harrje.

64. Axial Energy Distribution Studies in a Liquid Propellant Rocket Motor, *ICRPG Third Combustion Conference*, CPIA Publication No. 138, Vol. 1, February 1967, p. 523. With D. T. Harrje.

65. Longitudinal Shock Wave Combustion Instability in Liquid Propellant Rocket Engines, *ICRPG Third Combustion Conference*, CPIA Publication No. 138, Vol.1, February 1967, p. 395. With C. E. Mitchell and W. A. Sirignano.

66. Behavior of Supercritical Nozzles Under Three Dimensional Oscillatory Conditions, *AGARDograph* No. 117, 1967. With W. A. Sirignano (monograph).

67. Nonlinear Transversal Instability with a Simplified Droplet Vaporization Model, *Proceedings of the 4th Combustion Conference*, Chemical Propulsion Information Agency Publication No. 162, December 1967, p. 159.

68. Axial and Transversal Instability Using the η, τ Model, *Proceedings of the 4th Combustion Conference*, Chemical Propulsion Information Agency, Publication No. 162, December 1967, p. 169.

69. Theoretical Analysis of Nonlinear Transverse Combustion Instability in Liquid Propellant Rocket Motors Using a Droplet Evaporation Model, Presented at the 5th ICRPG Combustion Conference, The Johns Hopkins University, Silver Springs, Md., October 1–3, 1968, CPIA Publication No. 183, pp. 145–153, December 1968. With P. Tang and A. K. Varma.

70. Use of Shock Waves to Study Combustion Parameters, Presented at the 5th ICRPG Combustion Conference, The Johns Hopkins University, Silver Springs, Md., October 1–3, 1968, CPIA Publication No. 183, December 1968. With F. V. Bracco and D. T. Harrje.

71. Nonlinear Oscillations in Liquid Rocket Combustion Chambers, Presented at the International Colloquium on Gasdynamics of Explosions, Brussels, 19–21 September 1967. *Astronautica Acta*, Vol. 14, No. 5, pp. 409–410, June 1969. With C. E. Mitchell and W. A. Sirignano.

72. Periodic Finite-Amplitude Oscillations in Slowly Converging Nozzles, *Astronautica Acta*, Vol. 13, pp. 481–488, 1968. With B. T. Zinn.

73. The Nozzle Boundary Condition in the Nonlinear Rocket Instability Problem, *Astronautica Acta*, Vol. 13, pp. 489–496, 1968. With B. T. Zinn.

74. Nonlinear Longitudinal Instability in Rocket Motors with Concentrated Combustion, *Combustion Science and Technology*, Vol. 1, No. 1, pp. 35–63, July 1969. With C. E. Mitchell and W. A. Sirignano.

75. Nonlinear Periodic Oscillations in Rocket Motors with Distributed Combustion, *Combustion Science and Technology*, Vol. 1, No. 2, pp. 147–169, September 1969. With C. E. Mitchell.

76. Research on Combustion Instability in Liquid Propellant Rockets, *Twelfth Symposium (International) on Combustion*, The Combustion Institute, Pittsburgh, Pa., 1969.

77. *Early Rocket Research in Italy*, From the History of Rockets and Astronautics, Published by the Academy of Sciences of the U.S.S.R., Moscow, 1970.
78. L'Equazione dell'Eccesso di Energia in Meccanica dei Fluidi, *Missili e Spazio*, Vol. 1, 1971.
79. *Transversal Annular Waves and Multiple Scales*, Volume in Honor of Carlo Ferrari's 75th Birthday, Levrotto e Bella, Torino, 1972.
80. *Ecoulement du Troisième Ordre Autour d'une Lentille Biconvexe*, Publication Ecole Centrale, Paris, 1972.
81. Analytical Models of High Frequency Combustion Instability, Section 4.2 of *Liquid Propellant Rocket Combustion Instability*, NASA, 1972.
82. Intermediate Frequency Instability, Section 5.3.2 of *Liquid Propellant Rocket Combustion Instability*, NASA, 1972.
83. Coordinate Perturbation and Multiple Scales in Gasdynamics, *Philosophical Transactions of the Royal Society of London*, June 1972.
84. Flow Separation, *Proceedings of the AGARD Conference 168 on Flow Separation*, Supplement, Göttingen, May 1975.
85. *On Certain Spectral Models in Isotropic Turbulence, Part I*, Publications ONERA, Paris, 1982.
86. *On Certain Spectral Models in Isotropic Turbulence, Part II*, Publications ONERA, Paris, 1982.
87. *Cascade Spectral Models in Isotropic Turbulence.*
88. A Transformation of the Triatic Integral for Isotropic Turbulence, Report of Dipartimento di Meccanica ed Aeronautica, Università di Roma, 1983. With P. Orlandi.

I
MATHEMATICAL TECHNIQUES

1

Singular Perturbation for Stiff Equations Using Numerical Methods

S. H. Lam

ABSTRACT. The present paper explores the following question: can the number-crunching power of the computer be used not only for generating numerical solutions, but also for *deriving* alternative formulations for the given problems? In other words, can the traditional role of human theoreticians also be performed by digital computers? We shall limit our efforts here to stiff systems of ordinary differential equations. Our task is to translate the general singular perturbation procedures used by human theoreticians for this class of problems into a programmable set of computations; the output from the computations shall provide both the numerical solutions *and* the alternative formulation of the given problem.

1. INTRODUCTION

The general availability of modern high-speed digital computers has strongly influenced the methodology of modern theoretical research in engineering and applied science. Certain classes of problems can now be routinely solved by direct numerical computations.[1-5] For these problems there is apparently little need for the traditional activities such as "simplifying" the model or "neglecting" the unimportant terms. Frequently a modern theoretician can concentrate on the *formulation* of his problem,[6] including in his model any complication which may be important, and leave the task of constructing solutions to the numerical analysts and program-

S. H. Lam • Department of Mechanical and Aerospace Engineering, Princeton University, Princeton, NJ 08544, U.S.A.

mers. He must, however, pore over the voluminous printouts from the computer to try to discern trends, identifying causes and effects, determine the sensitivity of the results to input parameters, and generally *interpret* the computed behaviors of the problems at hand in terms of some physically meaningful statements. This new mode of theoretical research is clearly here to stay. Nevertheless, this brute-force direct computational approach is not without its shortcomings. First of all, there are problems which are *inherently difficult* to compute on the digital computer. Nearly all problems which are suitable for singular perturbation analysis are in this category.[7-9] The main difficulties are usually loss of significant figures and uneconomical use of the computer time. Secondly, the task of interpreting reams of computer printouts from a complicated simulation problem can itself be an overwhelming one. Frequently the computed solutions have to be further processed by additional computations before adequate "physical insights" of the computed behaviors can be achieved.

In the present paper, we shall concentrate on one special class of singular perturbation problems: stiff ordinary differential equations. A system of simultaneous first-order ordinary differential equations is said to be stiff if it contains vastly disparate time scales. The following is a typical example of a stiff problem:

$$\frac{dy_1}{dt} = \frac{(y_2^2 - y_1)}{\varepsilon} + 1 \tag{1.1a}$$

$$\frac{dy_2}{dt} = y_1(3 - 2y_2 - y_1) \tag{1.1b}$$

where ε is a small positive number, say 10^{-8}. A traditional theoretican would immediately take advantage of the smallness of ε, and would conclude after some analysis that y_2 remains approximately constant in a brief transient period $[t = O(\varepsilon)]$ while y_1 tends rapidly to y_2^2, and that thereafter the original set of stiff equations can be approximated by

$$y_1 \cong y_2^2 \tag{1.2a}$$

$$\frac{dy_2}{dt} \cong (3 - y_2)(1 + y_2)y_2^2 \tag{1.2b}$$

Single precision artithmetics and standard integration algorithms can be used to solve equations (1.2) on a digital computer. In contrast to this traditional approach, a modern theoretician would undertake to integrate equations (1.1) directly, using multiple-precision arithmetics and some sophisticated integration algorithm[10-12] to deal with the difficulties brought on by the smallness of ε. At least for this simple case, one would probably agree that the traditional approach is more satisfying and is therefore better in some sense. In both cases, however, the computer is used as a tool to construct numerical solutions. The traditional theoretician makes his contribution by *changing* the problem before giving it to the numerical analysts and programmers; the alternative problem he recommends is usually easier to solve and easier to interpret. The modern theoretician can focus his efforts on the formulation, since he has the option of giving the problem directly to the numerical analysts and programmers, and can rely on the tremendous number-crunching power of the computer to overpower the computational difficulties — loss of significant figures by multiple-precision arithmetics, numerical instabilities by smaller time-steps, or use of sophisticated integration algorithms. Because the recommending of an alternative problem is really only feasible for relatively simple problems, and because of the computational advantage of solving a simpler alternative problem decreases with increasing computer power, the modern approach is gaining and will continue to gain wider acceptance.

It is perhaps worthwhile at this point to emphasize one of the surviving virtues of the traditional approach in the modern age. In the process of arriving at the recommended alternative problem, the traditional theoretician must assess the "importance" of each of the contributing terms. In particular, he must identify all the "unimportant" terms. The computational advantage gained in neglecting these unimportant terms is *not* the main virtue of his developments. The knowledge of which terms are controlling the behaviors of the solutions and which terms are not important is this main virtue of the traditional approach and is precisely what is usually needed in interpreting numerical solutions and achieving "physical insights."

Let us consider another (admittedly contrived) example as follows:

$$\frac{dy_1}{dt} = \frac{y_2^2 - ay_1}{\varepsilon} \tag{1.3a}$$

$$\frac{dy_2}{dt} = -\frac{y_2^2 - y_1}{\varepsilon} + y_1(3 - 2y_2 - y_1) \tag{1.3b}$$

where $\varepsilon = 5.12345 \times 10^{-6}$, $a = 1.00001$. A direct-numerical integration package, probably using double-precision arithmetics, could readily generate solutions without comments. A traditional theoretician will point out that this problem as given is "poorly posed" and that its slow-time behavior is strongly affected by $(a - 1)/\varepsilon$. He would look into whether the given data for a can be written in the form $a = 1 + a'$ where a' is accurate to an acceptable number of significant figures, and recommend the following alternative problem in the slow-time period:

$$y_1 \cong y_2^2$$

$$\frac{dy_2}{dt} \cong \frac{1 - [(a'/\varepsilon) - 3]y_2^2 - 2y_2^3 - y_2^4}{2y_2 + 1} \qquad (1.4)^*$$

Such so-called "insights" on the given problem can only be obtained at the present time by exhaustive and careful sensitivity analysis of the computed solutions if the method of direct numerical integration is used.

In the present paper, we explore the following question: can the number-crunching power of the computer be used not only for generating numerical solutions but also for *deriving* alternative formulations for the given problems? In simpler terms, can the traditional role of deriving alternative problems also be performed by digital computers?

We shall limit our efforts here to stiff systems of ordinary differential equations. Our task is to express the general singular perturbation procedures used by theoreticians into a programmable set of computations. Our goal is to set the theoretical foundation for a computer program which can derive alternative formulations for any given stiff problems, without the usual limitations of human theoreticians who can only deal with relatively simple and "tractable" problems.

Stiff equations occur frequently in practical problems, particularly in complex chemical reactions. When a set of master kinetics equations is found to be stiff, it implies that some of the elementary reactions included in the model are much faster than others. These fast reactions frequently exhaust themselves and reach a state of *quasi-equilibrium* when the large forward and backward reaction rates become nearly equal. In such studies, the ability to correctly identify the fast reactions and to correctly apply the so-called "quasi-equilibrium approximations" are highly valued. At the present time, quasi-equilibrium theories are developed solely by human theoreticians, using singular perturbation analysis, order of magnitude estimates, experience, and intuition. Our present effort is to explore the possibility that these "quasi-equilibrium" theories can also be developed by digital computations.

2. PRELIMINARY DISCUSSIONS

A main difficulty of direct numerical computations for stiff equations is loss of significant figures. In finite precision arithmetics, the associative law

---

* The initial condition for y_2 should be the larger of the two roots of the equation $y_2^2(0) + y_2(0) - (Y_1 + Y_2) = 0$.

of algebra does not always hold. For example, consider the following relations:

$$A = C - B$$

$$X = (A + B) + 1 := C + 1$$

$$Y = A + (B + 1) := (C - B) + (B + 1)$$

We can easily verify, using a hand calculator, that X and Y are computationally not equal when B is sufficiently large in comparison to C. Large disparity of order of magnitude of additive terms is characteristic of stiff equations, and near-cancellations of these large terms is characteristic of quasi-equilibrium situations. To overcome the above difficulties, double or triple precision arithmetics is mandatory for direct numerical computation of stiff equations.

A traditional theoretician, upon recognizing the presence of near-cancellations, usually introduces a new variable *defined* as the difference and strives to compute it from its own equations.[12] For equations (1.1), he would define

$$F(y_1, y_2) \equiv y_2^2 - y_1 \tag{2.1}$$

and would caution himself that F can now be accurately evaluated by equation (2.1) under quasi-equilibrium conditions. In order to compute F, a differential equation can be obtained by differentiating equation (2.1) with respect to time:

$$\frac{dF}{dt} = \left(\frac{\partial F}{\partial y_1}\right)\frac{dy_1}{dt} + \left(\frac{\partial F}{\partial y_2}\right)\frac{dy_2}{dt} \tag{2.2a}$$

$$= -\frac{1}{\varepsilon}(F - \varepsilon Q) \tag{2.2b}$$

where

$$Q(y_1, y_2) = -1 + 2y_1 y_2(3 - 2y_2 - y_1) \tag{2.2c}$$

Equations (1.1) could now be written as

$$\frac{dy_1}{dt} = \frac{F}{\varepsilon} + 1 \tag{2.3a}$$

$$\frac{dy_2}{dt} = y_1(3 - 2y_2 - y_1) \tag{2.3b}$$

which, together with equations (2.2b,c), is a possible alternative formulation of the original problem. In this formulation, the near-cancellation of y_2^2 with y_1 no longer affects the computations.

To proceed further, the theoretician would normally apply the quasi-equilibrium approximation by recognizing from equation (2.2b) that F, while decaying rapidly, is expected to approach $\varepsilon Q(y_1, y_2)$ asymptotically. Using this approximation in equation (2.3a), he obtains

$$\frac{dy_1}{dt} \cong Q + 1 = 2y_1 y_2 (3 - 2y_2 - y_1) \tag{2.4a}$$

$$\frac{dy_2}{dt} = y_1 (3 - 2y_2 - y_1) \tag{2.4b}$$

He would recommend the use of equations (2.4) whenever severe loss of significant figures occurs in the use of equation (2.1) for the evaluation of F. It can readily be shown that equations (2.4) coupled with the fact that $F = 0$ initially is theoretically equivalent to equations (1.2), the conventional form of the quasi-equilibrium approximation, in the small-ε limit.

In the derivation immediately above, the traditional theoretician has "neglected" certain higher-order terms. These terms are expected on theoretical gounds to be small and therefore "unimportant." But in the modern age of computers, there is little need for neglecting small unimportant terms, *provided*, of course, *that these terms are not allowed to create havoc in the computational schemes.* Let us now re-examine equations (1.1) with the knowledge of equation (2.2b) and find out what can be done without neglecting terms.

Let us now simply add and subtract $Q(y_1, y_2)$ to the right-hand side of equation (1.1a). We can now rewrite equations (1.1) as follows:

$$\frac{dy_1}{dt} = (1 + Q) + \left(\frac{y_2^2 - y_1}{\varepsilon} - Q \right) \tag{2.5a}$$

$$\frac{dy_2}{dt} = [y_1 (3 - 2y_2 - y_1)] \tag{2.5b}$$

This new set of equations is theoretically identical to equation (1.1), using the usual associative law of algebra. There is but one crucial difference: the second term on the right-hand side of equation (2.5a) is now *known* to be theoretically small and unimportant *whenever its evaluation suffers severe loss of significant figures.* This luxury is not available to equation (1.1a); the term $(y_2^2 - y_1)/\varepsilon$ on its right-hand side is nonsmall and very important, *particularly* when its evaluation suffers loss of significant figures.

We claim that equations (2.5) coupled with the knowledge obtainable from equation (2.2b) is a superior formulation of the original problem.

Operationally, this alternative formulation simply involves addition and subtraction of certain terms to the right-hand side, and certain *intelligent* rearrangement of terms. The challenge is to make the above process *programmable* on a digital computer.

3. STATEMENT OF THE PROBLEM

We consider the class of initial-value problems given in the following form:

$$\frac{d\bar{y}}{dt} = \bar{G}(\bar{y}) \tag{3.1a}$$

$$\bar{y}(0) = \bar{Y} \tag{3.1b}$$

where $\bar{y}$ is an N-dimensional vector with components $\bar{y}_1(t), \bar{y}_2(t), \ldots, \bar{y}_N(t)$. Because the present work is strongly motivated by chemical kinetics, we shall in what follows adopt many of its terminologies. We shall call $\bar{y}$ the *state vector*, $\bar{Y}$ its initial value, and $\bar{G}(\bar{y})$ the *global reaction-rate vector*. Without loss of generality, $\bar{G}(\bar{y})$ is assumed to be given in the following form:

$$\bar{G}(\bar{y}) = \sum_{r=1}^{R} \bar{S}_r(\bar{y}) F_r(\bar{y}) \tag{3.2}$$

where $\bar{S}_r(\bar{y})$ and $F_r(\bar{y})$ are to be called the *stoichiometric vector* and the *reaction rate* of the *r*th *elementary reaction*, respectively. It is clear that for a given $\bar{G}(\bar{y})$, the form of equation (3.2) is nonunique. We shall assume that the given form has been arrived at with sound physical considerations so that each of the elementary reactions represents a *physically meaningful* physical process or mechanism to the researcher studying the problem. In chemical kinetics where each *elementary reaction* in fact represents a chemical reaction, the stoichiometric vector is normally a constant vector whose components are the stoichiometric coefficients of the chemical reaction equation, and the reaction rate is the algebraic sum of the forward and reverse rates. The number of such elementary reactions, R, may be greater, less than, or equal to N, the dimensionality of the state vector space.

The question is as follows: when the given problem is found to be stiff, can we find a vector function $\bar{G}^{00}(\bar{y})$ such that $\bar{G}(\bar{y})$ can be rewritten in the form

$$\bar{G}(\bar{y}) = \bar{G}^{00}(\bar{y}) + \bar{G}''(\bar{y}) \tag{3.3}$$

with the auxiliary knowledge that whenever severe loss of significant figures occurs in the evaluation of $\bar{G}^{00}(\bar{y})$, it is theoretically small and unimportant in comparison to $\bar{G}''(\bar{y})$?

4. FORMULATION OF THE EXTENDED PROBLEM

We shall assume that the problem under study is indeed stiff, i.e., it does involve some fast *and* some slow reactions. Fastness and slowness is usually measured against the desired time-resolution of the eventual computer printouts. If a reaction exhausts itself in one printout time-step or less, it *is* a fast reaction. We shall further assume that we are primarily interested in the slow-time period when the fast reactions are pretty well spent and have all reached quasi-equilibrium conditions.

We begin by dividing the N-dimensional global reaction-rate space into two subspaces: a fast subspace and a slow subspace. We tentatively resolve $\bar{G}$ into two components,

$$\bar{G} = \bar{G}^0 + \bar{G}' \tag{4.1}$$

where $\bar{G}^0$ and $\bar{G}'$ are the fast and the slow components, respectively.

We now need an expression for $\bar{G}^0(\bar{y})$. To do so we shall employ the concept of basis vectors. We shall formally span the as yet unknown fast subspace by a set of n linearly independent basis vectors $(\bar{a}_1^0(\bar{y}), \ldots, \bar{a}_n^0(\bar{y}))$ where $N > n > 0$. Since in general the fast subspace does not necessarily have a metric (i.e., $\bar{G}^0 \cdot \bar{G}^0$ is not physically meaningful), we shall need to associate $\bar{a}_i^0(\bar{y})$ with a set of dual basis vectors which satisfies the orthonormal relations

$$\bar{a}_i^0(\bar{y}) \cdot \bar{b}_j(\bar{y}) = \delta_{ij}, \qquad i, j = 1, 2, \ldots, n \tag{4.2}$$

where δ_{ij} is the Kronecker delta. Note that since $n < N$, equation (4.2) does not uniquely determine $\bar{b}_j^0(\bar{y})$s for a given set of $\bar{a}_i^0(\bar{y})$s. We shall in fact take advantage of this nonuniqueness later.

We can formally express $\bar{G}^0$ as follows:

$$\bar{G}^0 = \sum_{i=1}^{n} \bar{a}_i^0 F_i^0 \tag{4.3}$$

We can readily interpret $\bar{a}_i^0$ and F_i^0 as the stoichiometric vector and the net reaction rate of the ith fast-reaction group, and n as the number of such reaction groups. We shall define, tentatively, the slow subspace by the following requirement:

$$\bar{b}_j^0 \cdot \bar{G}' = 0, \qquad j = 1, 2, \ldots, n \tag{4.4}$$

Using equations (4.1)–(4.3) in equation (4.4), we obtain

$$F_i^0 = \bar{b}_i^0 \cdot \bar{G} \tag{4.5a}$$

$$= \sum_{r=1}^{R} [\bar{b}_i^0(\bar{y}) \cdot \bar{S}_r(\bar{y})] F_r(\bar{y}), \qquad i = 1, 2, \ldots, n \tag{4.5b}$$

It is to be emphasized here that any attempt to directly evaluate F_i^0 using (4.5b) under quasi-equilibrium conditions will suffer severe loss of significant figures.

We can now write $\bar{G}'$ as follows:

$$\bar{G}' \equiv \bar{G}^0 - \bar{G} \tag{4.6a}$$

$$= \left[\bar{\bar{I}} - \sum_{i=1}^{n} \bar{a}_i^0 \bar{b}_i^0 \right] \cdot \bar{G} \tag{4.6b}$$

where $\bar{\bar{I}}$ is the unit tensor. Alternatively, we have

$$\bar{G}'(y) = \sum_{r=1}^{R} \bar{S}_r' F_r(\bar{y}) \tag{4.6c}$$

$$\bar{S}_r'(\bar{y}) = \left[\bar{\bar{I}} - \sum_{i=1}^{n} \bar{a}_i^0 \bar{b}_i^0 \right] \cdot \bar{S}_r(\bar{y}) \tag{4.6d}$$

Note that $\bar{S}_r'(\bar{y})$ is simply the slow component of $\bar{S}_r(\bar{y})$, and is identically zero whenever $\bar{S}_r(\bar{y})$ lies entirely inside the fast subspace.

To avoid the use of equation (4.5b) for evaluation of F_i^0, we now derive its own differential equation by differentiating equation (4.5a) with respect to time. We obtain

$$\frac{dF_i^0}{dt} = \sum_{j=1}^{n} \Omega_{ij} F_j^0 + \bar{B}_i \cdot \bar{G}' \tag{4.7}$$

where

$$\bar{B}_i(\bar{y}) \equiv \nabla(\bar{G} \cdot \bar{b}_i^0) \tag{4.8a}$$

$$\Omega_{ij}(\bar{y}) \equiv \bar{a}_j^0 \cdot \bar{B}_i, \qquad i, j = 1, 2, \ldots, n \tag{4.8b}$$

and ∇ is the gradient operator in $\bar{y}$ space. The quantity $\bar{B}_i$ can alternatively be written as

$$\bar{B}_i = \frac{d\bar{b}_i^0}{dt} + (\nabla \bar{G}) \cdot \bar{b}_i^0 \tag{4.8c}$$

where $\nabla \bar{G}$ is simply the Jacobian of $\bar{G}$ with respect to $\bar{y}$. Note that Ω_{ij} is a

$\bar{y}$-dependent $n \times n$ matrix whose eigenvalues have the dimension of reciprocal time. In general, Ω_{ij} is not diagonal, and its eigenvalues vary with time because of its $\bar{y}$ dependence. However, we can require that the magnitude of each of the eigenvalues be larger than some specified value, say $(\Delta t)^{-1}$, where Δt is the desired printout time-resolution. This requirement guarantees that Ω_{ij} is nonsingular and therefore possesses an inverse $\tau_{ij}(\bar{y})$:

$$\sum_{i=1}^{n} \tau_{ij}\Omega_{ik} = \delta_{jk}, \quad \sum_{i=1}^{n} \tau_{ij}\Omega_{ki} = \delta_{jk}, \quad j, k = 1, 2, \ldots, n \quad (4.9)$$

Note that the eigenvalues of τ_{ij} have the dimension of time, and are all expected to be small in comparison to the desired printout time-resolution Δt. Equation (4.7) can now be written in the cleaner form

$$\frac{dF_i^0}{dt} = \sum_{j=1}^{n} \Omega_{ij}(F_j^0 - Q_j^0), \quad i = 1, 2, \ldots, n \quad (4.10)$$

where

$$Q_j^0 = \bar{b}_j^{00} \cdot \bar{G}' \quad (4.11)$$

and $\bar{b}_j^{00}(y)$, whose notation has been chosen advisedly, is given by

$$\bar{b}_j^{00}(\bar{y}) = \sum_{i=1}^{n} \tau_{ij}\bar{B}_i \quad (4.12)$$

Equation (4.10) is the generalized version of equation (2.2b) in the earlier example.

We can readily show that

$$\bar{a}_i^0 \cdot \bar{b}_j^{00} = \delta_{ij}, \quad i, j = 1, 2, \ldots, n \quad (4.13)$$

from equations (4.12), (4.9), and (4.8b). Hence $\bar{b}_j^{00}$ is eligible to serve as the set of dual basis vectors to $\bar{a}_i^0$. We shall presently show that $\bar{b}_j^{00}$ is in fact a superior set to use than $\bar{b}_j^0$. In fact, $\bar{b}_j^{00}$ is analytically the second approximation to $\bar{b}_j^0$.

Confining our attention to the case where all n eigenvalues of Ω_{ij} have sufficiently large negative real parts, we expect F_i^0 to approach Q_i^0 asymptotically in the quasi-equilibrium period. Hence G^0 approaches $\sum_{i=1}^{n} \bar{a}_i^0 Q_i^0$ but this contribution is *not*, in general, necessarily small or unimportant in comparison to $\bar{G}'$. In fact, usually this contribution *is* important. We are thus motivated to introduce F_j^{00} as follows:

$$F_j^{00} = F_j^0 - Q_j^0, \quad j = 1, 2, \ldots, n \quad (4.14)$$

Physically, F_j^{00} is the discrepancy between the exact F_j^0 and its value under the quasi-equilibrium approximations. Hence, F_j^{00} *is* now known to be small and unimportant.

Using equations (4.5a), (4.11), and (4.13) in equation (4.14), we can obtain

$$F_j^{00} = \bar{b}_j^{00} \cdot \bar{G}$$

$$= \sum_{r=1}^{R} (\bar{b}_j^{00} \cdot \bar{S}_r) F_r(\bar{y}) \tag{4.15}$$

Using equation (4.14) to eliminate F_j^0 in favor of F_j^{00}, we can rearrange the terms and rewrite equation (4.1) as follows:

$$\bar{G} = \bar{G}^{00} + \bar{G}'' \tag{4.16a}$$

where

$$\bar{G}^{00} = \sum_{i=1}^{n} \bar{a}_i^0 F_i^{00} \tag{4.16b}$$

$$\bar{G}'' = \sum_{r=1}^{R} \bar{S}_r''(\bar{y}) F_r(\bar{y}) \tag{4.16c}$$

and

$$S_r'' = \left[\bar{\bar{I}} - \sum_{i=1}^{n} \bar{a}_i^0 \bar{b}_i^{00} \right] \cdot \bar{S}_r(\bar{y}), \qquad r = 1, 2, \ldots, R \tag{4.16d}$$

Eliminating F_i^0 from equation (4.10), we have

$$\frac{dF_i^{00}}{dt} = \sum_{k=1}^{n} \Omega_{ik} (F_k^{00} - Q_k^{00}) \tag{4.17}$$

where

$$Q_k^{00} = \sum_{l=1}^{n} \tau_{lk} \frac{dQ_l^0}{dt}, \qquad i, k = 1, 2, \ldots, n \tag{4.18}$$

Since the eigenvalues of τ_{lk} are all expected to be small, Q_k^{00} and F_k^{00} and $\bar{G}^{00}$ are now theoretically small and are therefore unimportant. If $\bar{G}^{00}$ is simply neglected, we would recover the conventional quasi-equilibrium approximation, $\bar{G} \simeq \bar{G}''$.

In general, at least n, possibly more, elementary stoichiometric vectors can be completely contained in the fast subspace. These elementary stoichiometric vectors must be indentified, and their corresponding slow components $\bar{S}_r$ set explicitly to zero. We shall see that these considerations will later lead to the concept of "poorly posed" stiff systems.

5. THE REPRESENTATION OF THE FAST SUBSPACE

If the originally given problem were *linear* with respect to $\bar{y}$, the development in the subsequent sections is completely trivial. The Jacobian $(\nabla \bar{G})$ would be a *constant* $N \times N$ matrix, and one could choose constant vectors $\bar{a}_i^0$, $\bar{b}_j^0$ which span the subspace of its fast eigenvectors (with fastness appropriately specified). Determination of the fast subspace of a given $N \times N$ real matrix is a straightforward matter on digital computers. For the linear case, this fast subspace will be valid for all time and all initial conditions.

For general, nonlinear problems, particularly one involving a large number of elementary reactions, the Jacobian $\nabla \bar{G}$ is not a constant matrix, and its individual fast eigenvectors are also expected to vary with time. However, the fast subspace spanned by all the fast eigenvectors is, intuitively, expected to remain constant if the *identities* of the fast elementary reactions remain unchanged. The above intuitive reasoning provides a guide for what follows.

For nonlinear problems, the derivation of the alternative problem must be done in parallel with the computations for the solutions. One may, of course, choose to use the constant vectors $\bar{a}_i(\bar{Y})$, $\bar{b}_j(\bar{Y})$ that span the fast subspace of $\nabla \bar{G}(\bar{Y})$ at $t = 0$, and update these constant vectors only when necessary. But little physical insights or understanding would be gained from the output of these additional computations.

Since $\bar{a}_i^0(\bar{y})$ has the physical meaning of being the stoichiometric vector of the ith fast-reaction group, we propose to express it in terms of the physically meaningful elementary stoichiometric vectors $\bar{S}_r(\bar{y})$. We shall, without loss of generality, first assume that $R = N$. If the given $\bar{G}(\bar{y})$ has fewer elementary reactions than the number of components for $\bar{y}$, additional elementary reactions with zero reaction rate can be added to make $R = N$. We shall next assume, without loss of generality, that the N elementary stoichiometric vectors are linearly independent. Thus the set of vectors $\bar{S}_m(y) = [S_1(\bar{y}), S_2(\bar{y}), \ldots, S_N(\bar{y})]$ is a usable set of basis vectors spanning the global reaction-rate space.

Hence instead of using constant vectors $\bar{a}_i(\bar{Y})$, $\bar{b}_j(\bar{Y})$ and updating them as the computation proceeds, we may take advantage of the availability of this set of physically meaningful basis vectors to express $\bar{a}_i^0(\bar{y})$ as follows:

$$\bar{a}_i^0(\bar{y}) = \sum_{m=1}^{N} \alpha_{im} \bar{S}_m(\bar{y}), \qquad i = 1, 2, \ldots, n \tag{5.1}$$

where α_{im} is a (unknown) constant $n \times N$ matrix. We shall impose the following condition on $\bar{b}_j^0(y)$:

$$K_{jm} = \bar{b}_j^0(\bar{y}) \cdot \bar{S}_m(\bar{y}), \qquad j = 1, 2, \ldots, n, \quad m = 1, 2, \ldots, N \tag{5.2}$$

where K_{jm} is another (unknown) constant $n \times R$ matrix. These two matrices are not independent — they are related by equation (4.2), the orthonormal relation between $\bar{a}_i^0$ and $\bar{b}_j^0$:

$$\sum_{m=1}^{N} \alpha_{im} K_{jm} = \delta_{ij}, \qquad i, j = 1, 2, \ldots, n \tag{5.3}$$

It is intuitively clear that the use of these constant matrices (α_{im} and K_{jm}) to represent $\bar{a}_i^0(\bar{y})$ and $\bar{b}_j^0(\bar{y})$ is preferable to the use of constant basis vectors $\bar{a}_i(\bar{Y})$, $\bar{b}_i(\bar{Y})$ suggested earlier. Updating of these constant matrices would only become necessary if the physical processes controlling the fast reactions are changed.

6. THE INTERPRETATION AND DETERMINATION OF α_{im}, K_{jr}

Each of the constant matrices, α_{im} and K_{jr}, has clear physical interpretations. In fact we shall see presently that they contain most of the "physical insights" (if not more) in quantitative terms about the stiff problem at hand that traditional theoreticians claim they acquire as a consequence of their analysis and experience.

In general, the most interesting questions on highly complicated stiff systems are: (1) how many independent fast-reaction groups are there, (2) which elementary reactions are major participants in each of the fast-reaction groups, (3) what are the quasi-equilibrium equations of state for each of the fast-reaction groups, and finally, (4) is the data provided for the problem capable of yielding accurate answers. The answers to the above questions are of course not invariant with time and initial conditions.

In the present development, when the constant matrices α_{im} and K_{jr} are suitably determined (and updated as required), the answers to these questions are immediate. The number of independent fast-reaction groups is simply n, and the stoichiometric vector of each fast-reaction group is $\bar{a}_i^0$ as given by equation (5.1). It is seen that α_{im} is the "influence coefficient" of the mth elementary reaction on the ith fast-group stoichiometric vector. The net reaction rate F_i^0 of the ith fast group reaction is given by equation (4.5) and can be expressed in terms of K_{ir} as follows:

$$F_i^0(\bar{y}) = \sum_{r=1}^{N} K_{ir} F_r(\bar{y}), \qquad i = 1, 2, \ldots, n \tag{6.1}$$

Hence K_{ir} is the "influence coefficient" of the rth elementary reaction on the ith fast-group reaction rate. Under quasi-equilibrium conditions when the net fast-group reactions are nearly exhausted, the approximate quasi-equilibrium equations of state are given by *neglecting* F_i^0 from equations (6.1), yielding

$$\sum_{r=1}^{N} K_{ir} F_r(\bar{y}) \cong 0, \qquad i = 1, 2, \ldots, n \qquad (6.2)$$

Finally, the present development identifies the term $\bar{G}^{00}$ as being theoretically unimportant in the quasi-equilibrium period and equations (4.16b) and (4.17) are provided if it is to be kept in the computations. Therefore the issue of whether the originally posed problem is basically capable of generating accurate solutions reduces to asking whether $\bar{G}''$ as given by equation (4.16c) is capable of being accurately evaluated. We shall see that this consideration shall serve as a guide for the determination of the constant matrices.

A perceptive reader would have realized that $\bar{G}''$ and $\bar{G}$ are expressed in the same general format except that $\bar{S}''_r$ is used in place of $\bar{S}_r$. Obviously if all N of the $\bar{S}''_r$s are nonzero vectors, so that all N of the elementary reaction rates $F_r(y)$, fast as well as slow ones, participate in the evaluation of $\bar{G}''$, then whatever near-cancellation difficulties originally were encountered in the evaluation of G could remain. Hence, $\bar{G}''$ can only be accurately evaluated if all of the $\bar{S}''_r$s associated with the fast elementary reactions are *identically* zero in order to avoid loss of significant figures. Note that these fast $\bar{S}''_r$s being numerically small in *not* sufficient, since their contributions may be augmented by large numerical factors.

Thus we arrive at the important conclusion that the constant matrices must be chosen so that all of the fast $\bar{S}''_r$s be identically null vectors. If this requirement should prove impossible to meet, then *the originally posed problem is poorly posed and basically incapable of generating accurate solutions.* The problem posed by equations (1.3) is an example of a poorly posed problem — neither computer nor analysis can overcome such basic defects in the original problem.

From the physical interpretations of α_{im} and K_{ir}, these constant matrices are expected to remain constant so long as the mix of elementary reactions making up the fast-reaction groups remain unchanged. Thus no updating is required until the basic physics involved is altered. Thus even the updating process is physically meaningful.

7. A RECOMMENDED SCHEME FOR THE DETERMINATION OF α_{im}, K_{jr}

In the process of numerically integrating the originally posed problem, one may at any time (e.g., $t = 0$, $\bar{y} = \bar{Y}$) compute the eigenvalues and eigenvectors of $\nabla G(\bar{Y})$, and determine n and a set of constant basis vectors $\bar{a}_i(\bar{Y})$, $\bar{b}_i(\bar{Y})$ spanning the fast subspace at $t = 0$.

We can simply choose $\bar{b}_i^0(\bar{y})$ indirectly as follows:

$$K_{ir} = \bar{b}_i(\bar{Y}) \cdot \bar{S}_r(\bar{Y}) = \bar{b}_i^0(y) \cdot \bar{S}_r(\bar{y}) \quad i = 1, 2, \ldots, n, \quad r = 1, 2, \ldots, N \quad (7.1)$$

Thus K_{ir} is completely determined. Instead of determining α_{im} by looking solely at

$$\bar{a}_i(\bar{y}) = \sum_{m=1}^{N} \alpha_{im} \bar{S}_m(\bar{y}), \qquad i = 1, 2, \ldots, n \qquad (7.2)$$

we imposed the requirement of the fast $\bar{S}_r''$s as mentioned earlier. It can readily be shown that

$$\bar{S}_r'' = \left[\bar{\bar{I}} - \sum_{i=1}^{n} \bar{a}_i^0 \bar{b}_i^{0\infty} \right] \cdot \bar{S}_r' \qquad (7.3)$$

[compare this with equation (4.16d)]. Thus $\bar{S}_r''$ is null whenever $\bar{S}_r'$ is null. From equation (4.6d), we have

$$\bar{S}_r' = \bar{S}_r - \sum_{i=1}^{n} \bar{a}_i^0 K_{ir}, \qquad r = 1, 2, \ldots, N \qquad (7.4)$$

Using equation (5.1), we have

$$\bar{S}_r' = \bar{S}_r - \sum_{m=1}^{N} \left(\sum_{i=1}^{n} \alpha_{im} K_{ir} \right) \bar{S}_m, \qquad r = 1, 2, \ldots, N \qquad (7.5)$$

Hence, $\bar{S}_r'(y)$ is a null vector [and consequently $\bar{S}_r''(y)$ is null] if

$$\sum_{i=1}^{n} \alpha_{im} K_{ir} = \delta_{mr}, \qquad m = 1, 2, \ldots, N, \qquad \text{for selected } rs \qquad (7.6)$$

This equation in conjunction with equation (5.3) must be respected by α_{im}. Since there are $n \times n$ unknown elements in α_{im}, and each selected r in equation (7.6) provides N equations, the total number of $\bar{S}_r'$ that can be set to zero is thus the largest integer *below* $n \times (1 - n/N)$. Theoretically, the required number of null $\bar{S}_r'$ is n (since the set of $\bar{S}_r$s are assumed linearly independent). Thus in general, it is not theoretically possible to guarantee that *all* of the fast elementary reactions be removed from $\bar{G}''$. In other words, near cancellations of big terms in general remains in $\bar{G}''$.

However, when a problem is *properly posed*, some $\bar{S}'_r$ will be null vectors automatically, relieving the necessity of imposing equation (7.6) for that particular r. Physically, this means that some of the elementary reactions deemed meaningful to the investigator are, without complications, "pure" fast reactions. Only for such cases can all of the fast $\bar{S}''_r$s be set to zero, allowing accurate solutions to be generated.

The actual algorithms to be used in determining whether the originally posed problem is poorly posed, and the computations of α_{im}, K_{ir} are routine programming problems that are clearly solvable. Once the α_{im}, K_{ir}s are determined, updated, and outputed with the solutions, the "physical insights" available from these computer-generated numbers should be fully competitive with the traditional analytical solutions. Even though the additional computation required is considerable, once the problem is physically understood the simpler recommended problem

$$\frac{d\bar{y}}{dt} \cong \bar{G}''(\bar{y}) \tag{7.7}$$

can be used for further computations and studies.

CONCLUDING REMARKS

When I was working on my thesis under Professor Crocco, I once suggested an *ad hoc* assumption that could remove all the mathematical and computational difficulties then plaguing my problem. Professor Crocco looked at me gently, and said "Lam, a computed solution containing *ad hoc* assumptions is no better than one sketched by hand." I have remembered this lesson ever since.

REFERENCES

1. C. W. Gear, The automatic integration of ordinary differential equations, *Comm. ACM* **14**, 176 (1971).
2. C. W. Gear, *Numerical Initial Value Problems in Ordinary Differential Equations*, Prentice-Hall, Englewood Cliffs, New Jersey (1971).
3. L. Lapidus and J. H. Seinfeld, *Numerical Solution of Ordinary Differential Equations*, Academic Press, New New York (1971).
4. R. A. Willoughby, *Stiff Differential Systems*, Plenum Press, New York (1973).
5. R. W. Klopfenstein and C. B. Davis, ECE algorithms for the solution of stiff systems of ordinary differential equations, *Math. Comput.* **25**, 457 (1971).
6. C. K. Westbrook, J. Creighton, and C. Lund, A numerical model of chemical kinetics of combustion in a turbulent flow reactor, *J. Phys. Chem.* **81**, 23, 1977.
7. Heinz-Otto Kreiss, Numerical methods for singular perturbation problems, in: *Asympto-*

tic Methods and Singular Perturbations, SIAM-AMS Proceedings, Vol. X, American Mathematical Society (1976).

8. S. H. Lam, Singular perturbation problems with nearly resonant operators, *J. Appl. Math. Phys.* (ZAMP) **28**, 817 (1977).

9. R. E. O'Malley, Jr., *Introduction to Singular Perturbations*, Academic Press, New York (1974).

10. W. Liniger and C. A. Willoughby, Efficient integration methods for stiff systems of ordinary differential equations, *SIAM, J. Numer. Anal.* **7**, 1 (1970).

11. B. L. Ehle and J. D. Lawson, Generalized Runge Kutta processes for stiff initial value problems, *J. Inst. Math. Appl.* **16**, 11 (1975).

12. J. D. Ramshaw, Partial chemical equilibrium in fluid dynamics, *Phys. Fluids* **23**, 675 (1980).

2

Minimax Optimal Control and Its Application to the Reentry of a Space Glider

A. MIELE AND P. VENKATARAMAN

ABSTRACT. A transformation technique is employed in order to convert minimax problems of optimal control (also called Chebyshev problems) into Mayer–Bolza problems of the calculus of variations. The transformation requires the proper augmentation of the state vector $x(t)$, the control vector $u(t)$, and the parameter vector π, as well as the proper augmentation of the constraining relations. As a result of the transformation, the unknown minimax value of the performance index becomes a component of the vector parameter π being optimized.

The transformation technique is then applied to the following Chebyshev problems of interest in the reentry of a space glider: (Q1) minimization of the peak dynamic pressure; and (Q2) minimization of the peak heating rate.

Numerical results are obtained by means of the sequential gradient-restoration algorithm for solving optimal control problems on a digital computer. Reference is made to the hypervelocity regime, an exponential atmosphere, and a space glider whose trajectory is controlled by means of the angle of attack and the angle of bank.

1. INTRODUCTION

In recent years, considerable research has been done on the problem of optimizing a trajectory from the standpoint of an integral performance

A. MIELE AND P. VENKATARAMAN • Rice University, Department of Mechanical Engineering and Materials Science, P.O. Box 1892, Houston, TX 77251, U.S.A.

index. This problem can be formulated in the following form, called Problem (P) for easy identification.

PROBLEM (P). Minimize the functional

$$I = \int_0^1 f(x, u, \pi, t)dt + [h(x, \pi)]_0 + [g(x, \pi)]_1 \tag{1}$$

with respect to the state $x(t)$, the control $u(t)$, and the parameter π, which satisfy the following constraints:

$$\dot{x} = \phi(x, u, \pi, t), \qquad 0 \le t \le 1 \tag{2}$$

$$S(x, u, \pi, t) = 0, \qquad 0 \le t \le 1 \tag{3}$$

$$[\omega(x, \pi)]_0 = 0 \tag{4}$$

$$[\psi(x, \pi)]_1 = 0 \tag{5}$$

In the above equations, the functions f, h, g are scalar, and the functions ϕ, S, ω, ψ are vectors of appropriate dimensions. The independent variable t (the time) is a scalar, and the dependent variables x, u, π are vectors of appropriate dimension. The subscript 0 denotes the initial point, and the subscript 1 denotes the final point.

In the terminology of the calculus of variations, the above problem is called the Bolza problem and includes as particular cases the Lagrange problem and the Mayer problem. The former occurs when $h \equiv 0$, $g \equiv 0$, and the latter occurs when $f \equiv 0$. For the above Bolza problem, the necessary conditions for an extremum can be found, for example, in Bliss,[1] Hestenes,[2] and Pontryagin et al.[3] Computer algorithms of the first-order type can be found, for instance, in work by Miele and co-workers[4-6]; for recent surveys of gradient algorithms for the optimization of dynamic systems, see Miele.[7,8]

2. MINIMAX PROBLEMS

Within the formulation represented by Problem (P), an important class of problems has been omitted. These problems occur when the minimization of the integral performance index (1) is replaced with the minimization of a local performance index having the following form:

$$I = \max_t F, \qquad F = F(x, \pi, t), \qquad 0 \le t \le 1 \tag{6}$$

Thus, the following optimal control problem arises.

· PROBLEM (Q). Minimize the functional (6) with respect to the state $x(t)$, the control $u(t)$, and the parameter π, which satisfy the constraints (2)–(5).

Problem (Q) is a minimax problem: the objective is to minimize the maximum value achieved along the interval of integration by some function of the variables of the problem. Problem (Q) is a nonclassical problem of the calculus of variations, in that it is not a particular case of the Bolza problem. It is called the Chebyshev problem.

It must be noted that Chebyshev-type problems occur frequently in various branches of engineering. In aerospace engineering, the following Chebyshev problems are of interest for the reentry of a space glider: (Q1) minimization of the peak dynamic pressure; and (Q2) minimization of the peak heating rate.

For previous research on the analytical and/or numerical solution of Problem (Q), see elsewhere.[9-14]

3. TRANSFORMATION TECHNIQUE

In recent years, this writer and his associates have had considerable success in the development of transformation techniques that convert an optimal control problem into another.[7,8] More specifically, the idea is to bring a problem for which algorithms are not available into an equivalent problem that can be treated with some of the existing algorithms.

By suitable transformations, minimax problems of Type (Q) can be brought into the scheme of Problem (P). Hence, the sequential gradient-restoration algorithm developed by Gonzalez and Miele[6] can be applied[13,14]

More specifically, a minimax problem of Type (Q) can be converted into Problem (P) by exploiting the analogy with a bounded-state problem in combination with a transformation of the Jacobson type.[15] Generally speaking, this transformation results in an increase in the dimension of the state vector. However, in some cases, the increase in the dimension of the state vector can be limited or prevented by using a transformation of the Miele–Wu–Liu type.[16]

In order to illustrate the transformation technique, we refer to the problem represented by equations (2)–(5) and (6).

For any admissible choice of the state $x(t)$, the control $u(t)$, and the parameter π, let F_* denote the maximum value (or peak value) achieved by the function $F(x, \pi, t)$ along the interval of integration. With this understanding, the functional (6) can be rewritten as

$$I = F_*\tag{7}$$

$$F_* - F(x, \pi, t) \geq 0, \qquad 0 \leq t \leq 1\tag{8}$$

As a consequence, problem (2)–(6) is now replaced with problem (2)–(5) and (7)–(8). This is a Bolza problem, complicated by the fact that the state inequality constraint (8) must be satisfied everywhere along the trajectory.

The conversion of problem (2)–(5) and (7)–(8) into Problem (P) requires the proper augmentation of the state vector, the control vector, and the parameter vector, as well as the proper augmentation of the constraining relations. In this connection, an important element is the order of the state inequality constraint (8), i.e., the order of the minimax function $F(x, \pi, t)$.

A minimax function $F(x, \pi, t)$ is defined to be of order k if the kth total time derivative of $F(x, \pi, t)$ is the *first* to contain the control explicitly. As an example, if the minimax function is of order $k = 1$, we have

$$F = F(x, \pi, t), \qquad \dot{F} = G(x, u, \pi, t)\tag{9}$$

As another example, if the minimax function is of order $k = 2$, we have

$$F = F(x, \pi, t), \qquad \dot{F} = G(x, \pi, t), \qquad \ddot{F} = H(x, u, \pi, t)\tag{10}$$

We note that, in aerospace engineering, the case $k = 1$ occurs frequently, while the case $k = 2$ occurs rarely. Therefore, we limit ourselves to the case $k = 1$.

Case $k = 1$. This case is treated by introducing the auxiliary state variable $y(t)$ and the auxiliary control variable $w(t)$, defined by

$$F_* - F(x, \pi, t) = y^2, \qquad \dot{y} = w\tag{11}$$

with the implication that [see equations (9)]

$$G(x, u, \pi, t) + 2yw = 0\tag{12}$$

As a consequence, the inequality constrained problem (2)–(5) and (7)–(8) is replaced with the following equality constrained problem:

$$I = F_*\tag{13}$$

$$\dot{x} = \phi(x, u, \pi, t), \qquad 0 \leq t \leq 1\tag{14a}$$

$$\dot{y} = w, \qquad 0 \leqslant t \leqslant 1 \tag{14b}$$

$$S(x, u, \pi, t) = 0, \qquad 0 \leqslant t \leqslant 1 \tag{15a}$$

$$G(x, u, \pi, t) + 2yw = 0, \qquad 0 \leqslant t \leqslant 1 \tag{15b}$$

$$[\omega(x, \pi)]_0 = 0 \tag{16a}$$

$$[F_* - F(x, \pi, t) - y^2]_0 = 0 \tag{16b}$$

$$[\psi(x, \pi)]_1 = 0 \tag{17}$$

Next, we augment the state, the control, and the parameter as follows:

$$\tilde{x} = [x^{\mathrm{T}}, y]^{\mathrm{T}}, \qquad \tilde{u} = [u^{\mathrm{T}}, w]^{\mathrm{T}}, \qquad \tilde{\pi} = [\pi^{\mathrm{T}}, F_*]^{\mathrm{T}} \tag{18}$$

Also, we augment the constraining functions as follows:

$$\tilde{\phi} = [\phi^{\mathrm{T}}, w]^{\mathrm{T}}, \qquad \tilde{S} = [S^{\mathrm{T}}, G + 2yw]^{\mathrm{T}} \tag{19a}$$

$$\tilde{\omega} = [\omega^{\mathrm{T}}, F_* - F - y^2]^{\mathrm{T}}, \qquad \tilde{\psi} = \psi \tag{19b}$$

With this understanding, we see that problem (13)–(17) is identical with Problem (P). In (18)–(19), the superscript T denotes transposition of vector or matrix.

4. EQUATIONS OF MOTION

In this section, we consider the reentry of a space glider whose trajectory is controlled by means of the angle of attack α and the angle of bank σ. The following hypotheses are made: (i) the space glider is a particle of constant mass; (ii) the Earth is spherical and nonrotating; (iii) the acceleration of gravity g is assumed constant, $g = g_0$; (iv) the altitude above sea level h is negligible by comparison with the radius of the Earth r_0; (v) the path inclination γ is sufficiently small, so that the approximations $\sin \gamma \cong \gamma$, $\cos \gamma \cong 1$ hold; (vi) the drag D and the lift L have the form $D = D(h, V, \alpha)$, $L = L(h, V, \alpha)$, where V is the velocity.

With the above assumptions, and upon normalizing the flight time to unity, the equations of motion are given by[17]

$$\dot{h} = \tau V \gamma \tag{20a}$$

$$\dot{V} = -\tau(D/m + g_0 \gamma) \tag{20b}$$

$$\dot{\gamma} = \tau(L \cos \sigma/mV + V/r_0 - g_0/V) \tag{20c}$$

$$\dot{\theta} = \tau V \cos \psi/r_0 \cos \phi \tag{20d}$$

$$\dot{\phi} = \tau V \sin \psi / r_0 \tag{20e}$$

$$\dot{\psi} = \tau [L \sin \sigma / mV - (V/r_0)\cos \psi \tan \phi] \tag{20f}$$

In the above equations, the independent variable is the normalized time t, $0 \leq t \leq 1$. The dependent variables include six state variables $(h, V, \gamma, \theta, \phi, \psi)$, two control variables (α, σ), and one parameter (τ). Specifically, h is the altitude, V is the velocity, γ is the path inclination, θ is the longitude, ϕ is the latitude, ψ is the heading angle, α is the angle of attack, σ is the angle of bank, and τ is the total flight time from the initial point to the final point.

We examine the transfer of the space glider from a given initial point to a given final point. Therefore, we assume the following boundary conditions:

$$h(0) = C_1, \qquad V(0) = C_2, \qquad \gamma(0) = C_3 \tag{21a}$$

$$\theta(0) = C_4, \qquad \phi(0) = C_5, \qquad \psi(0) = C_6 \tag{21b}$$

and

$$h(1) = C_7, \qquad V(1) = C_8, \qquad \gamma(1) = C_9 \tag{22a}$$

$$\theta(1) = C_{10}, \qquad \phi(1) = C_{11}, \qquad \psi(1) = C_{12} \tag{22b}$$

where $C_1, C_2, \ldots, C_{12}$ are prescribed constants.

5. MINIMAX PROBLEMS

We consider the following minimax problems of interest in the reentry of a space glider: (Q1) minimization of the peak dynamic pressure; and (Q2) minimization of the peak heating rate. For these problems, the functional (6) takes the specific forms given below.

PROBLEM (Q1). Minimize the functional

$$I = \max_t F, \qquad F = \rho V^2, \qquad 0 \leq t \leq 1 \tag{23a}$$

subject to the dynamical constraints (20) and the boundary conditions (21)–(22). In (23a), ρ denotes the density, $\rho = \rho(h)$. Invoking (7)–(8), (23a) can be rewritten as

$$I = F_* \tag{23b}$$

$$F_* - \rho V^2 \geq 0, \qquad 0 \leq t \leq 1 \tag{23c}$$

PROBLEM (Q2). Minimize the functional

$$I = \max_t F, \qquad F = \sqrt{\rho}\, V^3, \qquad 0 \leqslant t \leqslant 1 \qquad (24a)$$

subject to the dynamical constraints (20) and the boundary conditions (21)-(22). Invoking (7)-(8), (24a) can be rewritten as

$$I = F_* \qquad (24b)$$

$$F_* - \sqrt{\rho}\, V^3 \geqslant 0, \qquad 0 \leqslant t \leqslant 1 \qquad (24c)$$

We note that, except for a proportionality constant, the minimax function F in (23) represents the dynamic pressure and the minimax function F in (24) represents an approximation to the stagnation-point heating rate.

6. TRANSFORMATION TECHNIQUE

Problems (Q1) and (Q2) involve a minimax function of order $k = 1$. As a consequence, the transformation technique of Section 3 can be employed in order to achieve the conversion of these problems into Bolza problems involving equality constraints.

PROBLEM (Q1). This problem is treated by introducing the auxiliary state variable $y(t)$ and the auxiliary control variable $w(t)$, defined by

$$F_* - \rho V^2 - y^2 = 0, \qquad \dot{y} = \tau w \qquad (25)$$

With this understanding and upon proceeding as in Section 3, Problem (Q1) is reformulated as follows:

$$I = F_* \qquad (26)$$

$$\dot{h} = \tau V \gamma \qquad (27a)$$

$$\dot{V} = -\tau(D/m + g_0\gamma) \qquad (27b)$$

$$\dot{\gamma} = \tau(L \cos \sigma/mV + V/r_0 - g_0/V) \qquad (27c)$$

$$\dot{\theta} = \tau V \cos \psi/r_0 \cos \phi \qquad (27d)$$

$$\dot{\phi} = \tau V \sin \psi/r_0 \qquad (27e)$$

$$\dot{\psi} = \tau[L \sin \sigma/mV - (V/r_0)\cos \psi \tan \phi] \qquad (27f)$$

$$\dot{y} = \tau w \tag{27g}$$

$$\rho' V^3 \gamma - 2\rho V(D/m + g_0\gamma) + 2yw = 0 \tag{28}$$

$$h(0) = C_1, \qquad V(0) = C_2, \qquad \gamma(0) = C_3 \tag{29a}$$

$$\theta(0) = C_4, \qquad \phi(0) = C_5, \qquad \psi(0) = C_6 \tag{29b}$$

$$(F_* - \rho V^2 - y^2)_0 = 0 \tag{29c}$$

$$h(1) = C_7, \qquad V(1) = C_8, \qquad \gamma(1) = C_9 \tag{30a}$$

$$\theta(1) = C_{10}, \qquad \phi(1) = C_{11}, \qquad \psi(1) = C_{12} \tag{30b}$$

Note that, in (28), the symbol ρ' denotes the derivative of the density with respect to the altitude, $\rho' = d\rho/dh$.

PROBLEM (Q2). This problem is treated by introducing the auxiliary state variable $y(t)$ and the auxiliary control variable $w(t)$, defined by

$$F_* - \sqrt{\rho}\, V^3 - y^2 = 0, \qquad \dot{y} = \tau w \tag{31}$$

With this understanding, and upon proceeding as in Section 3, Problem (Q2) is reformulated as follows:

$$I = F_* \tag{32}$$

$$\dot{h} = \tau V\gamma \tag{33a}$$

$$\dot{V} = -\tau(D/m + g_0\gamma) \tag{33b}$$

$$\dot{\gamma} = \tau(L \cos \sigma/mV + V/r_0 - g_0/V) \tag{33c}$$

$$\dot{\theta} = \tau V \cos \psi/r_0 \cos \phi \tag{33d}$$

$$\dot{\phi} = \tau V \sin \psi/r_0 \tag{33e}$$

$$\dot{\psi} = \tau[L \sin \sigma/mV - (V/r_0)\cos \psi \tan \phi] \tag{33f}$$

$$\dot{y} = \tau w \tag{33g}$$

$$\rho' V^4 \gamma/2\sqrt{\rho} - 3\sqrt{\rho}\, V^2(D/m + g_0\gamma) + 2yw = 0 \tag{34}$$

$$h(0) = C_1, \qquad V(0) = C_2, \qquad \gamma(0) = C_3 \tag{35a}$$

$$\theta(0) = C_4, \qquad \phi(0) = C_5, \qquad \psi(0) = C_6 \tag{35b}$$

$$(F_* - \sqrt{\rho}\, V^3 - y^2)_0 = 0 \tag{35c}$$

$$h(1) = C_7, \qquad V(1) = C_8, \qquad \gamma(1) = C_9 \tag{36a}$$

$$\theta(1) = C_{10}, \qquad \phi(1) = C_{11}, \qquad \psi(1) = C_{12} \tag{36b}$$

7. DECOUPLING OF THE SYSTEM EQUATIONS

With reference to Problem (Q1), assume that the system of dynamical equations and boundary conditions (27)–(30) is separated into two smaller systems: System (R1) includes equations (27a), (27b), (27c), (27g), (28), (29a), (29c), (30a); System (S1) includes equations (27d), (27e), (27f), (29b), (30b).

Next, consider System (R1), and assume that a feasible solution has been computed for the state variables $h(t)$, $V(t)$, $\gamma(t)$, $y(t)$, the control variables $\alpha(t)$, $\sigma(t)$, $w(t)$, and the parameters τ, F_*. Observe that the angle of bank $\sigma(t)$ is present only in equation (27c) through a cosine function. As a consequence, starting from a feasible solution of System (R1), one can obtain another feasible solution by replacing $\sigma(t)$ with $-\sigma(t)$ over the interval $0 \le t \le 1$ (see Figure 1). This property is called *total reflectivity* of the angle of bank.

In an analogous manner, starting from a feasible solution of System (R1), one can obtain another feasible solution by replacing $\sigma(t)$ with $-\sigma(t)$ over the subinterval $t_1 \le t \le t_2$ (see Figure 2). This property is called *partial reflectivity* of the angle of bank. Since the switching times t_1, t_2 are arbitrary, a double infinity of feasible solutions can be generated, starting from a feasible solution of System (R1).

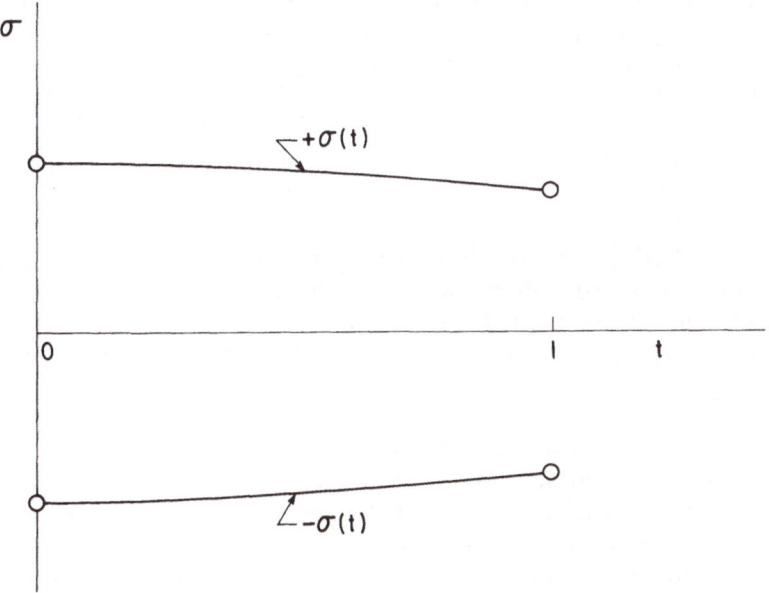

Fig. 1. Total reflectivity of the angle of bank.

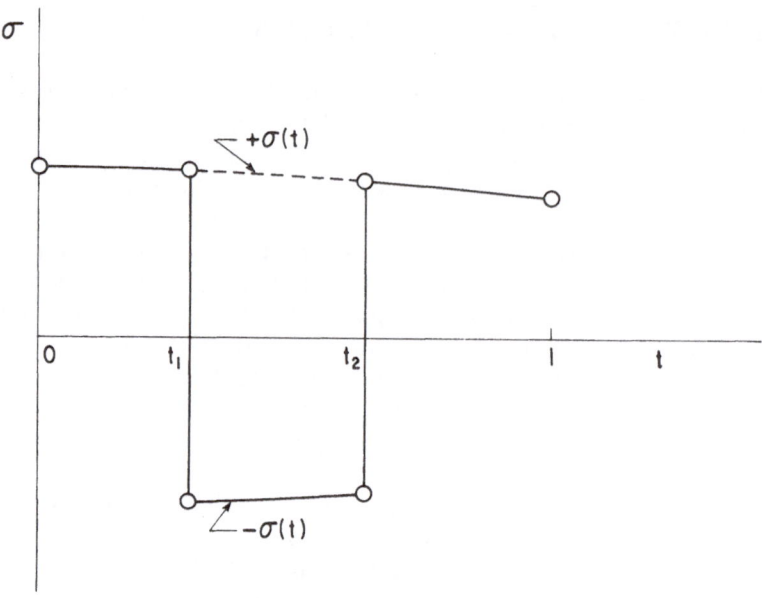

Fig. 2. Partial reflectivity of the angle of bank (two switches).

Of course, partial reflectivity of the angle of bank can be employed over several subintervals, for instance, the subintervals $t_1 \leqslant t \leqslant t_2$ and $t_3 \leqslant t \leqslant 1$ (see Figure 3). Since the switching times t_1, t_2, t_3 are arbitrary, a triple infinity of feasible solutions can be generated, starting from a feasible solution of System (R1). This allows one to decouple Problem (Q1) into a primary problem [Problem (R1)] and a secondary problem [Problem (S1)].

PROBLEM (R1). This is a Bolza problem: it involves the minimization of the functional (26), subject to the dynamical equations and boundary conditions of System (R1). Here, we compute the state variables $h(t)$, $V(t)$, $\gamma(t)$, $y(t)$, the control variables $\alpha(t)$, $\sigma(t)$, $w(t)$, and the parameters τ, F_*.

PROBLEM (S1). This is not a minimization problem, but a feasibility problem: once a solution to Problem (R1) is obtained, one employs the property of partial reflectivity of the angle of bank in order to solve System (S1). Here, we compute the state variables $\theta(t)$, $\phi(t)$, $\psi(t)$ and the switching times t_1, t_2, t_3 required to meet the final conditions (30b).

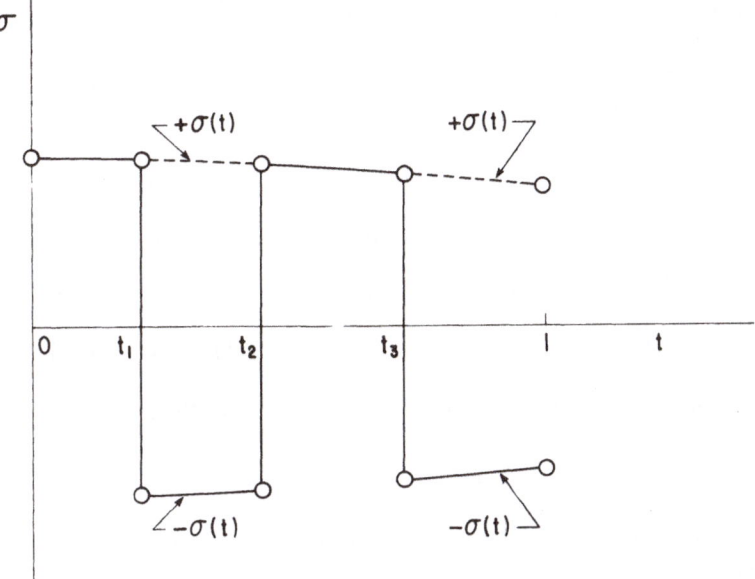

Fig. 3. Partial reflectivity of the angle of bank (three switches).

The same arguments can be extended to Problem (Q2). Here, the system of dynamical equations and boundary conditions (33)–(36) can be separated into two smaller systems: System (R2) includes equations (33a), (33b), (33c), (33g), (34), (35a), (35c), (36a); System (S2) includes equations (33d), (33e), (33f), (35b), (36b).

For System (R2), the properties of total reflectivity and partial reflectivity of the angle of bank hold. As a consequence, Problem (Q2) can be decoupled into a primary problem [Problem (R2)] and a secondary problem [Problem (S2)].

PROBLEM (R2). This is a Bolza problem: it involves the minimization of the functional (32), subject to the dynamical equations and boundary conditions of System (R2). Here, we compute the state variables $h(t)$, $V(t)$, $\gamma(t)$, $y(t)$, the control variables $\alpha(t)$, $\sigma(t)$, $w(t)$, and the parameters τ, F_*.

PROBLEM (S2). This is not a minimization problem, but a feasibility problem: once a solution to Problem (R2) is obtained, one employs the property of partial reflectivity of the angle of bank in order to solve System (S2). Here, we compute the state variables $\theta(t)$, $\phi(t)$, $\psi(t)$ and the switching times t_1, t_2, t_3 required to meet the final conditions (36b).

8. SUPPLEMENTARY ASSUMPTIONS

In solving the minimax problems of Sections 5–7, the following boundary conditions were assumed:

$$h(0) = 400,000 \text{ ft}, \quad V(0) = 24,500 \text{ ft s}^{-1}, \quad \gamma(0) = -0.020 \text{ rad} \quad \text{(37a)}$$

$$\theta(0) = 2.80 \text{ rad(E)}, \quad \phi(0) = 0.26 \text{ rad(N)}, \quad \psi(0) = 0.66 \text{ rad(N)} \quad \text{(37b)}$$

$$h(1) = 84,000 \text{ ft}, \quad V(1) = 2,500 \text{ ft s}^{-1}, \quad \gamma(1) = -0.085 \text{ rad} \quad \text{(38a)}$$

$$\theta(1) = 3.90 \text{ rad(E)}, \quad \phi(1) = 0.61 \text{ rad(N)}, \quad \psi(1) = 0.17 \text{ rad(N)} \quad \text{(38b)}$$

The drag D and the lift L were assumed to have the standard form

$$D = \tfrac{1}{2} C_D \rho S V^2, \quad L = \tfrac{1}{2} C_L \rho S V^2 \quad \text{(39)}$$

where S is a reference surface, C_D is the drag coefficient, and C_L is the lift coefficient.

An isothermal atmosphere was assumed, characterized by a constant speed of sound a. As a consequence, the density ρ in (39) was assumed to be an exponential function of the altitude,

$$\rho = \rho_0 \exp(- khg_0/a^2) \quad \text{(40)}$$

where ρ_0 is the density at sea level and $k = 1.4$ is the ratio of the specific heats.

A hypersonic drag polar was assumed, with coefficients C_D and C_L depending on the angle of attack α through the relations

$$C_D = k_1 + k_2 \sin^3 \alpha, \quad C_L = k_3 \sin^2 \alpha \cos \alpha \quad \text{(41a)}$$

where

$$k_1 = 0.07, \quad k_2 = 2.70, \quad k_3 = 3.00 \quad \text{(41b)}$$

The remaining physical constants appearing in the equations of motion were assumed to be

$$g_0 = 32.2 \text{ ft s}^{-2}, \quad r_0 = 20.9 \times 10^6 \text{ ft} \quad \text{(42a)}$$

$$a = 1,000 \text{ ft s}^{-1}, \quad \rho_0 = 2.38 \times 10^{-3} \text{ lb s}^2 \text{ ft}^{-4} \quad \text{(42b)}$$

$$S = 2,700 \text{ ft}^2, \quad mg_0 = 200,000 \text{ lb} \quad \text{(42c)}$$

9. NUMERICAL EXPERIMENTS

Problems (Q1) and (Q2) were solved employing the decomposition scheme outlined in Section 7. The primary problems [Problems (R1) and (R2)] were solved employing the sequential gradient-restoration algorithm of Gonzalez and Miele[6]; here, a single-subarc formulation was employed. The secondary problems [Problems (S1) and (S2)] were solved employing the restoration algorithm of the same authors,[6] modified to accommodate a multiple-subarc formulation (see, for instance, Miele[7,8]).

Details of the sequential gradient-restoration algorithm (SGRA) and the modified restoration algorithm (MRA) can be found elsewhere.[4-8,13,14] They are omitted here, for brevity. Both SGRA and MRA were programmed in FORTRAN IV, and the numerical results were obtained in double-precision arithmetic. Computations were performed at Rice University using an NAS-AS-9000 computer.

The interval of integration was divided into 100 steps. The differential systems were integrated using Hamming's modified predictor–corrector method, with a special Runge–Kutta starting procedure. Definite integrals were computed using a modified Simpson's rule. Linear-algebraic systems were solved using a standard Gaussian elimination routine. For both the gradient phase and the restoration phase, the linear two-point boundary-value problem was solved using the method of particular solutions.[4-8,13,14]

10. NUMERICAL RESULTS

The results presented here have a preliminary nature. They were obtained as follows: (i) the primary problems [Problems (R1) and (R2)] were solved by keeping the modulus of the angle of bank constant along the interval of integration; (ii) the secondary problems [Problems (S1) and (S2)] were solved by allowing three switching times (t_1, t_2, t_3) for the angle of bank.

Figures 4–7 refer to Problem (Q1). They show the distribution of altitude (Figure 4), velocity (Figure 5), dynamic pressure (Figure 6), and heating rate (Figure 7). Figures 8–11 refer to Problem (Q2). They show the distribution of altitude (Figure 8), velocity (Figure 9), dynamic pressure (Figure 10), and heating rate (Figure 11).

In Figures 4–11, the abscissa is the normalized time t. Concerning the ordinate, the following normalization factors were used for the altitude, velocity, dynamic pressure, and heating rate:

$$V_R = a, \qquad h_R = a^2/g_0 \tag{43a}$$

$$\rho_R V_R^2 = \rho_0 a^2, \qquad \sqrt{\rho_R} \, V_R^3 = \sqrt{\rho_0} \, a^3 \tag{43b}$$

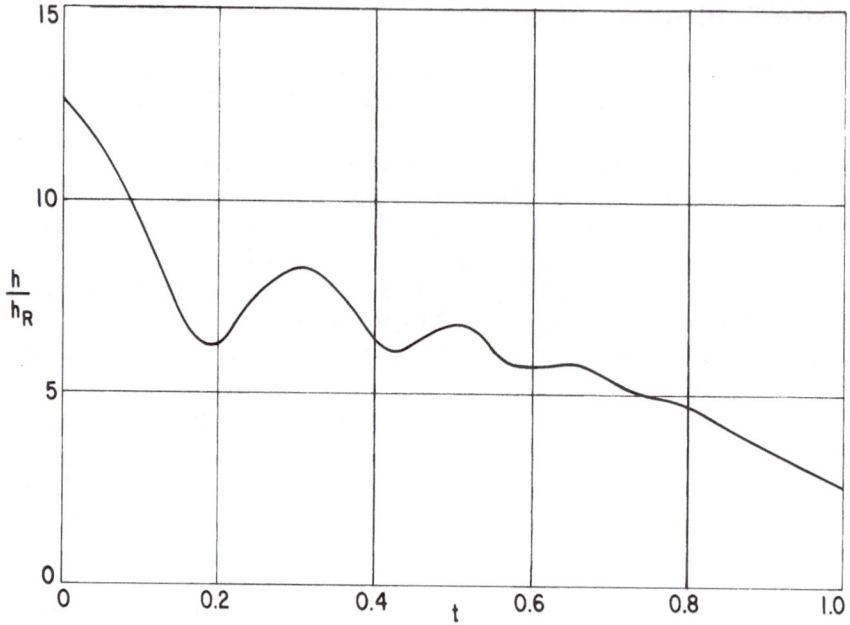

Fig. 4. Altitude distribution, Problem (Q1).

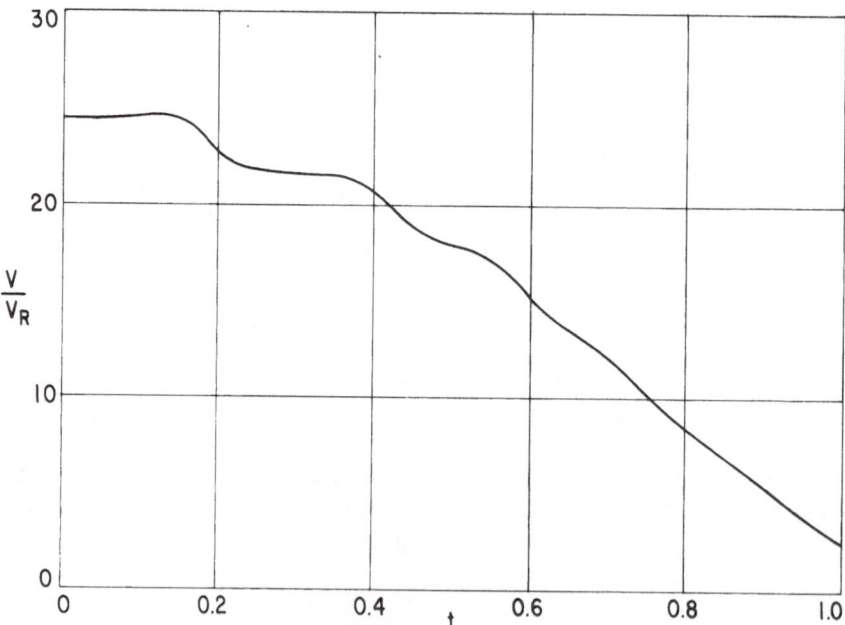

Fig. 5. Velocity distribution, Problem (Q1).

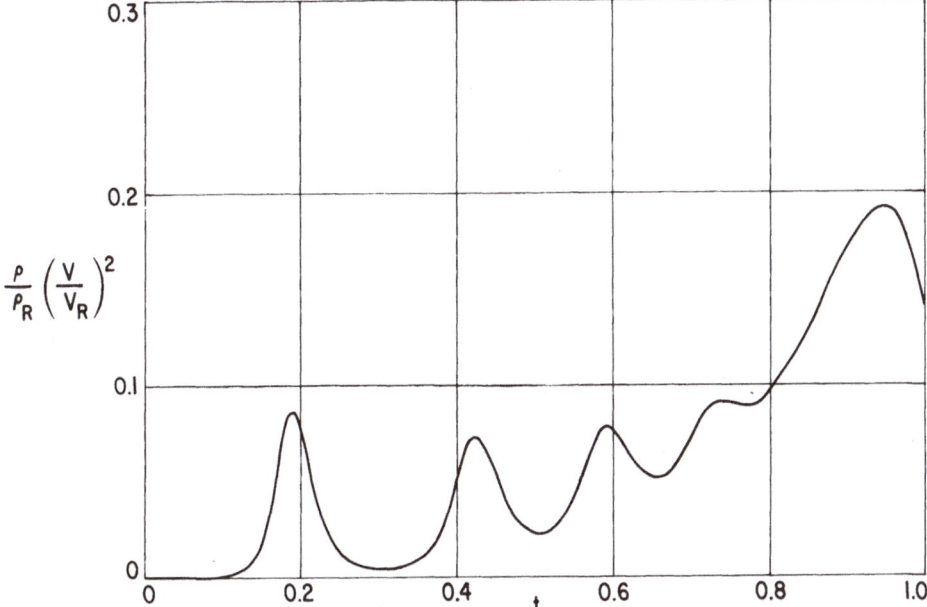

Fig. 6. Dynamic pressure distribution, Problem (Q1).

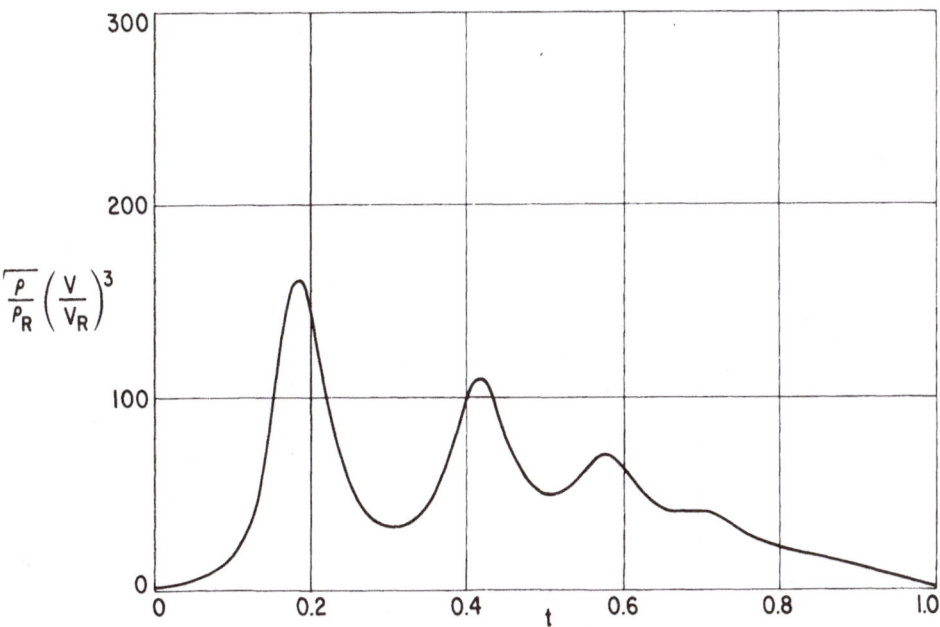

Fig. 7. Heating rate distribution, Problem (Q1).

A. Miele and P. Venkataraman

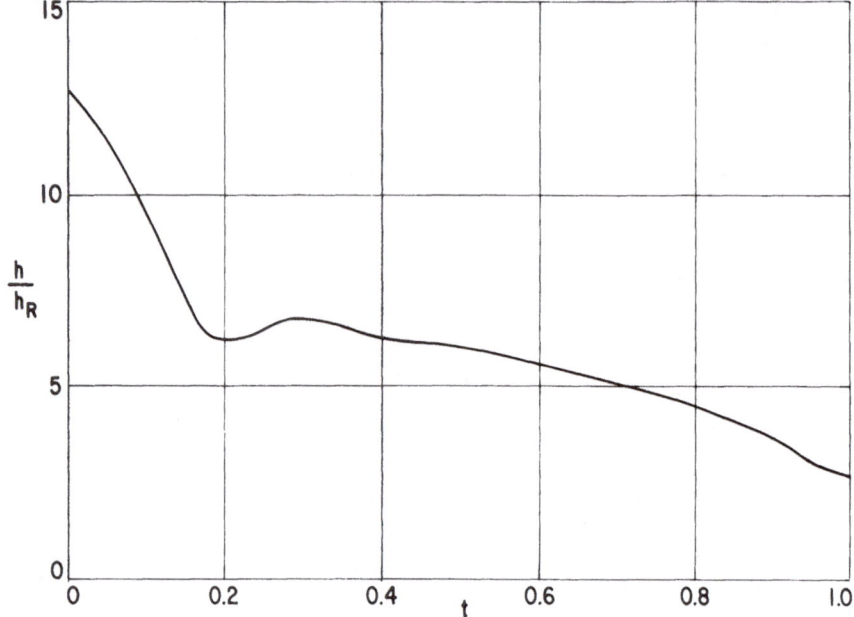

Fig. 8. Altitude distribution, Problem (Q2).

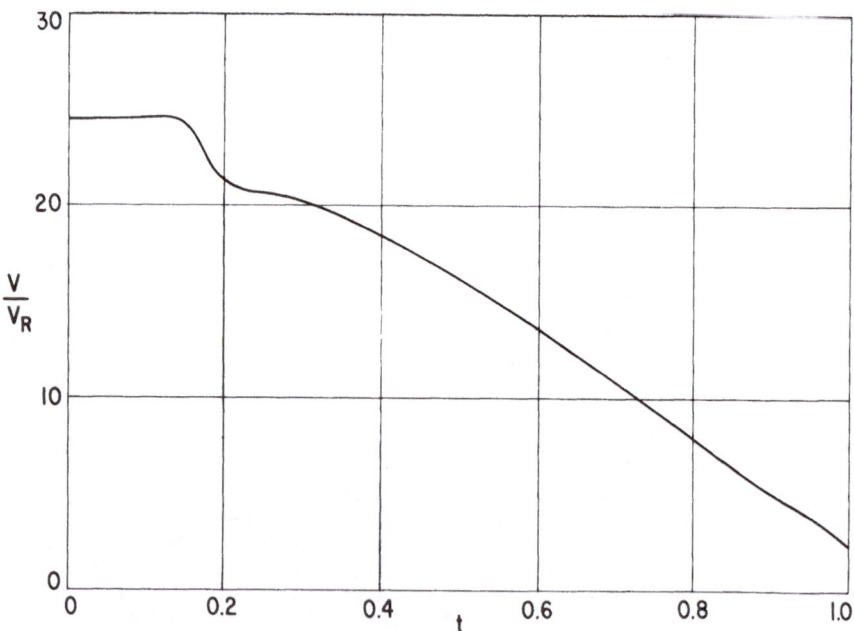

Fig. 9. Velocity distribution, Problem (Q2).

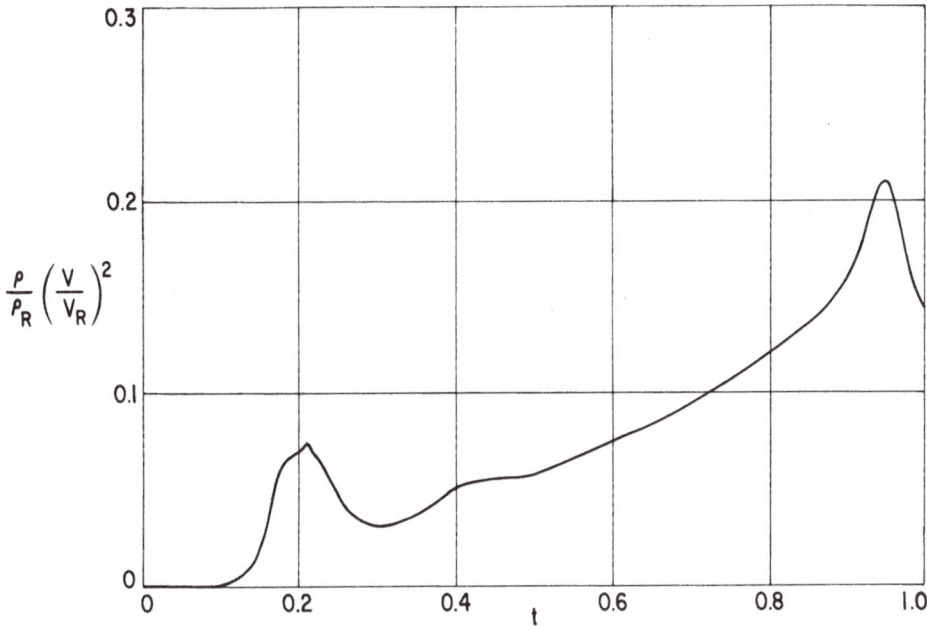

Fig. 10. Dynamic pressure distribution, Problem (Q2).

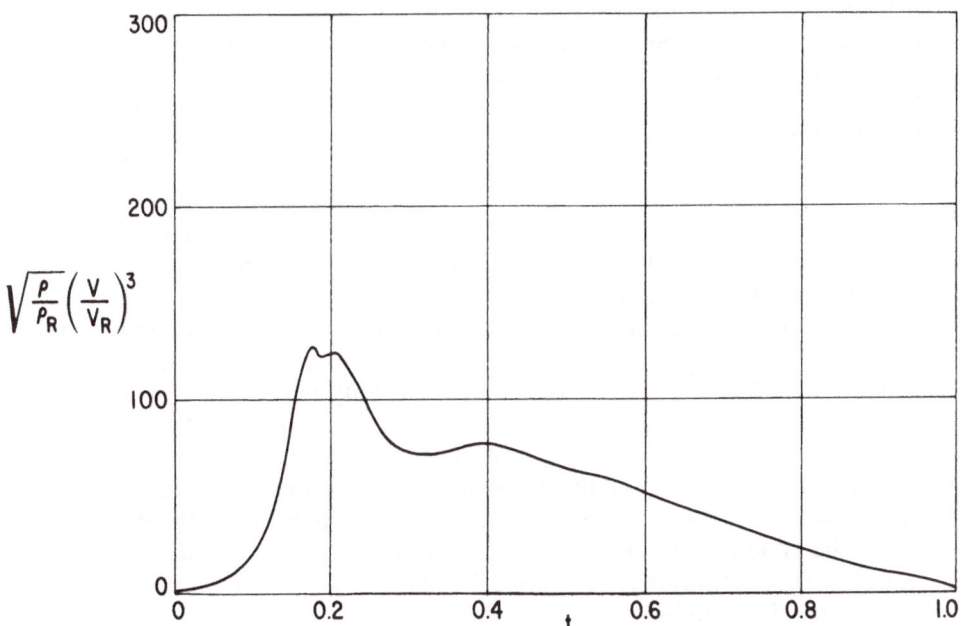

Fig. 11. Heating rate distribution, Problem (Q2).

For Problem (Q1), the optimal trajectory is a skipping trajectory, characterized by three skips. Each subsequent skip involves a diminishing altitude excursion (Figure 4). The dynamic pressure exhibits five relative maxima; among these, the peak value occurs in the vicinity of the final point ($t = 0.95$, Figure 6). The switching times for the angle of bank have the following values:

$$t_1 = 0.367, \qquad t_2 = 0.767, \qquad t_3 = 0.966 \tag{44}$$

For Problem (Q2), the optimal trajectory is a skipping trajectory, characterized by a single skip (Figure 8). The heating rate exhibits three relative maxima; among these, the peak value occurs in the vicinity of the initial point ($t = 0.18$, Figure 11). The switching times for the angle of bank have the following values:

$$t_1 = 0.332, \qquad t_2 = 0.769, \qquad t_3 = 0.971 \tag{45}$$

11. DISCUSSION AND CONCLUSIONS

A transformation technique is employed in order to convert minimax problems of optimal control (also called Chebyshev problems) into Mayer–Bolza problems of the calculus of variations. The transformation requires the proper augmentation of the state vector $x(t)$, the control vector $u(t)$, and the parameter vector π, as well as the proper augmentation of the constraining relations. As a result of the transformation, the unknown minimax value of the performance index becomes a component of the vector parameter π being optimized.

The transformation technique is then applied to the following Chebyshev problems of interest in the reentry of a space glider: (Q1) minimization of the peak dynamic pressure; and (Q2) minimization of the peak heating rate.

Numerical results are obtained by means of the sequential gradient-restoration algorithm for solving optimal control problems on a digital computer. Reference is made to the hypervelocity regime, an exponential atmosphere, and a space glider whose trajectory is controlled by means of the angle of attack and the angle of bank.

An important ingredient of the methodology employed here is the decomposition scheme outlined in Section 7. Indeed, it allows the decoupling of the system equations into a primary system (governing the distribution of angle of attack and angle of bank) and a secondary system (governing the switching times for the angle of bank).

The validity of the decomposition scheme is subordinated to having neglected the Coriolis acceleration terms and the transport acceleration terms in equations (20). If these terms are included, the decomposition scheme no longer holds. Therefore, Problem (Q1), governed by equations (26)–(30) suitably modified, must be solved as a single problem; analogously, Problem (Q2), governed by equations (32)–(36) suitably modified, must be solved as a single problem.

The authors feel that, if the Coriolis acceleration terms and the transport acceleration terms are included in equations (20), the optimal trajectories will not change appreciably. Therefore, they feel that the trajectories obtained here should constitute excellent nominal trajectories for the solution of the more complete problem.

ACKNOWLEDGMENT. This research was supported by the National Science Foundation, Grant No. ENG-79-18667. Portions of this paper were presented at the 33rd IAF Congress, Paris, France, September 26–October 2, 1982.

REFERENCES

1. G. A. Bliss, *Lectures on the Calculus of Variations*, University of Chicago Press, Chicago, Illinois (1946).
2. M. R. Hestenes, *Calculus of Variations and Optimal Control Theory*, John Wiley and Sons, New York (1966).
3. L. S. Pontryagin, V. G. Boltyanskii, R. V. Gamkrelidze, and E. F. Mishchenko, *The Mathematical Theory of Optimal Processes*, John Wiley and Sons (Interscience Publishers), New York (1962).
4. A. Miele, R. E. Pritchard, and J. N. Damoulakis, Sequential gradient-restoration algorithm for optimal control problems, *J. Optim. Theory Appl.* **5**, 235–282 (1970).
5. A. Miele, J. N. Damoulakis, J. R. Cloutier, and J. L. Tietze, Sequential gradient-restoration algorithm for optimal control problems with nondifferential constraints, *J. Optim. Theory Appl.* **13**, 218–255 (1974).
6. S. Gonzalez and A. Miele, Sequential gradient-restoration algorithm for optimal control problems with general boundary conditions, *J. Optim. Theory Appl.* **26**, 395–425 (1978).
7. A. Miele, Recent advances in gradient algorithms for optimal control problems, *J. Optim. Theory Appl.* **17**, 361–430 (1975).
8. A. Miele, Gradient algorithms for the optimization of dynamic systems, in: *Control and Dynamic Systems, Advances in Theory and Application* (C. T. Leondes, ed.), Vol. 16, pp. 1–52, Academic Press, New York (1980).
9. C. D. Johnson, Optimal control with Chebyshev minimax performance index, *J. Basic Eng.* **89**, 251–262 (1967).
10. G. J. Michael, Computation of Chebyshev optimal control, *AIAA J.* **9**, 973–975 (1971).
11. J. Warga, Minimax problems and unilateral curves in the calculus of variations, *SIAM. J. Control* **3**, 91–105 (1965).
12. W. F. Powers, A Chebyshev minimax technique oriented to aerospace trajectory optimization problems, *AIAA J.* **10**, 1291–1296 (1972).
13. A. Miele, B. P. Mohanty, P. Venkataraman, and Y. M. Kuo, Numerical solution of minimax problems of optimal control, Part 1, *J. Optim. Theory Appl.* **38**, 97–109 (1982).

14. A. Miele, B. P. Mohanty, P. Venkataraman, and Y. M. Kuo, Numerical solution of minimax problems of optimal control, Part 2, *J. Optim. Theory Appl.* **38**, 111–135 (1982).

15. D. H. Jacobson and M. M. Lele, A transformation technique for optimal control problems with a state variable inequality constraint, *IEEE Trans. Autom. Control*, **AC-14**, 457–464 (1969).

16. A. Miele, A. K. Wu, and C. T. Liu, A transformation technique for optimal control problems with partially linear state inequality constraints, *J. Optim. Theory Appl.* **28**, 185–212 (1979).

17. A. Miele, *Flight Mechanics, Vol. 1*, Addison-Wesley Publishing Company, Reading, Massachusetts (1962).

3

Pseudo-Unsteady Methods for
Inviscid or Viscous Flow Computation

R. Peyret and H. Viviand

ABSTRACT. This paper presents a review of various pseudo-unsteady techniques based upon different principles used for the numerical calculation of inviscid or viscous flows.

Two types of approaches to devise pseudo-unsteady methods are considered. The first, which relates to the mathematical formulation of the time-dependent problem, consists essentially in using artificial time-derivative terms associated with the steady equations, in order to obtain an unsteady (nonphysical) system with better properties (from the computational point of view) than the exact unsteady system. The second type, which relates to the numerical algorithm, consists in using numerical schemes that are not consistent with the unsteady equations but become consistent with the steady equations at convergence. Various criteria exist for such a choice and they are based on numerical considerations, such as stability, damping, and simplicity of implementation.

In the first part of the paper, the general concepts for constructing pseudo-unsteady systems are discussed and some examples presented for viscous flows. Then, some usual nonconsistent schemes are reviewed and their properties analyzed for model equations.

In the second part, a general family of pseudo-unsteady systems for the computation of inviscid compressible isoenergetic flows or incompressible flows is presented. These systems are required to be hyperbolic with respect to time. After a general discussion of this condition, examples are presented and it is shown, in particular, that there exist systems that are optimum for the stability condition of CFL type encountered with explicit numerical schemes.

R. Peyret • Département de Mathématiques, Université de Nice, 06034 Nice Cedex, France, and ONERA, 92320 Chatillon, France.   H. Viviand • ONERA, 92320 Chatillon, France.

1. INTRODUCTION

A common technique to compute steady flows is to use the evolutionary equations governing unsteady flows and to obtain the steady solution as the asymptotic limit of the transient solution when the time tends toward infinity. However, utilization of the physical unsteady equations could present drawbacks of various kinds:

1. The physical time needed to reach an asymptotic steady solution is excessive.
2. The numerical algorithm is complex due to the form of the evolutionary system.
3. The time steps allowed by the numerical stability condition become too small, depending on the flow properties.

For these reasons, it has been recognized that convergence to a steady state of the numerical solution could be accelerated by not attempting to compute the transient solution accurately, and even by calculating a transient solution without a physical interpretation provided the asymptotic limit exists and is a solution of the original steady problem. Such a procedure, called the "pseudo-unsteady approach," was suggested by Crocco[1] for the solution of the Navier–Stokes equations for compressible fluids and is now commonly used. In this approach we shall continue to speak of the time variable although it no longer has a physical meaning.

Ways to devise pseudo-unsteady methods fall into two large categories. The first concerns the mathematical formulation of the time-dependent problem and consists essentially in changing the time-derivative terms in order to construct a new unsteady system with better properties. The criterion for the choice of such a "pseudo-unsteady system" (PUS) depends on the problem under consideration.

The second category relates to the numerical algorithm. An example is the use of schemes with a large truncation error during the transient phase. The criterion for the construction of such schemes could be to have good stability properties or to get a high damping from the transient truncation error. Generally, these schemes are conditionally consistent with the unsteady equations, the consistency being verified for values of the time step small enough with respect to the mesh size. Practically, the transient solution can be calculated with much larger time steps so that it may bear no relation to the exact transient solution. Consequently, in this sense, the numerical scheme can be said to be "nonconsistent" with the unsteady equations. In this case only the consistency at convergence with the steady equations is required. Examples of such schemes are the well known Du Fort–Frankel[2] scheme, or the Evans *et al.*[3] scheme used by Crocco[1] for solving the Navier–Stokes equations.

In principle, a PUS belonging to the first category should be discretized by means of a consistent scheme, since the construction of the system is usually based on a consideration of the behavior of its transient solution. However, one can conceive of approximating the PUS with a nonconsistent scheme, a combination that can be very efficient in some special cases (see, e.g., Peyret and Viviand[4] and Raithby and Torrance[5]).

The aim of the present contribution is to first review some pseudo-unsteady methods used to solve the equations of fluid mechanics, then to analyze a general class of PUS for inviscid flows. In Section 2, the notion of PUS is introduced and some applications to viscous flows are described. Section 3 is devoted to a discussion of certain nonconsistent schemes. Finally, in Sections 4 and 5 a general class of PUS for inviscid, compressible or incompressible flows is introduced and examined.

2. PSEUDO-UNSTEADY SYSTEMS

2.1. General Concepts

Let us consider the solution of the following system of partial differential equations in a bounded domain D:

$$\mathscr{L}w = f \qquad \text{in } D \tag{2.1}$$

with the boundary conditions

$$\mathscr{B}w = g \qquad \text{on } S = \partial D \tag{2.2}$$

where $w = (w_1, \ldots, w_m)$ is a vector function of the space variable $x = (x_1, \ldots, x_N)$; f and g are given and $\mathscr{B}$ is the boundary operator. System (2.1) represents any nonlinear system of equations governing steady flows, compressible or incompressible, viscous or inviscid. The corresponding (physical) unsteady equations are of the general form

$$A\frac{\partial w}{\partial t} + \mathscr{L}w = f \tag{2.3}$$

where A is an $m \times m$ matrix.

If matrix A is invertible (this is the case for a compressible fluid), system (2.3) can be written as

$$\frac{\partial w}{\partial t} + A^{-1}\mathscr{L}w = A^{-1}f \tag{2.4}$$

System (2.4) is said to be in "normal form." It gives explicitly the time derivatives of all the dependent variables $w_1, \ldots, w_m$. Consequently, the Cauchy data at $t = 0$,

$$w(x, 0) = w^0(x) \qquad \text{in } D \tag{2.5}$$

permit the calculation of the solution $w(x, t)$ for $t > 0$. Then it is relatively easy to devise a numerical method to compute $w(x, t)$ step-by-step in time using an explicit scheme. The simplest example is based on the Taylor expansion

$$w(x, t + \Delta t) = w(x, t) + \Delta t \frac{\partial w}{\partial t}(x, t) + O(\Delta t^2)$$

$$\cong [w - \Delta t A^{-1}(\mathcal{L}w - f)]_{x,t} \tag{2.6}$$

A general PUS is obtained by replacing the exact system (2.3) by

$$B \frac{\partial w}{\partial t} + \mathcal{L}w = f \tag{2.7}$$

where the matrix B is supposed to be invertible and is chosen in order to eliminate drawbacks that might affect system (2.3), as mentioned in the previous section.

Another way to obtain a PUS analogous to (2.7) is by introducing new dependent variables $\tilde{w} = F(w)$, so that in this case $B = dF/dw$ and

$$\frac{\partial \tilde{w}}{\partial t} + \mathcal{L}w = f \tag{2.8a}$$

or, equivalently,

$$\frac{\partial \tilde{w}}{\partial t} + \tilde{\mathcal{L}}\tilde{w} = f \tag{2.8b}$$

When the matrix A is not invertible, system (2.3) cannot be written in the normal form (2.4), the initial conditions are not of Cauchy type (2.5), and a simple explicit method like (2.6) is no longer possible. This is the case of the Navier–Stokes equations for incompressible fluids. Consequently, it is of interest to consider a PUS of type (2.7). The steady solution is then obtained by computing the asymptotic solution of

$$\frac{\partial w}{\partial t} + B^{-1}\mathcal{L}w = B^{-1}f \tag{2.9}$$

with arbitrary initial conditions (2.5). In the following section, examples will be given for the Navier–Stokes equations in primitive-variable as well as in vorticity-potential formulations.

2.2. Pseudo-Unsteady Systems for Viscous Flows

2.2.1. Primitive-Variable Formulation for Incompressible Flow

The steady Navier–Stokes equations for incompressible fluids are

$$(\mathbf{U} \cdot \text{grad})\mathbf{U} + \frac{1}{\rho}\,\text{grad}\,p - \nu\Delta\mathbf{U} = 0 \tag{2.10}$$

$$\text{div}\,\mathbf{U} = 0 \tag{2.11}$$

In these equations, $\mathbf{U}$ is the velocity vector with cartesian components (u_1, u_2, u_3), p is the pressure, ρ is the (constant) density, and ν is the (constant) kinematic viscosity coefficient. Boundary conditions must be associated with the system (2.10) and (2.11), for example

$$\mathbf{U}\big|_s = \boldsymbol{\alpha} \qquad \text{with} \int_s \boldsymbol{\alpha} \cdot \mathbf{N} dS = 0 \tag{2.12}$$

where S is the boundary of the domain D in which the solution is calculated, and $\mathbf{N}$ is the unit normal to S.

The unsteady Navier–Stokes equations are

$$\frac{\partial\mathbf{U}}{\partial t} + (\mathbf{U} \cdot \text{grad})\mathbf{U} + \frac{1}{\rho}\,\text{grad}\,p - \nu\Delta\mathbf{U} = 0 \tag{2.13}$$

$$\text{div}\,\mathbf{U} = 0 \tag{2.14}$$

and the corresponding initial conditions are

$$\mathbf{U}\big|_{t=0} = \mathbf{U}^0 \qquad \text{with}\ \ \text{div}\,\mathbf{U}^0 = 0 \tag{2.15}$$

For system (2.13) and (2.14), the vector w and the matrix A defined by equation (2.3) are, respectively, given by

$$w = \begin{pmatrix} u_1 \\ u_2 \\ u_3 \\ p \end{pmatrix}, \quad A = \begin{pmatrix} 1 & & & \\ & 1 & & 0 \\ & 0 & 1 & \\ & & & 0 \end{pmatrix} \tag{2.16}$$

The matrix A is not invertible and the initial conditions (2.15) are not of Cauchy type (there is no condition for p). We have a situation where a simple explicit algorithm like (2.6) cannot be constructed. A common method of solving (2.13) and (2.14) with (2.12) and (2.15) is based on the utilization of a Poisson equation for the pressure. This equation is obtained by applying the divergence operator to equation (2.13), taking into account (2.14). Hence the simplest algorithm that can be conceived requires the solution of an elliptic equation at each time step. This kind of method, introduced in various forms by Harlow and Welsh,[6] Chorin,[7] or Temam,[8] is now currently used. We refer to Peyret and Taylor[9] for details concerning these methods.

A pseudo-unsteady method for the Navier–Stokes equations is the so-called "artificial compressibility method" based upon a modification of the continuity equation (2.14). Chorin[10,11] has proposed replacing equation (2.14) by

$$\delta \frac{\partial p}{\partial t} + \operatorname{div} \mathbf{U} = 0 \tag{2.17}$$

where δ is a positive constant. The pseudo-unsteady system of type (2.7) comprises equations (2.13) and (2.17), so that

$$w = \begin{pmatrix} u_1 \\ u_2 \\ u_3 \\ p \end{pmatrix}, \qquad B = \begin{pmatrix} 1 & & & \\ & 1 & & 0 \\ & 0 & 1 & \\ & & & \delta \end{pmatrix} \tag{2.18}$$

and the initial conditions

$$w(x, 0) = w^0(x) \tag{2.19}$$

are arbitrary, but it is recommended that the initial velocity $\mathbf{U}^0$ satisfies the boundary conditions in order to avoid singularities during the first time cycles. We note that the method can be generalized by replacing the diagonal entries $(1, 1, 1, \delta)$ of B by $(1, \kappa, \lambda, \delta)$, where κ and λ are positive constants. The role of κ and λ is to allow us to advance in time differently according to each velocity component. In other terms, the use of δ, κ, and λ amounts to using a different time step for each equation.

The artificial compressibility method of Chorin has been studied from a mathematical point of view by Temam[12,13] and applied, e.g., by Fortin et. al.,[14] Elsaesser and Peyret,[15,16] and Aubert and Deville[17] using various schemes.

Another type of perturbed continuity equation was introduced by Vladimirova, Kuznecov, and Yanenko in 1965 (see Yanenko[18]). This approach consists of using the new dependent variable q defined for plane flows, with $\mathbf{U} = (u_1, u_2)$, by

$$\rho q = p + \rho U^2/4 \qquad (2.20)$$

(U is the modulus of $\mathbf{U}$), so that equation (2.14) is replaced by

$$\delta \frac{\partial q}{\partial t} + \text{div}\, \mathbf{U} = 0 \qquad (2.21)$$

with $\delta > 0$ ($\delta = 1$ in Yanenko[18] where dimensionless variables are used). The momentum equation (2.13) is then written as

$$\frac{\partial \mathbf{U}}{\partial t} + (\mathbf{U}\cdot\text{grad})\mathbf{U} - \frac{1}{4}\,\text{grad}\, U^2 + \text{grad}\, q - \nu\Delta\mathbf{U} = 0 \qquad (2.22)$$

Equations (2.21) and (2.22) constitute a PUS of the form (2.8b) with $\tilde{w} = (u_1, u_2, q)$ as dependent variable. This system accepts arbitrary initial conditions

$$\tilde{w}(x,0) = \tilde{w}^0(x) \qquad (2.23)$$

The numerical method proposed by Yanenko[18] is a splitting method, where q is split in the form $q = (q_1 + q_2)/2$ with $\rho q_i = p + \rho u_i^2/2$. In three dimensions the definition of q should be $\rho q = p + \rho u^2/6$.

Taylor and Ndefo[19] reported difficulties in obtaining a steady-state solution with this method, possibly due to the fact that the value $\delta = 1$ (in dimensionless variables) used was not appropriate for their problem. Numerical applications[14] of the artificial compressibility method (2.13) and (2.17) have shown that the speed of convergence toward a steady state depends strongly on the choice of parameter δ. Criteria for this choice are dictated by the numerical scheme, but also by the mathematical properties of the PUS (see Section 5).

When the Reynolds number becomes large, the time required to reach a steady state can be long. To overcome this difficulty, Wirz[20] has proposed adding to the PUS based on equations (2.13) and (2.17) a supplementary dependent variable and a supplementary equation, in order to introduce damping of the transient solution independent of the viscous effect.

2.2.2. Vorticity-Potential Formulation for Incompressible Flow

One possible way of avoiding the difficulties associated with the continuity equation is to introduce the vorticity vector Ω, a scalar potential φ, and a solenoidal vector potential Ψ, so that

$$\Omega = \operatorname{curl} U, \qquad U = \operatorname{grad} \varphi + \operatorname{curl} \Psi \tag{2.24}$$

and the motion is governed by the equations

$$\frac{\partial \Omega}{\partial t} + \operatorname{curl}[\Omega \times (\operatorname{grad} \varphi + \operatorname{curl} \Psi)] - \nu \Delta \Omega = 0 \tag{2.25}$$

$$\Delta \Psi + \Omega = 0 \tag{2.26a}$$

$$\Delta \varphi = 0 \tag{2.26b}$$

We refer to Richardson and Cornish[21] for the appropriate boundary conditions associated with this formulation.*

Again system (2.25) and (2.26), where the dependent variables are $w = (\Omega, \Psi, \varphi)$, cannot be written in normal form: the time derivatives $\partial \Psi / \partial t$ and $\partial \varphi / \partial t$ do not appear in the equations. The initial condition is again (2.15), from which we deduce the initial value of Ω.

In the course of solving numerically equations (2.25) and (2.26), elliptic equations for the Ψ-components and for φ must be solved at each time step.

A PUS associated with the present formulation is obtained by adding $\partial \Psi / \partial t$ and $\partial \varphi / \partial t$ to equations (2.26a) and (2.26b), respectively, i.e.,

$$\delta \frac{\partial \Psi}{\partial t} - (\Delta \Psi + \Omega) = 0 \qquad (\delta = \text{const} > 0) \tag{2.27a}$$

$$\lambda \frac{\partial \varphi}{\partial t} - \Delta \varphi = 0 \qquad (\lambda = \text{const} > 0) \tag{2.27b}$$

so that system (2.25) and (2.27) can be easily solved, taking into account the initial conditions

$$\Omega \big|_{t=0} = \Omega^0, \qquad \Psi \big|_{t=0} = \Psi^0, \qquad \varphi \big|_{t=0} = \varphi^0 \tag{2.28}$$

---

* $\varphi \equiv 0$ if the normal velocity to the boundary vanishes.[21,22] If this condition is not satisfied the use of φ can be avoided, but at the price of a complication in the determination of the boundary conditions.[23]

These conditions can be arbitrary, however it is best to choose $\boldsymbol{\Psi}^0$ and φ^0 satisfying the boundary conditions, then to deduce $\boldsymbol{\Omega}^0$ from equation (2.26a).

The above pseudo-unsteady method, in the case where $\varphi \equiv 0$, has been proposed by Mallinson and De Vahl Davis[24] for calculations of two- and three-dimensional flows with heat transfer in a confined region. It has been also used by Bontoux[25] and Bontoux et al.[26] for analogous two-dimensional flows. In all these works, the equations are solved by means of alternating-direction implicit methods.[27] It should be pointed out that Burggraf[28] has solved the two-dimensional stream function–vorticity equations with an iterative method, which can be interpreted as a pseudo-unsteady method of type (2.25) and (2.27). Furthermore, in this method new values of the stream function and of the vorticity are used as soon as they are available (Gauss–Seidel technique). Hence, we have here an example of a PUS approximated with a "nonconsistent" scheme (Section 3).

2.2.3. Compressible Flow

The Navier–Stokes equations for compressible fluids can be written in normal form, so that the system of equations, written as equation (2.4), allows an explicit computation step-by-step in time as mentioned in Section 2.1. However, it could happen that the limitation on the time step due to the stability criterion is too stringent because of a peculiar behavior of the solution. In this case, one solution is to use a PUS for which such a difficulty does not appear. This problem was encountered by Peyret and Viviand[4] in the course of computing plane viscous flow around a parabolic finite body. The system is written in parabolic coordinates (ξ, η) and the unknown vector in dimensionless variables is given by

$$w = \begin{pmatrix} h^2\rho \\ h^3\rho\mathcal{U} \\ h^3\rho\mathcal{V} \\ h^2[p/(\gamma M_\infty^2) + (\gamma - 1)(\mathcal{U}^2 + \mathcal{V}^2)/2] \end{pmatrix} \qquad (2.29)$$

where $h^2 = \xi^2 + \eta^2$, $\mathcal{U}$ and $\mathcal{V}$ are the velocity components in the local (ξ, η) system, M_∞ is the free-stream Mach number, and γ is the adiabatic index. The discretization by means of an explicit scheme, which will be presented later [equation (3.5)], leads to the stability condition

$$\Delta t \le \frac{\text{Pr Re}}{2\gamma} h^2 \rho \frac{\Delta\xi^2 \Delta\eta^2}{\Delta\xi^2 + \Delta\eta^2} \qquad (2.30)$$

where Pr is the Prandtl number, Re the Reynolds number, and $\Delta\xi$ = const, $\Delta\eta$ = const. The maximal value of the time step Δt allowed by equation (2.30) depends in particular on the density ρ, and it turns out to be extremely small because of the very small values of the density in the base-flow region. This difficulty was solved by not using the exact Navier–Stokes equations but rather a PUS of the form (2.8a), where $\tilde{w}$ was chosen as

$$\tilde{w} = (\rho, h\mathcal{U}, h\mathcal{V}, T/\gamma M_\infty^2) \tag{2.31}$$

so that the time step required for stability no longer depended on $h^2\rho$ but was constant.

3. NONCONSISTENT SCHEMES

The discretization of unsteady equations by means of nonconsistent (with respect to time) schemes can be interpreted as a pseudo-unsteady approach. Generally, such schemes are constructed in order to possess good stability properties and to accelerate convergence of the corresponding transient solution toward a steady limit.

3.1. Mixing of Time Levels in Discretized Space Derivatives

In this section we outline some of the nonconsistent schemes used to solve the equations of fluid mechanics. For the sake of simplicity we describe these schemes for the heat equation

$$\frac{\partial w}{\partial t} = \sigma \frac{\partial^2 w}{\partial x^2} + f \qquad [\sigma > 0, f = f(x)] \tag{3.1}$$

Nonconsistent schemes can be generated by using a mixing of values at time levels $(n-1)\Delta t$, $n\Delta t$, and $(n+1)\Delta t$ in the approximation of the second-order derivative in equation (3.1). The basic idea of the technique is to introduce some degree of implicitness while maintaining the calculation really explicit. The resulting schemes are not unconditionally consistent in general, but can be consistent under certain conditions.

A first example is the well-known Du Fort–Frankel[2] scheme

$$\frac{1}{2\Delta t} (w_i^{n+1} - w_i^{n-1}) = \frac{\sigma}{\Delta x^2} [w_{i+1}^n - (w_i^{n+1} + w_i^{n-1}) + w_{i-1}^n] + f_i \tag{3.2}$$

where w_i^n refers to the numerical solution at $x_i = i\Delta x$, $t^n = n\Delta t$. The principal part of the truncation error of scheme (3.2) is $\sigma(\Delta t/\Delta x)^2 \partial^2 w/\partial t^2$.

The consistency during the transient stage would require $\Delta t / \Delta x \to 0$ when $\Delta x \to 0$. Scheme (3.2) is unconditionally stable and has been generalized by Gottlieb and Gustaffson[29] in the following manner:

$$\frac{1}{2\Delta t}(w_i^{n+1} - w_i^{n-1}) = \sigma D_{xx} w_i^n + f_i - \gamma(w_i^{n+1} - 2w_i^n + w_i^{n-1}) . \quad (3.3)$$

where $D_{xx} w_i^n$ is a consistent approximation of $\partial^2 w / \partial x^2$ at time $n\Delta t$, and γ is a positive parameter that must be greater than some value γ_0 for stability reasons. In the case of the original Du Fort–Frankel scheme (3.2), D_{xx} is the standard second-order accurate finite-difference operator and $\gamma = \sigma / \Delta x^2$. Other finite-difference approximations have been considered[29]; spectral approximations can also be used, such as Fourier series[29] or Chebyshev polynomials series.[30] It is noteworthy that the Du Fort–Frankel scheme (3.2) was used by Chorin[10,11] for approximating the PUS (2.13) and (2.17).

The nonconsistent scheme used by Crocco[1] for solving the one-dimensional Navier–Stokes equations for compressible fluid was originally proposed by Evans et al.[3] and is written for equation (3.1) as

$$\frac{1}{\Delta t}(w_i^{n+1} - w_i^n) = \frac{\sigma}{\Delta x^2}(w_{i+1}^n - 2w_i^{n+1} + w_{i-1}^n) + f_i \quad (3.4)$$

Here again, the scheme is unconditionally stable but its truncation error during the transient stage is of the order of $\Delta t / \Delta x^2$. A similar technique was introduced by Allen and Cheng[31] into the two-step scheme of Braïlovskaya[32] for the two-dimensional Navier–Stokes equations for compressible flows.

Peyret and Viviand[4,33] have used the nonconsistent scheme

$$\frac{1}{\Delta t}(w_i^{n+1} - w_i^n) = \frac{\sigma}{\Delta x^2}(w_{i+1}^n - 2w_i^n + w_{i-1}^{n+1}) + f_i \quad (3.5)$$

for computations of compressible flows about parabolic bodies. The truncation error of scheme (3.5) is of the order of $\Delta t / \Delta x^2$; the time step is limited by the stability condition $\Delta t < \Delta x^2 / \sigma$, which is half as restrictive as the usual condition for the explicit consistent scheme. We note that scheme (3.5) is similar to the one considered by Burggraf,[28] as mentioned in the previous section. It has also been used by Fortin et al.[14] combined with the artificial compressibility method, and by Raithby and Torrance[5] for the vorticity equation.

It should be noted that scheme (3.5) is none other than the successive overrelaxation (SOR) method for the solution of

$$\frac{\sigma}{\Delta x^2}(w_{i+1} - 2w_i + w_{i-1}) + f_i = 0 \tag{3.6}$$

with relaxation parameter α defined by

$$\alpha = 2\sigma\Delta t/\Delta x^2 \tag{3.7}$$

The condition of convergence $0 < \alpha < 2$ of the SOR method is comparable to the stability condition of scheme (3.5).

The analogy between some iterative methods and unsteady equations has been recognized by Garabedian[34] and since then has been widely exploited. We refer to works by Lomax and Steger[35] and Wirz[20] for a deeper analysis of relaxation methods and their relationship with unsteady equations.

Let us now consider in more detail some of the above nonconsistent schemes. By expanding in a Taylor series and retaining only the first significant term, we obtain the equation

$$K\frac{\partial w}{\partial t} = \sigma\frac{\partial^2 w}{\partial x^2} + f \tag{3.8}$$

with which these schemes can be considered as consistent, assuming α bounded. The coefficient K can be expressed as follows:

$$K = K_1 = 1 + \alpha \qquad \text{for scheme (3.4)}$$

$$K = K_2 = \frac{(1+\alpha)^2}{1+2\alpha} \qquad \text{for the Allen–Cheng scheme[31]}$$

$$K = K_3 = 1 - \alpha/2 \qquad \text{for scheme (3.5)}$$

where α is given by formula (3.7).

It is easily seen that $K_1 > 1$ and $K_2 > 1$; hence, for the corresponding schemes, the convergence toward a steady limit is slower than the one given by a consistent scheme (with the same Δt), but this relative slowness is balanced by the good stability that allows larger time steps.[36] For scheme (3.5), we have $0 < K_3 < 1$ if $\alpha < 2$. Consequently, in that case the convergence is faster than that obtained with the explicit consistent approximation, which can be identified with the iterative Jacobi procedure for the solution of (3.6) if $\Delta t = \Delta x^2/2\sigma$.

We note that equation (3.8), corresponding to scheme (3.5), degenerates when $\alpha = 2$. More precisely, if $\alpha = 2 - \lambda\Delta x$, $\lambda > 0$, it can be shown that scheme (3.5) is consistent with the hyperbolic equation

$$\lambda \frac{\partial w}{\partial \tau} - \sigma \frac{\partial^2 w}{\partial x^2} + \frac{\partial^2 w}{\partial \tau \partial x} = f \tag{3.9}$$

where the change of variable $t = \tau \Delta x$ has been introduced. This equation was derived by Garabedian,[34] who used it to determine an approximate value of the optimal relaxation parameter α_* of the SOR method.

3.2. Spatially Varying Time Step

Another way of constructing nonconsistent schemes with the purpose of accelerating convergence toward a steady state is to use a spatially varying time step rather than a constant one. The determination of this local time step may differ according to the problem under consideration.

Let us consider the scalar equation

$$\frac{\partial w}{\partial t} + A(w) \frac{\partial w}{\partial x} - \sigma \frac{\partial^2 w}{\partial x^2} = f \tag{3.10}$$

and assume that it is discretized in a nonuniform mesh. This can be done by a change of variable $x = H(\xi)$ in equation (3.10), then by discretization of the transformed equation in the (ξ, t)-plane using a uniform mesh $\Delta \xi =$ const. Another possibility is to discretize equation (3.10) directly in the (x, t)-plane by using unequally spaced points x_i. In this latter case, the change of variable

$$x = H(\xi) \quad \text{or} \quad \xi = h(x) \tag{3.11}$$

serves only to define the coordinates $x_i = H(i\Delta \xi)$ of the mesh points. We can write

$$x_{i+1} - x_i = \Delta x_i = H[(i+1)\Delta \xi] - H(i\Delta \xi)$$

$$= \Delta \xi H'(i\Delta \xi) + O(\Delta \xi^2) = \frac{\Delta \xi}{h'(x_i)} + O(\Delta \xi^2) \tag{3.12}$$

Consider then the discretization of equation (3.10) in the (x, t)-plane by means of an explicit scheme. For the sake of simplicity we consider an upwind approximation for the first-order spatial derivative:

$$w_i^{n+1} = w_i^n + \Delta t (a_i^n w_{i+1}^n + b_i^n w_i^n + c_i^n w_{i-1}^n + f_i) \tag{3.13a}$$

with

$$a_i^n = \frac{1}{2\Delta x_i} \left[(|A_i^n| - A_i^n) + \frac{4\sigma}{\Delta x_{i-1} + \Delta x_i} \right]$$

$$b_i^n = -\frac{1}{2\Delta x_{i-1}\Delta x_i} [(|A_i^n| + A_i^n)\Delta x_i + (|A_i^n| - A_i^n)\Delta x_{i-1} + 4\sigma]$$

(3.13b)

$$c_i^n = \frac{1}{2\Delta x_{i-1}} \left[(|A_i^n| + A_i^n) + \frac{4\sigma}{\Delta x_{i-1} + \Delta x_i} \right]$$

$(a_i^n + b_i^n + c_i^n = 0, \ a_i^n > 0, \ c_i^n > 0)$.

The stability criterion is

$$\Delta t \leq 1/(-b_i^n) \tag{3.14}$$

If a constant time step is used at each time level, the smallest Δt allowed by (3.14) must be employed, i.e.,

$$\Delta t \leq \Delta t_m = \min_i [1/(-b_i^n)] \tag{3.15}$$

In practical problems, this can be a very stringent restriction on Δt because in large regions Δt_m can be much smaller than the time step allowed by the local stability condition (3.14). Hence the idea arose of using at each point the time step corresponding to the local stability condition

$$\Delta t_i^n = \mu_i /(-b_i^n), \qquad 0 < \mu_i \leq 1 \tag{3.16}$$

where μ_i is a function of position x_i, and is introduced in order to avoid too large variations of the time step arising either from very large mesh variations or from small values of $|A_i^n|$ in the case $\sigma = 0$.

The resulting scheme is no longer consistent with equation (3.10). In that sense, the use of the local time step leads to a pseudo-unsteady method. The corresponding unsteady equation is determined by regarding the numerical solution w_i^n given by scheme (3.13), in which $\Delta t = \Delta t_i^n$ given by (3.16), as an approximation of a solution $W(x, \tau)$, where τ is a time variable discretized by $\tau^n = n\Delta\tau$, $\Delta\tau$ being an arbitrary constant time step: $w_i^n \cong W(x_i, \tau^n)$. This equation is obtained from Taylor expansions in equation (3.13) with (3.16), retaining only the first significant terms:

$$K\frac{\partial W}{\partial \tau} + A(W)\frac{\partial W}{\partial x} - \sigma\frac{\partial^2 W}{\partial x^2} = f \tag{3.17}$$

The neglected terms are $O(\Delta\tau, \Delta\xi)$; coefficient K is given by

$$K = \frac{\Delta\tau}{\Delta\xi^2} \frac{h'}{\mu} (2\sigma h' + |A| \Delta\xi) \qquad (3.18)$$

It is interesting to note that, if $\sigma = 0$, $f = 0$, and $A \neq 0$, equation (3.17) becomes simply

$$\varepsilon \frac{\Delta\tau}{\Delta\xi} \frac{h'}{\mu} \frac{\partial W}{\partial \tau} + \frac{\partial W}{\partial x} = 0, \qquad \varepsilon = \text{sign}(A) \qquad (3.19)$$

namely a piecewise linear equation (linear in regions where A maintains a constant sign), while the original equation is nonlinear. Moreover, in this case it can be shown that the numerical solution w_i^n given by scheme (3.13) with (3.16) can be considered as an approximation of the exact solution of the original equation

$$\frac{\partial w}{\partial t} + A(w) \frac{\partial w}{\partial x} = 0 \qquad (3.20)$$

at the point x_i and time t_i^n such that $t_i^n = g(x_i, \tau^n)$, where $g(x, \tau)$ must satisfy the equation

$$\varepsilon \frac{\Delta\tau}{\Delta\xi} \frac{h'}{\mu} \frac{\partial g}{\partial \tau} + \frac{\partial g}{\partial x} = \frac{1}{A} \qquad (3.21)$$

In other words, the solution w_i^n at a fixed level n represents the numerical solution of (3.20) on a curve $t = g(x, n\Delta t)$.

On the other hand, these peculiar properties are no longer true in the case of a system of equations. Assuming w is a vector and $\sigma = 0$, $f = 0$, the local time step is defined by the spectral radius $\rho(w)$ of the matrix $A(w)$. The equation analogous to (3.19) is then written as

$$\frac{\Delta\tau}{\Delta\xi} \frac{h'}{\mu} \rho(W) \frac{\partial W}{\partial \tau} + A(W) \frac{\partial W}{\partial x} = 0 \qquad (3.22)$$

which, in general, cannot be equivalent to a piecewise linear system. Also, the previous remark concerning the change of time variable $t = g(x, \tau)$ does not apply here.

The technique of the local time step has been used by Li[37,38] for the numerical solution of the Navier–Stokes equations for compressible fluids by means of the MacCormack scheme.[39] It is also used widely for calculating inviscid transonic flows.

Raithby and Torrance[5] made use of the same technique to solve the vorticity equation for incompressible viscous flows. Their numerical experiments show an important effect on the number of time cycles required to reach a steady solution. For example, if 500 time cycles are needed in the case where the exact transient solution is calculated, this number reduces to less than 100 for a scheme analogous to (3.13) with $\Delta t = \Delta t_i^n$ given by (3.16), and to less than 50 if the local time step is used in combination with the nonconsistent approximation (3.5).

Until now, we have considered explicit schemes so that the choice of the variable time step was based on the stability condition. However, the technique can be applied to implicit schemes also. In such schemes, matrices must be inverted and the use of a spatially varying time step amounts to a preconditioning of these matrices with a simple diagonal matrix. McDonald and Briley[40] have carried out a study in the case of the two-dimensional heat equation

$$\frac{\partial w}{\partial t} - \sigma\left(\frac{\partial^2 w}{\partial x^2} + \frac{\partial^2 w}{\partial y^2}\right) = 0 \tag{3.23}$$

and used a generalized alternating-direction method[27] in a nonuniform mesh. The scheme is written as

$$\left(1 - \frac{\Delta t}{2}\sigma D_{xx}\right)(w_{i,j}^* - w_{i,j}^n) = \sigma\Delta t(D_{xx} + D_{yy})w_{i,j}^n \tag{3.24a}$$

$$\left(1 - \frac{\Delta t}{2}\sigma D_{yy}\right)(w_{i,j}^{n+1} - w_{i,j}^n) = w_{i,j}^* - w_{i,j}^n \tag{3.24b}$$

where D_{xx} and D_{yy} approximate $\partial^2/\partial x^2$ and $\partial^2/\partial y^2$, respectively. We have

$$D_{xx}w_{i,j} = a_{i,j}w_{i+1,j} + b_{i,j}w_{i,j} + c_{i,j}w_{i-1,j} \tag{3.25a}$$

$$D_{yy}w_{i,j} = d_{i,j}w_{i,j+1} + e_{i,j}w_{i,j} + f_{i,j}w_{i,j-1} \tag{3.25b}$$

A first technique consists of using

$$\Delta t = \Delta t_{i,j}^{(1)} = \frac{\kappa}{|a_{i,j}| + |b_{i,j}| + |c_{i,j}|} \tag{3.26a}$$

for odd time cycles ($n = 1, 3, 5, \ldots$) and

$$\Delta t = \Delta t_{i,j}^{(2)} = \frac{\kappa}{|d_{i,j}| + |e_{i,j}| + |f_{i,j}|} \tag{3.26b}$$

for even time cycles ($n = 2, 4, 6, \ldots$). The parameter κ may be a constant or changed cyclically according to the Wachspress[41] rule.

A second technique makes use of the time step defined for each time cycle by

$$\Delta t = \Delta t_{i,j} = 2 \frac{\Delta t_{i,j}^{(1)} \Delta t_{i,j}^{(2)}}{\Delta t_{i,j}^{(1)} + \Delta t_{i,j}^{(2)}} \qquad (3.27)$$

Convergence tests reported elsewhere[40] show that the number of time cycles required to reach convergence, equal to 200 in the case of the consistent approximation, reduced to about 40 when (3.26) or (3.27) are used with optimized (constant) values of κ ($\kappa \approx 8$). Shamroth and Gibeling[42,43] have applied this technique of variable time steps to the solution of the Navier–Stokes equations for compressible fluids with the implicit method of Briley and McDonald.[44]

We shall now remark on the definition of the variable time step (3.26) or (3.27). Consider the following implicit version of the one-dimensional scheme (3.13):

$$w_i^{n+1} = w_i^n + \Delta t(a_i^n w_{i+1}^{n+1} + b_i^n w_i^{n+1} + c_i^n w_{i-1}^{n+1} + f_i) \qquad (3.28)$$

It is noteworthy that the time step defined by (3.26a) when used for scheme (3.28) is given by

$$\Delta t = \Delta t_i^n = \frac{\kappa}{|a_i^n| + |b_i^n| + |c_i^n|} = \frac{\kappa}{2(-b_i^n)} \qquad (3.29)$$

because $a_i^n > 0$, $c_i^n > 0$ and $b_i^n = -(a_i^n + c_i^n)$. Thus it is directly related to the time step (3.16) for the explicit scheme and based on the stability condition. Hence scheme (3.28) with Δt given by (3.29) is consistent with the unsteady equation (3.17), with μ replaced by $\kappa/2$ in the definition of K.

So when implicit schemes are used, the choice of a variable time step can be based on the local time step determined by the stability condition of the corresponding explicit scheme. The technique has been applied by Lerat et al.[45] to calculations of inviscid transonic flows using an implicit finite-volume method for the Euler equations. The local time step was 5.7 times greater than the value allowed by the Courant–Friedrichs–Lewy condition.

4. ANALYSIS OF A CLASS OF PSEUDO-UNSTEADY SYSTEMS FOR INVISCID FLOWS

4.1. Generalities

This section is concerned with a discussion of a general class of PUS for the computation of inviscid, compressible, or incompressible flows.

Various types of PUS are currently used for the calculation of inviscid transonic flows, a review of which has been given by Viviand[46] and is not repeated here. As a complement to this review, it is of interest to mention a new pseudo-unsteady technique, called enthalpy damping, which has been proposed by Jameson *et al.*[47] to accelerate the convergence. The aim of this section is to present a synthetic outlook of a class of PUS for the steady Euler equations that provides a generalization of existing PUS. A fuller exposition, including also a class of PUS systems for homentropic flows, is given by Viviand.[48]

Interestingly enough, we found that the incompressible case could be covered as well in this unified presentation first intended for compressible isoenergetic flows only. Although PUS do not seem to have been considered yet for the computation of incompressible inviscid flows, they might prove interesting for rotational flows. On the other hand, as discussed in Section 2.2, such methods are used for the solution of the incompressible Navier–Stokes equations, and the discussion of the inviscid counterpart of the corresponding PUS is of interest for these methods. Indeed, at moderate or high Reynolds numbers, the viscous terms are negligible in important regions of the flow where the solution of the Navier–Stokes equations behaves locally as a solution of Euler's equations. Thus it is necessary that the PUS used should also be suitable for inviscid flows.

4.2. Steady-Flow Equations

For steady inviscid flows, the continuity and momentum equations can be written in the following forms, which are equally valid for both incompressible and compressible fluids:

$$\operatorname{div} \mathbf{U} - \frac{U^2}{c^2} \mathbf{s} \cdot \operatorname{grad} U = 0 \tag{4.1}$$

$$(\mathbf{U} \cdot \operatorname{grad})\mathbf{U} + \frac{1}{\rho} \operatorname{grad} p = 0 \tag{4.2}$$

where $\mathbf{U}$ is the velocity vector, U its modulus, $\mathbf{s}$ its unit vector ($\mathbf{U} = U\mathbf{s}$), c is the speed of sound, ρ the density, and p the pressure. For an incompressible fluid, ρ is a constant and c is infinite. For a compressible fluid, we restrict the discussion to isoenergetic flows for which the energy equation can be integrated in the form of Bernoulli's relation:

$$h + \frac{1}{2} U^2 = H_0 \tag{4.3}$$

where the specific enthalpy h is a known function of p and ρ, and H_0 is the constant value of the total enthalpy. Relation (4.3) can be solved for ρ as a function of p and U. The speed of sound c is determined from the relations

$$c^2 = -(\gamma - 1)\rho \partial h / \partial \rho \tag{4.4}$$

$$\gamma = \rho \frac{\partial h}{\partial p} \left(\rho \frac{\partial h}{\partial p} - 1 \right)^{-1} \tag{4.5}$$

where γ reduces to the usual ratio of specific heats when these are constant (this assumption is not needed in what follows).

   With these definitions, the system of partial differential equations (4.1) and (4.2) is a closed system with respect to the basic dependent variables U and p. Of course, equations (4.1) and (4.2) need not be solved in this form; in particular, it may be appropriate to use the conservative forms in order to capture discontinuities (such as shock waves, slip surfaces, or vortex sheets). The form of equations (4.1) and (4.2) is convenient for studying the properties of associated PUS.

4.3. General Form of Pseudo-Unsteady Systems

   A general class of PUS for solving system (4.1) and (4.2) is obtained by equating the time derivatives of U and p to general linear combinations of the left-hand sides of equations (4.1) and (4.2). For convenience, we introduce the notation

$$\Psi \equiv \operatorname{div} U - \frac{U^2}{c^2} s \cdot \operatorname{grad} U \tag{4.6}$$

$$M \equiv (U \cdot \operatorname{grad})U + \frac{1}{\rho} \operatorname{grad} p \tag{4.7}$$

Hence the class of PUS to be studied is given by

$$\frac{\partial U}{\partial t} + aM + \beta U\Psi + eU(U \cdot M) = 0 \tag{4.8}$$

$$\frac{\partial p}{\partial t} + \rho\alpha U \cdot M + \rho d\Psi = 0 \tag{4.9}$$

where the scalar coefficients a, β, e, α, d are given functions of p and U. This system has an intrinsic form, i.e., it is independent of the choice of projection axis (which is why the vector coefficients are taken proportional to **U**, and the scalar coefficients depend on **U** only through its modulus).

Thanks to the use of p instead of ρ as basic dependent variable, this system applies to incompressible as well as to compressible flows. In the compressible case, the Bernoulli relation (4.3) gives ρ as a function of p and U, and the speed of sound c is obtained from relations (4.4) and (4.5).

A necessary condition to be satisfied by system (4.8) and (4.9) is that, at steady state, it yields a unique solution for $\varPsi$ and **M**, namely $\varPsi = 0$ and $\mathbf{M} = 0$. This is realized if and only if

$$\left.\begin{array}{l} a \neq 0 \\ (\alpha\beta - ed)U^2 - ad \neq 0 \end{array}\right\} \tag{4.10}$$

4.4. Hyperbolicity Condition

We now restrict the general class of PUS defined by equations (4.8) and (4.9) subject to the condition that such a system should be hyperbolic with respect to time. This property is a natural one to impose since it insures that perturbations will remain bounded in time, and since the existence of compatibility relations allows one to set up a systematic treatment at the boundaries.

There are different ways of expressing this condition. In the general three-dimensional case it is convenient, following Brochet,[49] to express it through the compatibility relations, especially as these relations are of direct use in the treatment of boundaries. Consider a linear combination of (4.8) and (4.9) with general form

$$\mu_0 \left(\frac{1}{\rho} \frac{\partial p}{\partial t} + \alpha \mathbf{U} . \mathbf{M} + d\varPsi \right)$$

$$+ \boldsymbol{\mu} . \left[\frac{\partial \mathbf{U}}{\partial t} + a\mathbf{M} + \beta \mathbf{U}\varPsi + e\mathbf{U}(\mathbf{U} . \mathbf{M}) \right] = 0 \tag{4.11}$$

Such a combination is called a compatibility relation if and only if it can be put into the special form

$$\mu_0 \frac{1}{\rho} \left(\frac{\partial p}{\partial t} + \lambda \frac{\partial p}{\partial \xi} \right) + \boldsymbol{\mu} . \left(\frac{\partial \mathbf{U}}{\partial t} + \lambda \frac{\partial \mathbf{U}}{\partial \xi} \right) = \mathcal{R} \tag{4.12}$$

where $\partial/\partial\xi = (\boldsymbol{\xi} . \mathrm{grad})$ and $\boldsymbol{\xi}$ is some unit vector in space, and where the right-hand side $\mathcal{R}$ depends on derivatives of p and U only in directions normal to $\boldsymbol{\xi}$. Hence the characteristic property of equation (4.12) is that the time derivative of any dependent variable combines with its space derivative in the ξ-direction to form a single transport operator

$$\frac{D}{D\xi} = \frac{\partial}{\partial t} + \lambda \xi \cdot \text{grad} \tag{4.13}$$

where λ is the wave speed for the direction ξ.

On projecting the gradient operator in directions parallel and normal to ξ,

$$\text{grad} \equiv \xi \frac{\partial}{\partial \xi} + \widetilde{\text{grad}} \qquad \text{with } \xi \cdot \widetilde{\text{grad}} = 0$$

we find that equation (4.11) will have the property of a compatibility relation (4.12) if and only if

$$\left.\begin{aligned}
\lambda \mu_0 &= a\boldsymbol{\mu} \cdot \xi + (\mu_0 \alpha + e\boldsymbol{\mu} \cdot \mathbf{U})U_\xi \\
\lambda \boldsymbol{\mu} &= aU_\xi \boldsymbol{\mu} + (\mu_0 \alpha + e\boldsymbol{\mu} \cdot \mathbf{U})U_\xi \mathbf{U} \\
&\quad + (\mu_0 d + \beta \boldsymbol{\mu} \cdot \mathbf{U})\left(\xi - \frac{1}{c^2}U_\xi \mathbf{U}\right)
\end{aligned}\right\} \tag{4.14}$$

where $U_\xi = \mathbf{U} \cdot \xi$. For given ξ and fixed λ, these relations constitute a linear homogeneous system for μ_0, $\boldsymbol{\mu}$, and the condition that there exists a nontrivial solution ($\mu_0^2 + \boldsymbol{\mu}^2 \neq 0$) is that the determinant vanishes. This condition yields an equation of fourth degree in λ, the solutions of which are the eigenvalues of system (4.14).

The hyperbolicity condition can then be stated as follows: for any direction ξ, compatibility relations should exist and should constitute a system of equations equivalent to system (4.8) and (4.9). It is also equivalent to say that, for any ξ, the four eigenvalues of system (4.14) should be real and that the corresponding eigenvectors ($\mu_0, \boldsymbol{\mu}$) should constitute a basis of the associated vector space.

Referring elsewhere[48] for a detailed discussion, we present the main results. First we find that an obvious eigenvalue (in 3D flow) is given by

$$\lambda^{(1)} = aU_\xi \tag{4.15}$$

To make the discussion tractable, we reduce the number of degrees of freedom by imposing that (aU_ξ) be at least a double eigenvalue, i.e.,

$$\lambda^{(2)} = aU_\xi \tag{4.16}$$

[this restriction is suggested by the fact that for the exact unsteady 3D

Euler's equations, U_ξ is a triple eigenvalue]. We find that this is the case if and only if

$$ed = \alpha\beta \tag{4.17}$$

The other two eigenvalues are then solutions of

$$\lambda^2 - A U_\xi \lambda - ad(1 - U_\xi^2/c^2) = 0 \tag{4.18}$$

with

$$A = a + eU^2 + \alpha + \beta(1 - U^2/c^2) \tag{4.19}$$

Taking into account relation (4.17), conditions (4.10) reduce to $(ad) \neq 0$. The number of boundary conditions to be imposed at a boundary point is equal to the number of negative eigenvalues associated with the vector ξ, equal to the unit outward normal (see, e.g., Viviand and Veuillot[50] and Brochet[49]). A study of the signs of the eigenvalues shows that this property implies the conditions

$$a > 0, \quad d > 0 \tag{4.20}$$

Then, taking into account relations (4.17) and (4.20), a rather lengthy discussion leads to the following conditions for the PUS (4.8) and (4.9) to be hyperbolic with respect to time. First define the quantities

$$\left. \begin{array}{l} D = d + \beta U^2 \\ E = \alpha U^2 + d(1 - U^2/c^2) \end{array} \right\} \tag{4.21}$$

Two cases must be considered

Case 1: $E \neq 0$,

$\quad\quad$ If $D \neq 0$, we must have $ED > 0$ $\hfill$ (4.22)

$\quad\quad$ If $D = 0$, we must have $a \neq \beta(1 - U^2/c^2)$ $\hfill$ (4.23)

Case 2: $E = 0$,

$\quad\quad$ If $a = \alpha$, we must have $D = 0$ $\hfill$ (4.24)

(since $a > 0$ and $d > 0$, this case implies $U > c$).

When the above conditions are satisfied, the eigenvectors are given by the following expressions, using an orthogonal basis with unit vectors $\mathbf{e}_1$, $\mathbf{e}_2$, $\mathbf{e}_3$ such that $\mathbf{e}_1 = \xi$ and $\mathbf{U} = U_\xi \xi + V\mathbf{e}_2$.

First, corresponding to the eigenvalue aU_ξ, we have:

if aU_ξ is of order 2 only,

$$\left.\begin{aligned}
\mu_0^{(1)} &= 0, \qquad \boldsymbol{\mu}^{(1)} = \mathbf{e}_3 \\
\mu_0^{(2)} &= \beta V, \qquad \boldsymbol{\mu}^{(2)} = \beta V\mathbf{U} - (d + \beta U^2)\mathbf{e}_2
\end{aligned}\right\} \tag{4.25}$$

if aU_ξ is of order 3 and not of order 4 (this implies $V = 0$),

$$\left.\begin{aligned}
\mu_0^{(1)} &= 0, \qquad \boldsymbol{\mu}^{(1)} = \mathbf{e}_3 \\
\mu_0^{(2)} &= 0, \qquad \boldsymbol{\mu}^{(2)} = \mathbf{e}_2 \\
\mu_0^{(3)} &= 1, \qquad \boldsymbol{\mu}^{(3)} = \mathbf{U} = U_\xi\boldsymbol{\xi}
\end{aligned}\right\} \tag{4.26}$$

if aU_ξ is of order 4, any vector is an eigenvector.

Then the eigenvector associated with an eigenvalue λ not equal to aU_ξ is:

$$\left.\begin{aligned}
\mu_0 &= \lambda - (A - \alpha)U_\xi \qquad (\lambda \neq aU_\xi) \\
\boldsymbol{\mu} &= (\alpha - d/c^2)U_\xi\mathbf{U} + d\boldsymbol{\xi}
\end{aligned}\right\} \tag{4.27}$$

5. SOME PARTICULAR PSEUDO-UNSTEADY SYSTEMS FOR INVISCID FLOWS

Here we consider particular PUS belonging to the class of PUS defined in the previous section.

5.1. Incompressible Flow

For an incompressible flow, it can be seen by setting $1/c = 0$ that the conditions given at the end of Section 4.4 reduce to

$$\left.\begin{aligned}
a &> 0, \qquad d > 0 \\
(\alpha U^2 &+ d)(\beta U^2 + d) \geq 0
\end{aligned}\right\} \tag{5.1}$$

In Section 2.2.1, PUS for the solution of the steady incompressible Navier–Stokes equations in primitive variables were reviewed and it is of interest (as mentioned in Section 4.1) to consider their inviscid counterpart. These systems are based on the exact unsteady momentum equation in either conservative or nonconservative form. Considering only the latter case, the exact inviscid momentum equation consists of equation (4.8) with

$$a = 1, \qquad \beta = 0, \qquad e = 0 \tag{5.2}$$

Therefore relation (4.17) is satisfied.

In Chorin's method, the pseudo-unsteady continuity equation (2.17) is used. This equation corresponds to equation (4.9) with

$$\alpha = 0, \qquad \rho d = 1/\delta \tag{5.3}$$

Condition (5.1) then reduces to $\delta > 0$, which is the condition imposed by Chorin.

The method of Yanenko *et al.* makes use of the pseudo-unsteady continuity equation (2.21) with $q = p/\rho + U^2/(2N)$, N being the number of spatial dimensions. This equation results from the PUS (4.8) and (4.9) in the case when

$$\left. \begin{array}{l} \alpha + (a + eU^2)/N = 0 \\ d + \beta U^2/N = 1/\delta \end{array} \right\} \tag{5.4}$$

Taking into acount (5.2), we obtain

$$\alpha = -1/N, \qquad d = 1/\delta \tag{5.5}$$

and conditions (5.1) give

$$\delta > 0 \tag{5.6a}$$

$$\delta U^2 \leqslant N \tag{5.6b}$$

In the original method of Yanenko *et al.*, the coefficient δ does not appear as a variable parameter but as a fixed numerical constant equal to 1 for dimensionless equations. In our notation, this translates into $\delta = 1/U_*^2$ where U_* is the reference velocity, and condition (5.6b) implies that U_* should not be taken smaller than the highest velocity in the flow field divided by $\sqrt{N}$ (at least in the region of quasi-inviscid flow). It will be shown in Section 5.3 that a PUS based on equation (2.21) with $q = p/\rho + U^2/2$ and $\delta = 1/U^2$ would be optimal from a numerical stability point of view.

5.2. Compressible Flow and Conservative Pseudo-Unsteady Systems

A now classical, pseudo-unsteady method for transonic flow calculation is based on the PUS consisting of the exact unsteady continuity and momentum equations in conservative form, together with the Bernoulli relation (4.3) (see, e.g., Viviand and Veuillot,[50] Brochet,[49] Rizzi and

Eriksson,[51] and Jameson et al.[47]). This is a particular case of system (4.8) and (4.9) corresponding to

$$a = 1, \qquad \beta = e = 0, \qquad \alpha = 1/\gamma, \qquad d = c^2/\gamma \qquad (5.7)$$

These coefficients satisfy relation (4.17) and conditions (4.20) and (4.22) (we are in case 1 with $E = D = c^2/\gamma > 0$).

Other particular PUS are suggested if we inspect the equivalent conservative forms of equations (4.8) and (4.9). Let us introduce the notation

$$\left. \begin{aligned} \rho \Theta &\equiv \mathrm{div}(\rho \mathbf{U}) \\ \rho \mathbf{D} &\equiv \mathrm{div}(\rho \mathbf{U} \otimes \mathbf{U} + p\mathbf{I}) \end{aligned} \right\} \qquad (5.8)$$

where $\mathbf{I}$ is the unit tensor and symbol $\otimes$ represents the dyadic product of vectors. Quantities Θ and $\mathbf{D}$ are related to their nonconservative counterparts Ψ and $\mathbf{M}$, defined in (4.6) and (4.7) respectively, by the following identities [making use of equation (4.3) in the case of a compressible fluid]:

$$\left. \begin{aligned} \Psi &= \left(1 + \gamma \frac{U^2}{c^2}\right) \Theta - \frac{\gamma}{c^2} \mathbf{U} . \mathbf{D} \\ \mathbf{M} &= \mathbf{D} - \mathbf{U}\,\Theta \end{aligned} \right\} \qquad (5.9)$$

Using these relations, the PUS (4.8) and (4.9) can be written in terms of Θ and $\mathbf{D}$, namely

$$\left. \begin{aligned} \frac{\partial \mathbf{U}}{\partial t} + a\mathbf{D} + \beta_1 \mathbf{U}\Theta + e_1 \mathbf{U}(\mathbf{U} . \mathbf{D}) &= 0 \\ \frac{\partial p}{\partial t} + \rho \alpha_1 \mathbf{U} . \mathbf{D} + \rho d_1 \Theta &= 0 \end{aligned} \right\} \qquad (5.10)$$

where the coefficients with subscript 1 are related to the previous coefficients by

$$\left. \begin{aligned} e_1 &= e - \gamma \beta/c^2, & \beta_1 &= \beta - a - e_1 U^2 \\ \alpha_1 &= \alpha - \gamma d/c^2, & d_1 &= d - \alpha_1 U^2 \end{aligned} \right\} \qquad (5.11)$$

The practical interest in system (5.10) is that it can be discretized in a conservative form with respect to Θ and $\mathbf{D}$, thus leading to conservative numerical solutions of the steady equations $\Theta = 0$ and $\mathbf{D} = 0$.

As regards the application of relation (4.17), it should be noted that

$$\alpha\beta - ed = \overline{\alpha_1\beta} - e_1 d \qquad (5.12)$$

This remark leads one to consider a subclass of simple PUS in conservative form defined by $\alpha_1 = e_1 = 0$, i.e.,

$$e = \gamma\beta/c^2, \qquad \alpha = \gamma d/c^2 \qquad (5.13)$$

By eliminating e and α, these simpler systems assume the form

$$\left. \begin{aligned} \frac{\partial \mathbf{U}}{\partial t} + a\mathbf{D} + (\beta - a)\mathbf{U}\Theta &= 0 \\ \frac{\partial p}{\partial t} + \rho d\Theta &= 0 \end{aligned} \right\} \qquad (5.14)$$

If relations (5.13) are taken into account, we obtain for E [equation (4.21)]

$$E = d[1 + (\gamma - 1)U^2/c^2] \qquad (5.15)$$

Thus we are in case 1 of Section 4.4, and conditions (4.22) and (4.23) yield

$$d + \beta U^2 \geqslant 0 \qquad \text{with } a \neq \beta(1 - U^2/c^2) \quad \text{if } d + \beta U^2 = 0 \qquad (5.16)$$

A system belonging to subclass (5.14) is selected in Section 5.3 from numerical stability considerations. It is noteworthy that the classical PUS corresponding to relations (5.7) does not belong to this subclass.

5.3. Selection of Pseudo-Unsteady Systems from Numerical Stability Considerations

Advantage can be taken of the existing arbitrariness in the choice of the coefficients to obtain the best stability criterion in the case when explicit time-integration schemes are used. The local CFL-type stability condition encountered with such schemes can be written in the general form

$$\Delta t \leqslant k\Delta x / V_{GM} \qquad (5.17)$$

where Δx is a measure of the local mesh size, k some positive constant of order unity, and V_{GM} is the largest of the modulus of the group velocities, i.e.,

$$V_{GM} = \max_{l,\xi} \{|\mathbf{V}_G^{(l)}(\xi)|\} \qquad (5.18)$$

$V_G^{(l)}$ being the group velocity associated with the eigenvalue (wave speed) $\lambda^{(l)}$ ($l = 1$ to 4) and given by

$$V_G^{(l)}(\xi) = \lambda^{(l)}\xi + \frac{\partial \lambda^{(l)}}{\partial \xi} - \xi\left(\xi \cdot \frac{\partial \lambda^{(l)}}{\partial \xi}\right) \qquad (5.19)$$

In general, the dependency of $\lambda^{(l)}$ with respect to ξ for $l = 3, 4$ is not very simple because the discriminant Δ of equation (4.18), namely

$$\Delta = (A^2 - 4ad/c^2)U_\xi^2 + 4ad \qquad (5.20)$$

depends itself on ξ through U_ξ. This is the case in particular for the classical PUS defined by relations (5.7).

However a remarkable case arises if Δ becomes independent of ξ, i.e., if

$$A^2 = 4ad/c^2 \qquad (5.21)$$

which is compatible with condition (4.20). Indeed, in this case the eigenvalues $\lambda^{(3)}$, $\lambda^{(4)}$ exhibit the simple type of dependency on ξ that exists for the exact unsteady (compressible) equations ($\lambda = U_\xi \pm c$), and more precisely we find

$$\lambda^{(3),(4)} = \frac{\sqrt{ad}}{c}(U_\xi \pm c) \qquad (5.22)$$

For compressible flow, in particular, by choosing $a = 1$ and $d = c^2$ [$A = 2$, hence $\alpha + eU^2 + \beta(1 - U^2/c^2) = 1$ and, using relation (4.17), $ED = c^4$], we obtain a subclass of PUS which satisfy (4.20) and (4.22), and which have exactly the same eigenvalues (hence the same characteristic cone) as the system of the exact unsteady Euler equations (the only difference is that aU_ξ is a double instead of a triple eigenvalue). In this subclass there is one system satisfying (5.13), hence of form (5.14), for β given by the relation

$$\beta = -\left(\frac{U^2}{c^2} + \frac{1}{\gamma - 1}\right)^{-1}$$

Let us therefore suppose that relation (5.21) is satisfied. Then

$$V_{GM} = \max_{l,\xi}\{|\lambda^{(l)}|\} = \max\left\{aU, \frac{\sqrt{ad}}{c}(U + c)\right\} \qquad (5.23)$$

and the largest time step will be obtained when the two quantities inside the brackets are equal, namely when

$$d = aU^2c^2/(U+c)^2 \tag{5.24}$$

in which case relation (5.22) becomes

$$\lambda^{(3),(4)} = aU(U_\xi \pm c)/(U+c) \tag{5.25}$$

and the stability criterion (5.17) yields

$$\Delta t \leq \Delta t_1 = k\Delta x/(aU) \tag{5.26}$$

We must check that conditions (5.21) and (5.24) lead to acceptable coefficients. From these relations we deduce $A = 2aU/(U+c)$ and $ED = 0$, which decomposes into:

1. $D = d + \beta U^2 = 0$, hence $\beta = -ac^2/(U+c)^2$, or
2. $E = \alpha U^2 + d(1 - U^2/c^2) = 0$, hence $\alpha = a(U-c)/(U+c)$.

The discussion in Section 4.4 then enables the following possibilities to be deduced:

Case 1:
$$\left. \begin{array}{l} \alpha \neq a(U-c)/(U+c) \text{ (hence } E \neq 0) \\ \text{then } \beta = -ac^2/(U+c)^2 = -dU^2 \end{array} \right\} \tag{5.27}$$

and condition (4.23) is satisfied.

Case 2:
$$\left. \begin{array}{l} \alpha = a(U-c)/(U+c) \\ = -(d/U^2)(1 - U^2/c^2) \quad (E=0) \\ \text{hence } a \neq \alpha, \text{ and } \beta \text{ remains free} \end{array} \right\} \tag{5.28}$$

Notice that, conversely, relations (5.24) and either (5.27) or (5.28) imply (5.21).

The stability condition (5.26) can be compared to the stability condition corresponding to the exact unsteady equations, namely

$$\Delta t \leq \Delta t_2 = k\Delta x/(U+c) \tag{5.29}$$

The comparison should be made with $a = 1$, since in fact coefficient a can be incorporated into the time variable by a suitable transformation, $d\tau = adt$. The ratio $\Delta t_1/\Delta t_2 = 1 + c/U$ is of order 2 at transonic speeds and becomes large for low Mach numbers. The fact that Δt_1 does not depend on the Mach number is an attractive feature for the case of flows which have extended low-speed regions.

Among the PUS satisfying relations (5.24) and (5.27) is one that also belongs to subclass (5.14) for which e and α are given by (5.13):

$$e = -\frac{a\gamma}{(U+c)^2}, \qquad \alpha = \frac{a\gamma U^2}{(U+c)^2} \tag{5.30}$$

The above discussion is valid for the case of incompressible flow ($1/c = 0$). Condition (5.21) reduces to $A = 0$, i.e.,

$$a + eU^2 + \alpha + \beta = 0 \tag{5.31}$$

and we have

$$\lambda^{(3),(4)} = \pm\sqrt{ad} \tag{5.32}$$

The optimum value of d, given by relation (5.24), becomes simply

$$d = aU^2 \tag{5.33}$$

and the corresponding stability criterion is still expressed by (5.26), since this expression does not depend on the speed of sound.

In the case of PUS (for incompressible flows) that make use of the exact unsteady momentum equation, i.e., with coefficients α, β, e given by equation (5.2), the maximum time step (5.26) is obtained when relations (5.31) and (5.33) are satisfied, which gives $\alpha = -1$ and $d = U^2$. This amounts to using an equation of type (2.21) but with $q = p/\rho + U^2/2$ and $\delta = 1/U^2$.

REFERENCES

1. L. Crocco, A suggestion for the numerical solution of the steady Navier–Stokes equations, *AIAA J.* **3**, 1824–1832 (1965).
2. E. C. Du Fort and S. P. Frankel, Stability conditions in the numerical treatment of parabolic differential equations, *Math. Tables Aids Comp.* **7**, 135–152 (1953).
3. G. W. Evans, R. Brousseau and R. Kierstead, Stability considerations for various difference equations derived from the linear heat conduction equation, *J. Math. Phys.* **34**, 267–285 (1956).
4. R. Peyret and H. Viviand, Calcul de l'écoulement d'un fluide visqueux compressible autour d'un obstacle de forme parabolique, in: *Lecture Notes in Physics*, Vol. 19, pp. 222–228, Springer-Verlag, Berlin–Heidelberg–New York (1973).
5. G. D. Raithby and K. E. Torrance, Upstream-weighted differencing schemes and their application to elliptic problems involving fluid flow, *Comput. Fluids* **2**, 191–206 (1974).
6. F. H. Harlow and J. E. Welsh, Numerical calculation of time-dependent viscous incompressible flow, *Phys. Fluids* **8**, 2182–2189 (1965).

7. A. J. Chorin, Numerical solution of the Navier–Stokes equations, *Math. Comp.* **22**, 745–762 (1968).
8. R. Temam, Sur l'approximation de la solution des équations de Navier–Stokes par la méthode des pas fractionnaires (II), *Arch. Ration. Mech. Anal.* **33**, 377–385 (1969).
9. R. Peyret and T. D. Taylor, *Computational Methods for Fluid Flow*, Springer-Verlag, Berlin–Heidelberg–New York (1983).
10. A. J. Chorin, Numerical Study of Thermal Convection in a Fluid Layer Heated from Below, New York University, A.E.C. Research and Development Report. No. NYO-1480-61 (August, 1966).
11. A. J. Chorin, A numerical method for solving incompressible viscous flow problems, *J. Comput. Phys.* **2**, 12–26 (1967).
12. R. Temam, Sur l'approximation de la solution des équations de Navier–Stokes par la méthode des pas fractionnaires (I), *Arch. Ration. Mech. Anal.* **32**, 135–153 (1969).
13. R. Temam, *Navier–Stokes Equations*, North-Holland Publishing Company, Amsterdam (1977).
14. M. Fortin, R. Peyret, and R. Temam, Résolution numérique des équations de Navier–Stokes pour un fluide incompressible, *J. Méc.* **10**, 357–390 (1971).
15. E. Elsaesser and R. Peyret, Méthodes hermitiennes pour la résolution des équations de Navier–Stokes, in: *Méthodes Numériques dans les Sciences de l'Ingénieur* (E. Absi and R. Glowinski, eds.), pp. 249–258, Dunod, Paris (1979).
16. E. Elsaesser and R. Peyret, Hermitian methods for the solution of the Navier–Stokes equations in primitive variables, in: *Recent Advances in Numerical Methods in Fluids*, Vol. 3: *Viscous Flows Computational Methods* (W. G. Habashi, ed.), Pineridge Press, Swansea (to appear).
17. X. Aubert and M. Deville, Steady viscous flows by compact differences in boundary-fitted coordinates, *J. Comput. Phys.* **49**, 490–522 (1983).
18. N. N. Yanenko, *The Method of Fractional Steps*, Springer-Verlag, Berlin–Heidelberg–New York (1971).
19. T. D. Taylor and E. Ndefo, Computation of viscous flow in a channel by the method of splitting, in: *Lecture Notes in Physics*, Vol. 8, pp. 356–364, Springer-Verlag, Berlin–Heidelberg–New York (1971).
20. H. J. Wirz, Relaxation methods for time-dependent conservation equations in fluid mechanics, in: *AGARD-VKI Lecture Series*, No. 86, Von Karman Institute for Fluid Dynamics, Rhode-St-Genèse (1977).
21. S. M. Richardson and A. R. H. Cornish, Solution of three-dimensional flow problems, *J. Fluid Mech.* **82**, 309–319 (1977).
22. G. J. Hirasaki and J. D. Hellums, Boundary conditions on the vector and scalar potentials in viscous three-dimensional hydrodynamics, *Q. Appl. Math.* **28**, 293–296 (1970).
23. G. J. Hirasaki and J. D. Hellums, A general formulation of the boundary conditions on the vector potential in three-dimensional hydrodynamics, *Q. Appl. Math.* **26**, 331–342 (1968).
24. G. D. Mallinson and G. I. De Vahl Davis, The method of false transient for the solution of coupled elliptic equations, *J. Comput. Phys.* **12**, 435–461 (1973).
25. P. Bontoux, Contribution à l'étude des écoulements visqueux en milieu confiné, Thèse Doctorat d'Etat, IMFM, Université d'Aix-Marseille (1978).
26. P. Bontoux, B. Gilly, and B. Roux, Analysis of the effect of boundary conditions on numerical stability of solutions of Navier–Stokes equations, *J. Comput. Phys.* **15**, 417–427 (1980).
27. J. Douglas and J. E. Gunn, A general formulation of alternating direction methods, *Numer. Math.* **6**, 428–453 (1964).
28. O. D. Burggraf, Analytical and numerical studies of the structure of steady separated flows, *J. Fluid Mech.* **24**, 113–151 (1966).

29. D. Gottlieb and B. Gustafsson, Generalized Du Fort–Frankel methods for parabolic initial-boundary-value problems, *SIAM. J. Numer. Anal.* **13**, 129–144 (1976).
30. D. Gottlieb and L. Lustman, The Du Fort–Frankel–Chebyshev Method for Parabolic Initial-Boundary-Value Problems, ICASE, Report No. 81-42 (December, 1981).
31. J. S. Allen and S. I. Cheng, Numerical solution of the compressible Navier–Stokes equations for the laminar near wake, *Phys. Fluids* **19**, 37–52 (1970).
32. I. Yu. Braïlovskaya, A difference scheme for numerical solution of the two-dimensional nonstationary Navier–Stokes equations for a compressible gas, *Sov. Phys. Dokl.* **10**, 107–110 (1965).
33. R. Peyret and H. Viviand, Calcul numérique de l'écoulement supersonique d'un fluide visqueux sur un obstacle parabolique, *La Rech. Aérosp.*, No. 1972-3, 123–131 (1972).
34. P. Garabedian, Estimation of the relaxation factor for small mesh size, *Math. Tables Aids Comp.* **10**, 183–185 (1956).
35. H. Lomax and J. L. Steger, Relaxation methods in fluid mechanics, in: *Annual Review in Fluid Mechanics*, Vol. 7, pp. 63–88, Annual Review Inc., Palo Alto (1975).
36. R. Peyret and H. Viviand, Computation of viscous compressible flows based on the Navier–Stokes equations, AGARDograph No. 212 (1975).
37. C. P. Li, Numerical solution of viscous reacting blunt body flows of a multicomponent mixture, AIAA paper No. 73-202 (1973).
38. C. P. Li, Hypersonic nonequilibrium flow past a sphere at low Reynolds numbers, AIAA paper No. 74-173 (1974).
39. R. W. MacCormack, The effect of viscosity in hypervelocity impact cratering, AIAA paper No. 69-354 (1969).
40. H. McDonald and W. R. Briley, Computational fluid dynamics aspects of internal flows, AIAA paper No. 79-1445 (1979).
41. E. L. Wachspress, *Iterative Solution of Elliptic Systems*, Prentice-Hall, Englewood Cliffs, New Jersey (1966).
42. S. J. Shamroth and H. J. Gibeling, The prediction of a turbulent flow field about an isolated airfoil, AIAA paper No. 79-1543 (1979).
43. S. J. Shamroth and H. J. Gibeling, A compressible solution of the Navier–Stokes equations for turbulent flow about an airfoil, NASA CR-3183 (October, 1979).
44. W. R. Briley and H. McDonald, Solution of the three-dimensional compressible Navier–Stokes equations by an implicit method, in: *Lecture Notes in Physics*, Vol. 35, pp. 105–110, Springer-Verlag, Berlin–Heidelberg–New York (1975).
45. A. Lerat, J. Sides, and V. Daru, An implicit finite-volume method for solving the Euler equations, *Lecture Notes in Physics*, Vol. 170, pp. 343–349, Springer-Verlag, Berlin–Heidelberg–New York (1982).
46. H. Viviand, Pseudo-unsteady methods for transonic flow computations, in: *Lecture Notes in Physics*, Vol. 141, pp. 44–54, Springer-Verlag, Berlin–Heidelberg–New York (1981).
47. A. Jameson, W. Schmidt, and E. Turkel, Numerical solution of the Euler equations by finite volume methods using Runge–Kutta time-stepping schemes, AIAA Paper No. 81-1259 (1981).
48. H. Viviand, Systèmes pseudo-instationnaires pour les écoulements stationnaires de fluide parfait, Publication ONERA No. 1983–4 (1983).
49. J. Brochet, Calcul numérique d'écoulements internes tridimensionnels transsoniques, *La Rech. Aérosp.*, No. 1980-5, 301–315 (1980).
50. H. Viviand and J. P. Veuillot, Méthodes pseudo-instationnaires pour le calcul d'écoulements transsoniques, Publication ONERA No. 1978-4 (1978).
51. A. Rizzi and L. E. Eriksson, Transfinite mesh generation and damped Euler equation algorithm for transonic flow around wing-body configurations, AIAA Paper, No. 81-0999 (1981).

II
GAS DYNAMICS

4

Boltzmann Equation and Rarefied Gas Dynamics

C. Cercignani

Abstract. A survey of the present state of the art in theoretical rarefied gas dynamics is presented.

1. INTRODUCTION

Considerable interest in rarefied gas dynamics developed during the late fifties and the sixties as a result of spectacular space programs and led to the study of a great number of theoretical and experimental problems concerning flows of neutral and ionized gases. This resulted in a rich harvest of original results. The pace of these activities has been slowing down considerably in recent years. In this paper we shall summarize the developments of the last decades and try to describe the theoretical research being undertaken at present, albeit at a reduced pace.

Rarefied gas dynamics is a subject easy to define, at least from a theoretical point of view. It involves all gas-flow phenomena for which the Navier–Stokes equations are not valid, at least in some significant region of the flow field. Quantitatively, the criterion of rarefaction is defined through the Knudsen number Kn, the ratio of a typical molecular mean free path to a typical length scale for the flow.[1,2] If this length scale is clearly defined *a priori*, which is not always the case, then the Knudsen number is inversely proportional to the density of the gas, as well as a function of temperature. The Chapman–Enskog theory, which relates the Navier–Stokes equations to the Boltzmann equation of kinetic theory, indicates that the length scale

C. Cercignani • Dipartimento di Matematica, Politecnico di Milano, Piazza L. da Vinci 32, 20133 Milano, Italy.

to be chosen must be based on the gradients occurring in the flow.[1,2] This scale may depend on the Mach number as well as on the geometric length scale, so that the Knudsen number may be a rather complicated function of Mach number, geometrical dimensions, density, and temperature but it increases with decreasing density, all other parameters remaining unchanged.

Thus a rarefied gas flow can be defined by the fact that the Knudsen number based on local gradients is not negligible compared to unity, or better, that the regions where the Knudsen number is not small are very thin layers (thin shocks, Knudsen layers with thickness negligible compared to the viscous boundary-layer thickness). If one recalls that the mean free path in air is of the order of 10^{-5} cm in normal conditions, it follows that rarefied gas flows occur typically at pressures of the order of a thousandth of the atmospheric pressure or less.

Many problems of rarefied gas dynamics arose in aerospace activities. Among the most significant were flows over reentering vehicles, with particular emphasis on kinetic heating, drag, and energy balance of satellites, the exhaust volume of rockets operating at extreme altitudes and in space, design problems of space simulators and their instrumentation, high-atmosphere sounding devices, production of gas clouds at extreme altitudes, etc. Other problems arose in other fields, such as astrophysics, the design of vacuum systems for chemical processing or particle accelerators, and so on.

Many of these problems of considerable practical importance indicated the need for a more thorough understanding of fundamental aspects of rarefied gas dynamics, such as gas–surface interactions, intermolecular forces, the foundations of kinetic theory, as well as basic flow phenomena such as the structure of Knudsen layers and shock waves, the free expansion of a gas into a vacuum, flow through pipes, orifices, and around bodies of simple geometric shapes, such as spheres, cones, and thin flat plates.

This presentation will be devoted to the description of selected results, concerning basic flow phenomena such as Knudsen layers, shock structure, flow near the leading edge of a flat plate, flows around bodies of simple geometric shapes, and free jets. For most of these flows there exist some theoretical results, usually based on approximation solutions of the Boltzmann equation or one of the model kinetic equations. It should be noted also that some of the basis flows mentioned involve boundaries, and therefore the interaction of gas molecules with solid surfaces.

For the basic concepts of the kinetic theory of gases we refer the reader to the author's books,[1,2] where one can also find a detailed treatment of some early results on some of the problems considered here.

2. INTERNAL FLOWS AND KINETIC LAYERS

There is an enormous amount of literature on Couette and Poiseuille flows, as well as on the similar problems of heat transfer between parallel plates and coaxial cylinders. Nowadays systematic, accurate, and relatively simple methods of solution are available for these flows, provided linearization is permitted. In general, the results obtained by these methods have produced predictions which are in spectacular agreement with experiment and have shed considerable light on the basic structure of transition-flow theory whenever nonlinear effects (in particular shock waves) can be neglected. For nonlinear problems, moment methods can be introduced, which lead to the solution of a system of nonlinear partial differential equations. The latter, in general, are harder to handle than the Navier–Stokes equations and, as a consequence, it is necessary to resort to numerical procedures for their solution. It may then be found convenient to try a direct numerical method, based on a discrete-ordinate technique, of the Boltzmann equation itself. Finally a great deal of research has been devoted to Monte-Carlo simulation methods.

In connection with all these methods, one can simplify the calculations by the use of suitable collision models, but it is noteworthy that the accuracy of kinetic models in nonlinear problems is less obvious than in the linearized ones. Detailed description and references may be found elsewhere.[2]

Very particular and important examples of flow regions having the features characteristic of the transition regime are the transition layers such as the "initial," "boundary," and "shock layers," in the terminology introduced by Grad.[3]

The shock layers arise in connection with sharp changes, such as those that occur in steady supersonic flows. Such shock layers or shock waves are well known in the inviscid-flow approximation and are described as discontinuity surfaces. If the Navier–Stokes equations are used, then the shock wave is a region across which physical quantities vary smoothly but rapidly, and the shock has a finite thickness generally of the order of a mean free path. This small thickness indicates that the Navier–Stokes equations, strictly speaking, should not be used. In order to obtain reliable results for the structure of shock waves, use has to be made of the Boltzmann equation.

The simplest case is offered by a steady normal plane shock wave: the gas flows parallel to the x-axis and all the quantities are independent of time or the remaining two space coordinates. The gas is in equilibrium at infinity upstream and downstream so that the corresponding distribution functions, f_- and f_+, are Maxwellians. The problem is to find the distribution function in between; this is the simplest problem in which the nonlinear structure of the Boltzmann equation plays an essential role.

If the shock wave is weak, i.e., if the downstream and upstream values of the density, velocity, and temperature are not very different from each other, then one may expect that the derivative $\partial f/\partial x$ is small, of the order of the strength of the shock (defined, e.g., by $\varepsilon = (p_- - p_+)/(p_- + p_+)$) and try to expand the solution as a series of powers of ε.[4,5] The result is similar to the Hilbert expansion[1,2] and the resulting equations describe a slight departure from a uniform state. To first order in ε, one is led to $\rho = \rho_0 + \varepsilon \rho_1$, where[5]

$$\rho_1/\rho_1^+ = \tanh\left[(x - x_0)/L\right] \tag{1}$$

Here the constant L is a measure of the shock thickness and is of the order of the ratio of the mean free path to the strength of the shock. The Navier–Stokes equations also lead to Taylor's equation (1) for weak shocks[6] and hence are correct to the lowest order in such a case.

The other extreme case is the infinitely strong shock profile. Grad[7] suggested that the limit of f for $\varepsilon \to \infty$ exists (for collision operators with finite collision frequency) and is given by a multiple of the delta function centered at the upstream velocity plus a comparatively smooth function for which it is not difficult to derive an equation. The latter seems more complicated than the Boltzmann equation itself, but the presumed smoothness of the solution should allow a simple approximate solution to be obtained. The simplest choice for the smooth remainder is a Maxwellian[7] whose parameters are determined by the conservation equations.

Grad's approach is one of the most recent analytical methods for dealing with the problem of shock-wave structure. In its simplest version it is strictly related to, and perhaps inspired by, the first approach to the same problem used by Mott-Smith[8] in 1951. The latter author represented the approximating distribution function in the form

$$f(x, \boldsymbol{\xi}) = a_+(x)f_+(\boldsymbol{\xi}) + a_-(x)f_-(\boldsymbol{\xi}) \tag{2}$$

where f_+ and f_- are the aforementioned Maxwellians, and a_+ and a_- are to be determined by suitable moment equations. The obvious choice for the first few moment equations are the conservation equations for mass, momentum, and energy. If, and only if, $a_+(x) + a_-(x) = 1$ then all the conservation equations are satisfied provided, of course, the upstream and downstream values satisfy the Rankine–Hugoniot relations. A possible choice is to take $\varphi = \xi_1^2$ in the moment equation

$$\frac{\partial}{\partial x} \int \xi_1 \varphi f \, d\boldsymbol{\xi} = \int \varphi Q(f, f) \, d\boldsymbol{\xi} \tag{3}$$

Then one is led to a hyperbolic tangent profile similar to that found for weak shocks, where the thickness L depends on the upstream mean free path l^- and Mach number M^-. For any power-law interaction l^-/L tends to zero when $M^- \to \infty$, but the same ratio tends to a finite limit for rigid spheres.

If we select another moment equation in place of that corresponding to $\varphi = \xi_1^2$, we are led to the same results except for a change in the expression and value of the thickness of the shock wave. Choosing $\varphi = \xi_1^3$ rather than $\varphi = \xi_1^2$, say, leads to a change of about 25% in the shock thickness. This fact indicates that the Mott-Smith method, although qualitatively correct, is not quantitatively adequate. In addition, the results for weak shocks ($M \to 1$) are not in agreement with the weak-shock theory or, equivalently, the Navier–Stokes results. Because of this fact several authors have presented modifications of the Mott-Smith approach.[9–11] A comparison with experimental data[12] indicates that the Navier–Stokes equations do not lead to results in agreement with experiments for $M_- > 2$. The situation is not improved by considering the Burnett equations or Grad's thirteen-moment equations. These sets of equations do not have a solution, in fact, for $M_- > 2.1$ and $M_- > 1.65$, respectively. In general, moment methods present difficulties for $M_- > 1.851$.

A general method extending Mott-Smith's approach would seem to use a linear combination of a number of Maxwellians. The main problem for the methods à la Mott-Smith remains the choice of moment equations; a way out of the difficulty would be to use a variational method, but the situation in this connection is not very encouraging for nonlinear problems, although Oberai[13] and Narasimha et al.[14] obtained interesting results for the shock structure by using the least-squares method.

An exact numerical solution of the problem for the BGK model was obtained by Liepmann et al.,[15] who used the integral form of the BGK model equation[12]; thus they obtained three integral equations for the three macroscopic quantities ρ, v, and T. Those equations were solved by the method of successive approximations with the Navier–Stokes solution as zeroth iterate.

The BGK solution was recalculated with greater accuracy by Anderson.[16] More recently, Segal and Ferziger[17] studied the problem using several models of the Boltzmann equations, including a new model, the trimodal gain function model, which was developed purposely also for the shock-structure problem.

The problem of shock-wave structure has been treated by several authors with Monte-Carlo methods.[18–21] The Monte-Carlo results as well as the numerical results for the trimodal model of Segal and Ferziger seem to agree in showing no evidence of a temperature overshoot found by some authors.[9,10]

Concerning comparison with experiments, it is not clear at present whether the experimental data (including those on the properties of the gas) are known to sufficient accuracy to discriminate between different results. In fact, as remarked by Bird,[20] a comparison based on either the shock-wave thickness or even the complete density profile does not provide a verdict on which of the methods provides the best description of the shock structure, and the prospects for a sufficiently accurate experimental determination of the higher moments are not hopeful.

Some recent developments of the problem of the shock wave structure appear worth mentioning. Nicolaenko[22] investigated the existence and uniqueness of shock-wave solutions for the nonlinear Boltzmann equation in the case of Grad's angular cutoff. He was able to show that the shock-wave solution arises by bifurcation from the constant Maxwellian distribution. The critical value of the bifurcation parameter corresponds to the sonic regime. These results had been previously demonstrated in a simpler but nonrigorous fashion.[5] The Chapman–Enskog solutions as well as the formal expansion using "stretched" variables are shown to fail to yield uniform asymptotics (in the x-variable) beyond the Navier–Stokes level. This fact is related to the circumstances that the Navier–Stokes and Fourier constitutive relations are not valid at the upstream and downstream points, even though $|f - f_\infty|$ becomes vanishingly small there, a fact discussed in detail by Elliot et al.[23,24] These authors proposed new closure relations for the thirteen-moment theory that model the upstream flow quite well. The picture downstream remains incomplete although superior to anything previously available. In addition, since the upstream singular point is the more critical one, the theory of Elliot et al. removes the difficulties previously encountered in computing shock-wave profiles at the thirteen-moment level.

Let us now consider the boundary layers, i.e., the regions where the matching between the boundary data on the one hand and the asymptotic continuum-like solution on the other takes place.

Solutions to the problem of finding the structure of a Knudsen layer were found by approximate methods in the last thirty years. In the last few years, experimental procedures have been developed[25,26] to measure velocity profiles in the Knudsen layer on a flat wall. In particular, Reynolds et al.[25] reported that in the Knudsen layer the deviation of the velocity defect (= deviation of the actual velocity from the hydrodynamic velocity profile) shows a behavior quantitatively different from the results obtained by the BGK model.[27]

Loyalka[28] pointed out that such a discrepancy could be due to a basic deficiency of the BGK model, in that it does not allow for the velocity dependence of the collision frequency, $\nu = \nu(\xi)$. With this in view, he carried out a detailed numerical study of a model of the Boltzmann equation which had been proposed and used to find an analytical solution for the structure of the kinetic layers as early as 1966.[28] The latter solution was considered by Loyalka not to be useful for the purpose of numerical evaluation, for which he preferred to use a direct numerical technique. His results show that the velocity dependence of $\nu(\xi)$ appropriate for rigid spheres does indeed have an important effect on the velocity defect and in fact the numerical solution practically coincides with the upper boundary of the region containing 80% of the experimental data reported by Reynolds *et al.*[25]

Another well-known deficiency of the BGK model is that it yields a Prandtl number Pr not appropriate for a monatomic gas (Pr = 1 instead of Pr = $\frac{2}{3}$). This unsatisfactory aspect was tackled by Tironi and the present author,[29] who presented a solution of the so-called ES model in 1966. When a trivial mistake is corrected the results are in reasonably good agreement with the experiments, in fact almost indistinguishable from the results presented by Loyalka.[28]

Gorelov and Kogan[30] reported the results of Monte-Carlo calculations which appear to fall on the lower boundary of the aforementioned region containing 80% of the data. These Monte-Carlo calculations were confirmed by results obtained by Bird.[31]

It was recently pointed out[32] that one can concoct a new model having the desirable properties of both a correct Prandtl number and variable collision frequency, while being amenable to a simple solution. The velocity profile computed by means of this model turns out to be in exceptionally good agreement[32] with the experiments of Reynolds *et al.*[25]

The discussion which has just been presented refers to the first-order effects (constant gradients, flat wall) in the kinetic layers. The second-order effects are more complicated to deal with but were successfully discussed by Grad[7] and Sone.[33–35] In particular, Sone[35] discovered the existence of a sublayer due to curvature.

An important feature of these problems is that they are amenable to a linear analysis. This is not true for the study of a completely different kind of kinetic layer, which shares the features of both a shock wave and a Knudsen layer and is met in the study of the evaporation of a surface into a low-pressure ambient. The problem of investigating the layer of evaporating gas close to the wall has been the subject of several researches because of its importance in various fields of physics, chemistry, and engineering. The aim of these researches is to improve on the approximate results obtained by Hertz[36] and Knudsen.[37]

Anisimov[38] in 1968 suggested the use of a trimodal ansatz for the molecular distribution function:

$$f(x, \xi) = a_L^+ f_L^+ + a_M^+ f_M^+ + a_M^- f_M^- \tag{4}$$

where x is the space coordinate normal to the evaporating surface, and f_L^+, f_M^+, f_M^- are half-range Maxwellians. He solved the conservation equations for one set of flow conditions (sonic) only and estimated the thickness in the corresponding Knudsen layer from the BGK model. Ytrehus[39,40] carried this method to completion in that he solved appropriate moment equations of the Boltzmann equation (Maxwell molecules) with a distribution function given by equation (4). The choice of equation (4) is such that the boundary conditions at infinity and at the evaporating boundary can be satisfied exactly (in the case of an absorption coefficient equal to unity). In addition to the three conservation equations, another moment equation is required in order to determine the three functions $a_L^+(x)$, $a_M^+(x)$, $a_M^-(x)$. Ytrehus was able to show that the problem is solvable if and only if M_M, the Mach number associated with the drift velocity of the downstream Maxwellian, is not larger than a critical value M_c, which turns out to be approximately unity ($M_c = 0.992$). This result is in agreement with the results obtained by Murakami and Oshima[41] by a Monte-Carlo method. These authors computed the solution and found that no steady state is reached when the downstream Mach number is larger than unity. It seems reasonable to conjecture that $M_c = 1$ exactly for the exact solution of the problem.[42] We remark that, as a consequence of the limitation of the downstream Mach number, no steady one-dimensional solution to the evaporation problem exists for a pressure ratio larger than (approximately) 4.8. In particular, no steady one-dimensional solution is possible for a gas evaporating into a vacuum.

The conjecture that $M_c = 1$ exactly has been partially confirmed by an analysis based on an exact solution of a one-dimensional BGK model linearized about a drifting Maxwellian.[43]

3. EXTERNAL AND EXPANDING FLOWS

In this section we shall survey the work on flows occurring in regions whose outer boundaries play a negligible role.

Mathematically, the simplest problems are pure heat transfer from a sphere[44] and flows past slowly rotating cylinders[45] and spheres.[46] When we pass to the more interesting problems of flow past a solid body, it is natural to study the low-Mach-number case first. Here we meet the analogue of Stokes' paradox,[47] which forbids using the simplest linearization for two-dimensional flows. We can, however, treat flows past three-dimensional bodies such as spheres.[48,49] Good agreement is then found between the solution of the linearized BGK model and Millikan's experimental data.[50] Such old experimental results are used for the purpose of comparison, because recent experiments deal with higher-Mach-number flows, i.e., with regimes for which the linearized solution does not apply.

When the Mach number is not small compared to unity, the problem of flow past a solid body becomes extremely difficult even for the simplest shapes. Apart from analytical studies of the far field[51,52] or for almost specularly reflecting surfaces, the approaches that have been used are discrete ordinate methods, Monte-Carlo simulation, and interpolation between the nearly free molecular and slip regimes. Particular attention has been paid to hypersonic flow: results in the slip regime are discussed in several basic references,[53,54] while results in the nearly free molecular regime are more scattered.[55-60]

A basic transition-flow problem arises in connection with the flow of a gas past a very sharp flat plate, parallel to the oncoming stream. When the Reynolds number $Re = \rho_\infty V_\infty L / \mu_\infty$ based on the plate length L is very large, the usual picture of a potential flow plus a boundary layer is valid everywhere except near the leading and trailing edges. According to estimates based on the work of Stewartson[61] and Messiter[62] kinetic theory is not needed to investigate the main behavior near the trailing edge. For the leading edge, the Knudsen number is of order M_∞; hence in supersonic or, even more, hypersonic flow, the description of the region about the leading edge is a typical problem in kinetic theory.

There are several methods based on simplified continuum models[63-69] which usefully predict surface and other gross properties in the so-called merged-layer regime, when the boundary layer and the outer flow are no longer distinct from each other, although a shock-like structure may still be identified.[70-72] Although there is good agreement between this theory and experiment, if we go sufficiently close to the leading edge the Navier–Stokes equations must be given up in favor of the Boltzmann equation. A full kinetic treatment, however, is very difficult and few papers deal with this problem. Kogan and Degtyarev[73] considered the flow over a very short plate, so that the whole flow is in the nearly free molecular regime, which makes it difficult to draw significant conclusions about the leading-edge effects. Charwat[74] assumed that the leading-edge region is in free-molecular conditions, an assumption which should be justified. Huang and co-workers[75-77] carried out extensive computations based on the discrete-ordinate technique for the BGK model and were able to exhibit the process of building up of the flow picture assumed in the simplified continuum models mentioned above. The Monte-Carlo computations of Vogenitz et al.[78] give a solution of the leading-edge problem for a reasonably long plate. All of these papers show evidence of a shear-flow region near the plate, shielded from the upstream pressure (or, rather, the normal stress), and a value at the leading edge larger than the free-molecule result for $5.5 < M_\infty < 30$, which is evidence of an upstream influence and appears to be in agreement with the experiments of Joss and Bogdonoff[79] at $M_\infty = 26$.

Another fundamental problem of rarefied gas dynamics is the free expansion of a gas into a vacuum, which occurs, e.g., in the discharge from an orifice into a low-pressure chamber. This problem embodies, within a simple framework, a transition from a continuum to almost free molecular flow without the usual complication of the effects of gas–surface interaction. In fact, according to Anderson and Fenn[80] and other experimenters, the increase in Mach number M predicted by the inviscid-gas theory does not continue indefinitely; M tends to a constant value. Theoretical progress on this problem was stimulated by the crucial remark by Ashkenas and Sherman[81] suggesting that the flow along the axis of the expanding jet can to some extent be simulated by the spherically symmetric expansion of a monatomic gas from a nearly continuum source. The study of spherical source flow based on the Boltzmann equation led to the explanation of the behavior of the jet in independent papers by Edwards and Cheng[82] and by Hamel and Willis.[83] These authors quantified, by means of suitable approximate solutions, the following intuitive picture of the flow: at large distances from the source the density will be so low that collisions will be unable to support the continuum expansion and the average energy due to random motion perpendicular to the stream lines will continually decrease and will be fed into the mean motion of the gas as well as the random motion parallel to the stream lines. Since random motion is connected with temperature, this circumstance leads to a very different behavior of the contributions to temperature from the motions parallel and transverse to the stream lines. This suggests the partitioning of kinetic temperature T into a parallel temperature $T_\parallel$ and a transversal temperature $T_\perp$ ($T = \frac{1}{3}T_\parallel + \frac{2}{3}T_\perp$). The picture above suggests that T tends to zero at large distances, but $T_\parallel$ (and hence T) reaches a constant value; the parallel temperature is said to "freeze." Since the speed of the gas is practically constant at large distances, the freezing of the temperature implies a freezing of the Mach number, in agreement with the experimental findings.

The above explanation implies that the transition from a collision-dominated toward collision-free flow occurs under hypersonic conditions (in order to have a constant flow speed). This remark led Hamel and Willis[83] as well as Edwards and Cheng[82] to use an expansion in powers of M^{-1}. This method, the so-called hypersonic approximation, produces a rational truncation of the infinite set of moment equations of the Boltzmann equation. A solution of these equations can be found that asymptotically approaches the isentropic flow solution for small values of a suitably scaled radial coordinate and exhibits a freezing of the parallel temperature at large distances. It is to be remarked that no such freezing is found for a cylindrical source.

Although the analysis confirms the freezing of the Mach number, it appears to contradict experimental findings concerning the behavior of T, which according to theory should fall like r^{-1} at large distances. Experiments seem to indicate clearly an r^{-2} dependence, but this contradiction was elucidated by Edwards and Cheng.[84] They illustrated that the distribution function of the perpendicular random velocity has, according to the BGK equation, a fairly narrow central spike with rather thick wings. The perpendicular temperature measurements are based on the spike width, which decreases as r^{-1}, and this explains why they indicate a temperature decay proportional to r^{-2}. Experiments measuring the energy content of the wings should indicate the r^{-1} behavior.

The mathematical theory of spherical expansion was formalized by Freeman,[85] who introduced an asymptotic expansion in powers of the source Knudsen number. At large distances from the source, the standard (Chapman–Enskog) expansion breaks down, but it is possible to rescale the Boltzmann equation in this outer region, and the resulting moment equations form a closed set (the same, of course, as found by previous authors). By using the same technique, Grundy[86] was able to support the assumption that the free jet can be approximated by a spherical source. In fact, he presented an analysis of the expansion of an axially symmetric jet which fully validates the basic remarks by Ashkenas and Sherman.[81]

The hypersonic approximation yields a valid solution for small values of the source Knudsen number Kn. As Kn increases, the method loses validity. This remark led Soga and Oguchi to analyze the source expansion flow by means of the BGK model.[87] According to their results the hypersonic approximation is very accurate for $Kn \leqslant 10^{-3}$, reasonably accurate for $Kn \simeq 10^{-2}$, while it is decidedly unacceptable for $Kn \geqslant 10^{-1}$. More recently, Abe and Oguchi[88] analyzed the problem by means of a kinetic model with a correct Prandtl number, but found no appreciable deviation from the results based on the BGK model.

The subject of jet expansions lends itself to a remark on the evaporation problem, discussed in Section 2. Since a steady ·one-dimensional solution is not possible for a gas evaporating into a vacuum, the question arises whether a steady solution for this problem is possible when the limitation to a one-dimensional geometry is suppressed. The answer is positive, as one can show by the method of matched asymptotic expansions.[89] The flow from, say an evaporating disk takes the form of an initially one-dimensional expansion, which reaches sonic conditions, then develops as an inviscid gas jet expansion from a sonic disk, and finally behaves as the frozen jet described above.

REFERENCES

1. C. Cercignani, *Mathematical Methods in Kinetic Theory*, Plenum Press, New York/Macmillan, London (1969).
2. C. Cercignani, *Theory and Application of the Boltzmann Equation*, Scottish Academic Press, Edinburgh/Elsevier, New York (1975).
3. H. Grad, Asymptotic theory of the Boltzmann equation, *Phys. Fluids* **6**, 147 (1963).
4. P. N. Hu, private communication.
5. C. Cercignani, Bifurcation problems in fluid-mechanics, *Meccanica* **5**, 7 (1970).
6. G. I. Taylor, The conditions necessary for discontinuous motions in gases, *Proc. R. Soc. London, Ser. A* **84**, 371 (1910).
7. H. Grad, Singular and nonuniform limits of solutions of the Boltzmann equation in transport theory (R. Bellmann *et al.*, eds.), SIAM-AMS Proc., Vol. I, p. 269, Am. Math. Soc., Providence, RI (1969).
8. H. M. Mott-Smith, The solution of the Boltzmann equation for a shock wave, *Phys. Rev.* **82**, 885 (1951).
9. H. Salwen, C. Grosch, and S. Ziering, Extension of the Mott-Smith method for a one-dimensional shock wave, *Phys. Fluids* **7**, 180 (1964).
10. L. H. Holway, Jr., Kinetic theory of shock structure using an ellipsoidal distribution function, in: *Rarefied Gas Dynamics* (J. H. de Leeuw, ed.), Vol. I, p. 193, Academic Press, New York (1965).
11. C. Muckenfuss, Some aspects of shock structure according to the bimodal model, *Phys. Fluids* **5**, 1325 (1962).
12. M. Linzer and D. F. Horning, Structure of shock fronts in argon and nitrogen, *Phys. Fluids* **6**, 1661 (1963).
13. M. M. Oberai, Mott-Smith distribution and solution of kinetic equations, *J. Mec.* **6**, 317 (1967).
14. R. Narasimha, S. M. Deshpande, and M. R. Ananthasayanam, Least squares methods for shock structure, in: *Rarefied Gas Dynamics* (L. Trilling and H. Wachman, eds.), Vol. I, p. 417, Academic Press, New York (1969).
15. H. W. Liepmann, R. Narasimha, and M. T. Chahine, Structure of a plane shock layer, *Phys. Fluids* **5**, 1313 (1962).
16. D. Anderson, Numerical solutions of the Krook kinetic equation, *J. Fluid Mech.* **25**, 271 (1966).
17. B. M. Segal and J. H. Ferziger, Shock-wave structure using nonlinear model Boltzmann equations, *Phys. Fluids* **15**, 1233 (1972).
18. J. K. Haviland and M. L. Lavin, Application of the Monte-Carlo method to heat transfer in a rarefied gas, *Phys. Fluids* **5**, 1399 (1962).
19. M. Perlmutter, Model sampling applied to the normal shock problem, in: *Rarefied Gas Dynamics* (L. Trilling and H. Wachman, eds.), Vol. I, p. 327, Academic Press, New York (1969).
20. G. A. Bird, Aspects of the structure of strong shock waves, *Phys. Fluids* **13**, 1172 (1970).
21. B. L. Hicks, S. M. Yen, and B. J. Railly, The internal structure of shock waves, *J. Fluid Mech.* **53**, 85 (1972).
22. B. Nicolaenko, Shock wave solutions of the Boltzmann equation as a nonlinear bifurcation problem from the essential spectrum, in: *Theories Cinetiques et Relativistes*, (G. Pichon, ed.), p. 127, CNRS, Paris (1975).
23. J. P. Elliot, D. Baganoff, and R. D. McGregor, Closure relations based on straining and shifting velocity space, in: *Rarefied Gas Dynamics* (J. L. Potter, ed.), Vol. II, p. 703, Academic Press, New York (1977).

24. J. P. Elliot and D. Baganoff, Solution of the Boltzmann equation at the upstream and downstream singular points in a shock wave, *J. Fluid Mech.* **65**, 603 (1974).
25. M. A. Reynolds, J. J. Smolderen, and J. F. Wendt, Velocity profile measurement in the Knudsen layer for the Kramers problem, in: *Rarefied Gas Dynamics* (M. Becker and M. Fiebig, eds.), Vol. I, p. A21, DFVLR Press, Porz-Wahn (1974).
26. W. Rixen and F. Adomeit, Simple moments of the molecular velocity distribution function in plane Poiseuille flow, in: *Rarefied Gas Dynamics* (M. Becker and M. Fiebig, eds.), Vol. I, p. B18, DFVLR Press, Porz-Wahn (1975).
27. C. Cercignani, Elementary solutions of the linearized gas-dynamics Boltzmann equation and their application to the slip-flow problem, *Ann. Phys.* **20**, 219 (1962).
28. S. K. Loyalka, Velocity profile in the Knudsen layer for the Kramer's problem, *Phys. Fluids* **18**, 1666 (1975).
29. C. Cercignani and G. Tironi, Some applications of a linearized kinetic model with correct Prandtl number, *Nuovo Cimento* **43B**, 64 (1966).
30. S. L. Gorelov and M. N. Kogan, Solutions of linear problems of rarefied gas dynamics by the Monte-Carlo method, *Izv. Akad. Nauk SSSR, Mekh. Zhidk. Gaza* **3**, 136 (1968); translated in: *Fluid Dynamics* **3**, 96 (1968).
31. G. A. Bird, Direct simulation of the incompressible Kramers problem, in: *Rarefied Gas Dynamics* (J. L. Potter, ed.), p. 323, Academic Press, New York (1977).
32. C. Cercignani, Knudsen layers. Theory and experiment, in: *Recent Developments in Theoretical and Experimental Fluid Mechanics* (U. Müller, K. G. Roesner, and B. Schmid, eds.), Springer-Verlag, Berlin (1979).
33. Y. Sone, Asymptotic theory of flow of rarefied gas over a smooth boundary I, in: *Rarefied Gas Dynamics* (L. Trilling and H. Wachman, eds.), Vol. I, p. 243, Academic Press, New York (1969).
34. Y. Sone, Asymptotic theory of flow of rarefied gas over a smooth boundary II, in: *Rarefied Gas Dynamics* (D. Dini, ed.), Vol. II, p. 737, Edizioni Tecnico Scientifiche, Pisa (1971).
35. Y. Sone, New kind of boundary layer over a convex solid boundary in a rarefied gas, *Phys. Fluids* **16**, 1422 (1973).
36. H. Hertz, Uber die Verdunstung der Flüssigkeiten, insbesondere des Quecksilbers, in luftleeren Raume, *Ann. Phys.* **17**, 177 (1882).
37. M. Knudsen, Die maximale Verdampfungsgeschwindigkeit des Quecksilbers, *Ann. Phys.* **47**, 697 (1915).
38. S. I. Anisimov, Vaporization of metal absorbing laser radiation, *Sov. Phys. JETP* **27**, 182 (1968).
39. T. Ytrehus, Kinetic theory description and experimental results for vapor motion in arbitrary strong evaporation, Von Karman Institute Technical Note 112 (1975).
40. T. Ytrehus, Theory and experiments on gas kinetics in evaporation, in: *Rarefied Gas Dynamics* (J. L. Potter, ed.), Vol. II, p. 1197, Academic Press, New York (1977).
41. M. Murakami and K. Oshima, Kinetic approach to the transient evaporation and condensation problem, in: *Rarefied Gas Dynamics* (M. Becker and M. Fiebig, eds.), Vol. II, p. F6, DFVLR Press, Porz-Wahn (1974).
42. C. Cercignani, A nonlinear criticality problem in the kinetic theory of gases, in: *Mathematical Problems in the Kinetic Theory of Gases*, Verlag Peter Lang D., Frankfurt (1980).
43. M. D. Arthur and C. Cercignani, Non-existence of a steady rarefied supersonic flow in a half-space, *ZAMP* **31**, 634 (1980).
44. C. Cercignani and C. D. Pagani, Variational approach to rarefied flows in cylindrical and spherical geometry, in: *Rarefied Gas Dynamics* (C. L. Brundin, ed.), Vol. I, p. 555, Academic Press, New York (1967).

45. P. Bassanini, C. Cercignani, and P. Schwendimann, The problem of a cylinder rotating in a rarefied gas, in: *Rarefied Gas Dynamics* (C. L. Brundin, ed.), Vol. I, p. 505, Academic Press, New York (1967).
46. C. Cercignani and G. Tironi, Some application to the transition regime of a new set of boundary conditions for Navier–Stokes equations, in: *Rarefied Gas Dynamics* (L. Trilling and H. Wachman, eds.), Vol. I, p. 281, Academic Press, New York (1969).
47. C. Cercignani, Stokes paradox in kinetic theory, *Phys. Fluids* **11**, 303 (1968).
48. C. Cercignani and C. D. Pagani, Flow of a rarefied gas past an axisymmetric body. I. General remarks, *Phys. Fluids* **11**, 1395 (1968).
49. C. Cercignani, C. D. Pagani and P. Bassanini, Flow of a rarefied gas past an axisymmetric body. II. Case of a sphere, *Phys. Fluids* **11**, 1399 (1968).
50. R. A. Millikan, The general law of fall of a small spherical body through a gas, and its bearing upon the nature of molecular reflection from surface, *Phys. Rev.* **22**, 1 (1923).
51. H. Grad, Equations of flow in a rarefied gas, New York University, NYO-2543 (1959).
52. L. Sirovich, On the kinetic theory of steady gas flows, in: *Rarefied Gas Dynamics* (L. Talbot, ed.), p. 238, Academic Press, New York (1961).
53. S. A. Schaaf, Mechanics of rarefied gas, in: *Handbuch der Physik* (S. Flugge, ed.), Vol. VII, p. 591, Springer-Verlag, Berlin (1963).
54. J. L. Potter, The transitional rarefied-flow regime, in: *Rarefied Gas Dynamics* (C. L. Brundin, ed.), Vol. II, p. 881, Academic Press, New York (1967).
55. D. R. Willis, A study of some nearly free-molecular flow problems, Princeton Univ. Aero. Engineering Lab. Report No. 440 (1958).
56. D. R. Willis, Mass flow through a circular orifice and a two-dimensional slit at high Knudsen numbers, *J. Fluid Mech.* **21**, 21 (1965).
57. D. R. Willis, Heat transfer and shear between coaxial cylinders for large Knudsen numbers, *Phys. Fluids* **8**, 1908 (1965).
58. H. Grad and P. N. Hu, Drag of a slender cone in hypersonic flow, in: *Rarefied Gas Dynamics* (L. Trilling and H. Wachman, eds.), Vol. I, p. 561, Academic Press, New York (1969).
59. B. B. Hamel and A. L. Cooper, A first collision theory of the hyperthermal leading edge problem, in: *Rarefied Gas Dynamics* (L. Trilling and H. Wachman, eds.), Vol. I, p. 433, Academic Press, New York (1969).
60. Y. S. Pan, Drag on a cylinder in hypersonic nearly-free molecular flow, in: *Rarefied Gas Dynamics* (L. Trilling and H. Wachman, eds.), Vol. I, p. 779, Academic Press, New York (1969).
61. K. O. Stewartson, On the flow near trailing edge of a flat plate II, *Mathematika* **16**, 106 (1969).
62. A. F. Messiter, Boundary-layer flow near the trailing edge of a flat plate, *SIAM J. Appl. Math.* **18**, 241 (1970).
63. H. Oguchi, The sharp leading edge problem in hypersonic flow, in: *Rarefied Gas Dynamics* (L. Talbot, ed.), p. 501, Academic Press, New York (1961).
64. M. Shorenstein and R. F. Probstein, The hypersonic leading-edge problem, *AIAA J.* **6**, 1898 (1968).
65. W. L. Chow, Hypersonic rarefied flow past the sharp leading edge of a flat plate, *AIAA J.* **5**, 1549 (1967).
66. W. L. Chow, Hypersonic slip flow past the leading edge of a flat plate, *AIAA J.* **4**, 2062 (1966).
67. S. Rudman and S. G. Rubin, Hypersonic viscous flow over slender bodies with sharp leading edges, *AIAA J.* **6**, 1833 (1968).
68. H. K. Cheng, S. Y. Chen, R. Mobly, and C. Huber, On the hypersonic leading-edge

problem in the merged-layer regime, in: *Rarefied Gas Dynamics* (L. Trilling and H. Wachman, eds.), Vol. I, p. 451, Academic Press, New York (1969).

69. S. S. Kot and D. L. Turcotte, Beam-continuum model for hypersonic flow over a flat plate, *AIAA J.* **10**, 891 (1972).
70. W. J. McCroskey, S. M. Bogdonoff, and J. C. McDougall, An experimental model for the sharp flat plate in rarefied hypersonic flow, *AIAA J.* **4**, 1580 (1966).
71. P. J. Harbour and J. H. Lewis, Preliminary measurements of the hypersonic rarefied flow field on a sharp flat plate using an electron beam probe, in: *Rarefied Gas Dynamics* (C. L. Brundin, ed.), Vol. II, p. 1031, Academic Press, New York (1967).
72. W. W. Joss, I. E. Vas, and S. M. Bogdonoff, Hypersonic rarefied flow over a flat plate, AIAA Paper 68–5 (January, 1968).
73. M. N. Kogan and L. M. Degtyarev, On the computation of flow at large Knudsen numbers, *Astronaut. Acta* **11**, 36 (1965).
74. A. F. Charwat, Molecular flow study of the hypersonic sharp leading edge interaction, in: *Rarefied Gas Dynamics* (L. Talbot, ed.), p. 553, Academic Press, New York (1961).
75. A. B. Huang and P. F. Hwang, Supersonic leading edge problem according to the ellipsoidal model, *Phys. Fluids* **13**, 309 (1970).
76. A. B. Huang, Kinetic theory of the rarefied supersonic flow over a finite plate, in: *Rarefied Gas Dynamics* (L. Trilling and H. Wachman, eds.), Vol. I, p. 529, Academic Press, New York (1969).
77. A. B. Huang and P. F. Hwang, Kinetic theory of leading edge flow, IAF Paper Re 63 (October, 1968).
78. F. W. Vogenitz, J. E. Broadwell, and G. A. Bird, Leading edge flow by the Monte-Carlo direct simulation technique, *AIAA J.* **8**, 504 (1970).
79. W. W. Joss and S. M. Bogdonoff, A detailed study of the flow around the leading edge of a flat plate in hypersonic low density flow, in: *Rarefied Gas Dynamics* (L. Trilling and H. Wachman, eds.), Vol. I, p. 483, Academic Press, New York (1969).
80. J. B. Anderson and J. B. Fenn, Velocity distributions in molecular beams from nozzle sources, *Phys. Fluids* **8**, 780 (1965).
81. H. Ashkenas and F. Sherman, The structure and utilization of supersonic free jets in low density wind tunnels, in: *Rarefied Gas Dynamics* (J. H. de Leeuw, ed.), Vol. II, p. 84, Academic Press, New York (1965).
82. R. H. Edwards and H. K. Cheng, Steady expansion of a gas into a vacuum, *AIAA J.* **4**, 558 (1966).
83. B. B. Hamel and D. R. Willis, Kinetic theory of source flow expansion with application to the free jet, *Phys. Fluids* **9**, 829 (1966).
84. R. H. Edwards and H. K. Cheng, Distribution function and temperatures in a monatomic gas under steady expansion into a vacuum, in: *Rarefied Gas Dynamics* (C. L. Brundin, ed.), Vol. I, p. 819, Academic Press, New York (1967).
85. N. C. Freeman, Solution of the Boltzmann equation for expanding flows, *AIAA J.* **5**, 1696 (1967).
86. R. E. Grundy, Axially symmetric expansion of a monatomic gas from an orifice into a vacuum, *Phys. Fluids* **12**, 2011 (1969).
87. T. Soga and H. Oguchi, Numerical analysis of the source expansion flow into vacuum, *Phys. Fluids* **13**, 317 (1972).
88. T. Abe and H. Oguchi, A kinetic-model analysis of the source expansion flow into vacuum, ISAS Report No. 554, Vol. 42, No. 9 (1977).
89. C. Cercignani, Strong evaporation of a polyatomic gas, in: *Rarefied Gas Dynamics* (S. S. Fisher, ed.), Vol. I, p. 305, AIAA, New York (1981).

5

On the Application of Light-Gas Guns to the Problem of Nuclear Fusion Refueling

A. Reggiori

ABSTRACT. The refueling of nuclear-fusion reactors will require the injection of solid hydrogen pellets at very high speeds. In present-day experiments single-stage light-gas guns are employed to inject 1-mm pellets at velocities up to 1000 m/s. In order to meet the future requirements, more sophisticated machines are required; one possible candidate is the two-stage light-gas gun. In the present work the main features and the basic theory of light-gas guns are presented. A simple analysis shows that the possible performance of these machines is well within the requirements of nuclear-fusion refueling.

1. INTRODUCTION

Light-gas guns have been extensively used in the last thirty years for research in the fields of ballistics and space flight.[1,2] Recently, a new application of these devices has been suggested[3,4] for accomplishing the task of refueling tokamak nuclear-fusion reactors.

For the operation of a tokamak reactor it is required that a fuel (hydrogen isotopes) be supplied in the innermost region of the hot plasma. The most practical method of doing this appears to be the injection of the fuel in the form of solid pellets (or possibly liquid droplets) at very high speed, in order to circumvent the problem of pellet ablation by the hot plasma while moving from the injector into the reaction zone.

A. REGGIORI • Dipartimento di Energetica, Politecnico di Milano, Piazza Leonardo da Vinci, n. 32, 20133 Milano, Italy.

The pellet-ablation theory that best describes the experiments performed up to now (known as the neutral gas-shielding model[5]) indicates that injection velocities in excess of 10 km/s may be needed.[6] Estimates of the mass required for refueling indicate pellet diameters of 1–5 mm and injection rates of 10–20 s^{-1}. Besides the possibility of efficient refueling, a high-speed injector of impurity pellets (made of carbon, iron, hydrocarbons, etc.) could also be very useful for research on the mechanisms of impurity transport and recycling in the tokamak plasma.[7]

Various solutions have been suggested for the purpose of obtaining the necessary high injection velocities, such as laser guns, electrostatic accelerators, centrifuges, and light-gas guns. Centrifuges are suitable for high delivery rates, but serious mechanical problems limit the injection velocity to about 2000 m/s; present-day machines can operate for several seconds at speeds up to 1000 m/s.

Practically, reliable experimental results have presently been obtained only with single-shot light-gas guns, which allow one to inject solid hydrogen pellets of approximately 1-mm diameter at velocities in the range of 1000 m/s.[8,9] In order to obtain much higher speeds more sophisticated machines, like the two-stage light-gas gun, are needed.

In this work the main features and the basic theory of light-gas guns are outlined and some considerations on the practical design of a pellet injector are presented.

2. SIMPLE THEORETICAL ANALYSIS

The basic configuration of the single-stage light-gas gun is shown in Figure 1. The pellet is accelerated along the barrel under the influence of the pressure acting on its base. Neglecting the effects of friction forces and downstream pressure, the motion of the pellet is simply described by the equation

$$mdu/dt = pA \qquad (1)$$

where m is the mass of the pellet, A the base area, p the pressure, t the time, and u the velocity.

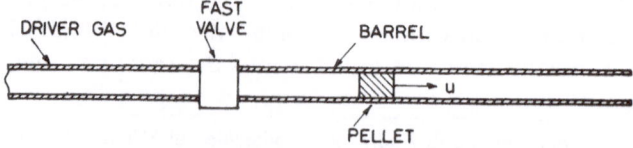

Fig. 1. Single-stage light-gas gun configuration.

If the pressure on the base could be kept constant, the velocity for a given length L of the barrel would be

$$u = (2pAL/m)^{1/2} \tag{2}$$

Unfortunately this is not the case because the pressure decreases rapidly during the acceleration, as shown by the elementary theory of unidimensional unsteady flow. With p_0 and a_0 the initial pressure and speed of sound of the driving gas, the pressure on the base of the pellet is given by

$$p/p_0 = \left(1 - \frac{\gamma - 1}{2} \frac{u}{a_0}\right)^{2\gamma/(\gamma-1)} \tag{3}$$

With the aid of equation (1) we obtain the barrel length and duration of acceleration needed to reach a given velocity:

$$L \frac{p_0}{a_0^2} \frac{A}{m} \frac{\gamma + 1}{2} = 1 + \frac{[(\gamma + 1)/2](u/a_0) - 1}{\{1 - [(\gamma - 1)/2](u/a_0)\}^{(\gamma+1)/(\gamma-1)}} \tag{4}$$

$$t \frac{A}{m} \frac{p_0}{a_0} \frac{\gamma + 1}{2} = \left(1 - \frac{\gamma - 1}{2} \frac{u}{a_0}\right)^{-(\gamma+1)/(\gamma-1)} - 1 \tag{5}$$

In particular, these equations show that the maximum speed attainable by this method is given by

$$u_M = \frac{2}{\gamma - 1} a_0 \tag{6}$$

Also, from equation (3) it can be seen that the base pressure decreases very rapidly for increasing values of u/a_0. Therefore, the maximum velocity indicated in equation (6) is only an asymptotic limit and practical speeds are well under this value.

For example, from equation (4) it appears that, using hydrogen at room temperature as a propellant ($a_0 = 1300$ m/s) with an initial pressure of 50 bar, solid hydrogen spherical pellets of 1-mm diameter could be launched at 1500 m/s with a barrel length of approximately 5 cm. On the other hand, in order to obtain a velocity of 4000 m/s with the same initial conditions, a barrel length of about 14 m would be necessary. (By comparison, with constant acceleration the same speed would be attained with a barrel length of 18 cm.)

In the simple analysis leading to equations (4)–(6) the effects of friction and gas leakage have been neglected. It is practically impossible to take these effects into account because they depend on unknown parameters, such as the degree of forcing and the deformation of the pellet. Some typical experimental results for solid hydrogen pellets of 1-mm diameter at speeds up to 1000 m/s are shown in Figure 2. The actual velocity is between 20 and 50% less than predicted by the simple theory, but how much of this reduction is due to friction and gas leakage is unknown.

From the above-mentioned considerations it appears that the possibility of reaching high velocities with a reasonable barrel length is essentially related to the speed of sound of the driver gas or, in turn, to its temperature. Moreover, a gradually increasing pressure would be desirable in order to compensate for expansion in the barrel.

Very high temperatures for short times can be obtained by means of adiabatic compression of the launching gas. This is the essential feature of the two-stage light-gas gun, which is shown schematically in Figure 3. The first stage consists of a piston tube in which a driver gas contained in volume (a) accelerates a piston along the tube, compressing and heating gas (b); the second stage, driven by hot gas (b), accelerates the pellet along the launching barrel.

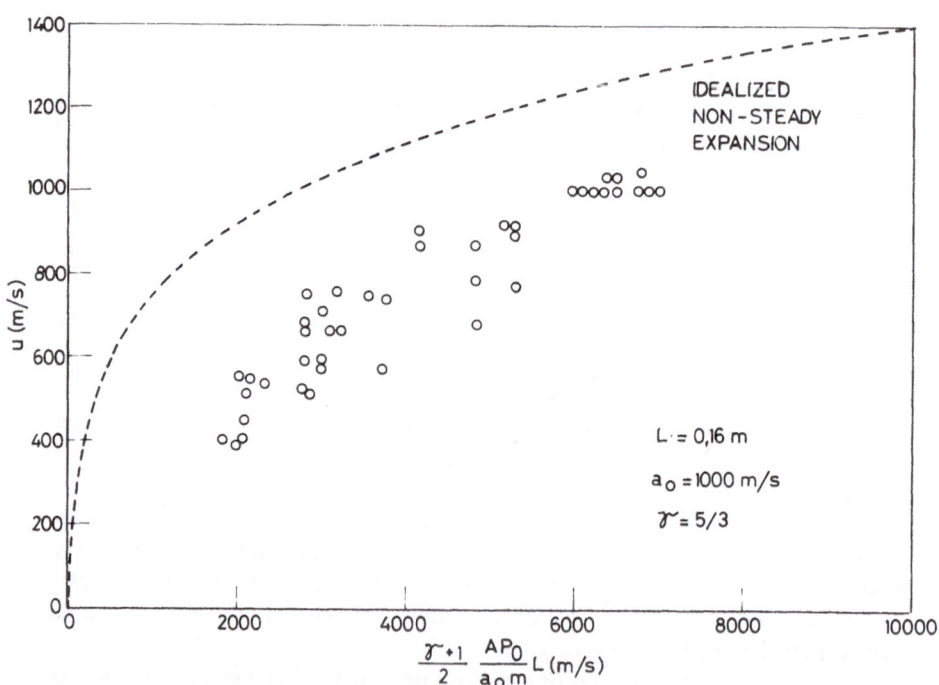

Fig. 2. Injection velocities of 1-mm hydrogen pellets obtained with a single-stage gun (after Milora and Foster[8]).

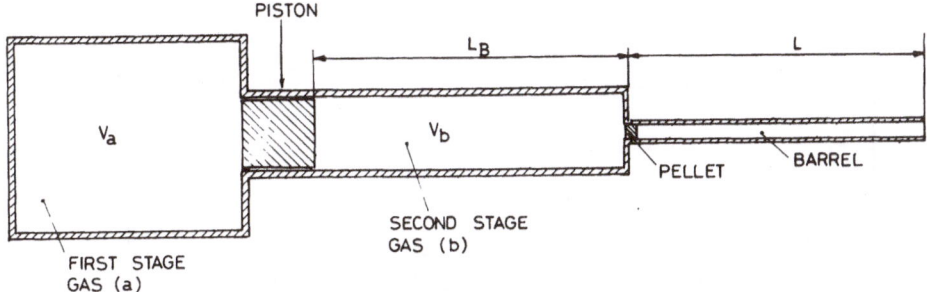

Fig. 3. Two-stage light-gas gun configuration.

Under some simplifying assumptions (such as perfect gas, no friction, no leakage, quasi-steady motion) it is easy to calculate the temperature and pressure of gas (b) during the motion of the piston.[10] In particular, if volume (a) is much larger than volume (b), the conditions of maximum temperature and pressure are given by

$$\frac{T_{bM}}{T_{b0}} = \frac{p_a}{p_{b0}} (\gamma - 1) + 1 \tag{7}$$

and

$$\frac{p_{bM}}{p_{b0}} = \left(\frac{T_{bM}}{T_{b0}}\right)^{\gamma/(\gamma-1)} \tag{8}$$

These equations show the theoretical possibility of obtaining unlimited pressures and temperatures. In practice it is easy to obtain short pressure impulses of thousands of bars and temperatures of thousands of degrees K. A typical pressure diagram obtained with the CNPM piston tube is shown in Figure 4.

The cross section of the piston tube is usually much larger than the cross section of the launching barrel. In this case it can easily be shown that the maximum velocity attainable with a two-stage gun is

$$u_M = \left(\frac{\gamma + 1}{2}\right)^{1/2} \frac{2}{\gamma - 1} a_{bM} \tag{9}$$

In practice, asymptotic velocity limits of the order of 20 km/s are obtainable with this method.

In order to verify the simple theory and to ascertain the practical feasibility of a two-stage injector, a series of preliminary tests were

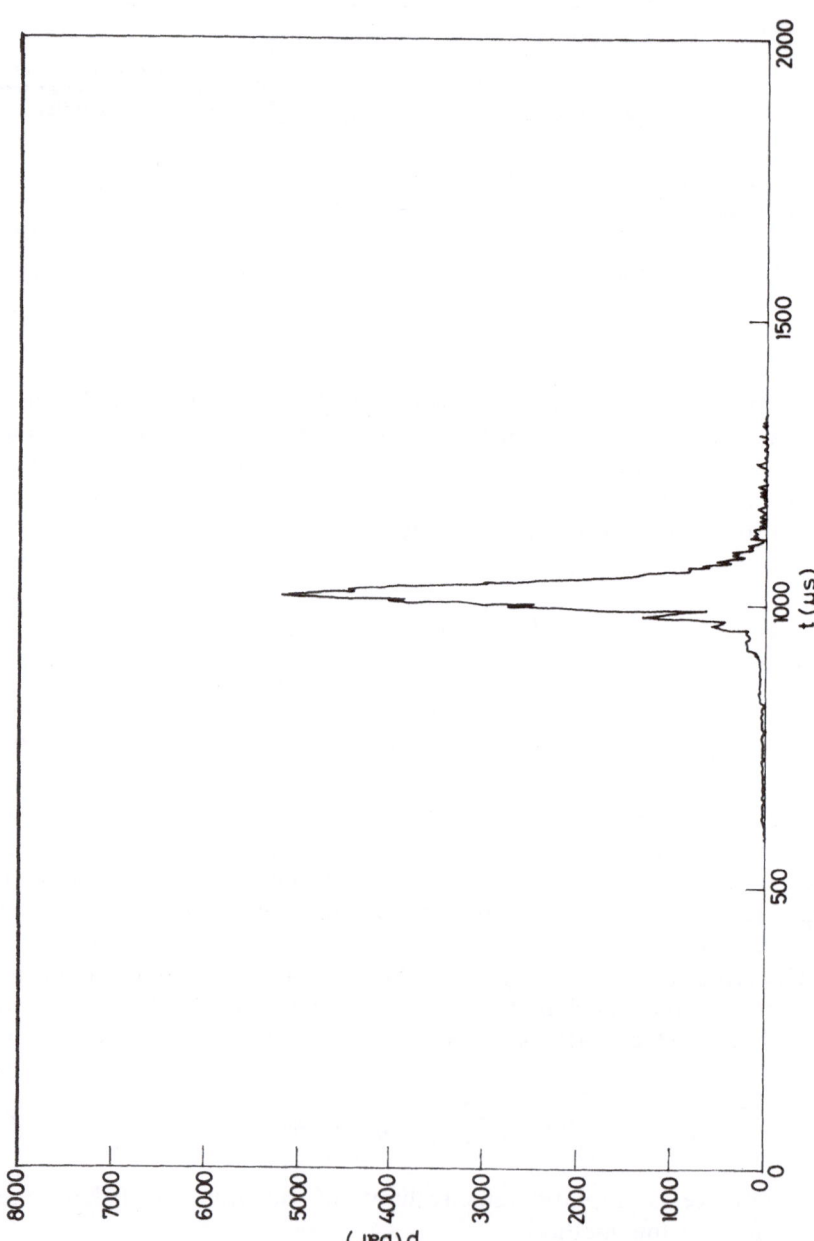

Fig. 4. Typical pressure–time diagram of a (CNPM) piston tube.

performed using as first stage the piston tube installed at CNPM. This tube has an internal diameter of 50 mm and a length L_b of 4 m; it is usually employed for research in the field of unsteady combustion and its dimensions are rather superabundant for the relevant application here. The second stage (launching barrel), a 15-cm-long tube with an internal diameter of 1.5 mm, was connected to the end of the piston tube. The pellets used in the tests were nylon spheres of 1.5-mm diameter. Pellet velocities in excess of 2000 m/s have been measured.[11]

Recently, another series of experiments were performed, still with 1.5-mm nylon spheres and a 20-cm-long barrel, using the same piston tube shortened to a length of 2 m. The propellant gas used in the tests was air at room temperature ($a_{bo} = 340$ m/s); the initial pressure p_{a0} in the first stage was varied between 20 and 40 bar and the initial pressure p_{bo} in the second stage between 0.05 and 0.1 bar. The results are shown in Figure 5. The velocities are of the same order as those obtained before, confirming that the present tube is larger than necessary. From the figure the advantage of the two-stage launching system is also evident, namely a single-stage gun, subject to the same initial conditions, would give a theoretical velocity of about 400 m/s.

In the preliminary tests no attempts were made to optimize the various parameters, such as the tube dimensions and the pressure of the propellant. Some basic design elements are given in the next section.

3. BASIC DESIGN FEATURES

An estimate of the correct dimensions of a two-stage gun for the injection of solid hydrogen pellets of given diameter at a prescribed velocity can be obtained with simple considerations based on an order-of-magnitude analysis.

If the driver gas in the second stage becomes very hot, the velocity of the pellet will be subsonic (except at very high speeds). Bearing in mind equation (3) and the fact that the driver pressure increases during the launching cycle, we may assume that the pressure acting on the pellet will remain almost constant along the barrel. If $\bar{p}$ is the average pressure and m/A the mass/area ratio of the pellet, the exit time and the barrel length are given by

$$t \simeq \frac{u}{\bar{p}} \frac{m}{A} \tag{10}$$

and

$$L \simeq ut/2 \tag{11}$$

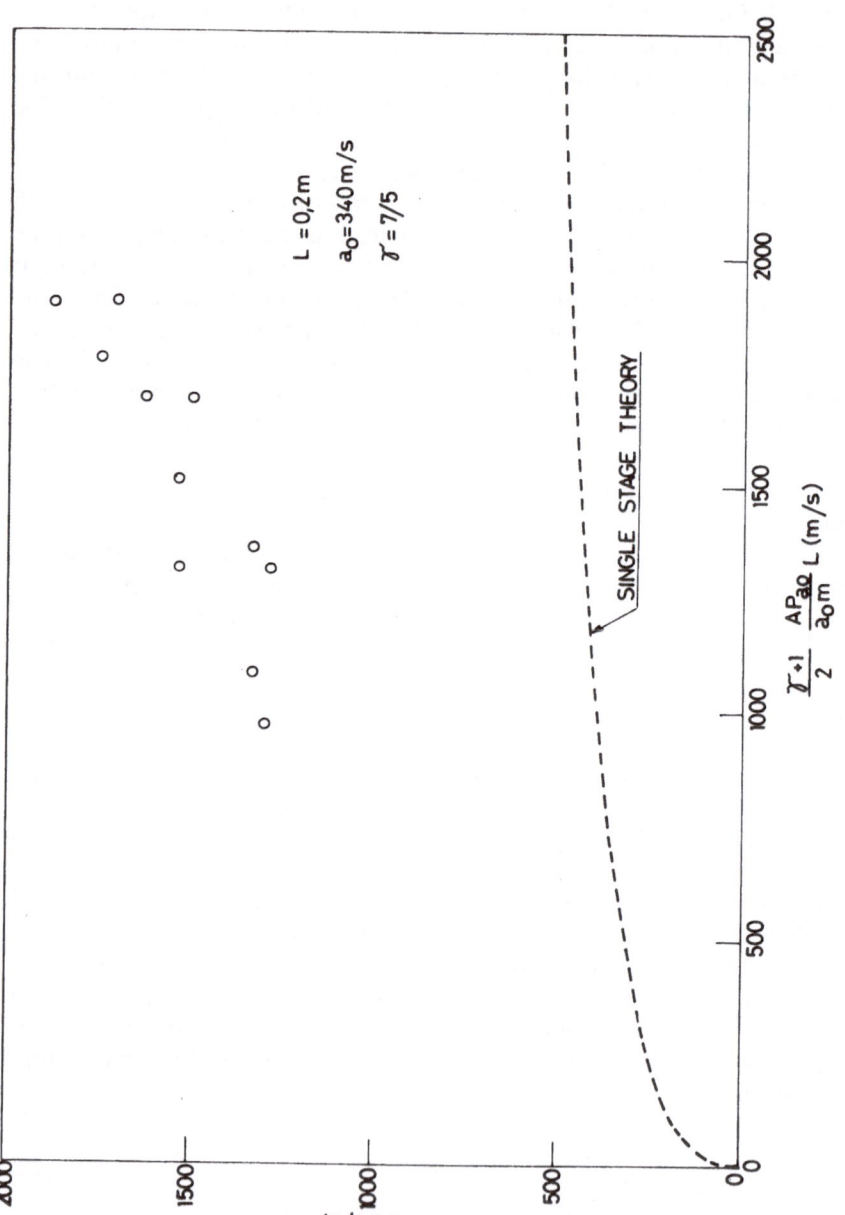

Fig. 5. Injection velocities of 1.5-mm nylon spheres obtained with the CNPM two-stage gun.

For the first stage we require that the rise time of the pressure pulse be of the same order of magnitude as the exit time of the pellet. It can be shown that the rise time is given approximately by

$$t_p \simeq \frac{p_a}{p_M} \left[\frac{2L_b}{p_a} (m/A)_p \right]^{1/2} \tag{12}$$

where $(m/A)_p$ is the mass/area ratio of the piston.

It follows from equations (10)–(12) that

$$L_b p_a (m/A)_p \simeq (um/A)^2 \tag{13}$$

In order to accelerate the pellet to the prescribed velocity, the first stage must supply an energy larger than the kinetic energy of the pellet. If V_b is the volume of the piston tube, the available energy is practically equal to $p_a V_b$ so we obtain the condition

$$p_a V_b > \tfrac{1}{2} m u^2 \tag{14}$$

Furthermore, the mass of gas initially contained in the piston tube must be larger than the mass leaving through the barrel. This is expressed by the condition

$$\frac{p_{b0}}{\bar{p}} > \left(\frac{V}{V_b} \right)^{\gamma} \tag{15}$$

where V is the internal volume of the launching barrel.

Conditions (13)–(15) can be satisfied with various combinations of the relevant parameters $[p_a, p_{b0}, L_b, V_b, (m/A)_p, \gamma]$. For example, consider an injector with a barrel of 2-mm diameter capable of launching hydrogen pellets with $m/A = 0.4$ kg/m² at velocity 5000 m/s. First we must choose a value for the accelerating pressure $\bar{p}$, which is limited by the mechanical strength of the pellet. For solid hydrogen this limiting value is not known; present-day launchers work successfully with pressures of around 50 bar. In the present example we shall assume $\bar{p} = 50$ bar.

With these data equations (10) and (11) immediately yield the barrel length,

$$L = 1 \text{ m}$$

Conditions (13)–(15) are then satisfied with a piston tube having the following characteristics: length $L_b = 1$ m, diameter $D_b = 50$ mm, piston mass $m_p = 100$ g, initial pressure $p_{b0} = 0.02$ bar, driving pressure $p_a = 1$ bar, and driver gas (helium) at room temperature ($\gamma = 5/3$, $a_{b0} = 1000$ m/s).

It is noteworthy that the first stage of this injector could be driven by the ambient air, thus simplifying the mechanical design of the apparatus. Alternatively, a shorter tube could be used with an increased driving pressure. For instance, with $p_a = 5$ bar, $L_b = 20$ cm, and $p_{bo} = 0.2$ bar, the prescribed conditions could still be satisfied.

This example shows that the required injection velocity could be obtained with a machine of rather small dimensions, thus allowing for the possibility of multiple injection simply by arranging several guns (possibly up to 10) in a cluster and firing them in sequence. This same arrangement was suggested elsewhere.[4]

Continuous operation at the repetition rates envisioned for future refueling conditions appears much more difficult to achieve. The development of a continuously operated single-stage injector was recently reported,[12] but no attempts to design a repeating two-stage injector have yet been made.

In conclusion, it appears from the above considerations that a rather simple two-stage injector should be able to meet the velocity requirements for the refueling of fusion reactors. Two main problems must be solved. The first is related to the mechanical strength of the pellet, and imposes a still unknown limit on the driving pressure; however, this problem is much less important for two-stage guns than for present-day single-stage guns because, for a given velocity, a smaller pressure is required. The second problem is posed by the technical difficulties in designing a two-stage gun capable of continuous operation. The practical feasibility of such a machine has yet to be demonstrated and much work in this direction will have to be invested.

NOMENCLATURE

a speed of sound
A cross section
L length
m mass
p pressure
t time
T temperature
u velocity
V volume
γ ratio of specific heats

Subscripts
0   initial conditions
a   first stage
b   second stage
M   maximum value

REFERENCES

1. A. E. Seigal, The Theory of High Speed Guns, *AGARDO graph 91* (May, 1965).
2. J. D. Watson, A Summary of the Development of Large Explosive Guns for Re-Entry Simulation, *PIFR-155*, Physics International Company, San Leandro, California (1970).
3. S. L. Milora, C. A. Foster, and G. D. Kerbel, A survey of possible pellet injection techniques for refueling tokamak reactors, *18th Annual Meeting of the Division of Plasma Physics of the American Physical Society*, San Francisco, California (November 15–19, 1976).
4. R. F. Flagg, A review of gas gun technology with emphasis on fusion fueling applications, *Proceedings of the Fusion Fueling Workshop*, pp. 123–130, Princeton, New Jersey (November 1–3, 1977).
5. S. L. Milora and C. A. Foster, A revised gas shielding model for pellet plasma interactions, *IEEE Trans. Plasma Sci.* **PS-6**, 578 (1978).
6. S. L. Milora, Review of pellet fueling, *J. Fusion Energy* **1**, 15–48 (1981).
7. S. M. Egorov, A. P. Zhilinsky, V. A. Krupin, B. V. Kuteew, V. A. Nikiforov, V. A. Roshansky, and L. D. Tsendin, Pellet diagnostics experiments in T-10 tokamak, *Proceedings of the 10th European Conference on Controlled Fusion and Plasma Physics*, Vol. 1, p. A-3b, Moscow (September, 1981).
8. S. L. Milora and C. A. Foster, Pneumatic hydrogen pellet injection system for the ISX tokamak, *Rev. Sci. Instrum.* **50**, 482–487 (1979).
9. C. Andelfinger, K. Büchl, R. S. Lang, H. B. Schilling, and M. Ulrich, Pellet injectors for JET, *Max Planck Institut für Plasmaphysik*, IPP 1/193 (September, 1981).
10. A. Reggiori and R. Carlevaro, Studio e sviluppo di un tubo d'urto a pistone per prove dinamiche ad alte pressioni, *XXXVI Congresso Nazionale ATI*, pp. 91–104, Viareggio (Ottobre, 1981).
11. G. Cima, A. Jacchia, and A. Reggiori, A shock tube device for pellet injection, preliminary experimental results, *Proceedings of the Workshop on Diagnostics for Fusion Reactor Conditions*, International School of Plasma Physics, Varenna (September, 1982).
12. S. L. Milora, S. K. Combs, C. A. Foster, W. A. Houlberg, J. T. Hogan, D. D. Schuresko, S. E. Attemberger, G. L. Schmidt, M. J. Greenwald, S. Wolfe, and J. Parker, Development of hydrogen pellet injectors and pellet fueling experiments at ORNL, *9th International Conference on Plasma Physics and Controlled Nuclear Fusion Research*, Baltimore, U.S.A. (1–8 September, 1982).

6

A More Modern Theory of Combustion Noise

W. C. STRAHLE

ABSTRACT. Some of the newer concepts that have emerged in the theory of turbulent reacting flows are applied to the problem of combustion-generated noise. A general theory is constructed for noise radiation from a turbulent combustion region in the practical limit of low frequency and low Mach number. The theory is specialized to simple cases of premixed and nonpremixed jet flames radiating to a free field. Comparison with experiment is given for premixed flames. A new interpretation is given for the frequency content of combustion noise, which is in accord with experiment.

1. INTRODUCTION

Noise production by the turbulent combustion process is a problem encountered in several combustor types. While it may be masked by other noise sources in some situations, such as in aeropropulsion engines, it may be a dominant noise source in others, such as in stationary gas-turbine power plants or furnaces. The noise emitted must be regarded as a pollutant to be minimized. As in other noise-control problems, the sound generated may be attenuated by dissipative shielding in many situations, but it would be desirable to have sufficient understanding of the noise source itself, in order to minimize the noise by combustor design. This paper is solely concerned with the source behavior.

The state of understanding of combustion noise through 1977 was reviewed by Strahle.[1] At the time of that review there was significant

---

W. C. STRAHLE • School of Aerospace Engineering, Georgia Institute of Technology, Atlanta, GA 30332, U.S.A.

progress being made in the theory of turbulent reacting flows that had not as yet found its way into combustion-noise theory. One of the purposes of this paper is to incorporate some of the more modern turbulent reacting flow concepts into the general theory and calculation of combustion noise. Moreover, it is now possible to treat the combustion-noise problem with more rigor, in one particular limiting case, than has previously been done. This will be the low frequency, low-Mach-number limit which eliminates discrepancies between two of the most widely used theories of combustion noise.[2-4] As a byproduct of the theory, a new interpretation of the frequency content of combustion noise will emerge. Adequate prediction of this frequency content has been notably lacking in prior work.[1]

Radiated combustion noise, as with all aerodynamic-induced sound, is enclosure-dependent because of reflections from walls and a change in source-radiation impedance. Enclosure effects greatly complicate the analytical problem, but do not, in general, change the physics of the noise-generation process. Consequently, all work here will be concerned with turbulent flames radiating to a free field (infinite surroundings). Another caveat to this work is that only what is called direct combustion noise will be considered. This is noise generated in, and radiated from, a region of turbulent combustion. The existence of direct combustion noise is indisputable[1] and is distinct from indirect effects, such as a change in jet noise when the flame is on as opposed to when it is off.

In the following, a general theory for flames radiating to a free field will first be developed. Then application will be made to two flame types. The first kind will be a premixed bunsen-type turbulent jet flame, for which there is substantial experimental information to test the theory. The second will be a turbulent jet diffusion flame for which there is only a paucity of experimental information. In the second case, therefore, the theory will remain untested.

2. GENERAL THEORY OF COMBUSTION NOISE FOR FREE FLAMES

2.1. The Low-Frequency Limit and the Radiation Field

Consider a general turbulent flame burning in free surroundings, as depicted in Figure 1. The flame is shown as a premixed flame such as may be stabilized on the mouth of a burner rim, but for analytical purposes it may be any flame type. Two coordinate vectors are shown in Figure 1, which may be called the field point variable, $\mathbf{r}$, and the source point variable, $\mathbf{r}_0$. Consider drawing a sphere, enclosing the entire flame, with surface S and volume $V = \frac{4}{3}\pi L^3$, where L is the sphere radius. The

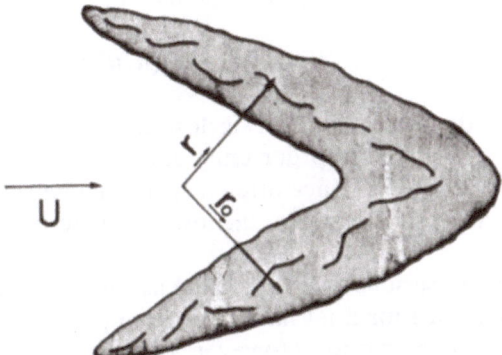

Fig. 1. Flame schematic and coordinate system.

following conditions are assumed satisfied: (i) $L \ll \lambda$ where λ is the sound wavelength emitted and (ii) the sphere surface lies on a region of fluid of constant mean density and speed of sound, $\bar{\rho}$ and $\bar{c}$, respectively, which are the ambient values of the free field. In this case it is known[5] that the first term of a multipole expansion of the sound field, which is a monopole field, is an adequate representation for the far-field pressure fluctuation, $p'(\mathbf{r}, t)$, that is,

$$p'(\mathbf{r}, t) = -\frac{\bar{\rho}}{4\pi r} \int_S \left[\frac{\partial \mathbf{w}'(\mathbf{r}_0, t)}{\partial t} \right]_{\mathbf{r}_0, \eta} \cdot \mathbf{n}_0 dS \tag{1}$$

In equation (1), t is time, w is the velocity vector, $r = |\mathbf{r}|$, and $\mathbf{n}_0$ is the unit normal on the sphere pointing toward the origin. The subscripts $\mathbf{r}_0$, η on the time derivative denote that the derivative is evaluated at position $\mathbf{r}_0$ and a retarded time $\eta = t - r_0/c$, which is the time at which a signal is generated to reach the observer at time t. Notice the monopole nature of the sound field; even if w varies with $\mathbf{r}_0$ over S, p' has no directionality.

The assumption $L \ll \lambda$ is the low-frequency assumption. Since combustion noise is usually broadband noise, consisting of all frequencies, there will be a portion of the frequency spectrum that will not be represented by equation (1). Fortunately, however, combustion noise is often dominated by low-frequency content so that the theory may apply to the portion of the spectrum carrying the majority of the sound power. There is a good reason for this low-frequency content. As will be seen, the combustion-noise frequency content tends to center about the frequency of the energy-containing eddies of the turbulence, u'/l, where u' is a measure of the rms velocity fluctuation and l is a measure of the turbulence integral scale. The low-frequency assumption entails

$$L \ll \lambda = \bar{c}l/u' = (\bar{c}/\bar{u})(\bar{u}/u')l$$

Consequently, if the Mach number is low and the relative turbulence intensity is low, L/l can be of order unity and above, while still satisfying the low-frequency criterion. Since often a typical flame size is of the order of a few integral scales, it is seen that the low-frequency assumption is often a good one.

Somewhat more serious in nature is the assumption that S is totally in the ambient fluid, since for a jet flame, for example, S must intersect the hot downstream part of the jet. However, the physical effect is primarily one of sound refraction and experimentally the effect seems weak. That is, most flames have only a weak directionality to the radiated-sound pattern.[1]

2.2. The Low-Frequency Limit and the Near Field

The region inside S contains, of course, the combustion field, which will be presumed to consist of thermally and calorically perfect gases. Strahle and Chandran[6] have shown that a particularly simple form emerges for the fluctuation in the energy equation, under rather weak assumptions. It is

$$\frac{1}{\bar{p}}\frac{\partial p'}{\partial t} + \gamma\nabla\cdot\mathbf{w}' + \frac{(\gamma-1)}{\bar{p}}(\nabla\cdot\mathbf{q}' - Q') = 0 \qquad (2)$$

Here γ is the specific-heat ratio, assumed constant, and q and Q are the heat-transfer vector and the heat-release rate per unit volume. The variables q and Q may in general be quite complex, involving heat transfer by conduction as well as Lewis-number differential diffusion effects and heat release by multiple reactions. However, these variables may always be clearly defined and physically identified; they are entropy-source terms. The other weak assumptions involved in deriving equation (2) are (i) low Mach number so that mean pressure gradients and convection effects are weak, (ii) second-order correlations between p' and w', which represent mechanical work, are negligible as compared with either thermal-energy fluctuations in the combustion region or the first-order acoustic quantities outside the combustion region, but still within S.

Of the four terms in equation (2), the following balances are appropriate in the stated regions and under the stated assumptions:

First + Second = 0   outside the combustion region and
                                      in the acoustic field.

Second $= 0$	under the low-frequency approximation and outside the combustion field.
Second + Third + Fourth $= 0$	in the combustion field and under the low-frequency approximation.
Third + Fourth $= 0$	under the low-frequency approximation and for very thin combustion regions, and then only in the time-dependent volume where combustion is occurring.

The low-frequency approximation will be invoked so that within S

$$\nabla \cdot \mathbf{w}' = \frac{\gamma - 1}{\gamma \bar{p}} (Q' - \nabla \cdot \mathbf{q}') \tag{3}$$

Outside the combustion region, but still within S, the incompressibility condition

$$\nabla \cdot \mathbf{w}' = 0$$

therefore holds. This is the approximation consistent with equation (1). What is being done, in effect, is the asymptotic matching[5] of the wave field to the incompressible near field in the limit $L \ll \lambda$.

The nice feature of equation (3) is, of course, that it is linear and may be solved as a Poisson equation for the fluctuating velocity potential, Φ',

$$\nabla^2 \Phi' = \nabla \cdot \mathbf{w}' = \frac{\gamma - 1}{\gamma \bar{p}} (Q' - \nabla \cdot \mathbf{q}') \tag{4}$$

presuming Q' and $\mathbf{q}'$ are somehow known. The solution is

$$\Phi' = \left(\frac{\gamma - 1}{\gamma \bar{p}} \right) \int_V G(\mathbf{r}, \mathbf{r}_0) [Q'(\mathbf{r}_0, t) - \nabla_0 \cdot \mathbf{q}'] dV(\mathbf{r}_0) \tag{5}$$

Here $G = 1/(4\pi R)$, $R = |\mathbf{r}_0 - \mathbf{r}|$, is the free-space Green's function. Then $\mathbf{w}' = \nabla \Phi'$ may be constructed and $\partial \mathbf{w}'/\partial t$ may be calculated and inserted in equation (1). Rigorous application of the theory of matched asymptotic expansion requires evaluation of G in the far field[5] (of the near-field solution) so that $|\mathbf{r}| \gg |\mathbf{r}_0|$ is considered. The result is that

$$p'(\mathbf{r}, t) = \left[\frac{(\gamma - 1)}{\gamma \bar{p}} \frac{\bar{p}}{4\pi r} \frac{\partial}{\partial t} \int_V Q' dV \right]_{\mathbf{r}_0, \eta} \tag{6}$$

Equation (6) recovers the original combustion-noise theory of this author.[1] However, it has now been done rigorously in the low frequency, low-Mach-number limit and with inclusion of q', which has always been incorrectly ignored in previous theories. The heat-transfer vector disappears from equation (6) because $\nabla \cdot \mathbf{q}'$ integrates to zero outside the combustion zone. It is indirectly included in equation (6) because Q is the source of all heat transfer, but $\mathbf{q}'$ does not explicitly appear in equation (6).

2.3. Frequency Content of Combustion Noise

If one were to decompose the pressure into its spectral content, the frequency information comes from the frequency content of

$$\frac{\partial}{\partial t} \int_v Q' dV = \int_v \frac{\partial Q'}{\partial t} dV \tag{7}$$

In order to investigate the behavior of this quantity it is instructive to consider a special case. Consider the Bray–Moss[6] model of premixed turbulent combustion where the actual combustion process takes place in thin laminar flamelets being whisked about by the turbulence. This particular model requires that the flame thickness be small compared with the Kolmogorov length scale.[7] In this case an observer either sees a combustion region at time t and position $\mathbf{r}_0$ or $Q = 0$. Thus, Q' is dominated by spikes approximately equal to Q_{max}, where Q_{max} is the maximum value of Q in a laminar flamelet. Instantaneously, therefore, the left side of equation (6) has the magnitude

$$Q_{max} \frac{\partial}{\partial t} V_Q(t)$$

where V_Q is the time-dependent volume containing the actual combustion. The frequency content of $\partial V_Q / \partial t$ is dominated by the distortions of the flame surfaces, which must be governed by the large-scale motions of the turbulence. It is for this reason that the frequency estimate above, u'/l, was made for combustion noise.

This observation is important, because if one uses the right side of equation (7) for an estimate, an erroneous answer is obtained. The quantity $\partial Q'/\partial t$ has some very high frequency content, and infinite if the flame thickness shrinks to zero. This quantity is dominated by the spikes one would see as a laminar flamelet is convected by a fixed observer. The characteristic distance here is δ, the flamelet thickness, not l, the "eddy" size. Evidently, there is strong cancellation of positive and negative high-frequency information in the right side of equation (7) to yield only the lower-frequency components estimated from the left side.

Of course there is still difficulty in choosing u' and l, because they will vary through a flame. There is, as yet, no theory by which they may be rigorously estimated. Nevertheless, the important point is that the large scale, energy-containing motions are the dominant ones that should determine the frequency content of combustion noise, at least in the limit where chemical times are much shorter than characteristic eddy times.

2.4. Sound Power from Combustion Noise

The sound power may be calculated from equation (6) by standard methods. It is

$$\mathcal{P} = \frac{\overline{p'^2}}{\bar{\rho}\bar{c}}(4\pi r^2) = \left[\frac{\bar{p}(\gamma-1)}{\gamma\bar{p}}\right]^2 \frac{1}{4\pi\bar{\rho}\bar{c}}\left(\frac{\partial}{\partial t}\int Q'\,dV\right)^2$$

Using the perfect-gas relation $\gamma\bar{p}/\bar{\rho} = \bar{c}^2$, and estimating the time derivative of the volume integral as a characteristic frequency f times the volume integral itself, the power estimate becomes

$$\mathcal{P} \approx \frac{1}{2\pi}\frac{(\gamma-1)^2}{\bar{\rho}\bar{c}^5}f^2\left(\overline{\int Q'\,dV}\right)^2 \tag{8}$$

The time average of the square of the volume integral of equation (8) may be estimated as[9]

$$\left(\overline{\int Q'\,dV}\right)^2 = \eta Q_{max}^2 V_f V_{cor} \tag{9}$$

Here η is the fraction of time an observer at $\mathbf{r}_0$ actually sees combustion, Q_{max} is the magnitude of maximum combustion rate, V_f is the measure of overall volume in which reaction may occur (macroscopic flame volume), and V_{cor} is the volume over which the combustion rates at two separated points are correlated. Evidently, however, $\eta \cdot Q_{max}V_f = m_F H$, where m_F is the fuel-flow rate and H is the fuel-heating value. Using this result in equation (9) and placing the overall result in equation (8), the resulting formula is

$$\mathcal{P} = \frac{\kappa}{4\pi}\frac{(\gamma-1)^2}{\bar{\rho}\bar{c}^5}f^2\dot{m}_f HQ_{max}V_{cor} \tag{10}$$

In equation (10) a factor κ has been inserted to correct for numerical errors made in the magnitude estimates. If the theory has validity, κ should turn out to be of order unity. Equation (10) is offered as the general theory of sound power radiated from flames in a free field. It is valid for any gaseous flame type, whether it is premixed, diffusion, or of hybrid type.

In equation (10) the acoustics of the medium enter through one factor of $\bar{\rho}\bar{c}$ in the denominator. This quantity is the characteristic acoustic impedance of the medium. The remaining factor in the denominator, $\bar{c}^4$, is basically the square of the ambient enthalpy, which shows that the radiated sound power depends upon the square of a heat-release to ambient-enthalpy ratio. The frequency f is expected to be dominated by fluid mechanics; but, since the combustion process and the turbulence may interact, exact estimation may be difficult. The estimation of Q_{max} will require a flame model as will the estimation of V_{cor}. If a characteristic width δ may be defined for the actual combustion process (such as a laminar flame thickness) and if Q' is at most correlated at space-separated points no larger than an integral scale,

$$V_{cor} \approx l^2 \delta \tag{11}$$

However, equation (11) is to be regarded as a special case.

3. APPLICATION OF THE THEORY

3.1. Premixed Jet Flames

The theory will first be applied to bunsen-type turbulent premixed flames. The data used for comparison is the fuel lean data reported by Strahle and Shivashankara.[9] For theoretical purposes the thin-flame theory of Bray and Moss will be accepted. In addition it will be demanded that the activation energy of the (assumed) global reaction is large so that the twin-zone structure of the laminar flame applies.[10] That is, there is a preheat zone in which there is a convective–diffusive balance and a reaction zone in which there is a diffusive–reactive balance. This is depicted in Figure 2 and in this case equation (11) applies. In this picture, the reaction-zone thickness is small compared with the preheat-zone thickness and analytically $\tilde{\delta} = \delta/\beta$. Here δ is approximately the thickness of the overall flamelet and is estimated by[11]

$$\delta = \nu_0/S_L \tag{12}$$

The quantity β is the dimensionless activation energy, ν_0 is the cold fluid kinematic viscosity, and S_L is the laminar flame speed. In equation (11), $\tilde{\delta}$ will replace δ, since reaction only occurs over $\tilde{\delta}$.

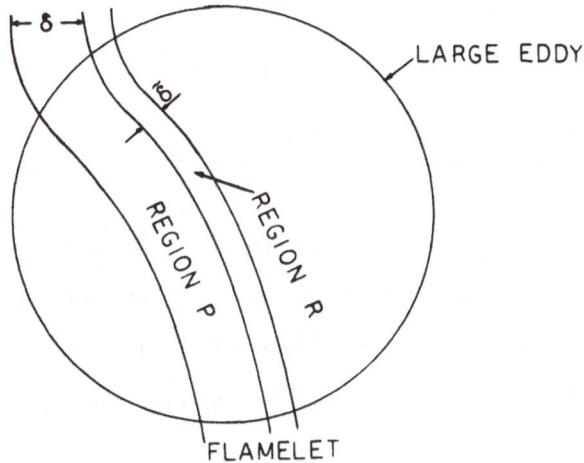

Fig. 2. Illustration of laminar flamelet imbedded in large-scale turbulent structure.

For this flame model, Q_{max} may be easily estimated as

$$Q_{max} = \frac{\bar{\rho} F S_L H}{\tilde{\delta}} \qquad (13)$$

where F is the fuel-mass fraction in the cold mixture. Since for these jet flames $\dot{m}_F = \frac{1}{4}\pi\bar{\rho} U D^2 F$, where U is the cold jet speed and D is the burner diameter, equation (8) becomes through use of equations (12) and (13)

$$\mathscr{P} = \frac{\kappa(\gamma-1)^2\bar{\rho}}{16\bar{c}^5}(fl)^2 F^2 S_L H^2 U D^2 \qquad (14)$$

It is presumed $f = u'/l$, but the data of Strahle and Shivashankara[9] suggest a complex estimate of u'. Using a combination of measured frequency data and estimates of the turbulence decay along the flame length, the result is that

$$u' \propto U^{.41} S_L^{.58} F^{-1.21} D^{.72} \qquad (15)$$

Proportionality (15) suggests complicated chemistry, heat release, and geometry effects upon u'. Using proportionality (15) in equation (14) yields

$$\mathscr{P} \propto F^{-.42} S_L^{2.16} U^{1.82} D^{3.44}$$

as compared with the experimental result[9]

$$\mathscr{P} \propto F^{-.40} S_{\mathrm{L}}^{1.83} U^{2.67} D^{2.78}$$

A numerical test case is chosen from the above-cited paper.[9] It is an ethylene–air flame at an equivalence ratio of 0.8. The following values apply:

$\bar{\rho} = 1.2$ kg/m^3	$c = 331$ m/s	$F = 0.0349$
$S_{\mathrm{L}} = 0.52$ m/s	$U = 91$ m/s	$D = 1.66$ cm
$u' = 3.7$ m/s	$H = 10{,}280$ J/g	$\gamma = 1.4$

The sound power for this case was 0.039 W so that from equation (14) and the numbers given

$$\kappa = 37 \tag{16}$$

which is the recommended value for use in equation (14). Note, however, that this should only be used for jet flames.

Experiments have only tested the F, U, S_{L}, and D variations. The theory, in addition, predicts the $\bar{\rho}$, $\bar{c}$, H, and β dependence; notice that no variation with respect to β remains and the sole chemical effect comes through S_{L}. Also noteworthy is that no direct effect of molecular transport appears.

3.2. Jet Diffusion Flames

In the case of a fuel jet burning in still air or with a coflowing stream of air, there are fairly accurate methods of flow-field prediction, at least for the H_2–air system.[12] However, there are no noise data. The theory will be specialized here for the diffusion-flame case, but no comparison with experiment will be attempted.

Again, a fast chemistry limit will be assumed that here implies local equilibrium of species. Flames for which the conserved scalar approach of Bilger[13] applies will also be assumed. In this approach the mass fraction of any species is uniquely related to a conserved scalar (the H-element mass fraction, for example in an H_2–air flame). The instantaneous reaction rate ω_i of any species is then exactly given by

$$\omega_i = -\rho \mathscr{D} \frac{\partial \xi}{\partial x_k} \frac{\partial \xi}{\partial x_k} \frac{d^2 Y_i^e}{d\xi^2} \tag{17}$$

Here ξ is the conserved scalar of interest, $\mathscr{D}$ is the diffusion coefficient (assumed the same for all species), x_k is the coordinate in the kth direction, and Y_i^e is the equilibrium mass fraction of species i.

The quantity $2D(\partial\xi/\partial x_k)(\partial\xi/\partial x_k) = \chi$ is the scalar dissipation. Taking i to be the fuel, therefore

$$Q_{max} = -\omega_{F_{max}}H = (\rho\chi)_{max}\frac{d^2 Y_F^e}{d\xi^2} H \tag{18}$$

The estimate of δ follows from the χ definition if the actual flame zones are thin, that is,

$$\delta \approx (\mathscr{D}/\chi)^{1/2} \tag{19}$$

A typical estimate of χ is[14] u'/l, if it is assumed that the mean-square conserved scalar fluctuation is of the order of unity. A coarse estimate of u' in the case of coflowing stream of fuel and oxidizer with a velocity difference of ΔU is $u' \propto \Delta U$. Finally, if the simple notion[15] is used that $l \propto$ (flame height) $\propto D$, the result for the sound power is

$$\mathscr{P} \propto \frac{(\gamma-1)^2\bar{\rho}F\mathscr{D}^{1/2}}{\bar{c}^5} U(\Delta U)^{5/2}D^{3/2}H^2\left(\frac{d^2 Y_F^e}{d\xi^2}\right) \tag{20}$$

Although equation (20) differs in detail from the premixed flame equivalent, equation (14), many of the same effects are present. Chemical kinetics and molecular transport occur through the product

$$\mathscr{D}^{1/2}d^2 Y_F^e/d\xi^2$$

in equation (20). For infinitely fast reaction rates, the Y_F^e derivative has a delta-function property.[13] The analogue of this behavior for the premixed flame would be $S_L \to \infty$.

4. COMMENTS ON NOISE REDUCTION

It is unfortunately true that reference to the general equation (10) or the special equations (14) or (20) leads one to some pessimism in the case of noise reduction. This negative view holds because considerations other than noise usually determine most of the factors. That is, one could not seriously suggest altering the chemical kinetic rates or the total firing rate. Moreover, the characteristic impedance is not generally at the designer's command, nor is the fuel-heating value. However, there is one factor in equation (10) that can be controlled and it is V_{cor}. A successful technique in both gas turbines[16] and furnaces[1] is to increase the number of injection points while holding the total firing rate fixed. This presumably breaks up larger correlated regions into smaller ones. This has the detrimental effect of raising f, usually, but the net effect appears beneficial in some cases.

The only other viable method of noise reduction at the source appears to be in a drop of turbulence level, which should also decrease f. This, of course, is often not desirable because of combustor-size constraints.

ACKNOWLEDGMENTS. This work was performed while the author was a visitor at the Mechanical Engineering Department in the University of Sydney, Australia. Useful discussions were held with Prof. Robert W. Bilger. Partial financial support for this work was supplied by the National Science Foundation under Grant No. CME 80-22366.

REFERENCES

1. W. C. Strahle, Combustion noise, *Prog. Energy Combust. Sci.* **4**, 157–176 (1978).
2. W. C. Strahle, On combustion generated noise, *J. Fluid Mech.* **49**, 399–414 (1971).
3. W. C. Strahle, Some results in combustion generated noise, *J. Sound Vib.* **23**, 113–115 (1972).
4. H. H. Chiu and M. M. Summerfield, Theory of combustion noise, *Astronaut. Acta* **1**, 967–984 (1974).
5. A. D. Pierce, *Acoustics*, McGraw-Hill, New York (1981).
6. W. C. Strahle and S. B. S. Chandran, The pressure-velocity correlation in a reactive turbulent flow, *AIAA J.* **120**, 129–135 (1982).
7. K. N. C. Bray and J. B. Moss, A unified statistical model of the premixed turbulent flame, *Astronaut. Acta* **4**, 291–319 (1977).
8. W. C. Strahle, Estimation of some correlations in a premixed reactive turbulent flow, *Combust. Sci. Technol.* (in press).
9. W. C. Strahle and B. N. Shivashankara, in: *Fifteenth Symp. (Int.) on Combust.*, pp. 1379–1385, The Combustion Institute, Pittsburgh, PA (1974).
10. W. B. Bush and F. E. Fendell, Asymptotic analysis of laminar flame propagation for general Lewis number, *Combust. Sci. Technol.* **1**, 421–428 (1970).
11. F. A. Williams, *Combustion Theory*, Addison-Wesley Publishing Company, Reading, Massachusetts (1965).
12. J. H. Kent and R. W. Bilger, in: *Sixteenth Symp. (Int.) on Combust.*, pp. 1636–1643, The Combustion Institute, Pittsburgh, PA (1976).
13. R. W. Bilger, in: *Turbulent Reacting Flows*, pp. 65–114, Springer-Verlag, Berlin (1980).
14. D. B. Spalding, Concentration fluctuations in a round free jet, *Chem. Eng. Sci.* **26**, 95–114 (1971).
15. I. Glassman, *Combustion*, Academic Press, New York (1977).
16. D. C. Mathews, N. F. Rekos, and R. R. Nagel, Combustion Noise Investigation, U.S. Federal Aviation Agency Report No. FAA-RD-77-3 (1977).

III
TURBULENCE

7

On the Structure and Morphology of Turbulent Premixed Flames

R. BORGHI

ABSTRACT. Turbulent combustion occurs in a majority of industrial devices. Many prediction methods have been proposed in recent years, and at the same time much experimental investigation has been performed. However, the physical structure of turbulent flames cannot be regarded as completely understood at present due to the complexity of the relevant experimentation.

This study aims at a synthesis between experimental findings and theoretical reasonings concerning the problem. A brief survey of previous discussions is presented first, with emphasis on three points around which controversy has been concentrated. The morphology of turbulent premixed flames in different conditions is described following a theoretical point of view; the presentation here synthesizes the discussions of earlier workers from different countries and different periods. The result is a comprehensive classification of turbulent premixed flames. The theoretical description is then compared with available experimental observations of the detailed structure of turbulent flames. Good agreement is achieved; however, experiments are very few and more research is required in this direction.

1. INTRODUCTION

Turbulent flames are present in a majority of industrial devices involving combustion, so prediction methods are of great interest for engineering purposes. Numerous experimental studies on turbulent com-

R. BORGHI • Laboratoire de Thermodynamique — L.A.C.N.R.S. No. 230, Faculté des Sciences et des Techniques de Rouen, B.P. 67, 76130 Mont-Saint-Aignan, France.

bustion have been performed in recent years, and numerical computations based on turbulent-combustion models now exist, leading to results of interest for engineers. However, the physical and chemical phenomena occurring in turbulent flames are not known in sufficient detail; their physical structure, in particular, has been the subject of many discussions, from 1940 to the present time, and the problem has not been solved to the extent that a common approach has emerged.

One reason is probably that the detailed study of the fine-scale fluctuating structure of turbulent flames is very difficult to investigate by experimental means, and consequently theoretical studies, based on numerous physical assumptions, suffer from a lack of validation. During the last few years, experimental research has become more powerful due particularly to laser devices, while theoretical studies on turbulence or the physical structure of laminar flamelets have been refined.

The time now appears ripe to synthesize past discussions on the basis of new experimental and theoretical results. We shall limit ourselves here to perfectly premixed flames. On the other hand, although we believe our approach is correct and substantiated when possible by experimental facts, our presentation and proposal result from personal thinking and may eventually be modified by new experimental results.

2. THE ESSENCE OF PAST DISCUSSIONS

2.1. The Wrinkled versus Distributed Flame

During the fifties, discussions on turbulent-flame structure concentrated on the distinction between the model of a wrinkled flame and the "distributed reaction zone" model. The former model, supported by Karlovitz and Schelkin, considered the turbulent flame as essentially constructed from a wrinkled laminar flame, or a set of laminar flames, moving randomly. The latter model, supported by Summerfield and Longwell, regarded the turbulent flame as a large zone where no flame fronts were present, i.e., without too large temperature and concentration fluctuations. This second model was in fact proposed jointly with the wrinkled-flame model by Damköhler, but was finally considered irrelevant by the supporters of the first model. Both these models were regarded at that time as being completely opposed to each other and irreconcilable. One can sense the mood of these discussions particularly in the first paper of Summerfield[1] and in the papers cited therein.

A few years later, however, the works of Schelkin and Russian researchers concluded that turbulent flames can be represented by the first or second model, according to the intensity and scale of turbulence with respect to some scale and velocity characterizing the laminar flame (see Schelkin[2] and Talantov *et al.*[3]). In addition, the need for one intermediate model was emphasized and, more recently, the study of Zimont[4] has allowed one to quantify, in some sense, this new model and identify it as a regime of thickened flamelets, wrinkled again by the largest scales of the turbulence.

On the other hand, new experimental and theoretical studies have been devoted these past ten years to laminar flames experiencing strain from the flow field, and it was recognized very early that these studies can greatly help our understanding of turbulent flames.[5,6] The main result of these studies is the possibility of extinction for the flamelets of a wrinkled flame, if it experiences too large a stretching. Can this phenomenon lead to a general extinction of the turbulent flame, as appears to be the case in the experiments of Ganji and Sawyer[18]? This question also simulated some discussion and is still largely open.

2.2. Flame-Generated Turbulence

Another heatedly discussed topic in the framework of turbulent combustion is so-called "flame-generated turbulence." This phenomenon appears to have been "invented" by Karlovitz *et al.*,[7] who saw it as a small-scale effect directly related to the expansion of gases due to heat release. At almost the same time, Scurlock and Grover[24] noticed that high, large-scale velocity gradients induced by combustion due to the particular geometric design of the burners could actually produce a large amount of turbulence. This topic has been the subject of much controversy in the past, even though correct measurements of turbulence kinetic energy in flames have become available only very recently, and even though the correct definition of which turbulence is actually "generated" by the flame is lacking. It seems clear now that a good approach to this problem is based on a balance equation for the turbulence kinetic energy, and that causes external to the combustion itself, but related to the experimental devices, do play a large role. For these reasons we do not intend to discuss this point here.

2.3. The Turbulent-Flame Velocity

The main purpose of turbulent-combustion studies was originally to determine the relevant law for the turbulent-flame velocity. A well-defined propagation velocity existed in the laminar case, so it was thought that a similar quantity could be defined for the turbulent case. However, theoretical work has not succeeded in defining this quantity as clearly as for the laminar case. In addition, experimental work showed that since the turbulent flame is much thicker than the laminar one, the turbulent-flame velocity was difficult to measure without arbitrariness. Finally, the very different, if not contradictory, experimental results originally cast doubt that the turbulent propagation velocity was strongly dependent on the experimental device. In the light of what we know today on turbulence kinetic energy, we acknowledge that a turbulent-flame velocity is actually a nonintrinsic quantity.

For these reasons, the following discussions on the turbulent-flame structure will not be extended further to some new proposal on a flame-speed law. Our intention here is to help to construct turbulent-combustion models that will describe in a more convenient way turbulent flows with combustion.

3. A SYNTHETIC PRESENTATION OF THE MORPHOLOGY OF TURBULENT PREMIXED FLAMES

3.1. A Classification of Turbulent Premixed Flames

We shall base this discussion on the work of Schelkin[2] and try to incorporate additional knowledge; a first attempt in this direction was made elsewhere.[8]

The discussion is clarified if we compare the turbulence characteristics to those of the laminar flames by the use of two quantities. The first, $k^{1/2}/u_L$, is the ratio of the square root of the turbulence kinetic energy to the laminar-flame speed of the considered mixture; k is intended to be the actual turbulence kinetic energy, including the influence of combustion on turbulence. The second quantity, l_t/e_L, is the ratio of the integral length scale of the turbulence, defined classically, to e_L, the laminar-flame thickness. We know from laminar-flame studies that $u_L \sim (a/\tau_c)^{1/2}$, where a is the heat diffusivity (we assume that the Lewis numbers are close to unity) and τ_c is a chemical characteristic time definable in several ways; also, we know that

$$e_L \sim u_L \tau_c \sim (a\tau_c)^{1/2}$$

Let us consider the plane $(l_t/e_L, k^{1/2}/u_L)$, in which the laminar flame is represented by the l_t/e_L axis, where $k^{1/2}/u_L = 0$. We can distinguish in this plane between the two previous models of turbulent combustion: if $e_L < \eta$ (where η is the Kolmogorov microscale defined by $\eta = \varepsilon^{-1/4}\nu^{3/4} \sim k^{-3/8}l_t^{1/4}\nu^{3/4}$) the turbulent flame is formed from wrinkled flamelets; if $l_t < e_L$ we are confronted with a distributed combustion.

If Kolmogorov equilibrium exists, the relation $e_L = \eta$ is written in terms of our variables as $l_t/e_L \sim (k^{1/2}/u_L)^3$. We shall assume in the following that Kolmogorov equilibrium occurs if Re_T is large, but we shall use $\eta \sim l_t$ if Re_T is small; in our plane, $Re_T = k^{1/2}l_t/\nu = 1$ gives the hyperbola $(k^{1/2}/u_L) \cdot (l_t/e_L) = 1$.

Between the line $\eta = e_L$ and $e_L = l_t$ there is room for an intermediate type of combustion. This has been emphasized by Zimont.[4] Following Zimont, the flame within this zone is formed from flamelets too, but the flamelets are not laminar and are thickened by the turbulence. One can understand this in two ways. First, as soon as $\eta < e_L$, some fluctuations enter the flame front, mainly in the preheat zone, if the global activation energy of chemical reactions is high enough. The heat diffusivity increases, the flame velocity grows, and its thickness increases. In the second approach, a more physical picture of the phenomenon can be given. If $\eta < e_L$, the radii of curvature of laminar fronts become of the same order as their thicknesses, and consequently the fronts interact. Even without exact calculation of this phenomenon, one can say that the resulting flame front will be thicker and, if the thickness of the reaction zone itself is always less than η, the reaction zone is not thickened but just wrinkled. In fact, these two pictures are in agreement: the physical phenomenon of interacting preheat zones just causes the mathematical effect of increasing heat diffusivity.

The structure of this new type of turbulent flame is such that it can be called a thickened-wrinkled flame.

The previous discussion must, however, be extended further. Indeed, if the thickness of flamelets is larger than e_L, more fluctuations have a scale less than it, and consequently the flame thickens even more, and so on. Zimont[4] argues that this process will stop when the equilibrium values e^*, u^*, and a^* of the flame thickness, flame velocity, and heat diffusivity inside the flame, respectively, will be such that e^*u^*/a^* is unity, as in laminar flames. If $a^* \sim e^*(k^*)^{1/2}$ (where k^* is the turbulence kinetic energy of fluctuations with scale less than e^*), and the k spectrum is assumed to be given by the well-known $n^{-5/3}$ Kolmogorov law, then

$$e^* \sim (k^{3/2}\tau_c^3 l_t^{-1})^{1/2}, \qquad u^* \sim (k^{3/2}\tau_c l_t^{-1})^{1/2}$$

It follows that this flame structure is changed into a distributed combustion when $e^* = l_t$, and not only when $e_L = l_t$. If the previous relation is taken into account, $e^* = l_t$ gives a very simple relationship: $\tau_t = \tau_c$ or $l_t/k^{1/2} = e_L/u_L$, which is easily plotted in the $(l_t/e_L, k^{1/2}/u_L)$ plane.

These three regimes are outlined roughly in Figure 1. In fact, we have assumed here that the proportionality constants in the previously obtained relations are unity, but the limits must not be taken as very precise. Figure 1 strongly resembles the turbulent-flame classification of Ballal and Lefevre.[13]

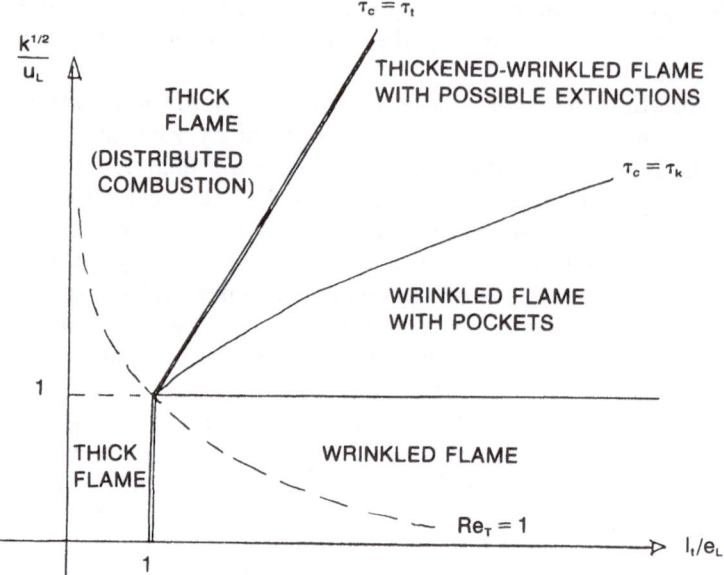

Fig. 1. The different regimes of premixed turbulent combustion.

3.2. More Details on the Physical Phenomena

In examining wrinkled flames, Schelkin[2] also discussed whether the wrinkled flamelet was continuous, or whether some pockets of burned gases existed in unburned environment (or the opposite). It is quite possible that, due to the wrinkling of a laminar flame, the oscillations are large enough with a nonsinusoidal shape such that two close parts of the flame front can interact; such a situation is displayed in Figure 2. Roughly

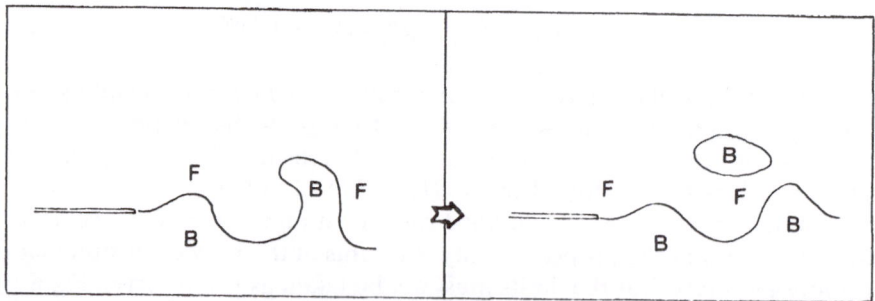

Fig. 2. The formation of "pockets" in a turbulent premixed flame. F denotes fresh mixture, B burnt gases.

speaking, the size of the burned or unburned pockets is of the order of l_t. A characteristic time for their birth is then $l_t/k^{1/2}$, while a characteristic time for their collapse by propagation of the flame that constitutes their surface is l_t/u_L. If $l_t/u_L < l_t/k^{1/2}$, there is therefore virtually no possibility of seeing such islands, even if the Reynolds number of the turbulence is large. The line $k^{1/2}/u_L$ in Figure 1 divides the wrinkled-flame domain into two subdomains: wrinkled flames with pockets can exist if $k^{1/2}/u_L > 1$, but not if $k^{1/2}/u_L < 1$.

The picture of turbulent-flame structure given here may now appear complete, but this is not the case. We must now discuss another physical phenomenon, the strain that the flamelets experience due to the turbulence velocity field. It is known that intense velocity gradients exist in a turbulence field, and their characteristic measure is not a mean velocity gradient. In the case of equilibrium turbulence, a characteristic strain rate is $(\varepsilon/\nu)^{1/2} = \tau_k^{-1}$.

Since the works of Klimov,[5] detailed studies have been devoted to laminar stretched flames. Lebedev and Klimov[9] studied the case where the Lewis number of a weakly reacting mixture is unity and showed that a premixed flame with intense stretching can experience extinction. Sivashinsky[10] found that this effect can be amplified if the Lewis number is larger than unity. Libby and Williams[11] more recently studied the particular case of a flame in the vicinity of a stagnation point and, taking into account the effect of heat release on the flow field, have shown that this effect could make the extinction phenomenon less frequent. Nevertheless, one can expect that any type of stretch can be imposed on the flame front by the turbulence, and it follows that the extinction phenomenon is likely to occur in a turbulent flame.

It is noteworthy that, from theoretical studies, the extinction limit is given by $\tau_c\gamma > W$, where W is a constant depending on the Lewis number and γ is the stretch rate. A characteristic value of γ is τ_k, so it follows that extinctions are possible up to line $\tau_c = W\tau_k$, i.e., a line similar to the thickened-wrinkled flame limit. This last finding is very important: if the turbulence thickens the flame, this flame experiences less stretch, because only gradients with a scale larger than the flame thickness can stretch the flame, and the gradients continue to decrease as the turbulence scale increases.

Hence thickening and stretching of the flame (both effects being produced by the same cause, namely the turbulence) are competing with each other. The outcome in some cases probably favors the stretching, and in others the thickening. It is to be expected that the larger the Reynolds number and the lower the Lewis number, the smaller the likelihood of extinction, and on the other hand if the Reynolds number is low and the Lewis number large, extinctions are probably more frequent.

If extinctions occur in the flamelets, it is clear that re-ignitions would also occur elsewhere and at other times, when the stretch is low enough. So the structure of the turbulent flame will be composed of segments of thickened-wrinkled flamelets whose length and position are continuously fluctuating.

Before concluding this description of the morphology of turbulent flames, one point remains to be discussed. Until now, we have implicitly assumed that flamelets are in a quasi-steady regime of propagation. This means that the frequencies of the turbulence are low enough with respect to some characteristic frequency of the flamelet. In the two regimes of wrinkled and thickened-wrinkled flames, the characteristic frequency of the flamelet itself is again none other than $1/\tau_c$; similarly, the highest frequency of the turbulence in a frame moving with mean velocity is just $1/\tau_k$.* It follows that the quasi-steady-state assumption is probably verified for wrinkled flames, but poorly justified in thickened-wrinkled flames, and probably out of order in distributed combustion. Therefore, if our description of the wrinkled flame is probably true, the discussions relative to thickened-wrinkled flames, particularly the calculations of e^*, u^*, and a^*, are probably rough approximations only. As regards distributed combustion, we can only state that it constitutes a medium with fluctuating temperature concentration and velocity, and consequently the reaction rate fluctuates between zero (extinction) and its laminar maximum value.

3.3. Some Consequences of the Flame Structure on Modeling

Modeling turbulent premixed flames gives rise to three particularly interesting quantities. The first is the probability density function (pdf) of the fluctuations of species concentrations and temperature, the second is the destruction rate of these fluctuations by the well-known scale reduction leading to small-scale molecular diffusion, and the third is the turbulent diffusion flux. The above picture of the structure of the different types of turbulent flame can help us to obtain a deeper understanding of the first two quantities.

It is clear that in the case of wrinkled flames, the probability density of a reactant, or a product, must exhibit two peaks: one corresponding to fresh gases, and one to burnt gases (at chemical adiabatic equilibrium). The question to be answered is whether the probability is zero or nonzero between these peaks.

Let us consider a small volume of a wrinkled flame, of the order of the integral scale of the turbulence. The probability of finding gases in an

---

* We see in the paper of Saitoh and Otsuka[12] that a frequency larger than one or two times $1/\tau_k$ leads immediately to a strong drop in the response of the flame.

intermediate state is equal to the ratio of the volume of flame fronts to the considered volume. In Figure 3a, which shows a section of such a volume, this probability is represented by the percentage of the flame area. Hence this probability will be nonzero if the thickness e_L is not infinitely small with respect to l_t. However, even if e_L/l_t is very small, this probability can exist if the surface of the flame per unit volume is infinite. Such a situation can occur if the surface is highly corrugated. Concerning the destruction rate ε, a recent study[14] based on the structure of wrinkled flames gives the result that, if the pdf has two peaks only, ε is controlled by the turbulent reduction of the scales prior to combustion of the reactant, that is related to an integral time scale of the turbulence. In addition, the surface per unit volume, Σ, and the destruction rate ε of the fluctuations of the reactant are related; it is found that $\varepsilon \sim u_L \Sigma$, and therefore $u_L \Sigma \sim 1/\tau_t = k^{1/2}/l_t$. It follows that $u_L/k^{1/2} \cdot \Sigma l_t$ is of the order of unity and consequently Σ remains of the order of $1/l_t$ if $u_L/k^{1/2}$ is finite. We can then infer that the pdf of the reactant has only two peaks when $u_L/k^{1/2}$ is finite and e_L/l_t infinitely small, i.e., τ_c/τ_t is infinitely small ($e_L/l_t = \tau_c/\tau_t \cdot u_L/k^{1/2}$). On the other hand, if $u_L/k^{1/2}$ tends to zero together with e_L/l_t, or if e_L/l_t remains finite, whatever the value of $u_L/k^{1/2}$ in the limit of the wrinkled-flame zone, then the pdf displays some probability in between the two peaks. In addition, it has been found[14] that in this case ε involves a contribution related to combustion that can be determined quantitatively.

For thickened-wrinkled flames, it is clear also from Figure 3b that the pdf must have the two peaks, and that they are smaller with the probability in between larger. For distributed combustion, the peaks are no longer

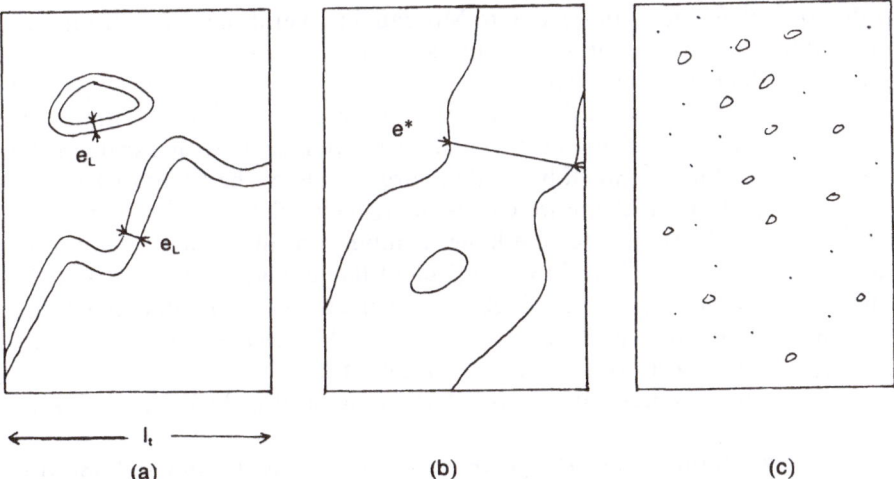

(a) (b) (c)

Fig. 3. Schematized structure for (a) wrinkled flame, (b) thickened-wrinkled flame, (c) thick flame.

present as the flame thickness becomes larger than the integral scale (Figure 3c). However, in nonhomogeneous flows and when we are considering a point located to one side of the mean combustion zone, we find a pdf with only one peak.

We can expect our hypothesis[14] to fail for wrinkled flames due to the more complex structure of the flamelets when they are thickened. The evaluation for ε is probably far from reality, and it is presently unclear how to improve it. Nonetheless, we can remark that as the turbulent flame evolves to a distributed-combustion zone, the implication for ε is that the contribution related to flame fronts decreases, and that estimates based on balance equations, such as those usually used in turbulence mathematical modeling, would probably be more relevant (see, e.g., our earlier paper[15]).

4. EXPERIMENTAL OBSERVATIONS

4.1. Visualization

The experimental demonstration of the structure of turbulent flames is not easy, especially for a high turbulence level. Photography with direct exposure or with a shadowgraph method can be used, but the information obtained is not relative to a well-defined cross section of the flame and cannot yield instantaneous visualization of the flamelets (if they exist).

It is interesting, however, to compare two recent photographs obtained for highly turbulent premixed flames. Figure 4 shows a shadowgraph of a premixed flame in an air–CH_4 high-velocity stream (approximately 50 m/s, equivalence ratio 0.85), due to Moreau[16]; several successive frames of a 4000-frames-per-second movie film are presented. The flame appears to be of the thickened-wrinkled type. The exposure time is about 30 μs and the apparent thickness of the flamelet is of the order of a few centimeters. Inside the flamelet a granular irregular structure is seen, with a scale of the order of 1 or 2 mm; a wrinkling with a larger scale of the order of 10 cm is also present. For this flame we can estimate a chemical time of about 1 ms from chemical studies of methane combustion in similar conditions. Measurements of $k^{1/2}$ within the turbulent flame gave values of about 20 m/s, and an integral scale of about 5 cm can be estimated from the geometrical dimensions of the large wrinkles. All this leads to $Re_T \sim 2 \cdot 10^4$, giving η of the order of 0.14 mm, τ_t of about $2 \cdot 10^{-3}$, and τ_k about 10^{-5} s. These rough estimates correspond to the regime of a thickened-wrinkled flame.

Two additional remarks can be made concerning this flame. First, it is difficult to see local extinctions, but it may be argued that these could be scrambled due to the velocity of the flow, even with an exposure time as

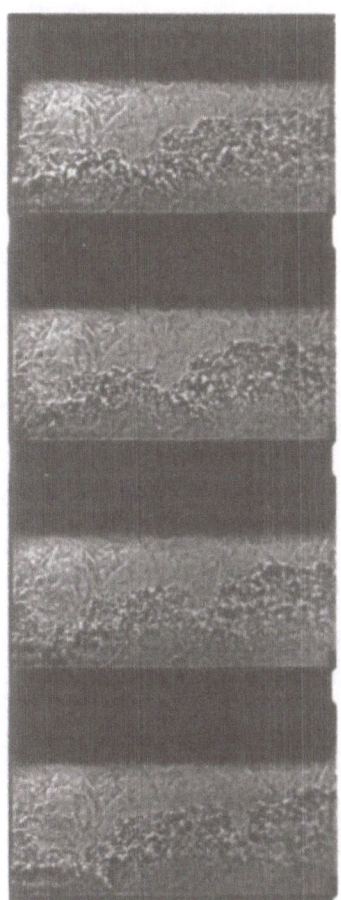

Fig. 4. Cinematographic shadowgraph record of the flame. Time interval between frames is 0.25 ms (after Moreau and Boutier[16]).

short as 30 μs. Second, it is noteworthy that large-scale undulations seem to display some coherence, not only in geometrical shape but also in crossing frequency. Numerous discussions have very recently been devoted to the presence of so-called "coherent structures." It is now clear that they are related to shear instability, and some researchers in combustion are attributing to them primary importance in turbulent combustion (see Yule et al.[17]). They can be seen in Figure 4, but they clearly appear to vary in shape similar to mixing without reaction of two streams possessing different velocity and density (Figure 5). On the other hand, a more detailed study of the crossing frequency of these structures seems to indicate that its value of about 800 Hz correponds to an acoustic frequency of the experimental device which, we suspect, could efficiently sustain the coherent structures.

Figure 6 shows photographs of another turbulent premixed flame by

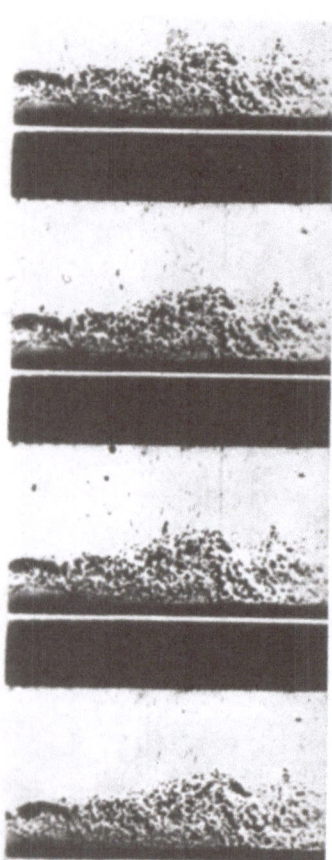

Fig. 5. Cinematographic shadowgraph record of the air–helium mixing layer. Same conditions as Figure 4 (after Moreau and Boutier[16]).

Ganji and Sawyer[18]; the photographs are from a high-velocity movie (5000 frames per second) employing the Schlieren method. One can see the so-called coherent structures much more clearly. The flamelets are again thickened with respect to laminar flames, as their thickness is about 1 cm. The flamelets do not show, as previously, internal granulation but, on the contrary, display elongated and nearly parallel streaks; many local extinctions seem to take place between them. The fuel here is propane and the equivalence ratio 0.6. The chemical time can again be estimated to be of the order of 1 ms; from measurements, $k^{1/2}$ is about 3 m/s and the integral scale about 1 cm. These values give $\tau_t \sim 3 \cdot 10^{-2}$ s, $\mathrm{Re}_T \sim 3 \cdot 10^3$, $\eta \sim 0.2$ mm, and $\tau_k \sim 10^{-4}$ s. The conditions here appear to differ slightly from the previous ones; η is larger and approaches e_L, Re_T is smaller, and the large structures are probably more "coherent." These results correspond to the photographs.

As distinct from the previous flame, this one clearly presents local

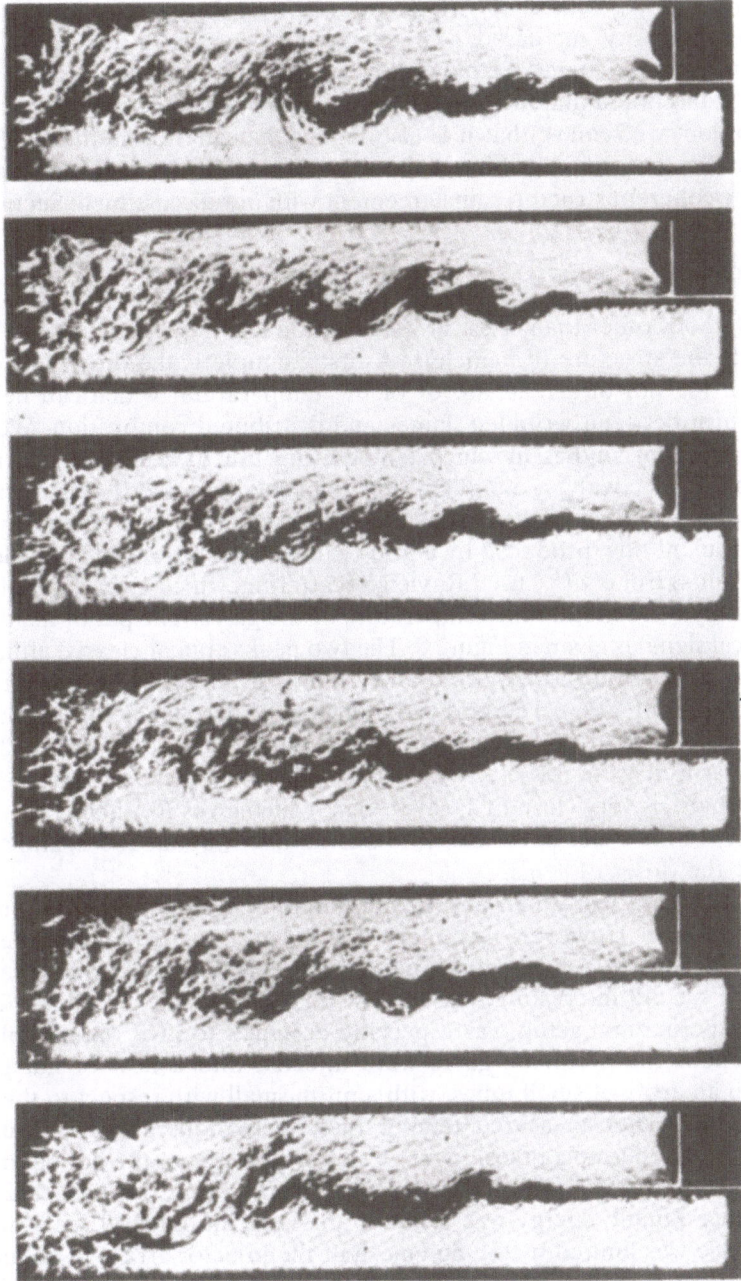

Fig. 6. Cinematographic Schlieren record of stable combustion. Time interval between frames is 5 ms (after Ganji and Sawyer[18]).

extinctions; more remarkably, total extinction of the flame followed when the flow velocity increased. An example is shown in Figure 7. This phenomenon has been related to stretching, as discussed in Section 3. Peters[19] has noted that the Lewis number of the propane was high enough. In addition, we remark that it is also higher than that of methane. Peters argues also that the stretching responsible for this extinction is associated with the coherent structures, in agreement with our discussion in Section 3.

4.2. Pdf Measurements

Methods other than classical visualization techniques can now be used to study the structure of flamelets. A first example is the direct measurement of the pdf of a reactant, or of the temperature, which can help to distinguish between wrinkled flames and distributed combustion. We can cite two recent studies in which temperature pdf have been performed, showing respectively a wrinkled (or thickened-wrinkled) flame and a distributed reaction. In their study of a premixed flame of methane and air in a turbulent flow produced by a grid (equivalence ratio 0.75 and velocity about 7 m/s) Bill $et\ al.$[20] used Rayleigh scattering, which gives a signal that can mainly be related to the temperature of the gases. The pdf of the direct electrical signal is given in Figure 8. The two peaks appear clearly, although they are not as thin as expected and there is a noticeable probability in between (about 40%). The grid turbulence used here leads to a moderate Reynolds number Re_λ between 10 and 30, with λ approximately 0.2 cm. The measured value of $k^{1/2}$, with flame, is about 30 cm/s. We can then expect that l_t is very close to λ, say 0.3 cm, which gives Re_T about 90 and η probably of the order of 0.1 cm, τ_k about 1.5 ms, τ_c about 10^{-2} s, and again τ_c is of the order of 1 ms.

In this case we obtain a wrinkled flame, as e_L is less than η and τ_c smaller than τ_k. However, we remark that the probability between the two peaks indicates that τ_c/τ_t cannot be considered infinitely small.

The second interesting study in this respect is that of Bellet $et\ al.$[21] Their experimental setup was especially designed to give rise to volume combustion: the premixed gases were injected into a flow of hot gases through an array of small tubes, with a mesh small with respect to the size of the duct. They measured temperature fluctuations and their pdf by means of fine thermocouples. Figure 9 shows a sample of the pdf obtained, without any peak and with fairly small fluctuations. In this case the turbulence kinetic energy was $k^{1/2} \simeq 20$ m/s and the integral scale of the turbulence was limited by the fine mesh of the injector arrays to about 0.3 cm. The length scale for temperature fluctuations was also estimated from optical measurements[22] to be of the same order. There is no doubt that a well-distributed zone with very small τ_t/τ_c has been achieved here.

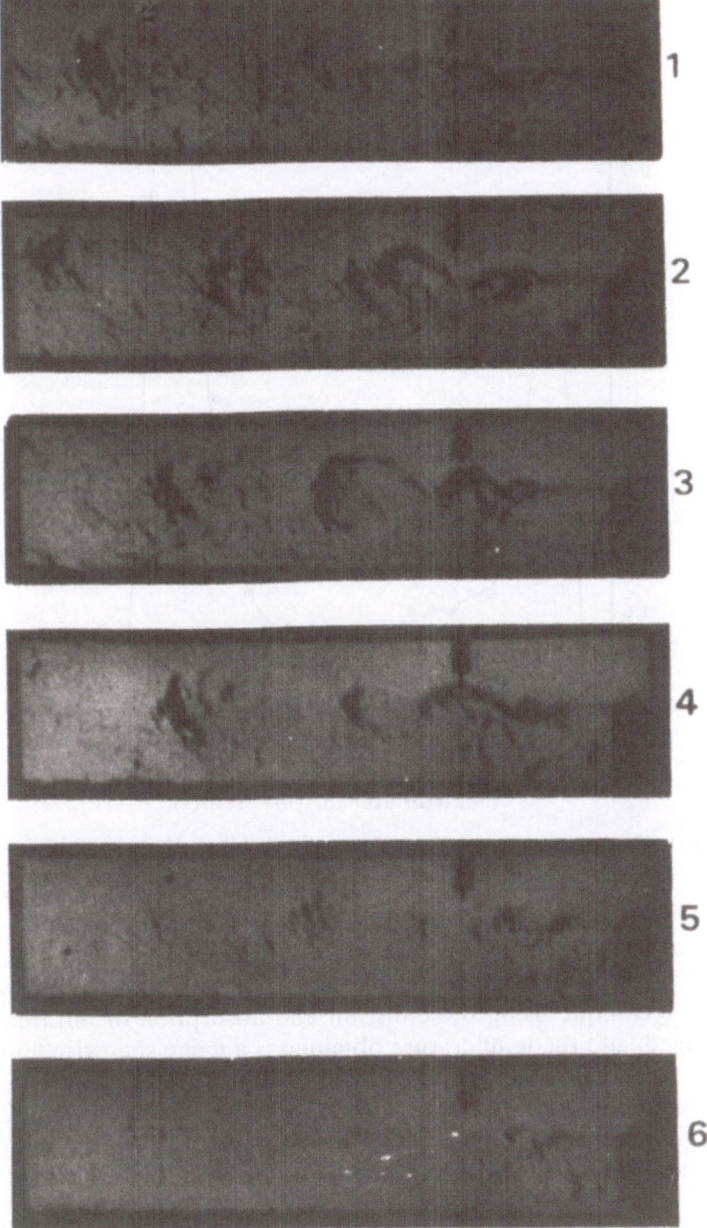

Fig. 7. Cinematographic Schlieren record of blowing-out flame. Flow from right to left. Time interval between frames is 25 ms (after Ganji and Sawyer[18]).

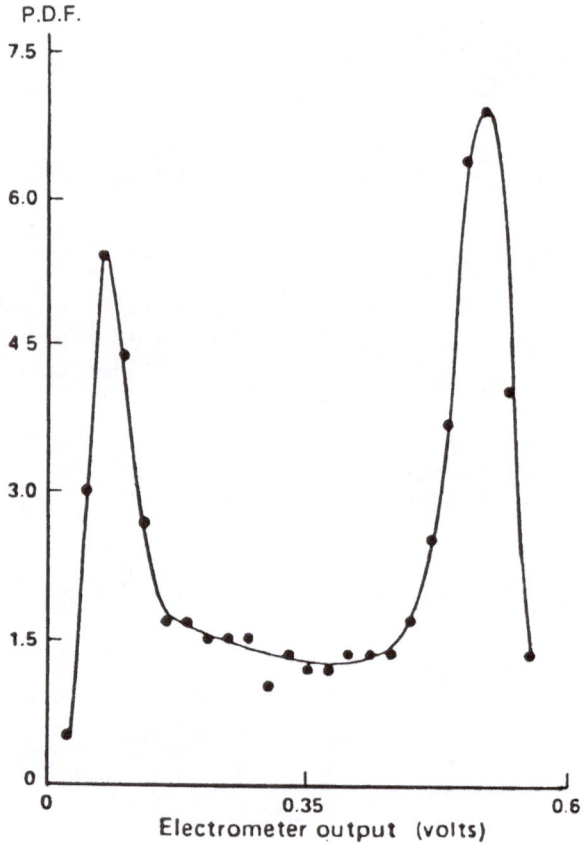

Fig. 8. Probability density function of a signal related to local temperature. Experiment of Bill *et al.*[20] in a V-shaped turbulent premixed flame.

An attempt to measure the temperature-fluctuation pdf has also been made within the flame of Figure 4, and is reported elsewhere.[24] However, a nonlocal technique using the emission and absorption of infrared light has been used, and the temperature obtained is a mean spanwise temperature. For this reason, the fluctuations obtained are clearly the two-dimensional fluctuations. Following the pdf obtained (Figure 10), they obey a strongly non-Gaussian pdf and exhibit two large peaks in the middle of the flame and only one on either side of it.

4.3. Direct Measurements of Flamelet Shapes

Finally, a very fine study has been performed by Smith[23] for the difficult case of a turbulent flame propagating in the cylinder of a

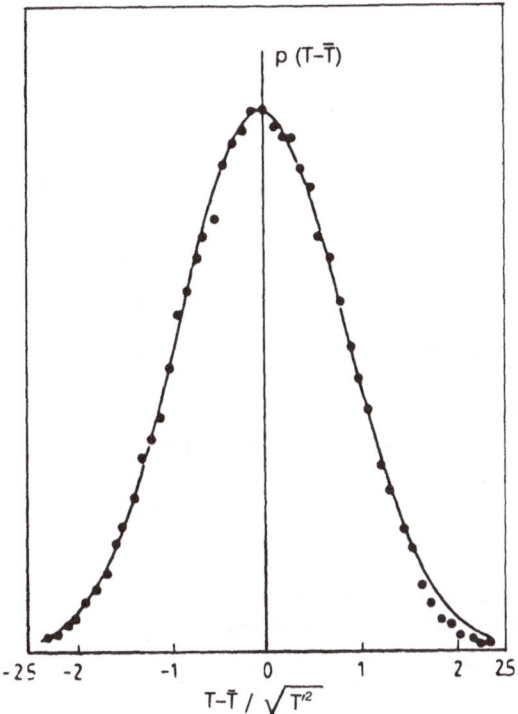

Fig. 9. Probability density function of temperature. Experiment of Chauveau *et al.*[22] in a reaction zone in small-scale turbulence: ●, experimental data; ——, Gaussian fit.

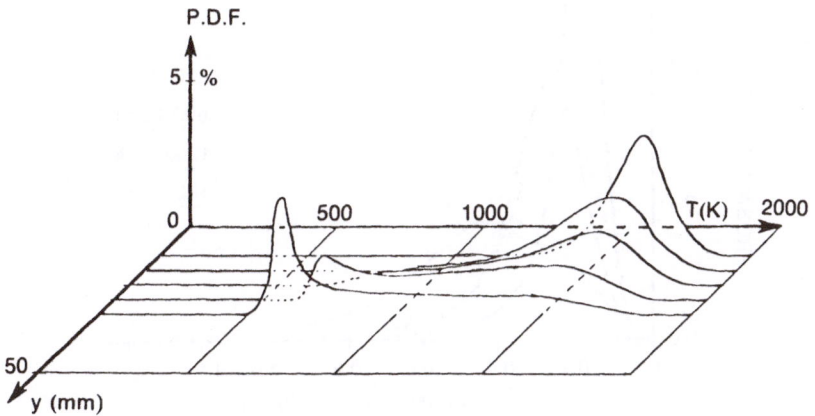

Fig. 10. Probability density function of temperature. Experiment of Moreau and Boutier[16] in a highly turbulent premixed flame. Same conditions as Figure 4.

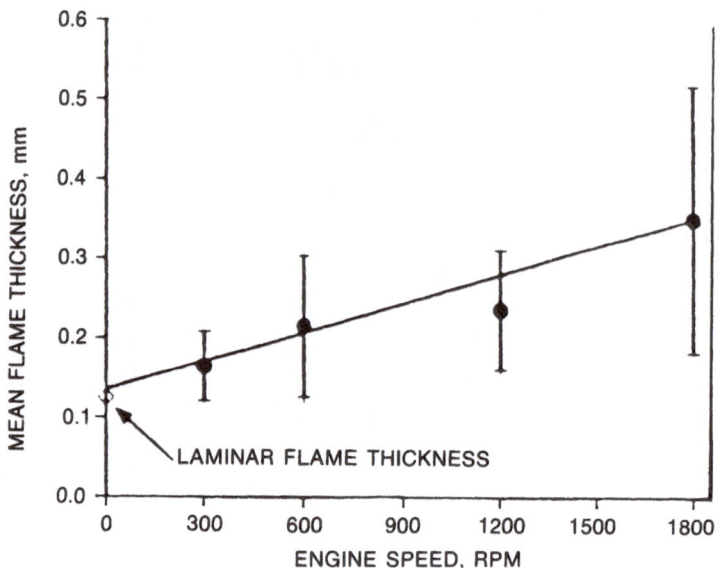

Fig. 11a. Mean flame thickness vs engine speed. Experiment of Smith[23] in a reciprocating engine. Bars indicate one standard-deviation value.

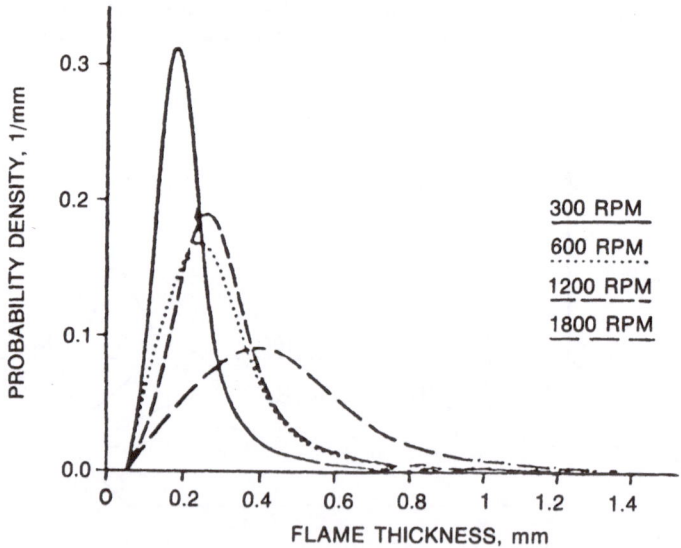

Fig. 11b. Probability density function of flame thickness for four engine speeds. Experiment of Smith.[23]

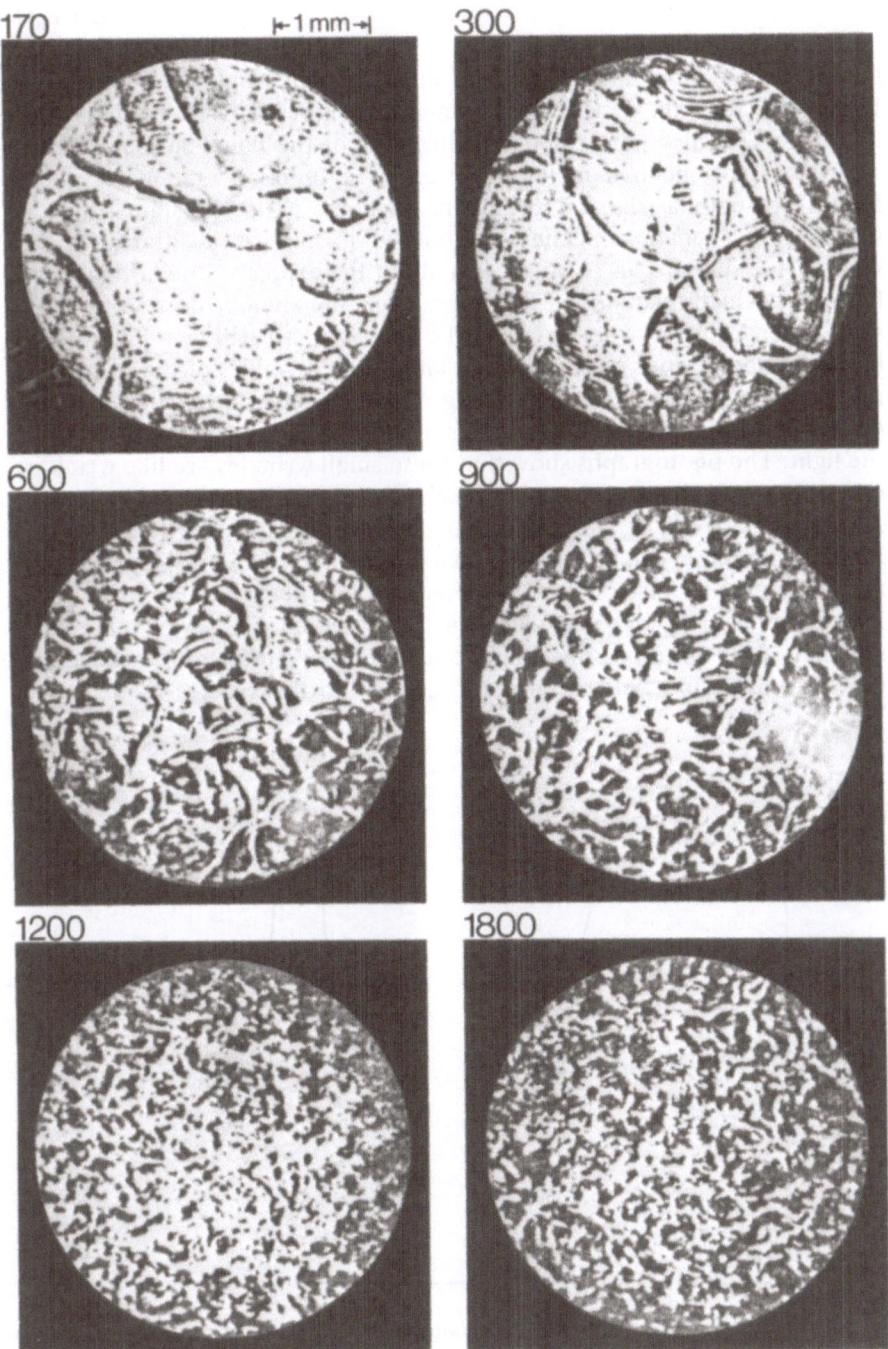

Fig. 12. Microshadowgraphs of flame propagating toward the viewer. Experiment of Smith.[23] Engine rpm is indicated in the upper left corner.

laboratory reciprocating engine. He measured the flamelet thickness by Rayleigh scattering, using the instantaneous temperature-profile measurements, and he also performed visualizations of the flame surface. Figure 11a shows the mean flamelet thickness as a function of engine speed. Figure 11b emphasizes that the flame thickness is not constant: there is scatter in the flamelet thickness of about 100%, and the scatter increases with engine speed. This scatter is just about the expected one in thickened flames, in which fluctuations with a scale smaller than e_L do not have the same intensity everywhere. Figure 12 shows microshadowgraphs of the flame surface. It does not display the large-scale wrinkles of the flame, but rather the small-scale ones, having radii of curvature small enough (probably of the order of a few times e_L) to produce sensible deviation of the light. The photographs show that these small wrinkles are like wrinkles on a sheet, and that their number increases as the engine speed increases. At the highest engine speed a granulation, very similar to that found in Figure 4, is all that can be detected. An interesting feature displayed in the measurements is also shown on Figure 13, where the record of the instantaneous Rayleigh signal reveals the presence of two very close flamelets. It is quite possible that we have here experimental evidence of the presence of pockets of burned gases within this type of flame.

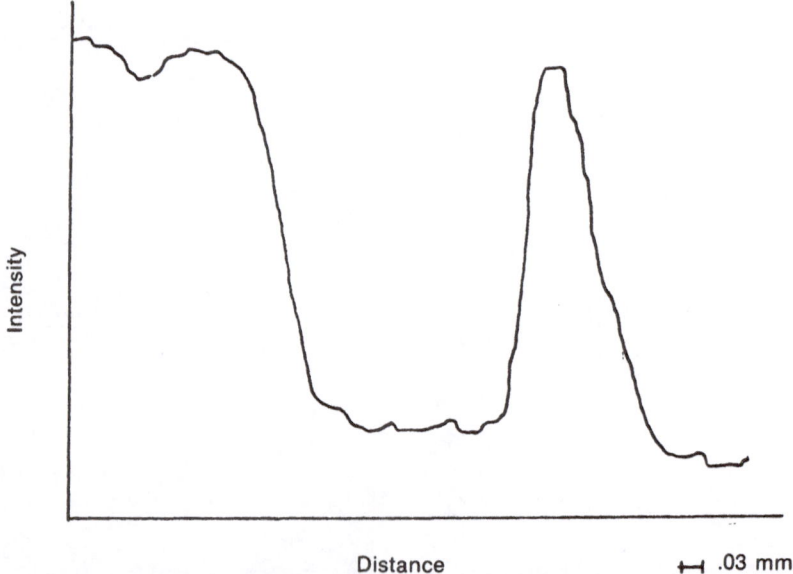

Fig. 13. Rayleigh intensity vs distance in a propagating flame. Experiment of Smith[23] at 600 rpm. Low intensity corresponds to low density and high temperature. An "island" or pocket of unburned mixture behind the flame front is shown at left.

5. CONCLUDING REMARKS

In conclusion, it is noteworthy that a wrinkled flame with a pdf having only two peaks has not been observed, but the majority of experiments have shown a pdf corresponding to wrinkled or thickened-wrinkled flames. Volume combustion has also been observed in some experiments, but special care must be taken to maintain this type of turbulent combustion.

Unfortunately, there are very few experiments where the results can be interpreted in terms of flame structure without any ambiguity. The reason is probably that the study of flame structure requires spatial information, not only one-point measurements (except for pdfs), and this complicates the measuring devices. On the other hand, many studies (too many, in our opinion) have measured turbulent flame speed, or gross properties only, without any attempt to obtain more physical information.

In summary, the picture presented here of the structure of different types of turbulent flames appears in agreement with both existing experimental evidence and up-to-date theoretical knowledge. New experiments must be conducted in the near future in order to verify it in more detail. We hope that their results will not contradict our synthesis.

REFERENCES

1. M. Summerfield, S. H. Reiter, V. Kebely, and R. W. Mascolo, The structure and propagation mechanism of turbulent flames in high speed flow, *Jet Propulsion* **25**, 377–384 (1955).
2. K. I. Schelkin, Combustion hydrodynamics, *Fiz. Goreniya Vzryva* **4**, 455–468 (1968).
3. A. Talantov, V. M. Ermolaev, V. K. Zotin, and E. A. Petrov, Laws of combustion of a homogeneous mixture in a turbulent flow, *Combust., Explos. Shock Waves* **5**, 73–75 (1969).
4. V. L. Zimont, Theory of turbulent combustion of a homogeneous fuel mixture at high Reynolds numbers, *Fiz. Goreniya Vzryva* **15**, 23–32 (1979).
5. A. M. Klimov, Flame propagation under conditions of strong turbulence, *Sov. Phys. Dokl.* **20**, 168–170 (1975).
6. F. A. Williams, A review of some theoretical considerations of turbulent flame structure, in: *AGARD Conf. Proc.* no. 164, on: Analytical and numerical methods for investigation of flows fields with chemical reactions, especially related to combustion, Liège (1974).
7. B. Karlovitz, J. R. Denniston, D. H. Knapschaefer, and F. E. Wells, Studies on turbulent flames, in: *Fourth Symp. (Int.) on Combust.*, pp. 613–635, The Williams and Wilkins Co., Baltimore (1953).
8. R. Borghi and E. Pourbaix, On the coupling of complex chemistry with a turbulent combustion model, *Physicochemical Hydrodynamics* **2**, 65–71 (1981).
9. V. N. Lebedev and A. M. Klimov, Elongation of a non-adiabatic flame during combustion of mixed gases and volatile explosive, *Fiz. Goreniya Vzryva* **17**, 68–73 (1981).
10. G. I. Sivashinsky, On a distorted flame front as a hydrodynamic discontinuity, *Acta Astronaut.* **3**, 889–918 (1976).

11. P. A. Libby and F. A. Williams, Structure of laminar flamelets in premixed turbulent flames, *Combust. Flame* **44**, 287–303 (1982).
12. T. Saitoh and Y. Otsuka, Unsteady behavior of diffusion flames and premixed flames for counter flow geometry, *Combust. Sci. Technol.* **12**, 135–146 (1976).
13. D. R. Ballal and R. H. Lefevre, The structure and propagation of turbulent flame, *Proc. R. Soc. London. Ser. A* **344**, 217–234 (1975).
14. R. Borghi, Turbulent premixed combustion: further discussions on the scales of fluctuations, Paper presented at the 9th International Colloquium on Dynamics of Explosions and Reactive Systems, Poitiers, France (3–8 July, 1983).
15. R. Borghi and D. Dutoya, On the scales of the fluctuations in turbulent combustion, in: *Seventeenth Symp. (Int.) on Combust.*, pp. 235–244, The Combustion Institute, Pittsburgh, PA (1978).
16. P. Moreau and A. Boutier, Laser velocimeter measurements in a turbulent flame, in: *Sixteenth Symp. (Int.) on Combust.*, pp. 17–47, The Combustion Institute, Pittsburgh, PA (1977).
17. A. J. Yule, N. A. Chigier, S. Ralph, R. Boulderstone, and J. Ventura, Combustion-transition interaction in a jet flame, *AIAA J.* **19**, 752–760 (1981).
18. A. R. Ganji and R. F. Sawyer, An experimental study of flow field and pollutant formation in a two dimensional, premixed, turbulent flame, AIAA Paper 79–0017 (1979).
19. N. Peters, Local quenching due to flame stretch and non-premixed turbulent combustion, *Combust. Sci. Technol.* **30**, 1–17 (1983).
20. R. G. Bill, I. Namer, L. Talbot, R. K. Cheng, and F. Robben, Flame propagation in grid-induced turbulence, *Combust. Flame* **43**, 229–242 (1981).
21. J. C. Bellet, M. Champion, Y. Chauveau, and J. Merlin, Expériences sur la combustion d'un hydrocarbure stabilisé par un mélange de gaz brûlés avec les gaz frais, *C. R. Hebd. Séances Acad. Sci. de France* **293**, série II, 259–263 (1981).
22. Y. Chauveau, P. Cambray, M. Champion, and J. C. Bellet, Experimental study and modelling of the reactive flow in a constant section reactor, *Physicochemical Hydrodynamics* **4**, 231–242 (1983).
23. J. R. Smith, Turbulent flame structure in a homogeneous-charge engine, SAE Paper 820043 (1982).
24. A. C. Scurlock and J. H. Grover, Propagation of turbulent flames, in: *Fourth Symp. (Int.) on Combust.*, pp. 645–658, The Williams and Wilkins Co., Baltimore (1953).

8

Model of Low-Reynolds-Number Wall Turbulence for Equilibrium Layers

P. ORLANDI

ABSTRACT. A one-equation turbulence model simulating the wall region is applied to study the equilibrium boundary layers. In these layers a universal mixing-length distribution can be assumed, as shown in the experiments of East and Sawyer. Particular emphasis is placed on representing accurately the turbulence-energy balance in the viscous and buffer regions, where an explicit model of the pressure-work term is shown to be necessary. Comparisons with experimental results of the mean quantities in the near-wall region are presented. Analysis of the turbulence-energy balance shows a behavior in qualitative agreement with that described by Shubauer. The same initial conditions have been assumed for the whole range of favorable and adverse pressure gradients considered. The free-stream pressure gradient influences the mean quantities and the turbulence-energy balance in a manner that is in agreement with experiment.

1. INTRODUCTION

Equilibrium boundary layers have been a subject of study from both theoretical and experimental points of view since the first theoretical study of Rotta.[1] He assumed that the boundary layer can be schematized by the thin layer near the wall, affected only by the kinematic viscosity, and the viscosity-independent outer part, in which the flow is determined by the edge velocity distribution. Rotta showed that self-preserving turbulent

P. ORLANDI • Scuola di Ingegegneria Aerospaziale, Istituto di Aerodinamica, Università degli Studi di Roma, Via Eudossiana 16, 00184 Rome, Italy.

boundary layers can be obtained with the same distribution of external velocity which gives self-similar laminar boundary layers (i.e., $U_e \propto X^m$) and that nonseparating flows in the turbulent case occur for values of exponent m larger than those in the laminar case. Clauser[2] studied experimentally the equilibrium layers by measuring the velocity profiles for two different external adverse pressure gradients and introduced the thickness Δ, scaling the normal direction in order to obtain universal velocity defect profiles. He also defined the acceleration parameter

$$\beta = \delta_1 \frac{dp}{dx} \Big/ \tau_w$$

which has to be held constant to have an equilibrium layer. Clauser examined the question whether only one or two different equilibrium layers can be obtained for the same exponent m. A definite answer to this question has not yet been given. While Clauser, and later Townsend,[3] showed that two equilibrium layers develop for large values of m, Mellor,[4] taking into account the viscous layer, gave a unique relationship between m and β. Bradshaw,[5] while leaving the question open, was in favor of a single equilibrium layer. Recently, Head[6] used an integral method to show that a wide range of equilibrium layers are possible for $m = -0.255$ depending on the initial conditions, and that it is impossible to achieve equilibrium at some values of the initial Reynolds momentum thickness.

Equilibrium boundary layers possess a very interesting turbulence-structure behavior. Negative values of m give rise to a turbulence structure very different from the structure at $m = 0$, while for positive values of m the turbulence structure is very similar to that of the flat-plate case. Accurate knowledge of the turbulence-energy balance may be a useful tool in understanding the turbulence structure of equilibrium layers. Bradshaw[5,7] measured some terms of the turbulence-energy balance in the log region and in the outer region for two flows at $m = -0.155$ and $m = -0.25$. More recently, East and Sawyer[8] conducted experimental studies of the equilibrium layers in a more systematic way, going from favorable to adverse pressure gradients while measuring turbulence production and diffusion. Measurements in the laminar sublayer and in the buffer region, however, are very difficult to carry out; experimental data in these regions have been obtained only in the case of a zero pressure-gradient boundary layer by Klebanoff[9] and for pipe flow by Laufer.[10]

The use of turbulence models, taking into consideration the near-wall region, may give some insight toward understanding mechanisms of the turbulence-energy balance in regions where accurate measurements are difficult to carry out. The large-eddy turbulence model is the best available model for this purpose, but it requires large computers and an enormous computational time. This model has been applied to solve the flow in a two-dimensional channel by Moin et al.,[11] showing a capability to simulate the coherent structures that have been visualized by Kim et al.[12] and measured by Blackwelder.[13] Moin et al.[11] employed the instantaneous velocity and pressure field to calculate the pressure-work term, which plays an important role in the turbulence-energy balance, especially in the near-wall region. This calculation, together with the qualitative behavior depicted by Laufer[10] for the pressure-work term, are very useful in producing improved Reynolds-averaged models, which require more reasonable computational time and can be used for engineering purposes.

The one-equation turbulence model, which takes into consideration the low-Reynolds-number effect, is adequate to simulate equilibrium boundary layers, and represents the turbulence-energy balance across the whole boundary layer provided the distribution of the length scale is known. In the near-wall region, usually not modeled, all the terms of the turbulence-energy balance except the advection term are of the same order of magnitude. In this region, maximum production and maximum dissipation occur due to the inrush–ejection cycle. The production term has been measured by Klebanoff[9] and Kim et al.,[12] thus a quantitative comparison can be made between the numerical and experimental results. For dissipation only, some terms can be measured while others can be estimated by making some questionable assumptions that can lead to erroneous results. Measurement of the turbulent diffusion requires very sophisticated electronic equipment, while the pressure-work term cannot be measured at all. The aim of this paper is to model explicitly, in the near-wall region, the pressure-work term that is usually included in the turbulent-diffusion term. It should represent a large diffusion of turbulent energy against the turbulent energy gradient, as pointed out by Laufer[10] and Moin et al.[11]

The turbulence-energy balance calculated by the one-equation turbulence model shows that the viscous sublayer and the buffer region behave as self-similar regions, while the log region and the advection and diffusion terms become more important as exponent m increases. The behavior of the calculated production and turbulent-diffusion terms is in good qualitative agreement with values measured by East and Sawyer[8] at low values of m; at large values of m there are discrepancies. The opinion of this author is that these discrepancies arise because the flow of East and Sawyer[8] was only approximately two-dimensional. Moreover, East and Sawyer obtained large adverse pressure gradients passing through a favorable pressure gradient, and each investigated flow had different initial conditions. Head[6] has pointed out that initial conditions can influence the flow in a sensitive

manner. The same initial conditions have been imposed in the numerical
solution over the whole range of external velocity distributions, to analyze
the effect of the pressure gradient on the turbulence structure. As a result
transition from an equilibrium to a nonequilibrium condition is not so
drastic, as shown by experiments.

2. TURBULENCE MODEL

When the local low turbulence Reynolds-number effect is considered
the viscous stresses cannot be neglected, and the turbulent momentum
boundary-layer equation is

$$\frac{\partial U}{\partial t} + U\frac{\partial U}{\partial x} + V\frac{\partial U}{\partial y} = \frac{\partial U_e}{\partial t} + U_e\frac{\partial U_e}{\partial x} + \nu\frac{\partial^2 U}{\partial y^2} - \frac{\partial \overline{U'V'}}{\partial y} \tag{1}$$

The Reynolds stress is obtained by multiplying the local mean rate of strain
by the turbulent viscosity ν_T of the fluid, which can be expressed in terms of
a velocity scale, the turbulent kinetic energy, and a length scale. The length
scale is related to the dimension of the eddies carrying energy in the
boundary layer. In the outer region the eddies possess the dimension of the
boundary-layer thickness, and in the near-wall region they have dimensions proportional to the distance from the wall. Earlier turbulence models
take into account the effect of the wall modifying the mixing-length
distribution by Van Driest's damping factor; this approach assumes that
near the wall the mixing length is no longer proportional to the distance
from the wall. Norris and Reynolds[14] argued that Van Driest's assumption
does not correctly represent the physics of the phenomenon in the wall
region, because the length scale should behave like $l = ky$ right down to
the wall. This point of view was supported by Ueda and Hinze[15] who
analyzed the fine structure of the turbulence in the near-wall region. Their
measurements of the dissipation length scale, which is related to the mixing
length, showed that the dissipation length increases linearly with distance
from the wall. Norris and Reynolds[14] introduced a "damping" factor in the
turbulent viscosity to model the wall suppression of turbulent transport and
proposed the following expression:

$$\nu_T = C_2 Q^{1/2} l \left[1 - \exp\left(-C_6 \frac{Q^{1/2}y}{\nu} \right) \right] \tag{2}$$

The dynamics of the turbulent kinetic energy $Q = \overline{U'^2} + \overline{V'^2} + \overline{W'^2}$ is
given by a transport equation that, subject to the boundary-layer simplifi-

cation, is written as

$$\frac{\partial Q}{\partial t} + U\frac{\partial Q}{\partial x} + V\frac{\partial Q}{\partial y} = -2\overline{U'V'}\frac{\partial U}{\partial y} + \nu\frac{\partial^2 Q}{\partial y^2} - \frac{\partial \overline{Q'V'}}{\partial y} - 2\frac{\partial \overline{P'V'}}{\partial y} - 2\mathcal{D}$$

| advection | production | viscous | turb. | press. | dissipation |
| | | diff. | diff. | work | |

$$(3)$$

where $\mathcal{D}$ is the "isotropic dissipation," which must be modeled as a function of turbulent kinetic energy and mixing length. The rate of energy dissipation at a high turbulent Reynolds number does not depend on the viscosity, and in this case dimensional analysis leads to the expression $\mathcal{D} = C_3 Q^{3/2}/l$. On the other hand, near the wall the viscosity is a dissipative agent for the largest eddies, and the dissipation depends on the viscosity. By dimensional analysis, this condition yields the expression $\mathcal{D} \simeq \nu Q/l^2$. In order to obtain a nonzero constant value of the dissipation at the wall, where Q^2 behaves as y^2, it is necessary that $l \simeq y$. This is further evidence that Van Driest's damping factor cannot be used in this turbulence model. The final expression for $\mathcal{D}$ is assumed here to have the form

$$\mathcal{D} = C_3 \frac{Q^{3/2}}{l}\left(1 + C_5\frac{\nu}{Q^{1/2}l}\right) \qquad (4)$$

The turbulent kinetic-energy diffusion has been treated by a gradient-diffusion model in which the pressure-work term has not been included, as is usual in one- and two-equation turbulence models. The pressure-work term should play a significant role mostly in the near-wall region. Thus it is important to model explicitly the pressure-work term only when the turbulence model has to take into consideration the low-Reynolds-number effect. This term cannot be measured at all, but a qualitative picture can be drawn from the literature.

Laufer[10] measured all the other terms and hence derived the pressure work; he showed that this term plays an important role only in the viscous and buffer regions. Shubauer,[16] analyzing the results of Laufer[10] and Klebanoff,[9] emphasized that the near-wall region is a region where dissipation must be greater than production and excess energy must be diffused from the outer region. Since the turbulent and laminar energy-diffusion terms cannot bring energy from the outer region into the near-wall region due to the negative gradient of the turbulent energy, the only term able to go against the adverse energy gradient is the pressure-work term. Moin et al.[11] confirmed the behavior depicted by Laufer[10] in their large-eddy simulation of the flow in a two-dimensional channel. From the instantaneous velocity and pressure distribution they were able to compute the pressure-work term. The results of Moin *et al.* indicate a large gain of energy in the buffer region, with the maximum situated at about the same value of y^+ where the maximum turbulent-energy production occurs. The pressure-work term attains zero at the wall and assumes low positive and negative values in the outer region. In channel flow the pressure-work term, integrated across the channel width, has to be zero. The pressure-work term in the near-wall region of boundary layers behaves as in two-dimensional channel flow, and its integral across the whole layer must be negative, $\overline{P'V'}$ being undoubtedly positive in the irrotational field. This implies that in the region away from the wall the pressure-work term behaves like a loss of energy. Lilley[17] imposed the condition that the production is completely balanced by the local dissipation and suggested that, in the inner layer, $2\overline{P'V'}$ has the opposite sign of $\overline{Q'V'}$. Here an assumption of constant $\overline{P'V'}$ in the log and outer region will be made, as it should not lead to large errors. The above arguments have enabled the pressure-work term to be modeled by dimensional considerations in the form

$$-\frac{\partial \overline{P'V'}}{\partial y} = C_3 \frac{Q^{3/2}}{l} \exp\left[-\left(\frac{Q^{1/2}l}{C_8 \nu}\right)^2\right] \tag{5}$$

The pressure work as given by equation (5) vanishes at the wall and at the end of the buffer region (around $y^+ \simeq 30$). Equation (5) is capable of representing the diffusion of energy from the outer region against the adverse turbulent-energy gradient, but cannot give a positive value of $\overline{P'V'}$ at the boundary layer's edge. However, it is this author's opinion that the simulation of the pressure-work-term behavior in the near-wall region is much more important. In fact, in this region the pressure-work term is comparable with the other terms, while in the equilibrium region it is negligible with respect to the production and rate of dissipation.

Equation (5) is similar to the general form of the wall-proximity effect on the pressure strain suggested by Launder *et al.*[18] This is consistent with the fact that by splitting the pressure-work term into pressure strain and $\overline{V'\partial p'/\partial y}$, the latter assumes values much smaller than $\overline{p'\partial V'/\partial y}$, as shown by Moin *et al.*[11]

The values of the constants in the turbulent-energy equation, in the eddy viscosity expression, and in the length-scale relationship

$$l/\delta = C_0[1 - (1 - y/\delta)^n]$$

can be evaluated following the procedure described by Norris and Reynolds[14] and Orlandi and Reynolds.[19] In Orlandi's paper the constant

C_6 of the damping factor in the eddy-viscosity expression has been calculated by imposing

$$\left(-2\overline{U'V'}\frac{\partial U}{\partial y}\frac{\nu}{U_\tau^4}\right)_{y^+\simeq 11}=\frac{1}{2}$$

The influence of this constant on the results is much greater than the influence of all the other constants. The effect consists in translating the log region "side by side" with respect to the law

$$U^+=\frac{1}{k}\ln y^++4.9$$

The velocity profile in wall coordinates moves upward if the value of C_6 is too small and downward if it is too large. The other constants were determined in the previous paper for the zero-pressure-gradient boundary layer by imposing conditions at the wall and at $y^+=100$. From equilibrium-layer experimental investigations one can see that the external pressure gradients do not affect these regions. The values of the constants obtained for the zero-pressure-gradient case have then been retained for the whole range of equilibrium layers considered.

3. SOLUTION METHOD

The following coordinate transformation in the normal direction with respect to the boundary-layer thickness has been assumed in order to solve the system of continuity, momentum, and turbulent energy-conservation equations:

$$y = \delta(x, t)\eta \qquad (6)$$

If the governing equation is integrated down to the wall, some computational points must lie inside the viscous layer, extending from $y^+=0$ to $y^+\simeq 5$. Since the boundary-layer thickness in wall coordinates is of the order of thousands, a finite-difference scheme with uniform mesh would require a large number of grid points. To avoid enormous computational times, a further coordinate transformation

$$\eta = \eta_\infty\left[1-\frac{\tanh(1-x_2)\alpha}{\tanh(\alpha)}\right] \qquad (7)$$

has been introduced, where x_2 is the "new" coordinate with respect to which the governing equation will be discretized. Large values of the transformation parameter α give more computational points near the wall. The value $\eta_\infty = 1.25$ has been assumed, so that the whole boundary layer is situated inside the computational domain.

Transformations (6) and (7) applied to the momentum and turbulent energy-transport equations yield

$$\frac{\partial U}{\partial t} + AU\frac{\partial U}{\partial x_1} + BV\frac{\partial U}{\partial x_2} + C\frac{\partial U}{\partial x_2} + DU\frac{\partial U}{\partial x_2}$$

$$= \frac{\partial U_e}{\partial t} + U_e\frac{\partial U_e}{\partial x} + F\frac{\partial}{\partial x_2}\left[E(1+\nu_T)\frac{\partial U}{\partial x_2}\right] \tag{8}$$

$$\frac{\partial Q}{\partial t} + AU\frac{\partial Q}{\partial x_1} + BV\frac{\partial Q}{\partial x_2} + C\frac{\partial Q}{\partial x_2} + DU\frac{\partial Q}{\partial x_2}$$

$$= 2F\nu_T E\left(\frac{\partial U}{\partial x_2}\right)^2 + F\frac{\partial}{\partial x_2}\left[E(1+\nu_T)\frac{\partial Q}{\partial x_2}\right]$$

$$- 2\mathcal{D} + \text{pressure work term} \tag{9}$$

where A, B, C, D, E, and F are functions of the coordinate transformation.

The continuity equation is differentiated with respect to y such that the matrix corresponding to the vertical velocity possesses diagonal dominance after finite-difference discretization. Coordinate transformations (6) and (7) introduced into the differentiated continuity equation give

$$G\frac{\partial^2 U}{\partial x_1 \partial x_2} + H\left[\eta\frac{\partial}{\partial x_2}\left(E\frac{\partial U}{\partial x_2}\right) + \frac{\partial U}{\partial x_2}\right] + L\frac{\partial}{\partial x_2}\left(E\frac{\partial V}{\partial x_2}\right) = 0 \tag{10}$$

where G, H, and L are functions of the coordinate transformation.

The boundary conditions governing system of equations (8), (9), and (10) are:

$$\text{at } x_2 = 0, \qquad U = V = Q = 0$$

$$\text{at } x_2 = 1 \qquad U = U_e, \qquad E\left(\frac{\partial V}{\partial x_2}\right) = -\frac{\partial U_e}{\partial x}, \qquad \frac{\partial Q}{\partial x_2} = 0 \tag{11}$$

The condition $\partial Q/\partial y = 0$ at $y = 1.25\delta$ yields free-stream turbulence due only to the boundary-layer effect.

The solution of the steady boundary layer requires the boundary-layer-thickness distribution along the downstream coordinate. Satisfactory empirical expressions are available for the zero-pressure-gradient case, while if a pressure gradient is imposed it is better to predict the boundary-layer-thickness distribution and then compare it with available experimental results. The steady equilibrium boundary layer has been simulated numerically using an unsteady code, as follows. At $t = 0$ and $x = x_0$, the horizontal-velocity profile measured by Wieghardt[20] at Re $\delta_1 = 6754$ has been prescribed, and the turbulent kinetic-energy profile has been calculated by an iterative procedure, namely solving the nonlinear equation (9) with the assumption of zero advection. This assumption gives a satisfactory distribution in the inner region, where the advection term is negligible. Instead, the advection term plays a very important role in the outer region $(y/\delta > 0.6)$, and in this region the distribution of the turbulent energy is then incorrect. The same velocity and turbulent-energy profiles have been assigned upstream as boundary conditions ($x = x_0$ and $t > 0$) for the whole range of the free-stream velocities considered.

The boundary-layer-thickness distribution given by the empirical law $\delta \propto x^{4/5}$ has been assigned and system of equations (8), (9), and (10) solved in the steady and zero-pressure-gradient case. The solution obtained has been taken as the initial condition ($t = 0$ and $x \geqslant x_0$). Then an external velocity profile extending from a constant value to the distribution for which the equilibrium layer is expected to exist has been imposed. Such a velocity distribution is given by

$$U_e = 1 - f(t)\left[1 - \left(\frac{x - x_0}{x_0}\right)^m\right] \tag{12}$$

Function $f(t)$ gives a smooth transition from $t = 0$ to $t = \tau$, where τ is chosen *a priori*. For $t \geqslant \tau$ the edge velocity does not change in time. It has been checked that the steady solution does not depend on the assumed value of τ. Equation (12) at $x = x_0$ and $t > 0$ gives $\partial P/\partial x \neq 0$, which is not in agreement with Wieghardt's velocity profile. Between $x = x_0$ and the first streamwise computational point there is then a large variation in pressure gradient, but only a short region close to $x = x_0$ should be influenced. The solution in this region is affected also by the assumption of zero vertical velocity and by the fact that the advection term in the turbulent-energy calculation at $x = x_0$ and $t = 0$ has been neglected.

The nonlinear system of unsteady boundary-layer equations, together with the initial and boundary conditions described above, have been solved by an implicit procedure. This is worthwhile for two reasons: first to avoid stability conditions on the mesh size in the streamwise direction; second and more important, to have rapid coupling between the velocity field and the turbulent quantities. Implicit methods applied to nonlinear equations require iterative procedures whose number of iterations depends on the complexity of the boundary layer (large pressure gradients).

All the details of the method, of the finite-difference scheme, and of the methodology of solution of the algebraic system of equations are reported in Orlandi and Ferziger[21] and Orlandi and Reynolds[19] and will not be repeated here.

4. RESULTS AND DISCUSSION

4.1. Zero Pressure Gradient

The zero-pressure-gradient boundary layer has been the most studied, and in this case measurements in the near-wall region are available. The universally validated measurements of velocity profiles, turbulent intensities, and Reynolds stress were obtained by Laufer[10] and Klebanoff.[9] The behavior of the boundary layer is analogous to the behavior of the flow in a two-dimensional channel or pipe in the near-wall region. Laufer[10] could measure only the energy balance in the thick layer he considered, and for low-speed flow, for which good conditions are difficult to obtain. Moreover, questionable assumptions had to be introduced in order to determine the nine components of the dissipation term. Laufer himself pointed out that the large differences between the measured triple velocity correlation term observed at the two considered Reynolds number are probably due to experimental errors. Since the pressure-work term was derived from these measurements, its behavior should be considered only from a qualitative point of view. Klebanoff's[9] measurements in this region are less exact than Laufer's since he disregarded the pressure-work term and derived the diffusion term by subtraction. Due to the inaccuracy of the measurements for some terms of the energy balance, the turbulent-energy balance obtained by the numerical model has been compared from a qualitative point of view, while a quantitative comparison with experimental results is possible for velocity, turbulent-energy, Reynolds-stress, and eddy-viscosity profiles.

Figure 1 shows the nondimensional profiles of the turbulent energy, Reynolds stress, and eddy viscosity in the viscous sublayer and in the buffer region, and also Shubauer's[16] data obtained by analyzing measurements of the boundary-layer and pipe flows. The calculated and experimental results appear to be in very good agreement. The value of the maximum turbulent energy calculated is in very good agreement with the experimental value but is located at a position closer to the wall, in better agreement with the position found by Kim et al.[12] than with the position found by Klebanoff[9] and Laufer.[10] For the limiting condition $y^+ \to 0$ theory predicts that the turbulent energy decreases quadratically, however this could not be verified experimentally due to the impossibility of taking meaurements very close to the wall. According to Laufer's[10] data, the eddy-viscosity

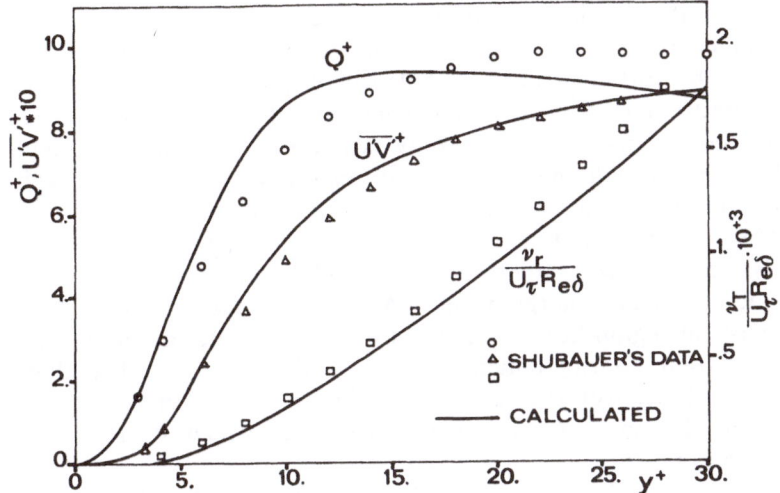

Fig. 1. Distribution of turbulent energy, Reynolds stress, and eddy viscosity in the near-wall region.

distribution in the viscous sublayer varies as $(y^+)^n$ with $n \simeq 4$, and the term $\nu/U_\tau \mathrm{Re}\delta$ approaches a linear behavior as the distance from the wall is increased.

Figure 2 shows the turbulent-energy balance in the region between $y^+ = 0$ and $y^+ = 30$. This region should be studied in detail, since it is the region where intense turbulence takes place. The one-equation turbulence model shows that dissipation at the wall is balanced by viscous diffusion. On moving away from the wall the dissipation increases and becomes larger than the production, reaching a maximum about the same position at which maximum production occurs. The the viscous diffusion in the viscous sublayer moves energy toward the wall, with an effect much greater than that of turbulent diffusion; it also moves turbulent energy toward the wall, but at a lower rate. In the first part of the buffer layer turbulent diffusion moves energy toward the wall, while viscous diffusion moves energy away from the wall. As y^+ increases further, turbulent diffusion also begins to move energy away from the wall. The turbulent and viscous diffusion are equal when the eddy viscosity equals the kinematic viscosity, since in this region the turbulent energy is almost constant.

Numerical simulation shows almost the same behavior of the turbulence balance yielded by the measurements; however, the measurements show a stronger effect of turbulent diffusion, a situation in which turbulent and viscous diffusion are not equal, and a maximum dissipation that is lower than the maximum production. The greater measured turbulent diffusion gives rise to a large pressure work, as was derived by subtraction. The turbulence-energy balance of Figure 2a probably gives a better representation of the real mechanism in this very important region. A final check may be made only when experimental equipment is developed to provide measurements in regions very close to the wall and when apparatus is designed to measure the pressure-work term.

This region is not usually modeled using the Reynolds averaged models; attempts were made by Jones and Launder[22] but they did not yield the turbulence-energy balance. Mellor and Herring[23] have modeled the near-wall region but they did not explicitly model the pressure-work term, so their energy balance shows a dissipation smaller than the production. Moreover, they did not plot the viscous and turbulent diffusion separately, thus making it impossible to discern which of these contributions is larger.

In the layer between $y^+ = 30$ and $y/\delta = 0.6$ there is equilibrium between production and dissipation. For $y/\delta > 0.6$, Klebanoff[9] and Brad-

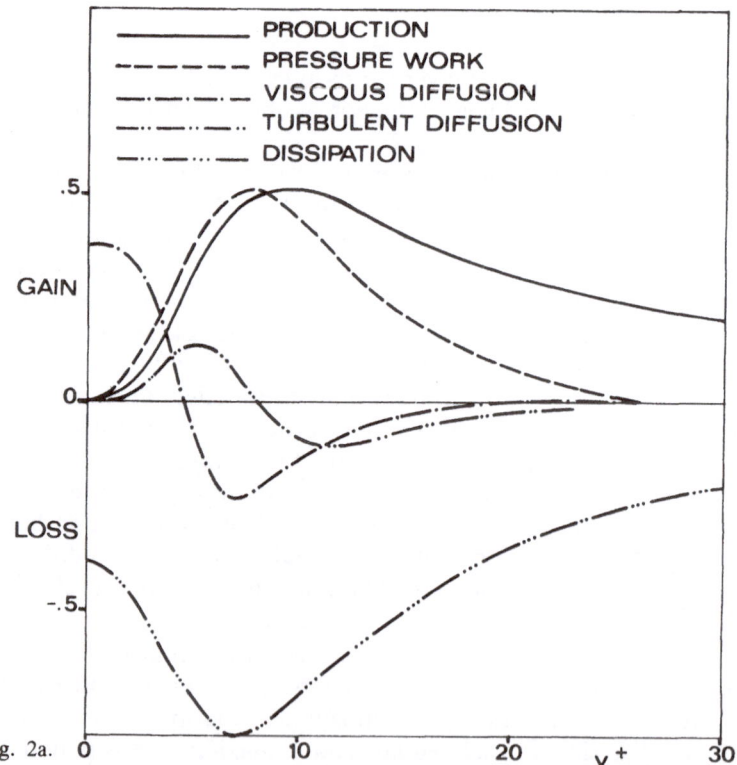

Fig. 2a.

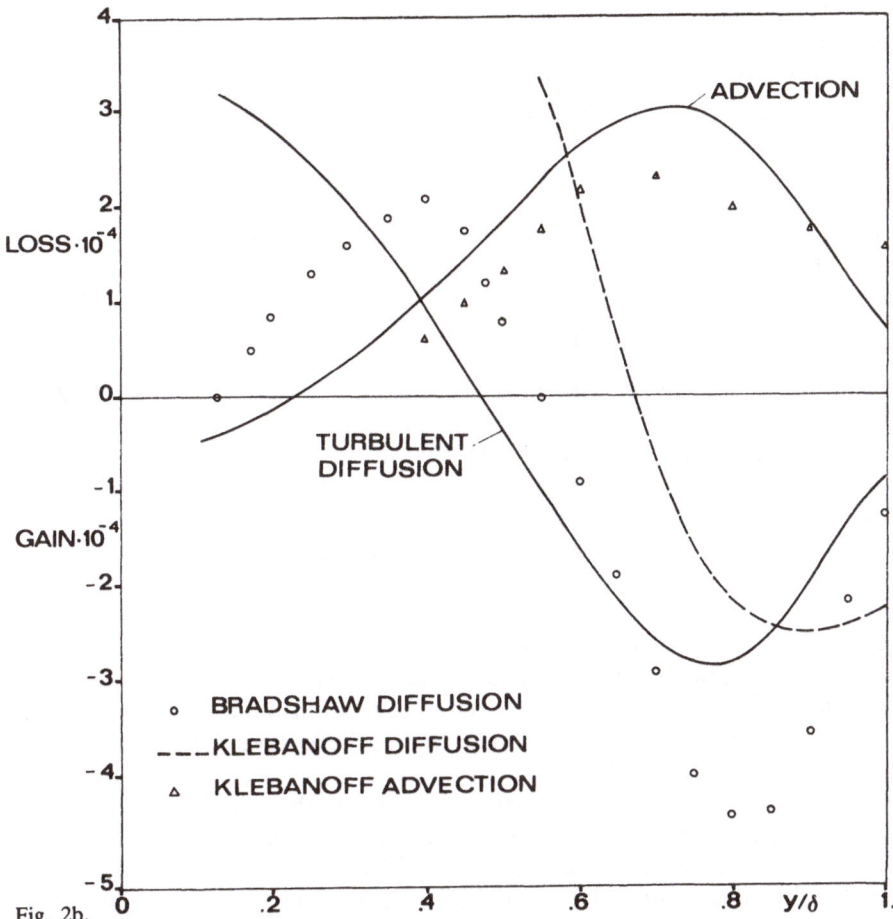

Fig. 2b.

Fig. 2. Turbulent-energy balance, dimensionless with respect to Re U_τ^4. (a) Near-wall region. (b) Outer region.

shaw[5] have pointed out that the turbulence-energy balance is due mostly to the contribution of the advection and turbulent-diffusion terms, namely the large-eddy structure is responsible for the turbulence mechanism. Bradshaw[5] criticized Klebanoff's turbulent-diffusion distribution[9] obtained by subtraction using the measured dissipation because it does not integrate to zero and shows a large loss by diffusion from the inner half, which is not correct. Bradshaw[5] preferred to measure turbulent diffusion rather than dissipation, although it required sophisticated linearizing electronic circuits. Figure 2b shows Bradshaw's[5] and Klebanoff's[9] distributions of turbulent diffusion compared with the one obtained here numerically. Good agreement between calculated data and Bradshaw's results is observed for $y/\delta > 0.4$. The computed distribution integrates to zero (while Bradshaw's distribution does not), and for $y/\delta < 0.4$ it yields a gain in the buffer region, in disagreement with Laufer's[10] behavior. Good modeling of the region close to the edge is very important, because there nonturbulent fluid is entrained by the very "corrugated" external edge of the boundary layer; this entrainment becomes much more important for boundary layers with adverse pressure gradient.

4.2. Equilibrium Layer with $\partial P/\partial x \neq 0$

Equilibrium layers with pressure gradient have not been studied as extensively as the zero-pressure-gradient case (e.g., turbulent-energy-balance measurements are available only in the region $y/\delta > 0.1$). This is mainly a consequence of the difficulty in setting up experimental apparatus for equilibrium layers, especially for large adverse pressure gradients. Thus a numerical solution using an accurate turbulence model can help in understanding how the turbulence mechanism is affected by the external pressure gradients. The turbulence model, which extends right to the wall, also allows one to understand the turbulence mechanisms in those regions where it is difficult to take measurements. The present one-equation turbulence model may be used to study the equilibrium layers, in line with the observation by East and Sawyer[8] that in these flows the mixing length has a universal distribution.

The computation has been carried out for both favorable and adverse pressure gradients with exponent m varying from $m = 0.2$ to $m = -0.4$. Skin friction has been compared with empirical laws given by Kader and Yaglom.[24] They divided the boundary layer into several regions and derived a general family of velocities, which were compared with the experimental profiles of a large number of flows in conditions of moving equilibrium. This family was used to obtain the following skin-friction law:

$$\left(\frac{2}{C_f}\right)^{1/2} = 2.44 \ln \Gamma - \frac{15}{\Gamma^{1/2}} - \frac{6}{\Gamma} + F \tag{13}$$

where

$$\Gamma = 6 \frac{\text{Re } \delta_1 U_\tau}{\beta U_e} \bigg/ \frac{50\delta_1}{\beta\delta} + 5$$

This law allows the skin-friction coefficient to be computed at any nonnegative value of the pressure gradient when the values of Re δ and of the modified acceleration parameter $\beta\delta/\delta_1$ are known. Kader and Yaglom[24] plotted the values of

$$F = 10\left[3 + 0.16\left(\frac{\beta\delta}{\delta_1}\right)^2\right]^{1/4} \tag{14}$$

vs. the quantity $\beta\delta/\delta_1$, together with the experimental results. Figure 3 shows quantity F calculated by equation (13) from the numerical solution at different m, compared with Kader and Yaglom's empirical law. The agreement is satisfactory at high values of $\beta\delta/\delta_1$; at intermediate values the computed values of F, like the experimental data, lie slightly above the empirical law given by equation (14).

In order to evaluate where the computed boundary layers are in equilibrium, one may verify how accurately the following conditions, defining equilibrium layers, are satisfied:

1. The momentum thickness must increase linearly with the streamwise distance.
2. The shape parameter $H = \delta_1/\delta_2$ must be constant or decrease slowly as the Reynolds number increases.
3. The similarity parameter β must be constant throughout the streamwise distance.

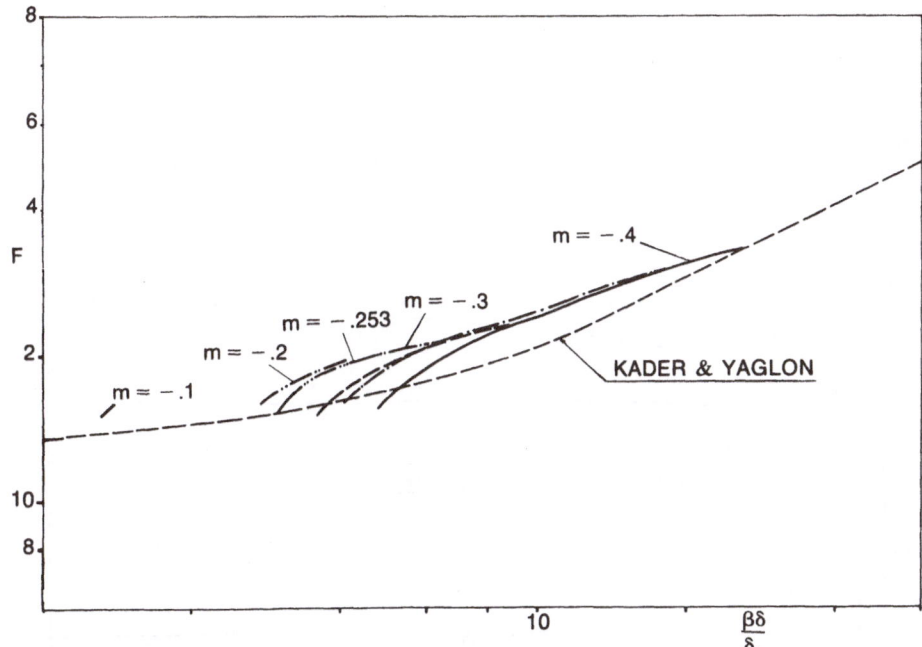

Fig. 3. Dependence of parameter F on $\beta\delta/\delta_1$ for the calculated equilibrium layer, compared with the empirical law of Kader and Yaglom.

Figures 4a–c show the profiles of the above-mentioned parameters vs. the dimensionless streamwise coordinate for several values of exponent m. The plot of the shape factor vs. exponent m at streamwise locations larger than $(x - x_0)/x_0 = 2$ shows that condition 2 above is satisfied up to $m \geq -0.23$, i.e., the same value predicted by Mellor[4] to obtain equilibrium boundary layers. The sharp increase in H, β, and δ_2 observed near $x = x_0$ is due to the sharp transition between the initial condition of zero pressure gradient and the condition $\partial P/\partial x \neq 0$. The assumption of the same zero-pressure-gradient initial condition for all the considered flows causes the layer to go from an equilibrium to a moving-equilibrium condition more weakly than if an initial equilibrium condition was assumed for each flow. Figures 4a–c indicate that on increasing the negative value of m and increasing the distance from the origin, conditions 1–3 become less accurately satisfied and the layers tend to a nonequilibrium condition.

Let us now examine the influence of the adverse pressure gradient on some important flow features. The free-stream pressure gradient does not affect the velocity in the near-wall region, and the width of the log region decreases as the adverse pressure gradient becomes larger (Figure 5a). As for turbulent energy and Reynolds stress, Figures 5b and c show the same

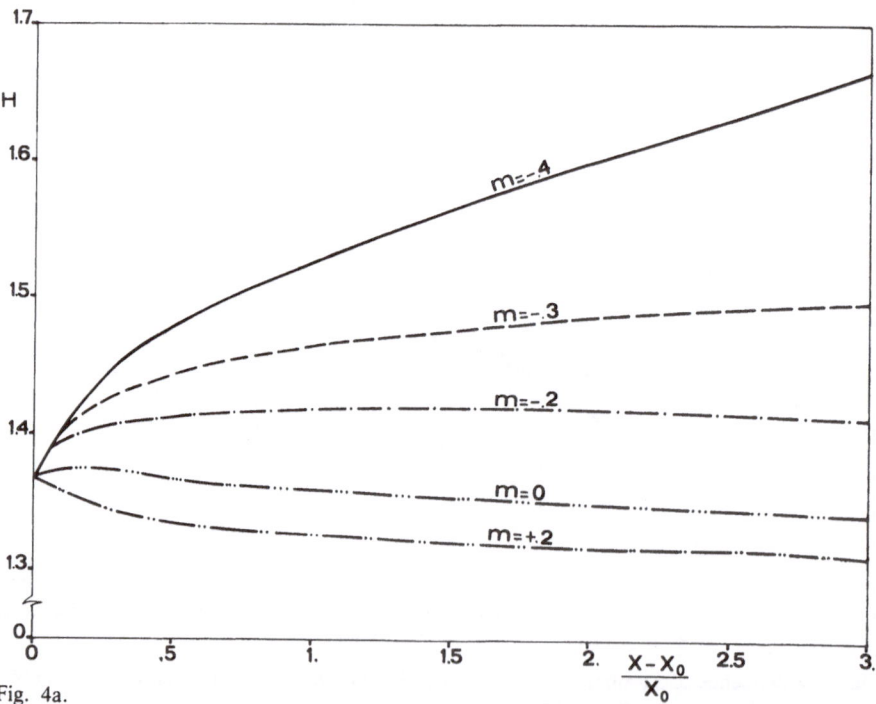

Fig. 4a.

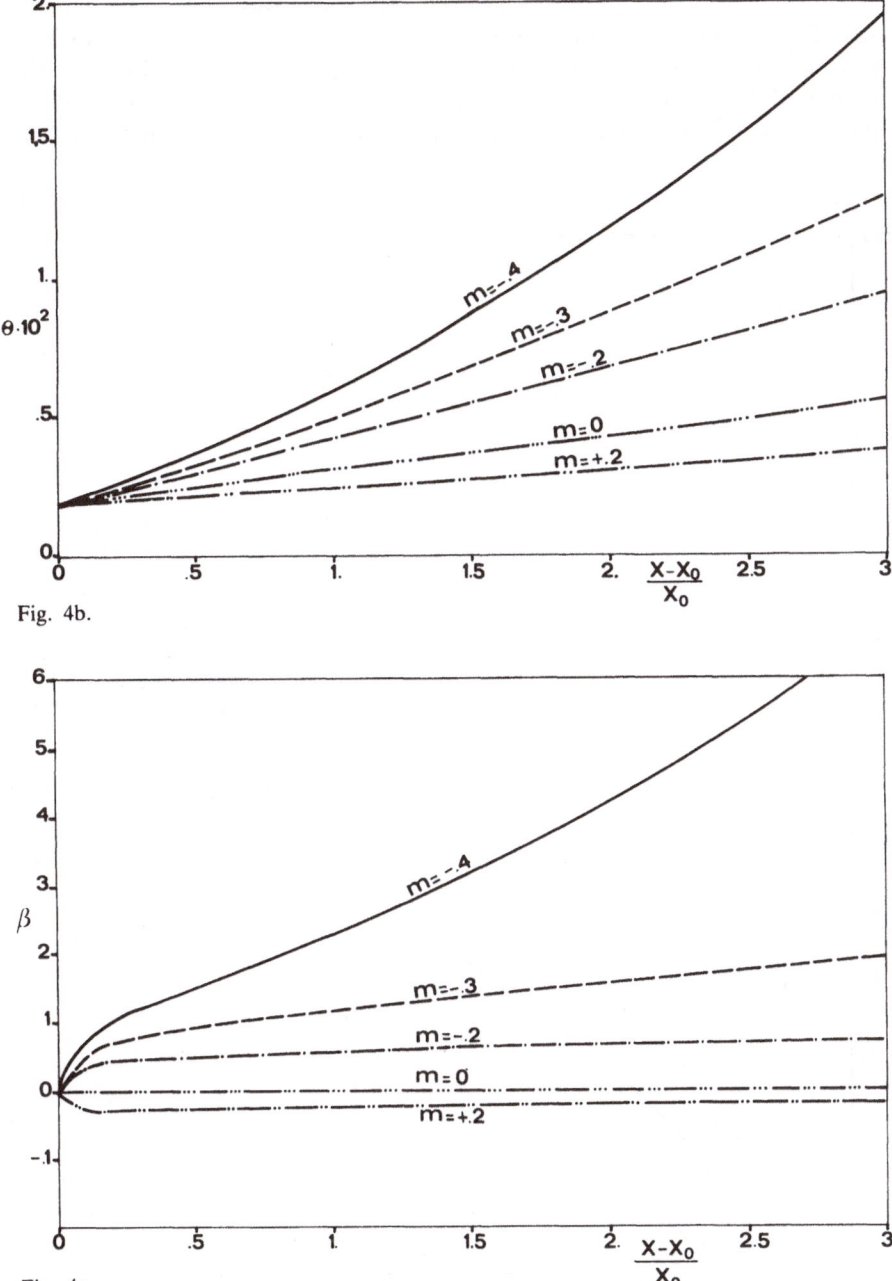

Fig. 4b.

Fig. 4c.

Fig. 4. Distribution of global quantities for all flows. (a) Shape parameter. (b) Momentum thickness. (c) Similarity parameter β.

behavior observed experimentally by Bradshaw[5] and East and Sawyer,[8] i.e., if the adverse pressure gradient is increased, then the position at which the turbulent quantities attain maxima moves in the outer region, and the values of these maxima are larger. Numerical simulation shows a point where, for different m, the turbulent quantities assume the same value and

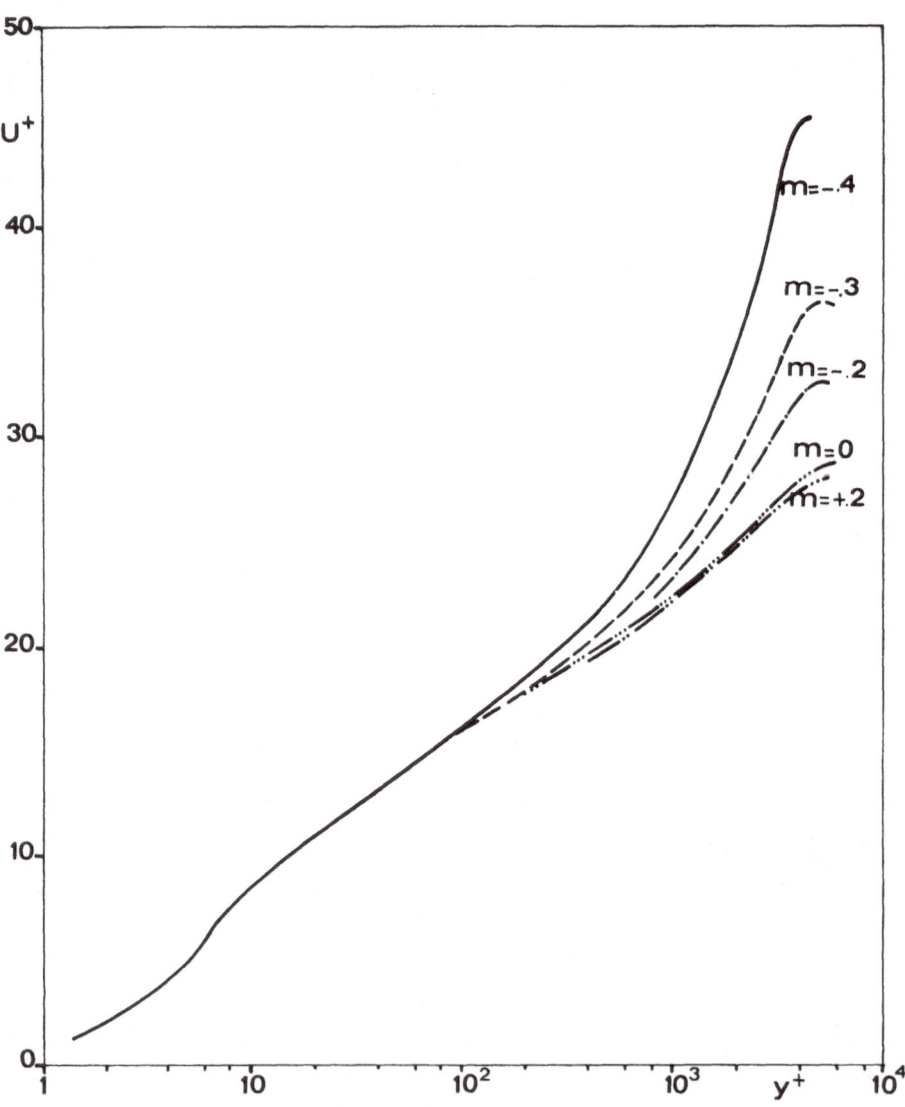

Fig. 5a.

this point moves toward greater y^+, increasing the downstream distance. This point also seems to occur in the experimental Reynolds-stress profiles of East and Sawyer,[8] but in this case it is not so well defined due to the difficulty of measuring close to the wall. Figures 5b and c show the pressure gradient influencing the turbulent quantities in the near-wall region, but, if

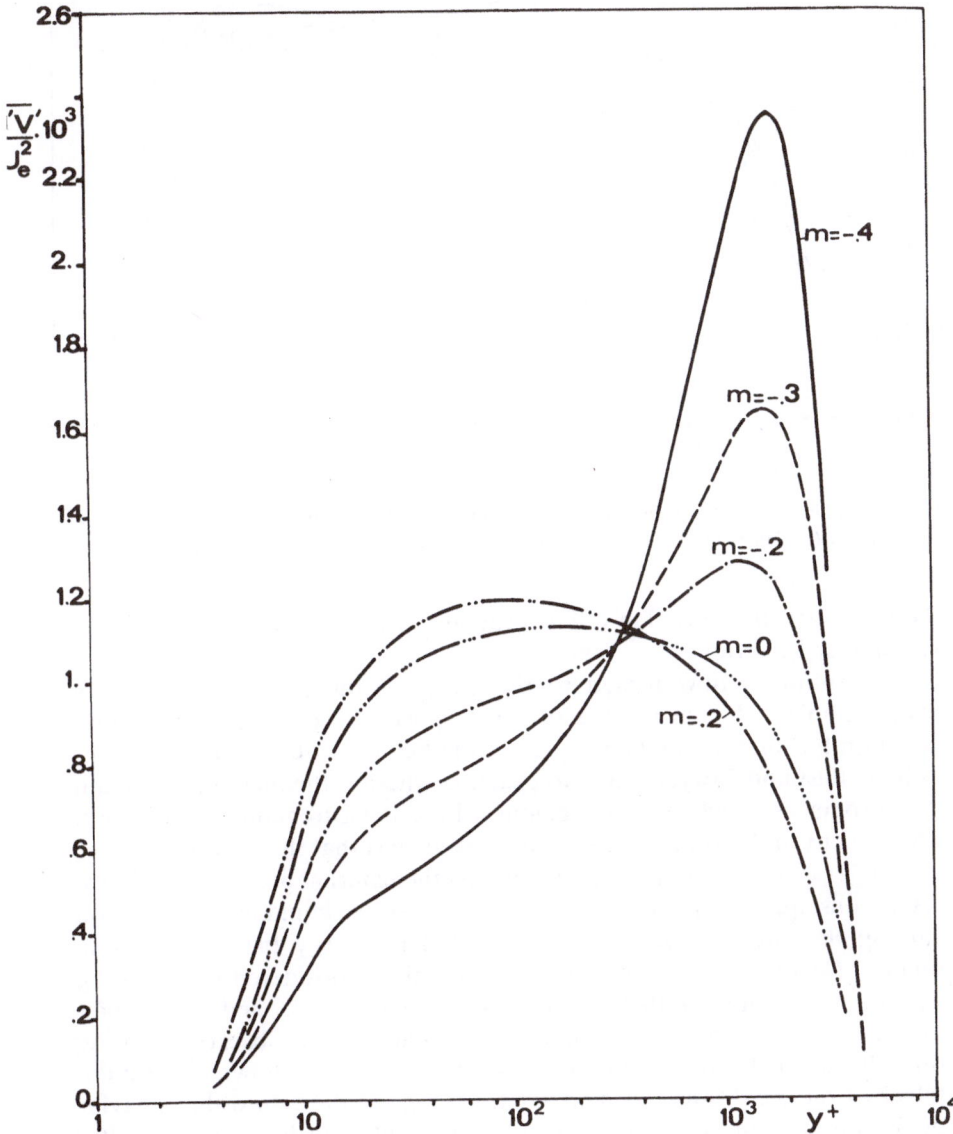

Fig. 5b.

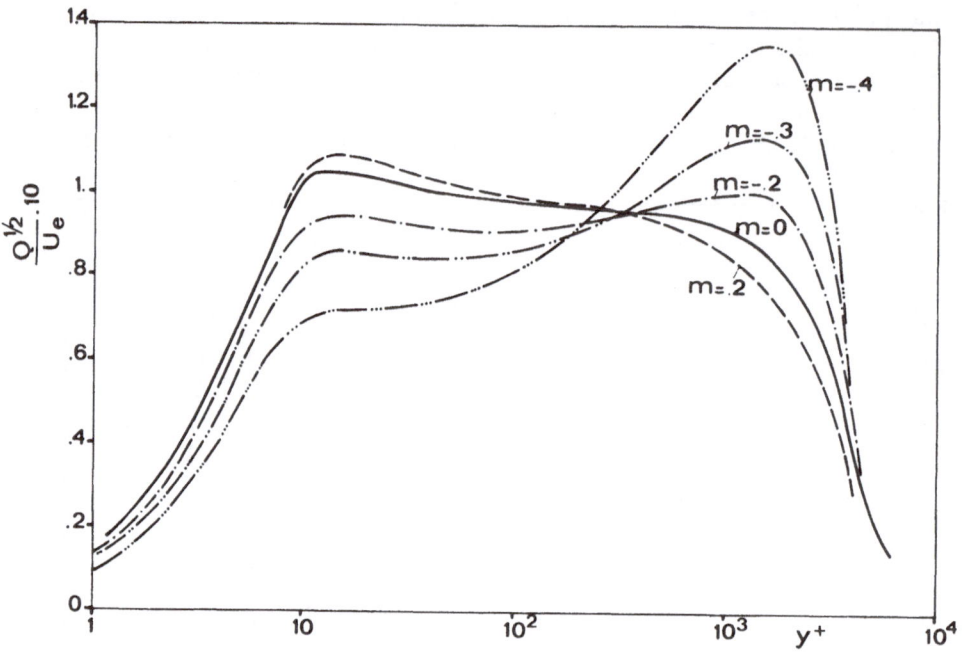

Fig. 5c.

Fig. 5. Profiles of velocity and turbulent quantities for all flows. (a) Velocity in wall coordinates. (b) Turbulent energy. (c) Reynolds stress.

the quantities are plotted in wall coordinates, the near-wall region exhibits self-similar behavior.

Attention is now turned to the effect of the free-stream velocity distribution on the turbulence-energy balance, to see whether the one-equation turbulence model can predict the behavior observed experimentally by East and Sawyer,[8] and to examine whether the model can explain the turbulence mechanism in regions where measurements are difficult. The velocity and turbulent-energy profiles suggest that the viscous sublayer and buffer layer are not greatly affected by the external pressure gradients. As a consequence the turbulent-energy balance should not change on varying the pressure gradient. The turbulent-energy balance in these regions, for the larger adverse pressure gradient considered ($m = -0.4$), shows a trend similar to that obtained at $m = 0$ (Figure 2a) and only a slight increase in the maxima of the nondimensional terms is observed. On the other hand, the balance in the outer region changes, as shown experimentally by Bradshaw[5] and East and Sawyer.[8] Figures 6a–c show the behavior of the production, advection, and turbulent-diffusion terms in the outer region. The rate of dissipation has not been reported, but can be easily obtained from the other terms, since the viscous-diffusion and pressure-work terms are negligible in this region. Figure 6a indicates that a stronger adverse pressure gradient increases and modifies the production term, giving rise to a second maximum. The same qualitative trend has been measured elsewhere.[8]

The turbulent-diffusion term plays a very important role in the turbulent-energy balance, because by this process turbulence is transferred from one streamline to another; in particular, turbulent diffusion controls the growth by the entrainment of boundary layers. Figure 6b shows that, on increasing the adverse pressure gradient, greater diffusion occurs at the

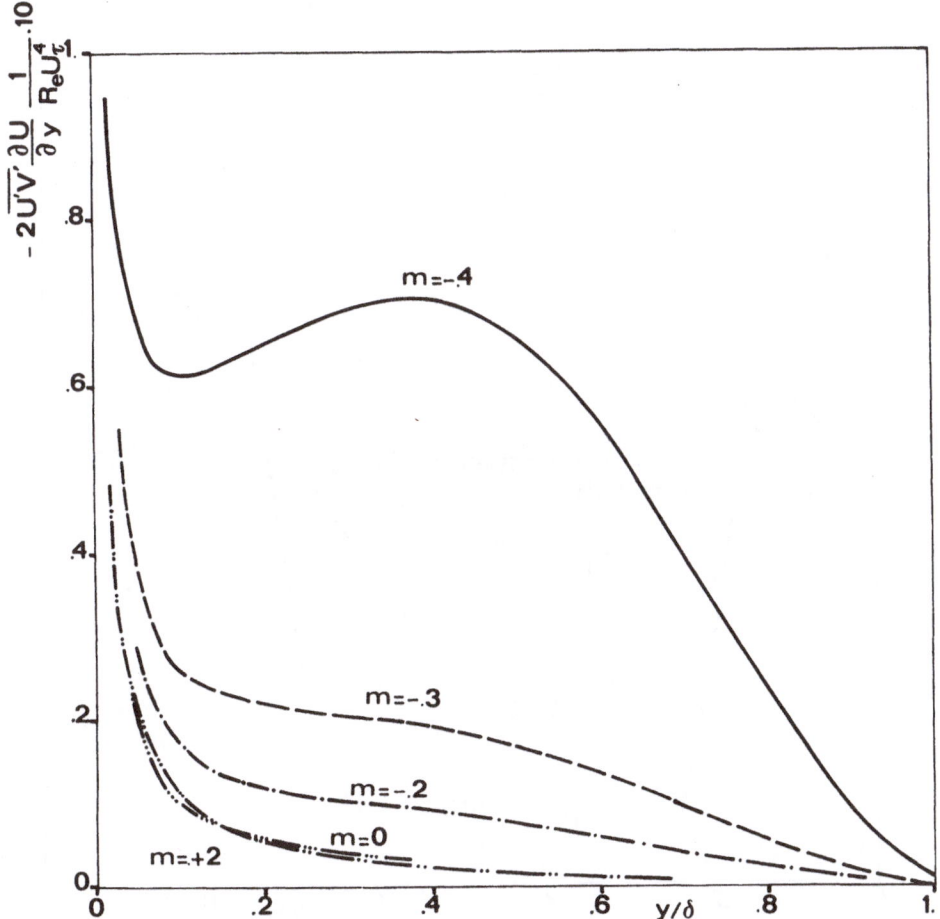

Fig. 6a.

outer edge, while in the central part of the layer turbulent diffusion increases much more than the production term. The inversion point moves outward as exponent m is reduced; the position of this point is in good agreement with measurements[8] up to $m = -0.2$. Numerical results show the occurrence of two points where diffusion is independent of the pressure gradient. These points are similar to the fixed points seen in the turbulent-energy and Reynolds-stress profiles. In the outer-edge region the diffusion is balanced by the advection term, as shown in Figure 6c. Also, the advection-term profile presents a point where advection does not change,

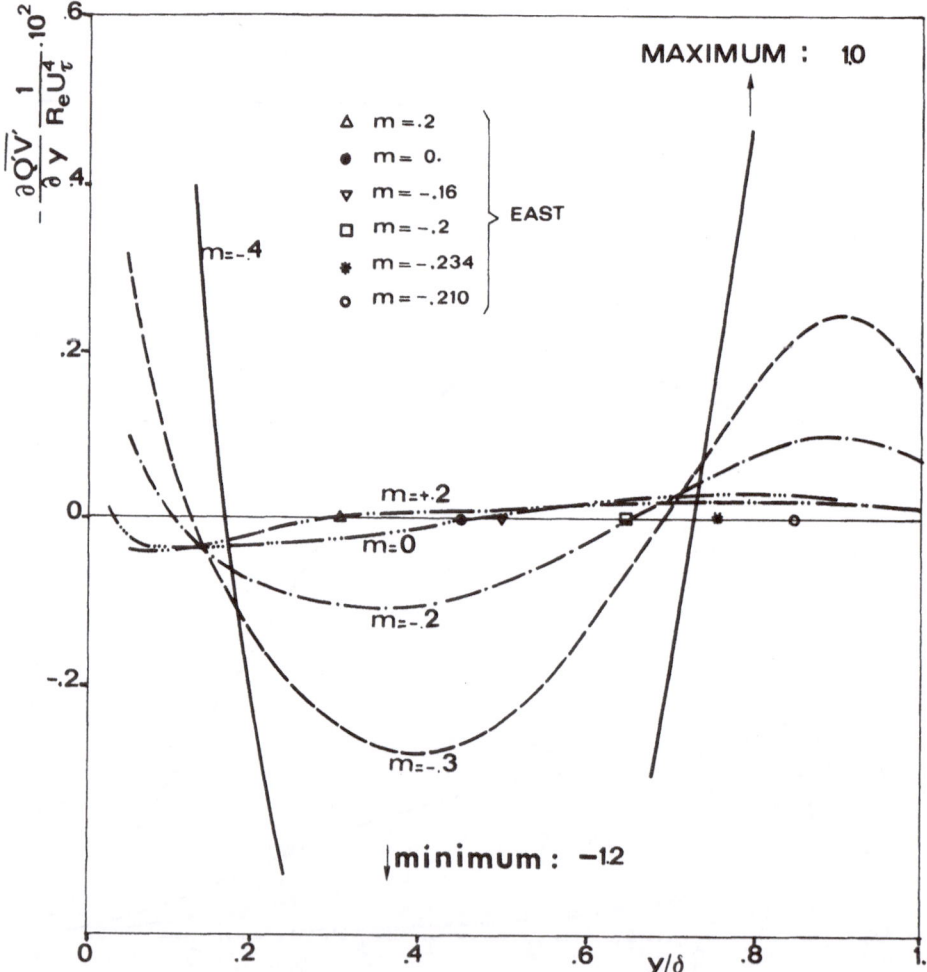

Fig. 6b.

so increasing the adverse pressure gradient. Computation has shown that, in the advection term, the contribution of $u\partial Q/\partial x$ is much larger than that of $V\partial Q/\partial y$.

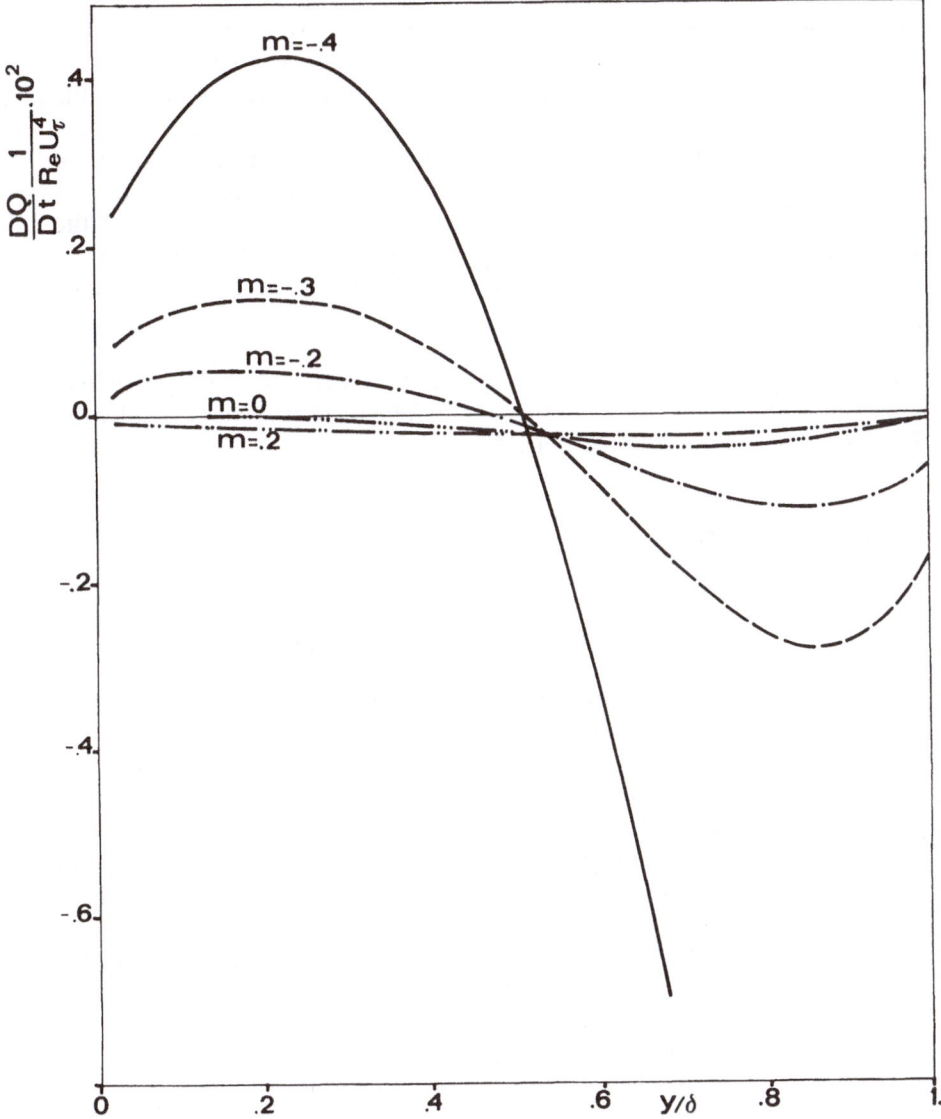

Fig. 6c.

Fig. 6. Turbulent-energy-balance profiles in the outer region for all flows. (a) Production. (b) Turbulent diffusion. (c) Advection.

5. CONCLUSIONS

The one-equation turbulence model presented here is able to predict accurately the turbulence structure throughout the boundary layer. Emphasis has been laid on the viscous and buffer regions, which are very difficult to analyze experimentally. The pressure-work term has been modeled explicitly. This term makes its greatest contribution in the near-wall region and its introduction yields a realistic representation of the turbulent-energy transfer process in this region. This term was not modeled explicitly in the outer region, where it contributes less to the turbulence structure. This one-equation turbulence model can predict only the behavior of nonseparated boundary layers. Work is in progress to expand the applicability of this model, and to describe more complex boundary layers, such as boundary layers with curvature, and boundary layers with blowing or suction.

ACKNOWLEDGMENTS. I am grateful to Prof. F. Sabetta for his constructive criticism of the original draft of this paper. The work was supported by the Italian "Consiglio Nazionale delle Ricerche."

NOMENCLATURE

A	$(\partial x/\partial x_1)^{-1}$, function of coordinate transformation
B	$(d\eta/dx_2 - 1)\delta$, function of coordinate transformation
C	$-\eta(\partial\delta/\partial t)B$
$C_2, \ldots, C_8$	constants in the turbulent model
C_f	skin friction coefficient
D	$-\eta(\partial\delta/\partial x)B$, function of coordinate transformation
$\mathcal{D}$	isotropic dissipation
E	$(d\eta/dx_2)^{-1}$
F	$B/(\mathrm{Re}\,\delta)$, function of coordinate transformation, defined in equation (14)
G	$(dx/dx_1)^{-1}B$, function of coordinate transformation
H	$-(\partial\delta/\partial x)B/\delta$, function of coordinate transformation; also shape factor $= \delta_1/\delta_2$
k	Von Karman constant
l	mixing length
L	B/δ, function of coordinate transformation
m	pressure-gradient parameter
P	pressure
Q	turbulent kinetic energy
Re	Reynolds number

t	time
U	mean velocity component in the x-direction
U_e	free-stream velocity
U_τ	friction velocity
U^+	U/U_τ
V	mean velocity component perpendicular to the wall
x	streamwise coordinate
y	coordinate perpendicular to the wall
y^+	yU_τ/ν, wall coordinate

Greek Symbols

α	stretching parameter
β	equilibrium parameter
γ	intermittency factor
Γ	parameter defined in equation (13)
δ	boundary layer thickness
δ_1	displacement thickness
δ_2	momentum thickness
Δ	Clauser thickness
η	stretched coordinate, defined in equation (7)
ν	kinematic viscosity
ν_D	turbulent viscosity
τ_w	wall friction

REFERENCES

1. J. Rotta, On the Theory of the Turbulent Boundary Layer, NACA TM 1344 (1950).
2. F. H. Clauser, Turbulent boundary layers in adverse pressure gradients, *J. Aero. Sci.* **21**, 91–108 (1954).
3. A. A. Townsend, Equilibrium layers and wall turbulence, *J. Fluid Mech.* **11**, 97–120 (1961).
4. G. L. Mellor, The effects of pressure gradients on turbulent flow near a smooth wall, *J. Fluid Mech.* **24**, 255–274 (1966).
5. P. Bradshaw, The turbulent structure of equilibrium layers, *J. Fluid Mech.* **29**, 625–645 (1967).
6. M. R. Head, Equilibrium and near-equilibrium turbulent boundary layers, *J. Fluid Mech.* **73**, 1–8 (1976).
7. P. Bradshaw, Inactive motion and pressure fluctuations in turbulent boundary layers, *J. Fluid Mech.* **30**, 241–258 (1967).
8. L. F. East and W. G. Sawyer, An Investigation of the Structure of Equilibrium Turbulent Boundary Layers, AGARD CP 271 (1979).
9. P. S. Klebanoff, Characteristics of Turbulence in a Boundary Layer with Zero Pressure Gradient, NACA Report 1247 (1955).
10. J. Laufer, The Structure of Turbulence in Fully-Developed Pipe Flow, NACA TN 2954 (1953).

11. P. Moin, W. C. Reynolds, and J. H. Ferziger, Large Eddy Simulation of Incompressible Channel Flow, Report No. TF-12, Mechanical Engineering Department, Stanford University (1978).
12. T. H. Kim, S. J. Kline, and W. C. Reynolds, The production of turbulence near a smooth wall in a turbulent boundary layer, *J. Fluid Mech.* **50**, 133–160 (1971).
13. R. F. Blackwelder and R. E. Kaplan, On the wall structure of the turbulent boundary layer, *J. Fluid Mech.* **76**, 89–112 (1976).
14. H. L. Norris and W. C. Reynolds, Turbulent Channel Flow with a Moving Wavy Boundary, Report No. TF-7, Mechanical Engineering Department, Stanford University (1975).
15. H. Ueda and J. O. Hinze, Fine-structure turbulence in the wall region of a turbulent boundary layer, *J. Fluid Mech.* **67**, 125–143 (1975).
16. G. B. Shubauer, Turbulent processes as observed in boundary layer and pipe, *J. Appl. Phys.* **25**, 188–196 (1954).
17. G. M. Lilley, A review of pressure fluctuations in turbulent boundary layers at subsonic and supersonic speeds, *Arch. Mech. Stos.* **16**, 301–330 (1964).
18. B. E. Launder, G. J. Reece, and W. Rodi, Progress in the development of a Reynolds-stress turbulence closure, *J. Fluid Mech.* **68**, 537–566 (1975).
19. P. Orlandi and W. C. Reynolds, A provisional model for unsteady turbulent boundary layers (in press).
20. K. Wieghardt, in: *Proceedings of the Conference on Computation of Turbulent Boundary Layers* (D. E. Coles and E. A. Hirst, eds.), Vol. 2, pp. 98–124, AFOSR-IFP-Stanford Conference (1968).
21. P. Orlandi and J. H. Ferziger, Implicit non-iterative schemes for unsteady boundary layers (in press).
22. W. P. Jones and B. E. Launder, The calculation of low-Reynolds-number phenomena with a two-equation model of turbulence, *Int. J. Heat Mass Transfer* **16**, 1119–1130 (1973).
23. G. L. Mellor and H. J. Herring, A survey of mean turbulent field closure models, *AIAA J.* **5**, 590–599 (1973).
24. B. A. Kader and A. M. Yaglom, Similarity treatment of moving-equilibrium turbulent boundary layers in adverse pressure gradients, *J. Fluid Mech.* **89**, 305–342 (1978).

9

The Relationship between Temperature and Velocity in Turbulent Boundary Layers with Injection

L. C. SQUIRE

ABSTRACT. The relationship between temperature and velocity in turbulent boundary layers has been of considerable interest ever since the publication of the pioneering papers by Crocco on the corresponding results in laminar flow. As a result a number of authors have derived a simple algebraic form for the relation, the Crocco relation. This simple form has been found to be in close agreement with experimental results over a wide range of conditions although there are a large number of inexplicable discrepancies. This paper reviews the various derivations of the Crocco relation to see if the approximations used break down under certain circumstances. In fact no obvious reasons for the discrepancies have been found. However, it has been possible to extend some of the derivations to boundary-layer flows over porous surfaces and the results are found to be in good agreement with experiment, even in highly nonequilibrium layers, for example, in the boundary layer downstream of a porous surface.

1. INTRODUCTION

The relationship between velocity and temperature in compressible turbulent boundary layers has been of considerable interest for at least the past thirty years. Recently there has been a renewed interest in the relationship due partly to the development of computational methods for viscous flow, since the assumption of such a relationship removes the need

L. C. SQUIRE • Cambridge University Engineering Department, Trumpington Street, Cambridge CB2 1PZ, England.

to solve the energy equation. However, the main reason for the interest
arises from the relative ease of obtaining density profiles in boundary
layers by use of holographic interferometry, since if the static pressure is
constant across the boundary layer then velocity profiles can be obtained
directly from the density profiles by use of the velocity–temperature
relation. All algebraic forms of this relation follow directly from the early
pioneering papers by Crocco[1,2] on compressible laminar boundary layers.
Essentially Crocco showed that in gases with a Prandtl number of one, the
variation of temperature across a boundary layer in zero pressure gradient
was such that d^2T/du^2 was constant across the whole layer. Thus the actual
temperature profile could be expressed as a quadratic in u with the
coefficients determined from the wall and free-stream temperatures.
Exactly the same result can be obtained for a turbulent boundary layer if
both the laminar and turbulent Prandtl numbers have values of unity.
Crocco[2] also considered the case of laminar flows with Prandtl numbers
not equal to one and, as will be described below, a number of authors have
extended this work to turbulent layers. As a result various authors have
suggested the following simple relation between velocity and temperature
in a compressible turbulent boundary layer:

$$\frac{T}{T_1} = \frac{T_w}{T_1} + \frac{(T_r - T_w)}{T_1}\left(\frac{u}{u_1}\right) + \left(1 - \frac{T_r}{T_1}\right)\left(\frac{u}{u_1}\right)^2 \tag{1}$$

where T is the static temperature, T_w the wall temperature, T_r the recovery
(or adiabatic) temperature, u the velocity, and suffix 1 corresponds to
conditions in the free stream. Many workers who use this formula refer to
it as the Crocco relation, and we shall follow this notation throughout the
paper.

A number of authors have attempted to compare the predictions of
equation (1) with experimental results, but most of the early attempts were
hampered by the lack of suitable data. Recently Fernholz and Finley[3–5]
published a large compilation of compressible boundary-layer data and
have used these data to make an exhaustive study of the applicability of
equation (1) for flows over solid surfaces. Fernholz and Finley find
reasonable agreement between experiment and equation (1) for a wide
range of conditions, but there are a large number of cases in which there
are apparent inexplicable discrepancies. Clearly some of these discrepancies must be due to experimental errors, but it is also possible that they
arise from the approximation used in the derivation of the Crocco relation.
For example, it is well known that equation (1) is definitely in error near
the outer edge of a boundary layer over an adiabatic surface, since it
cannot produce the peak (or overshoot) in total temperature needed to
compensate for the low total temperature near the wall as required by the

condition of constant enthalpy flux along the layer. Equation (1) also gives a Reynolds analogy factor $2S_t/c_f$ of $1/Pr$ ($= 1.4$ for a laminar Prandtl number Pr of 0.71) while most of the experimental data suggest a value in the range $1 < 2S_t/c_f < 1.2$, thus some discrepancies should be expected in conditions with high heat-transfer rates. However, it is easy to see that these two effects are not sufficient to explain some of the gross discrepancies noted by Fernholz and Finley. Thus it appears profitable to look at the various derivations of the Crocco relation to see if any of the approximations used break down in certain regions and hence explain the discrepancies. The result of such a survey is carried out in Section 2 of this paper, while in Section 3 the work is extended to flows over porous surfaces with injection and comparisons made with experimental results, including flows in nonequilibrium conditions.

2. DERIVATION OF THE CROCCO RELATION

2.1. General Derivation

Most derivations of this relation start from the energy equation written in Crocco coordinates (i.e., with x and u as independent variables) so that the equation assumes the form

$$\frac{\partial u}{\partial y}\left[\frac{\partial h}{\partial u}\cdot\frac{\partial \tau}{\partial u} - \tau - \frac{\partial q}{\partial u}\right] = -\rho u \frac{\partial h}{\partial x} + \left(\frac{\partial h}{\partial u} + u\right)\frac{dp}{dx} \tag{2}$$

where h is static enthalpy, τ is shear stress ($= \mu \partial u/\partial y - \overline{\rho v'u'} \equiv \tau_L + \tau_t$), and q is heat flux toward the wall [$= (k/c_p)\partial h/\partial y - \overline{\rho v'h'} \equiv q_L + q_t$].

It is then assumed that the longitudinal pressure gradient is zero and that h is independent of x, so that equation (2) reduces to

$$\frac{dh}{du}\cdot\frac{\partial \tau}{\partial u} - \tau - \frac{\partial q}{\partial u} = 0 \tag{3}$$

The laminar and turbulent heat flux are then related to the corresponding shear stresses by the introduction of the laminar and turbulent Prandtl numbers, that is,

$$q_L = \frac{\tau_L}{Pr}\frac{dh}{du} \quad \text{and} \quad q_t = \frac{\tau_t}{Prt}\frac{dh}{du} \tag{4}$$

so that equation (3) becomes

$$\frac{dh}{du}\left[\frac{\partial \tau_L}{\partial u} + \frac{\partial \tau_t}{\partial u}\right] - \tau_L - \tau_t - \frac{\partial}{\partial u}\left[\frac{\tau_L}{Pr}\frac{dh}{du} + \frac{\tau_t}{Prt}\frac{dh}{du}\right] = 0 \tag{5}$$

Thus if the shear stresses are known h can be found in terms of u, and so for a perfect gas the relationship between T and u can be obtained.

Clearly if $\mathrm{Pr} = \mathrm{Prt} = 1$, equation (5) reduces to

$$\frac{d^2h}{du^2} = -1 \tag{6}$$

With boundary conditions $h = h_w$ at $u = 0$, and $h \to h_1$ as $u \to u_1$, equation (6) has the solution

$$h = h_w + (h_{01} - h_w)\frac{u}{u_1} - (h_{01} - h_1)\left(\frac{u}{u_1}\right)^2 \tag{7}$$

where h_{01} is the stagnation enthalpy in the free stream, h_1 the free-stream static enthalpy, and u_1 the free-stream velocity. This solution gives zero heat transfer at the wall when the wall temperature is equal to the stagnation temperature in the free stream, whereas in general, the adiabatic wall temperature is less than the stagnation temperature for gases with a Prandtl number below one. Thus it has been argued that for Prandtl numbers near unity equation (7) can be modified by replacing h_{01} by h_r, the recovery enthalpy, in which case equation (7) reduces to the Crocco relation [equation (1)] for a perfect gas. Of course such a simple approach does not provide any means of finding the recovery temperature.

In order to derive expressions for the recovery factor and for the heat transfer at the wall Van Driest[6] and Walz[7] both introduced the concept of a mixed Prandtl number defined by

$$q = \frac{\tau}{\mathrm{Prm}}\frac{dh}{du}$$

In this case equation (5) reduces to the laminar form and, following Crocco, can be integrated by use of the integrating factor.

$$\exp\left[-\int_{\tau_w}^{\tau}\frac{(1-\mathrm{Prm})}{\tau}\,du\right]$$

so that in the case of a constant mixed Prandtl number the solution is given in the form of integrals of the form

$$\int_0^u \left(\frac{\tau}{\tau_w}\right)^{(\mathrm{Prm}-1)} du' \tag{8}$$

and

$$\int_0^u \left(\frac{\tau}{\tau_w}\right)^{(\mathrm{Prm}-1)}\left[\int_0^{u''}\left(\frac{\tau}{\tau_w}\right)^{(1-\mathrm{Prm})} du'\right] du'' \tag{9}$$

By making use of suitable assumptions for the shear-stress distribution through the layer, both Van Driest and Walz were able to find expressions for the recovery factor and the heat transfer at the wall. However, Van Driest did not attempt to derive an explicit relationship between temperature and velocity. On the other hand, Walz suggested that since Prm is close to one the integrals given in equations (8) and (9) are approximately proportional to u and u^2, respectively, in which case the relationship between temperature and velocity again reduces to the Crocco relation [equation (1)], but with the addition of a means of finding the recovery temperature.

In a slightly different approach Spence[8] rewrote equation (5) as

$$\frac{d}{du}\left\{\left[\frac{\tau}{\text{Prt}} + \left(\frac{1}{\text{Pr}} - \frac{1}{\text{Prt}}\right)\tau_L\right]\frac{dh}{du}\right\} - \frac{\partial\tau}{\partial u}\frac{dh}{du} + \tau = 0 \qquad (10)$$

where τ is now the total shear stress. He was then able to integrate this equation explicitly for constant Pr and Prt by use of the integrating factor mentioned above together with the following assumptions:

1. In the inner part of the layer where the laminar shear stress is important, the total shear stress is constant and is equal to the wall value.

2.
$$\tau = \tau_w[1 - (u/u_1)^{2+n}] \qquad (11)$$

   where $7 < n < 11$.

3. The laminar shear stress can be found from a correlation of mean-flow velocity profiles based on the Howarth transformation[9] between compressible and incompressible flows.

Using this approach Spence was able to obtain expressions for the recovery factor and the Reynolds analogy factor. His results can be combined to produce an explicit form for the enthalpy–velocity relation as

$$(h - h_w) = (h_r - h_w)A(u) + (h_1 - h_r)B(u) \qquad (12)$$

where $A(u)$ and $B(u)$ are given in Appendix 1. However, Spence did not produce this result explicitly, but instead suggested that in relating enthalpy and velocity it was permissible to ignore the contribution from the laminar shear stress and to replace the integrals based on total shear stress by u and u^2 so that equation (12) again reduces to the Crocco relation. In the course of the present work it has been found that the use of the exact expressions for $A(u)$ and $B(u)$ (as given in Appendix 1) do reproduce some of the measured results, in particular total temperature profiles based on equation (12) do show an overshoot near the outer edge of the layer.

Results given by Spence for the recovery factor and the Reynolds analogy factor are dependent on the frictional velocity, while in the literature it is usually assumed that the recovery factor for turbulent layers is constant at a value of about 0.89. It was found that this dependence on Reynolds number was slightly reduced if the laminar shear-stress distribution in the buffer and logarithmic layer is represented by more realistic forms based on the current knowledge of turbulent boundary layers. Even without this modification the calculated values of the recovery factor do, in fact, follow the general trend of the experimental results in that they fall toward the laminar value of $(Pr)^{1/2}$ at low Reynolds numbers.

In the most recent work on this topic Whitfield and High,[10] and Laderman[11] have derived algebraic expressions for the enthalpy–velocity relation on solid surfaces based on a constant mixed Prandtl number by expanding the solution in terms of $(1 - Prm)$ and then ignoring second-order terms in this small parameter. For adiabatic layers on solid surfaces the results again reproduce the overshoot in total temperature near the outer edge of the layer. Laderman was essentially interested in obtaining a temperature–velocity relationship for use in the analysis of experimental data, so his analysis is not written in the most general form and in particular he does not derive expressions for the recovery factor and Reynolds analogy factor. In fact such a derivation is relatively straightforward and is given in Appendix 2 for any shear-stress distribution in a boundary layer in zero pressure gradient conditions. In particular the derivation holds for layers with suction or injection at the surface, although in this case the various integrals have to be evaluated numerically, while Laderman used an exponential form for the shear-stress distribution so that the integrals could be evaluated in closed form. It is also shown that in all cases the enthalpy–velocity relation can be recast in the form given by equation (12) and that as Prm tends to one the original Crocco relation is recovered. Results from Laderman's method are compared with results obtained from Spence and from the Crocco relation in Section 2.2, while in Section 3 his method, as extended in Appendix 2, is used to produce results for layers with injection that are then compared with experiment.

2.2. Comparison of Various Forms of the Temperature–Velocity Relationship

Calculations of the temperature–velocity relationship as given by equation (12) were made for a range of Mach numbers, Reynolds numbers, and heat-transfer rates, using the formulas for $A(u)$ and $B(u)$ given in Appendixes 1 and 2. In making these calculations it was decided to use the total shear-stress distributions recommended by the original authors; these shear-stress distributions are compared in Figure 1. As will be seen from

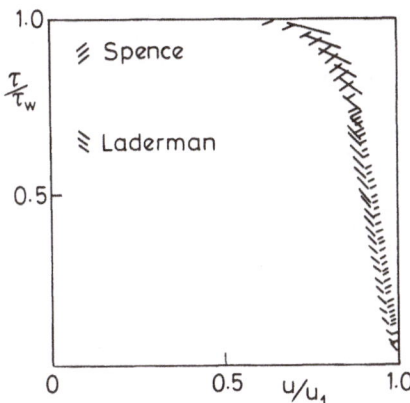

Fig. 1. Comparison of total shear-stress distributions.

this figure the shear stress is taken as equal to the wall value for $0 < u < 0.6\ u_1$, but it then falls rapidly to zero in the outer part of the layer. In fact it was found that the shape of the curve of temperature against velocity was virtually independent of the shape of the shear-stress profile provided the value of the (constant) Prandtl number was allowed to vary from case to case in order to achieve a recovery factor of 0.89 at high Reynolds numbers. As a caveat to this statement it should be pointed out that if the results are plotted in terms of the parameter $(T_0 - T_w)/(T_{01} - T_w)$ (where T_0 is the local stagnation temperature and T_{01} the stagnation temperature in the free stream) then the effects of the different forms of the shear-stress profiles do show up, particularly in zero-heat-transfer conditions. However, as pointed out by Fernholz and Finley and by Laderman, in zero-heat-transfer conditions $(T_{01} - T_w)$ is small, so apparent large differences in the parameter $(T_0 - T_w)/(T_{01} - T_w)$ correspond to relatively small differences in the actual temperature.

Typical results for three heat-transfer rates at $M = 8.0$ are shown in Figure 2 for two Reynolds numbers. Figure 2a shows results obtained by Spence's method using $\tau/\tau_w = 1 - (u/u_1)^{12}$ with Prt = 0.88, while Figure 2b shows the corresponding results obtained by Laderman's method with $\tau/\tau_w = \exp[-4(u/u_1)^{20}]$ with Prt = 0.86; both figures include results calculated from equation (1) with T_r based on a recovery factor of 0.89. The three cases of heat transfer correspond to wall temperatures equal to T_r, $0.6\,T_r$, and $0.2\,T_r$ where, again, T_r is based on a recovery factor of 0.89. As will be seen both sets of calculations give temperatures above those calculated from equation (1) near the outer edge of the boundary layer for zero-heat-transfer conditions. This corresponds to the overshoot in total temperature required to produce zero enthalpy flux along the layer. It will also be noted that near zero-heat-transfer conditions there are some effects

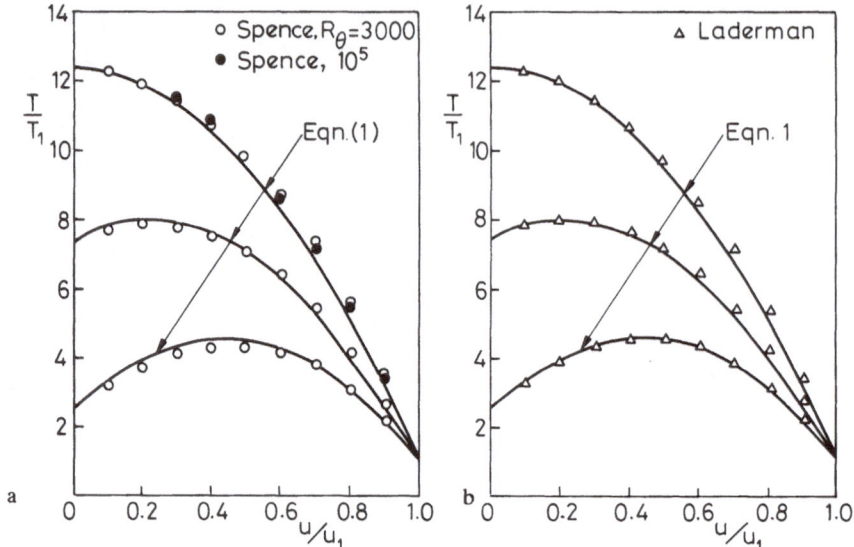

Fig. 2. Temperature–velocity relationships.

of Reynolds number on the temperatures as calculated by Spence's method. However, these effects are relatively small and they were found to be negligible at the higher heat-transfer rates and so only results for the lower Reynolds number are plotted in these cases. At moderate heat-transfer rates ($T_w = 0.6T_r$) the results from both methods are virtually the same as given by equation (1) except near the outer edge of the layer. In making all these assessments it should be noted that the only way of deciding which of the various methods is the most accurate is by comparison with experimental results, and the differences between the various methods shown for $T_w = 0.6T_r$ are probably less than the experimental accuracy.

At the higher heat-transfer rate, $T_w = 0.2T_r$, the results calculated by Laderman's method coincide with equation (1), but the results calculated by Spence's method are significantly lower near the wall. In fact, this follows directly from the Reynolds analogy factor given by the various methods, since equation (1) gives a value of this factor of 1.41, while Laderman's method gives a value of about 1.35 and Spence's method a value near 1.2. Since most of the experimental evidence points to a value for the Reynolds analogy factor between 1 and 1.2 it would appear that the results obtained by Spence's method are the more realistic.

As noted in the introduction a very detailed comparison of results based on equation (1) with experimental results has been made by Fernholz and Finley. In many cases they found very close agreement between the experimental results and equation (1) with the discrepancies being within the experimental error bands. A typical result of such a comparison in shown here by the zero pressure gradient, zero-blowing result plotted in Figures 6 and 7, but for a fuller comparison the reader is referred to Figures 2.5.1 to 2.5.21 in the paper by Fernholz and Finley.[4] From these figures it would appear that most of the experimental results tend to lie slightly below equation (1) while both sets of results shown in Figure 2 suggest that equation (1) is low in the outer edge of the layer. In some cases the measured results are significantly lower than equation (1); see Figure 2.5.13 and Figures 5.1.7 and 5.1.8 of the aforementioned paper,[4] for example. In these cases the temperatures measured within the boundary layer, that is near $u = 0.5u_1$, may be up to $2T_1$ below those calculated from equation (1). As pointed out by Fernholz and Finley the worst discrepancies occur at relatively low Reynolds numbers with high heat-transfer rate and with slight pressure gradients. There is certainly nothing in the present analysis to suggest the reasons for the discrepancies, although the calculations based on Spence's method should be able to deal with both low Reynolds-number effects and high heat transfer. The possibility that the discrepancies may be due to upstream influences will be considered later when the effects of sudden changes in blowing rate through a porous surface are considered.

3. EXTENSION TO LAYERS WITH INJECTION

3.1. Derivation of Temperature–Velocity Relation

The results presented in Figure 2 show that, except at high heat-transfer rates, the temperature–velocity relationship, as given by the methods of Spence and Laderman, are similar and neither can be judged better than the other by reference to the experimental data. Thus it was decided to use the generalization of Laderman's method as given in Appendix 2 to study the effects of injection. The shear-stress profiles required in the various integrals were obtained from the experimental work of the present author,[12,13] and the actual profiles, together with the functions $(1/\tau)d\tau/du$ that occur in the integrals, are plotted in Figure 3. These profiles were obtained from mean-flow profiles measured at various blowing rates ($F \equiv \rho_w v_w / \rho_1 u_1$) for a Mach number of 2.5 at relatively high Reynolds numbers so that the Reynolds number based on momentum thickness was always greater than 10,000. As will be seen the function $(1/\tau)d\tau/du$ is virtually independent of F for $u > 0.5u_1$, but is strongly

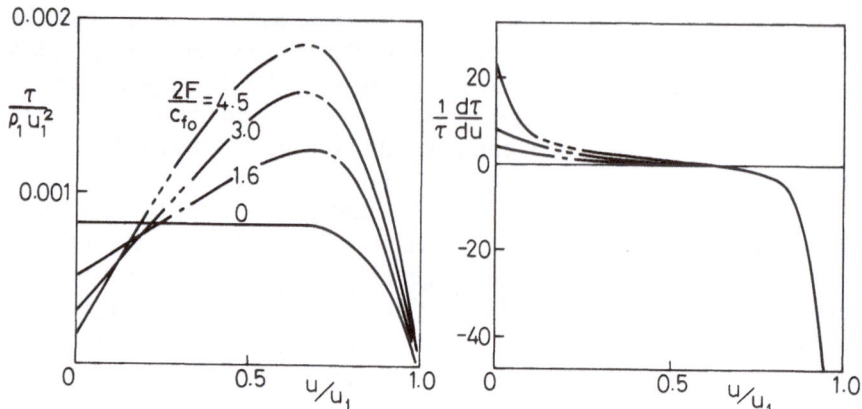

Fig. 3. Shear-stress profiles and functions $(1/\tau)d\tau/du$.

dependent on F near the wall. At the wall the momentum equation reduces to

$$\rho_w v_w \frac{\partial u}{\partial y} = \frac{\partial \tau}{\partial y}$$

so $\tau = \tau_w + \rho_w v_w u$; hence

$$\frac{1}{\tau}\frac{d\tau}{d(u/u_1)} = \frac{\rho_w v_w u_1}{\tau_w + \rho_w v_w u} = \frac{F}{c_f/2 + F(u/u_1)} \tag{13}$$

At the wall

$$\frac{1}{\tau}\frac{d\tau}{d(u/u_1)} = \frac{2F}{c_f} = \frac{2F}{c_{f0}} \cdot \frac{c_{f0}}{c_f} \tag{14}$$

where c_{f0} is the skin-friction coefficient in the absence of injection. Most experimental[14] and theoretical[15] results with injection show that the ratio c_f/c_{f0} may be correlated in terms of the parameter $2F/c_{f0}$, so that near the wall the integrand depends on this parameter only. The results shown in Figure 3 were used to evaluate the various integrals occurring in Appendix 2 and the results were recast in the form of equation (12); the functions $A(u)/u$ and $B(u)/u^2$ are plotted in Figure 4. Since, as noted above, the integrands are functions of $2F/c_{f0}$, it is assumed that the functions $A(u)$ and $B(u)$ may be used for any Mach number at the corresponding value of $2F/c_{f0}$. The corresponding values of the recovery factor as given in Appendix 2 are shown in Figure 5, where they are compared with the experimental results of Leadon and Scott[16] and of Bartle and Leadon.[17]

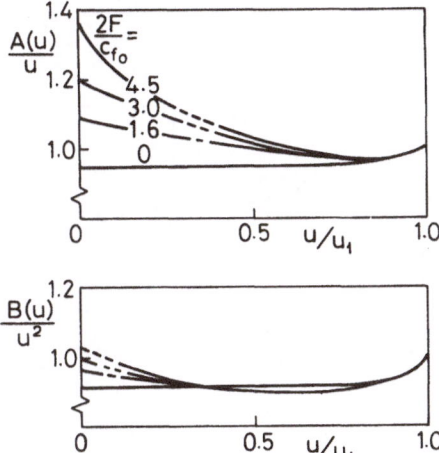

Fig. 4. Plots of functions $A(u)/u$ and $B(u)/u^2$ vs. u/u_1.

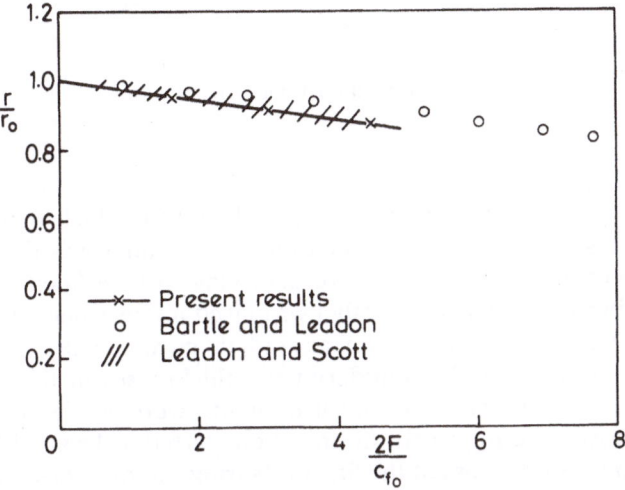

Fig. 5. Values of the recovery factor.

(In plotting the experimental results it has been necessary to estimate values of c_{fo}, and these estimates have been based on the charts published by Hopkins.[18]) As will be seen the trend of the calculated recovery factors are in excellent agreement with the experimental results.

The calculated results for the actual temperature–velocity relationship are compared with the experimental results of Thomas,[19] Marriott,[20] and Danberg[21] in Figures 6–10. Figure 6 shows results obtained by Thomas in the blowdown supersonic tunnel at Cambridge. These results were ob-

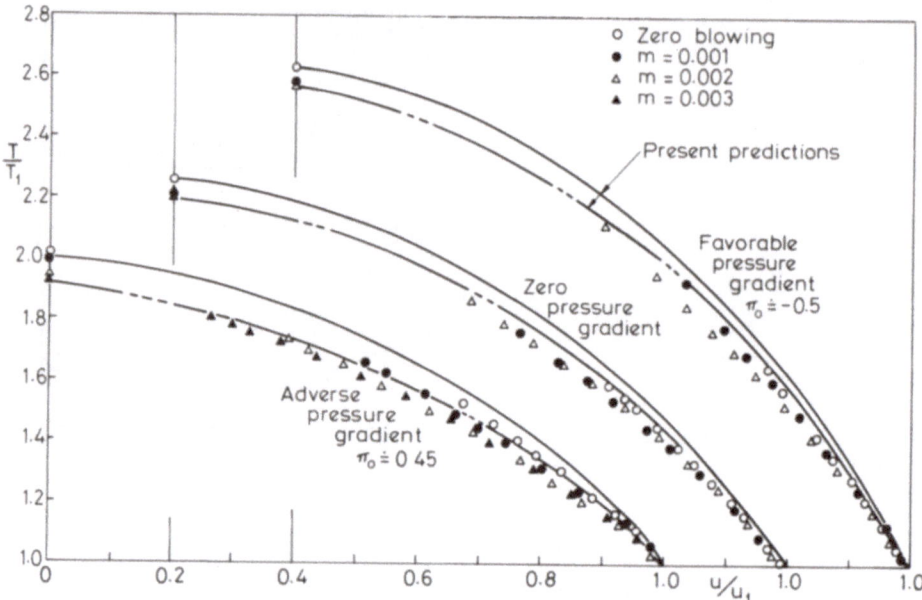

Fig. 6. Temperature–velocity relationship.

tained at high Reynolds numbers ($R_\theta > 14{,}000$) for Mach numbers near $M = 2.5$ using two specially designed liners. One liner was designated to produce a constant Mach number of 2.5 followed by a linear increase in Mach number to 2.9, while the other produced a linear decrease in Mach number to 2.2. In both cases the linear gradient covered a distance of about 130 mm, that is, about 25 boundary-layer thicknesses in the absence of injection. The boundary-layer measurements were made on a porous surface opposite the liner and the injection started at least 20 boundary-layer thicknesses upstream of the first measuring station. The results shown in Figure 6 were obtained in the zero-pressure gradient conditions upstream of the pressure-gradient region and at approximately 100 mm downstream of the start of the pressure gradient for both the favorable and the adverse gradients. In all cases results are presented for zero injection and for the two, or three, injection rates used by Thomas. Since the porous plate used in this experiment was backed by a single plenum chamber the actual injection rate F varied along the surface; hence the results presented in Figure 6 are labeled by the parameter m as used by Thomas, where m is the average mass flow per unit area divided by the mass flow per unit area in the free stream ahead of the gradient. Thomas does give estimates of the local values of F and these were used, together with the measured values of

c_{f0}, to find the appropriate value of $2F/c_{f0}$ for the calculation of the predicted temperature–velocity relationship as shown in Figure 6 for zero injection and for the highest injection rate.

In considering these experimental results it should be noted that the boundary layer in the absence of injection is about 6 mm thick while the equilibrium temperature probe used is of 1.6 mm diameter, so that the first temperature reading was taken at 0.8 mm from the wall. Thus in the absence of injection temperatures were only obtained for $u > 0.7u_1$. However, with injection the velocity profile becomes less full, and the overall boundary layer thicker, so that temperatures could be measured at lower values of u/u_1. From Figure 6 it will be seen that temperatures measured with injection are slightly lower than those in the absence of injection, although the total temperatures of the injected and free-stream air are the same. The same characteristic is shown by the predicted temperatures, but in all cases the measured temperatures are slightly below the predicted values. However, when the experimental results are compared (Figure 7) with the predictions of equation (1), with the recovery factor taken from Figure 5, the agreement is excellent and is well within the measurement accuracy of $\mp 2°$ ($\pm 0.015 T_1$).

Turning to the pressure-gradient results it will be seen that again the temperatures drop slightly with increasing injection and that the results are

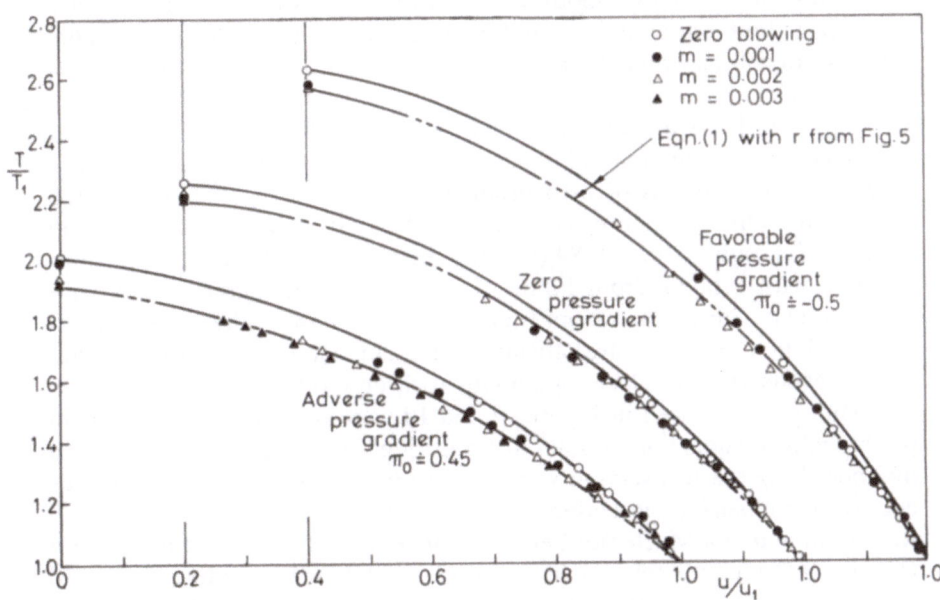

Fig. 7. Temperature–velocity relationship.

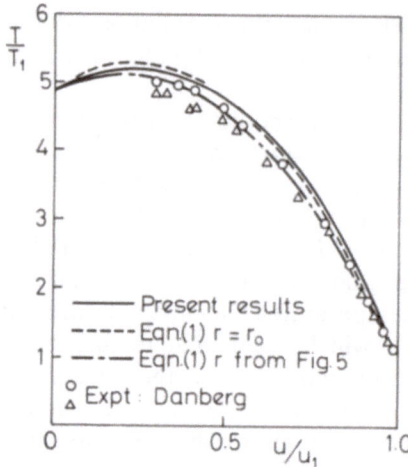

Fig. 8. Temperature–velocity relationship.

all slightly below the current predictions, while they are in close agreement with the predictions of equation (1). However, it should be noted that the present predictions are based on the assumption of the absence of pressure gradients, while in the present flows the pressure gradients are comparatively strong. Thus in the absence of injection the parameter π_0 $[=(\theta/\tau_w)dp/dx]$, where θ is the momentum thickness and τ_w the wall shear stress, has values of about ∓ 0.5. At the highest injection rates θ has more than doubled, while τ_w has fallen to at least a quarter of its value without injection so that the pressure-gradient parameter now has values nearer ∓ 4.

Figure 8 shows a set of experimental results obtained by Danberg at $M \doteq 6.35$ with injection ($F \doteq 0.0023$, $2F/c_{f0} \doteq 4$) and heat transfer ($T_w/T_r \doteq 0.6$). Two sets of measurements are shown together with three predictions. One prediction is based on the current theory and two on equation (1); one with $r = 0.89$ (i.e., the recovery factor without injection) and one with r taken from Figure 5. Again the best agreement is with equation (1), with r taken from Figure 5, but in view of the experimental scatter all the predictions are satisfactory and, in fact, Danberg showed that all his results were in close agreement with equation (1).

The results plotted in Figures 9 and 10 were obtained by Marriott in the blowdown supersonic tunnel at Cambridge. For these flows the Mach number along the test section was constant but there was a discontinuous change in the surface injection rate. Thus the results shown in Figure 9 were obtained at a Mach number of 1.8 on a porous surface with injection ($F = 0.0044$) immediately downstream of a solid surface, while Figure 10 shows the reverse case of the boundary-layer development at $M = 3.6$ on a solid surface downstream of a porous surface with injection ($F = 0.00294$).

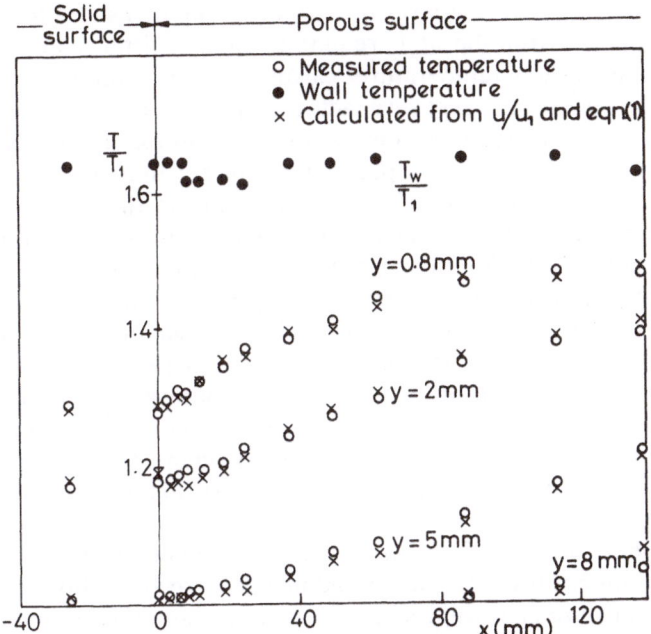

Fig. 9. Temperature-velocity relationship.

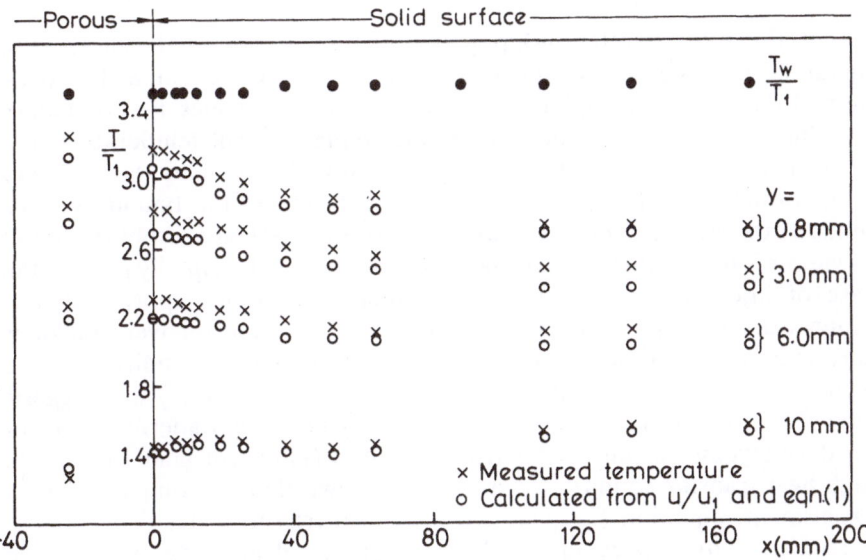

Fig. 10. Temperature–velocity relationship.

To illustrate the boundary-layer development the measured temperatures at fixed distances from the wall are plotted against distance along the wall. The figures also show temperatures calculated from equation (1), with the appropriate value of the recovery factor, and the measured values of u/u_1. The measured and predicted temperatures are in excellent agreement at $M = 1.8$ (Figure 9) and it is particularly interesting to note the close agreement just downstream of the discontinuity in the surface condition where the boundary layer is clearly in a nonequilibrium condition. (This nonequilibrium state is enhanced by a pressure disturbance from the surface discontinuity, which is shown in Figure 9 by an increase in T_1, that is, a fall in T_w/T_1, at $x \doteq 10$ mm.) This close agreement with equation (1) was also found by Marriott in his original paper, although he only compared results well away from the discontinuity. On the other hand Marriott found that the agreement was less good at $M = 3.6$ and this is confirmed by the results shown in Figure 10, where the measured results are all less than the predictions. More recent tests in the tunnel suggest that the discrepancy may be associated with three-dimensional effects at this Mach number. In spite of this discrepancy the shape of the measured and predicted temperatures are similar and in particular both show a significant fall in temperature near the wall at $x \doteq 15$ mm where the boundary layer is again in a nonequilibrium state.

4. CONCLUDING REMARKS

If the Crocco relationship [equation (1)] is regarded simply as a quadratic interpolation formula, then its overall agreement with experimental data is not completely surprising since it satisfies two boundary conditions at the wall (fluid temperature equal to wall temperature and zero heat transfer when the wall temperature is equal to the recovery temperature) and it ensures that the temperature in the boundary layer tends to the free-stream temperature smoothly. Looked at from this point of view it is to be expected that the agreement would be equally good in the case of injection, since the same boundary conditions apply and it is noteworthy that the agreement is improved when the recovery factor is corrected for injection because in this case the boundary conditions at the wall are more closely satisfied. This approach to equation (1) also suggests why the more sophisticated methods of Spence and Laderman do not produce significant differences from equation (1) except near the wall in high heat-transfer conditions, where it is known that, in comparison with experiment, equation (1) gives too high heat-transfer rates. This overall agreement with experiment might be improved if an additional term in $(u/u_1)^3$ was added to equation (1), or if the coefficient of (u/u_1) was adjusted, so as to bring the heat-transfer rate at the wall into closer agreement with experiment as in fact was done by Crocco[22] in order to account for pressure-gradient effects in his work on transformations of compressible turbulent boundary layers with heat transfer. With a Reynolds analogy factor of 1.2 such modifications produce a temperature–velocity relationship similar to that calculated by Spence's method. Clearly, if a Reynolds analogy factor of unity is used the temperatures near the wall are much lower. However, before continuing along this path it is necessary to obtain more reliable data on heat transfer and skin friction at high heat-transfer rates.

The most interesting and significant aspect of the current review is the close agreement between temperatures predicted by the Crocco relationship, and those measured by Thomas with strong pressure gradients and those measured by Marriott downstream of a discontinuous change in surface boundary condition, since the theories developed in Section 2 specifically imply that dp/dx and $\partial h/\partial x$ are zero. Of course equation (1) still satisfies the correct local boundary conditions so it would not be expected to be greatly in error for small pressure gradients or small changes in external or wall conditions. However, the pressure gradients in the flows studied by Thomas are not small and they increase with injection (since the same absolute pressure change occurs in less boundary-layer thicknesses) while there is no significant change in the agreement between equation (1) and experiment with increasing injection. The good agreement in these cases is at variance with some of the work of Fernholz and Finley, who found that the largest discrepancies between the Crocco relationship and experiment occurred in flows with pressure gradients, although these pressure gradients were usually combined with high heat-transfer rates at relatively low Reynolds numbers, while Thomas's results were obtained at high Reynolds numbers in near-adiabatic conditions. Thus the results presented here should not be taken to indicate that the Crocco relationship holds for all flows with pressure gradients, but it would certainly appear to apply for many flows of aeronautical interest.

The good agreement with Marriott's results essentially implies that the term $\partial h/\partial x$ is small so the conditions of the theory are satisfied. Since h at constant distance from the surface changes rapidly this implies a close link between the diffusion of heat and vorticity within the boundary layer, and it is interesting to note that numerical solutions[15] of the boundary-layer equations together with the energy equation do show this link, although in general the numerical solutions tend to make the boundary layer respond too quickly to disturbances. Thus, for example, numerical experiments have shown that if temperature disturbances are introduced into the flow the actual temperature–velocity curve returns to that given by equation (1) within one or two boundary-layer thicknesses. Clearly this is a subject for further study but at the moment it is hampered by the lack of turbulence measurements in compressible boundary layers with heat transfer. (A survey of the available data has been published by Fernholz and Finley.[5]) Thus while Marriottt's experiments do provide some evidence that the Crocco relationship holds in certain nonequilibrium conditions, care should be taken before extending this conclusion to extreme conditions such as, for example, shock/boundary-layer interactions. Much more experimental work is needed on this topic and a joint program to study the boundary-layer development downstream of a step change in surface temperature has just started in Cambridge and Berlin.

Together the work presented in this paper shows that the simple Crocco relationship as given by equation (1) holds over a wide range of conditions, including flows over porous surfaces with air injection through the surface (this includes discontinuous changes in injection rate, but excludes foreign gas injection[23]), with and without pressure gradients. In fact, in all the flows studied here the differences between measured temperatures and those predicted by the Crocco relationship are similar in magnitude to the estimated errors in measurement. Thus it would appear that for most calculations, and particularly those in which the main interest is in the velocity field, it is unnecessary to solve the energy equation since results can be obtained directly from equation (1). Of course this approach must be used with care in flows with very rapid changes in the stream direction, but even in these cases it is doubtful if our current knowledge of turbulence is sufficient to provide solutions of the energy equation which are more reliable than those given by the straightforward use of the Crocco relationship.

ACKNOWLEDGMENTS. The author would like to thank his former research students, D. I. A. Dunbar, L. O. F. Jeromin, P. G. Marriott, and G. D. Thomas, for the use of their results and for many discussions on the prediction and measurement of temperatures in compressible turbulent boundary layers with injection.

APPENDIX 1. Velocity–Temperature Relation after Spence

Using the assumptions listed in Section 2.1 Spence obtains a solution of equation (10) which, in the present notation, can be written as

$$h - h_w = \frac{u_1}{\text{Pr}} \left(\frac{dh}{du}\right)_w \left[I(u) - (\text{Prt} - \text{Pr}) \left(\frac{u_\tau}{u_1}\right) L(g) \right]$$
$$- u_1^2 \left[K(u) - (\text{Prt} - \text{Pr}) \left(\frac{u_\tau}{u_1}\right)^2 N(g) \right] \tag{15}$$

where u_τ is the frictional velocity based on the density evaluated at

Eckert's intermediate temperature, $I(u)$ and $K(u)$ are the integrals of total shear stress given in equations (8) and (9), respectively, using the shear-stress distribution given by equation (11), while $L(g)$ and $N(g)$ are integrals of the laminar shear stress,

$$L(g) = \int_0^g \frac{\tau_L}{\tau_w}\, dg', \qquad N(g) = \int_0^g \frac{\tau_L}{\tau_w} g'\, dg' \qquad (16)$$

where $g = u/u_\tau$. By applying the boundary condition $h = h_1$ at the outer edge of the layer and by noting that $(dh/du)_w = 0$ at $h_w = h_r$, Spence was able to obtain expressions for h_r and $(dh/du)_w$ in terms of the integrals $I(u_1)$, $K(u_1)$, L_0, and N_0 where L_0 and N_0 are the limiting values $L(g)$ and $N(g)$, respectively. Using the same approach it is possible to recast equation (15) into the form given by equation (12) with

$$A(u) = \frac{I(u) - (\mathrm{Prt} - \mathrm{Pr})(u_\tau/u_1)L(g)}{I(u_1) - (\mathrm{Prt} - \mathrm{Pr})(u_\tau/u_1)L_0}$$

and

$$B(u) = \frac{K(u) - (\mathrm{Prt} - \mathrm{Pr})(u_\tau/u_1)^2 N(g)}{K(u_1) - (\mathrm{Prt} - \mathrm{Pr})(u_\tau/u_1)^2 N_0} \qquad (17)$$

To evaluate the integrals $I(u)$ and $K(u)$ Spence used the shear-stress distribution given by equation (11), while to evaluate the integrals $L(g)$ and $N(g)$ he assumed that the velocity distribution in the buffer layer was the same as in incompressible flow provided u_τ was based on the density evaluated at the intermediate temperature and the distance from the wall was replaced by the Howarth variable $\eta = \int_0^y (\rho/\rho_0)dy'$, where ρ_0 is again the density at the intermediate temperature. In the present calculations it was decided to base u_τ on the wall density and to evaluate the laminar shear stress from the correlation of mean-flow profiles based on the Van Driest transformation, since it was felt that most of the experimental data favor this transformation as against the Howarth transformation for turbulent layers. In the event it was found that for the same conditions of Mach number and Reynolds number the differences between the laminar shear stress as derived from the two transformations had only a minor effect on the shape of the temperature–velocity relationship.

APPENDIX 2. Temperature–Velocity Relationship after Laderman

In developing the work of Laderman it is convenient to work in terms of a nondimensional enthalpy H $(= h/h_1)$ and a nondimensional velocity

U $(= u/u_1)$. Using this notation, together with the assumption of a constant, mixed, Prandtl number, equation (5) can be written as

$$\frac{d^2H}{dU^2} + (1 - \text{Prm})\frac{1}{\tau}\frac{d\tau}{dU}\frac{dH}{dU} + \text{Prm } A = 0 \tag{18}$$

where $A = (\gamma - 1)M_1^2$.

Putting $\text{Prm} = 1 - \varepsilon$, and expanding $H(U)$ as

$$H(U) = H_0(U) + \varepsilon H_1(Y) + O(\varepsilon^2) \tag{19}$$

equation (18) can be integrated to give

$$H - H_w = -\frac{A}{2}U^2 + BU + \varepsilon\left[DU + \frac{A}{2}U^2 + AF_1(U) - BF_2(U)\right] \tag{20}$$

where

$$F_1(U) = \int_0^U \int_0^{U'} \frac{U''}{\tau}\frac{d\tau}{dU''}\,dU''dU'$$

$$F_2(U) = \int_0^U \int_0^{U'} \frac{1}{\tau}\frac{d\tau}{dU''}\,dU''dU'$$

$$B = 1 - H_w + A/2$$

$$D = -A\left[\tfrac{1}{2} + F_1(1)\right] + BF_2(1) \tag{21}$$

so that $H = H_w$ at $U = 0$ and $H = 1$ at $U = 1$.

Using the condition that zero heat transfer occurs at $H_w = H_r$, the recovery factor is found as $r = 1 - \varepsilon[1 + 2F_1(1)]$ and with this result in equation (20) the Reynolds analogy is found as $2S_t/c_f = [1 + \varepsilon F_2(1)]/\text{Pr}$. Finally equation (20) can be recast in the form of equation (12), where

$$A(U) = U\{1 + \varepsilon[F_2(1) - F_2(U)/U]\}$$

and

$$B(U) = U^2\{1 + 2\varepsilon[F_1(1) - F_1(U)/U^2]\} \tag{22}$$

REFERENCES

1. L. Crocco, Su di un valor massimo del coefficiente di trasmissione del calore da una lamina piana a un fluido scorrente, *L'Aerotecnica* **12**, 181–197 (1932).
2. L. Crocco, Lo strato limite laminare nei gas, Monografie Scientifiche d'Aeronautica, No. 6 (1947).

3. H. H. Fernholz and P. J. Finley, A critical compilation of compressible boundary layer data, AGARD AG-223 (1977).
4. H. H. Fernholz and P. J. Finley, A critical commentary on mean flow data for two-dimensional compressible turbulent boundary layers, AGARD AG-253 (1980).
5. H. H. Fernholz and P. J. Finley, A further compilation of compressible boundary layer data with a survey of turbulence data, AGARD AG-263 (1981).
6. E. R. Van Driest, in: *50 Jahre Grenzschichtforschung*, pp. 257–271, Friedr. Vieweg & Sohn, Braunschweig (1955).
7. A. Walz, Compressible turbulent boundary layers with heat transfer and pressure gradient in the flow direction, *J. Res. Natl. Bur. Stand., Sect. B* **63**, 53–70 (1959).
8. D. A. Spence, Velocity and enthalpy distributions in the compressible turbulent boundary layer on a flat plate, *J. Fluid Mech.* **8**, 368–387 (1960).
9. L. Howarth, Concerning the effects of compressibility on laminar boundary layers and their separation, *Proc. R. Soc. London* **194**, 16–41 (1948).
10. D. L. Whitfield and M. D. High, Velocity-temperature relations for turbulent boundary layers with non unity Prandtl numbers, *AIAA J.* **15**, 431–434 (1977).
11. A. J. Laderman, Total temperature-velocity relation in turbulent compressible boundary layers, *Int. J. Heat Mass Transfer* **24**, 1990–1992 (1981).
12. L. C. Squire, Further experimental investigations of compressible turbulent boundary layers with air injection, ARC R & M 3627 (1970).
13. L. C. Squire, Eddy viscosity distributions in compressible turbulent boundary layers with injection, *Aeronaut. Q.* **23**, 169–182 (1971).
14. L. O. F. Jeromin, The status of research in turbulent boundary layers with fluid injection, *Prog. Aeronaut. Sci.* **10**, 65–189 (1970).
15. L. C. Squire and V. K. Verma, The calculation of compressible turbulent boundary layers with fluid injection, ARC Current Paper No. 1265 (1973).
16. B. M. Leadon and C. J. Scott, Transpiration cooling experiments in a turbulent boundary layer at $M = 3$, *J. Aeronaut. Sci.* **23**, 798–799 (1956).
17. E. R. Bartle and B. M. Leadon, Experimental evaluation of heat transfer with transpiration cooling in a turbulent boundary layer at $M = 3.2$, *J. Aeronaut. Sci.* **27**, 78–80 (1960).
18. E. J. Hopkins, Charts for predicting turbulent skin friction from Van Driest method (II), NASA TN D-6945 (1972).
19. G. D. Thomas, Compressible turbulent boundary layers with combined air injection and pressure qradient, ARC R & M 3779 (1976).
20. P. G. Marriott, Compressible turbulent boundary layers with discontinuous air transpiration: An experimental and theoretical investigation, ARC R & M 3780 (1977).
21. J. E. Danberg, Characteristics of the turbulent boundary layer with heat and mass transfer at $M = 6.7$, NOLTR 64-99 (1964) [see also NOLTR 67-6 (1967)].
22. L. Crocco, Transformations of the compressible turbulent boundary layer with heat exchange, *AIAA J.* **1**, 2723–2731 (1963).
23. D. I. A. Dunbar and L. C. Squire, Correlations of concentration, temperature and velocity profiles in compressible turbulent boundary layers with foreign gas injection, *Int. J. Heat Mass Transfer* **14**, 27–40 (1971).

IV
LIQUID SPRAYS

10

Structure of High-Speed Full-Cone Sprays

F. V. BRACCO

ABSTRACT. A better understanding and characterization of the formation and propagation of high-velocity sprays from single-hole cylindrical nozzles is of importance both fundamentally and practically. The
steady and transient structure of these sprays is qualitatively similar to
that of incompressible jets but the breakup of the liquid column into
drops and the presence of drops introduce substantial quantitative
differences. Measurements of the angle of the spray and of the size of
the drops near the nozzle suggest that the breakup of the outer surface
of the liquid jet is due to aerodynamic forces that lead to the rapid and
selective growth of surface perturbations generated within the nozzle.
The state and mechanism of disruption of the inner part of the liquid
jet is less clear but sufficiently downstream only individual drops are
present. Recent LDV drop velocity measurements and detailed mul-
tidimensional computations have shown that at distances of the order
of hundreds of nozzle diameters so much ambient gas has been
entrained by the spray that the subsequent structure of the jet is
dominated by the entrained ambient gas and the fully developed
incompressible jet structure and drop-gas equilibrium are approached.

1. INTRODUCTION

Sprays exhibit a large variety of geometrical, dynamic, and thermodynamic configuratons.[1-3] The general features of these configurations
are understood but the details are not and predictions remain elusive. In
this field, as in many others, the stability of complex interfaces, nonequilib-

F. V. BRACCO • Department of Mechanical and Aerospace Engineering, Princeton
University, Princeton, NJ 08544, U.S.A.

rium thermodynamics, intricate reactions, and turbulence ultimately set
the limits of our knowledge.

This is true even for the simplest of configurations, i.e., the one of
interest in this review, in which a liquid is injected into a gas through a
single straight hole of circular cross section, the two media have negligible
angular momenta, and their thermodynamic states are such that vaporization and chemical reactions can be ignored. Only dynamic forces control
the field and the evolution of interfaces and turbulence present the greatest
difficulties.

A general idea of the structure of our sprays can be obtained by
considering the similar and better-understood family of incompressible
jets.[4-6]

Figure 1 shows the initial propagation of a turbulent incompressible
jet[7] and of a spray.[8] The propagation of the incompressible jet is marked
by the advancement and growth of a head vortex that is fed, from its

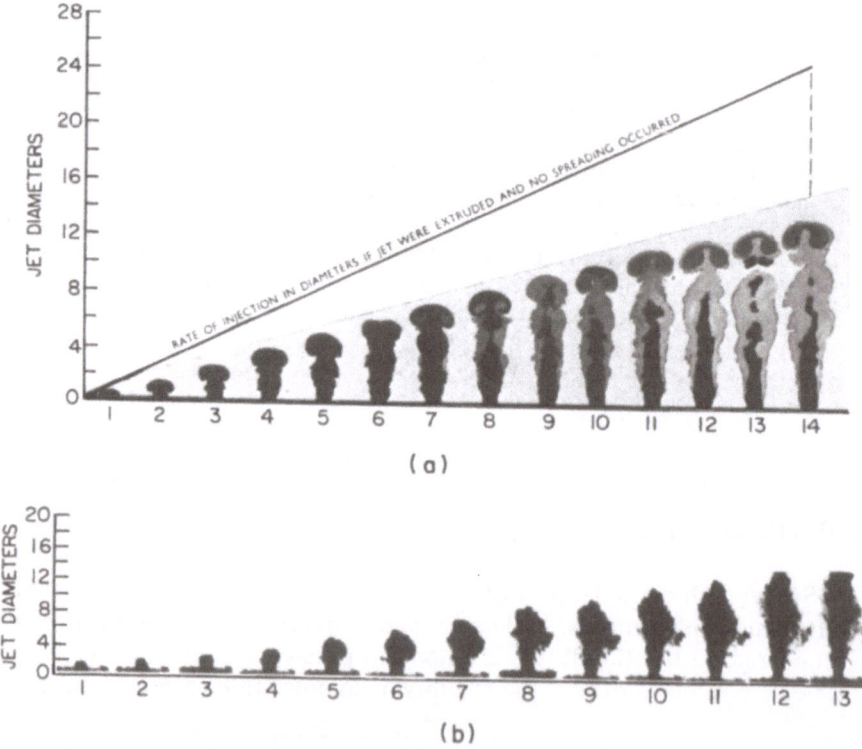

Fig. 1. Development of (a) a turbulent incompressible jet[7] (water in water, $Re_D = 4 \cdot 10^4$), and
(b) an atomizing spray[8] (50% glycerol + 50% water into N_2).

downstream side, by the injected medium and the ambient medium that was entrained in the region between the vortex head and the nozzle. Practically, this intermediate region is in its steady-state configuration, i.e., the head vortex leaves a turbulent, steady, incompressible jet behind itself. In the laboratory frame, the vortex moves at a fraction of the local steady-state centerline velocity. Within the steady part, the shear layer, the potential core, and its end can be seen.

The corresponding picture of a spray raises several questions. Its internal structure cannot be seen and is not known precisely because no technique has been found to probe it without altering it significantly. There is a head structure but its details are not obvious. It is expected that when the head structure is at sufficient distance from the nozzle, enough ambient gas has been entrained to set up a flow field similar to that of incompressible jets behind itself. But closer to the nozzle, the liquid core, liquid ligaments, and drops impose different configurations on the shear layer and on the head of the jet due to their different modes of momentum transfer. The length of the transition region is not known. Behind the head, however, the spray divergence angle, which is the only quantity that can be measured with relative ease, rapidly achieves its steady-state value.[8] This indicates that also in sprays the adjustment to steady state occurs primarily within their head region. In the steady part of the spray of Figure 1, as in the corresponding incompressible jet, we expect the mean axial velocity to point downstream and the mean radial velocity to point toward the axis always. Thus we call them full-cone sprays in contrast to hollow-cone sprays that exhibit mean recirculation flows along the axis and correspondingly larger spray angles.

In Figure 1b the spray appears to start diverging immediately at the nozzle exit. Although this is the mode of breakup of interest in this review, there are other modes and some are shown[8] in Figure 2. All other parameters being the same, at very low injection velocities a jet breaks up many diameters downstreams and forms large drops. Surface tension is the disruptive force (Figure 2a). As the velocity is increased, the displacement of the gas by the moving undulated liquid surface generates a pressure distribution that aids surface tension in amplifying surface waves (Figure 2b). This process, which is called aerodynamic interaction, eventually becomes the main destabilizing force and leads to the formation of very small drops while being opposed by surface tension (Figure 2c). All along, an intact surface is visible and the growth of unstable surface waves is detectable. These are classical regimes of breakup of circular liquid columns and major contributions to our understanding of them were made[9] by Lord Rayleigh, C. Weber, and G. I. Taylor. They continue to be the subject of current research[10] but are not the ones of interest to us here.

The regime of interest to us is obtained when the injection velocity is

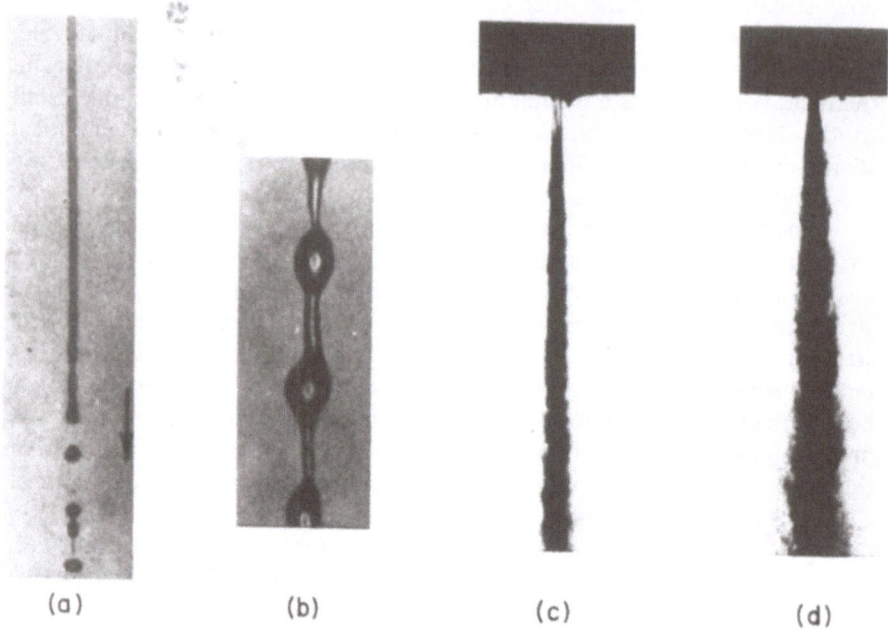

(a)   (b)   (c)   (d)

Fig. 2. Some regimes of breakup of liquid jets.[8]

so high that the intact length of the outer jet surface seems to disappear and the configuration of Figures 2d and 1b is obtained. This regime is called the atomization regime[11] and the forces that control it are more complex and less known even though it has been the subject of extensive research because of its considerable practical importance.

Thus we will concentrate on nonvaporizing, nonreactive, isothermal, atomizing, full-cone sprays, such as those of Figures 1b and 2d, and discuss their possible structure. We will consider first the steady far field and then the breakup and development regions. For liquids such as water and hydrocarbon fuels injected at room temperature into standard or compressed air or similar gases, such sprays are found at injection velocities of the order of 10^2 m/s. The high velocity allows large mass flow rates to be obtained even with small nozzles. Their diameter is often of the order of 10^2 μm so that their initial characteristic time is of the order of 10^{-6} s. Drop diameters are of the order of 10 μm. The combination of large velocity and small size makes detailed measurements difficult in the breakup and development regions. Thus their initial transient was observed for the first time only recently.[8]

2. THE STEADY FAR FIELD

The classical point source, boundary layer, similarity solution of Tollmien[4] for incompressible jets is a satisfactory guide to the understanding of the global structure of the far field of full-cone sprays. In principle, only one constant, such as C_t that relates the effective turbulent diffusivity D_t to the specific momentum flow M, is left undetermined by this solution and must be evaluated from measurements. Then the width increases as x, the centerline velocity decays as $1/x$, and the ratio of the entrained mass to the injected mass increases as x:

$$M = C_m \pi \bar{u}_{0,\mathrm{CL}}^2 d^2/4 \tag{1}$$

$$Q_0 = C_q \pi \bar{u}_{0,\mathrm{CL}} d^2/4 \tag{2}$$

$$D_t = C_t [C_m \pi \bar{u}_{0,\mathrm{CL}}^2 d^2/4]^{1/2} \tag{3}$$

$$r_{0.5} = 8[\pi(2^{1/2}-1)/3]^{1/2} C_t(x-x_0) \tag{4}$$

$$\bar{u}_{x,\mathrm{CL}} = (3C_m^{1/2}/16\pi^{1/2}C_t)d\bar{u}_{0,\mathrm{CL}}/(x-x_0) \tag{5}$$

$$Q_e/Q_0 = (16\pi^{1/2}C_tC_m^{1/2}/C_q)(x-x_0)/d \tag{6}$$

In the above equations the constants C_m and C_q are used to relate the specific momentum flow rate and the volumetric flow rate to the injection centerline velocity through the injection velocity profile, and x_0 is the virtual origin. (In practice the determination of C_m, x_0, and of an additional needed constant x_2 is nontrivial even for incompressible jets. We will return to this subject later.)

At this point the growth of the entrained mass and local characteristic convection and diffusion times are of interest to us. Schlichting[4] gives $C_t = 0.0161$ and equation (6) shows that for $C_m = C_q = 1$, the entrained mass is already 10 times the injected one at 22 nozzle diameters from the virtual origin. Local convection and diffusion times can be obtained dividing $(x-x_0)$ by $\bar{u}_{x,\mathrm{CL}}$ from equation (5) and $r_{0.5}^2$ by D_t from equations (4) and (3). The two times are of the same order and increase as $(x-x_0)^2$.

For our sprays the injected medium is a liquid and the entrained medium is a gas, so that the initial entrainment process can be expected to differ from that of incompressible jets. But at sufficient distance from the injector, the injected mass must become negligible in comparison to the entrained one because past a certain distance their ratio grows as x. Downstream of such a region, the momentum that leads to the entrainment of more ambient gas resides mostly with the ambient gas that was entrained earlier, i.e., the structure of fully developed incompressible jets is recovered.

Imbedded in this far field there are drops that move in equilibrium with the gas. This is because the time for the drop velocity to relax to the local gas velocity, $\rho_l d_i^2/18\mu_g$, has an upper limit while the fluid time continues to grow as x^2. Since there is a distribution of drop sizes and of eddy times, equilibration will be selective and dependent on conditions, but, at some appropriate distance from the injector, all sprays become incompressible jets dominated by the entrained ambient medium, and drops move within them as markers of the motion of the ambient medium.

Having accepted the existence of this limit, equations (1)–(6) can be modified for direct application to the far field of sprays. Following Thring and Newby[12] and Kleinstein,[13] we can equate the axial momentum evaluated at the nozzle exit, where the density is ρ_l, to that evaluated in the far field, where the density is ρ_g and the velocity profile is fully developed, and conclude that, as regards the far field, a liquid spray is an incompressible jet with an equivalent specific momentum flow rate of $M(\rho_l/\rho_g)$. Then equations (1)–(6) become

$$M^* = M(\rho_l/\rho_g) \tag{7}$$

$$Q_0^* = Q_0\rho_l \tag{8}$$

$$D_t^* = D_t(\rho_l/\rho_g)^{1/2} \tag{9}$$

$$r_{0.5}^* = r_{0.5} \tag{10}$$

$$\bar{u}_{x,CL}^* = \bar{u}_{x,CL}(\rho_l/\rho_g)^{1/2} \tag{11}$$

$$Q_e^*/Q_0^* = (Q_l/Q_0)(\rho_g/\rho_l)^{1/2} \tag{12}$$

$$d^* = d(\rho_l/\rho_g)^{1/2} \tag{13}$$

Parallel interpretations are that in the far field a spray is an incompressible jet from an equivalent nozzle diameter[12] of $d(\rho_l/\rho_g)^{1/2}$ or with an equivalent turbulent diffusivity[13] of $D_t(\rho_l/\rho_g)^{1/2}$. Notice however that the change in the virtual origin is not identified by this theory since this quantity is determined by the structure of the development region. The ratio of the turbulent diffusion time to the drop relaxation time becomes

$$\frac{\tau_{\text{diff}}^*}{\tau_{\text{drop}}} = 18\left(\frac{r_{0.5}}{d_l}\right)^2\left(\frac{\mu_g}{\rho_l D_t^*}\right) \tag{14}$$

Several authors have realized the existence of this limit and taken good advantage of it.[14] But only recently detailed drop-velocity measurements within sprays of this family have provided direct evidence of it. We shall review these measurements briefly, but first a note of caution. In specific applications, particularly to time-varying and/or closed volumes, the development region is often the one of prime interest. Indeed in many cases no more volume or time are provided than those necessary to complete some specific degree of mixing of the injected and ambient media. In the far-field limit, the injected medium is but a trace within the ambient one. Thus we are not saying or implying that the local equilibrium limit is necessarily achieved in applications. We are saying that these limits exist and are conceptually useful.

The data of Figure 3 were taken in the sprays of Table 1 under steady injection, at distances greater than $300d$, and with the gas contained in a vessel of such size that a wall-free environment was approached.[15] These sprays looked very much like those of Figures 1b and 2d. Up to $r = r_{0.5}$, the radial profiles of the mean axial drop velocity and of the amplitude, skewness, and flatness of the fluctuation of the drop axial velocity are seen to fall within the range of the corresponding fluid quantities measured in incompressible jets by Wygnanski and Fiedler[16] by hot-wire anemometry and by others[4–6,17,18] with hot-wire anemometry and laser Doppler velocimetry. The same parallel was also found for the radial component of the drop velocity, for the centerline velocity decay, and for the width. Also, the drop velocity was measured with various laser power levels, thus weighing the measurements in favor of drops of different sizes, and the results were found to be independent of it for $x \geq 300d$. If the drops had not been in equilibrium with the gas, their velocity would have depended on their size. Thus indications are that for the full-cone sprays of Table 1 the incompressible jet and drop-gas equilibrium limits are achieved around $x = 300d$ and for $r \leq r_{0.5}$. For spray E, at $x = 200d$ the measurements indicate that drops are not in equilibrium with the gas and for $x \geq 300d$ but $r > r_{0.5}$ the evidence is inconclusive because of large errors in both HWA and LDV data.[15]

There is no reason to expect that all sprays of Table 1 should reach equilibrium at the same axial distance, but the data were not sufficiently numerous or sufficiently accurate to differentiate and the $300d$ location should be considered indicative. By extrapolating backward the centerline velocity decay, and also checking the growth of the jet width, it was determined that $x_0 = 30d$ for $\rho_l/\rho_g = 13.7$ and $x_0 = 100d$ for $\rho_l/\rho_g = 39.1$. If the $300d$ range and the two values of x_0 are compared with the corresponding $50d$–$70d$ and $5d$–$10d$ values for incompressible jets,[16] the reasonable conclusion is reached that full-cone sprays develop into incompressible jets but require a longer distance and that such a distance is likely to be an

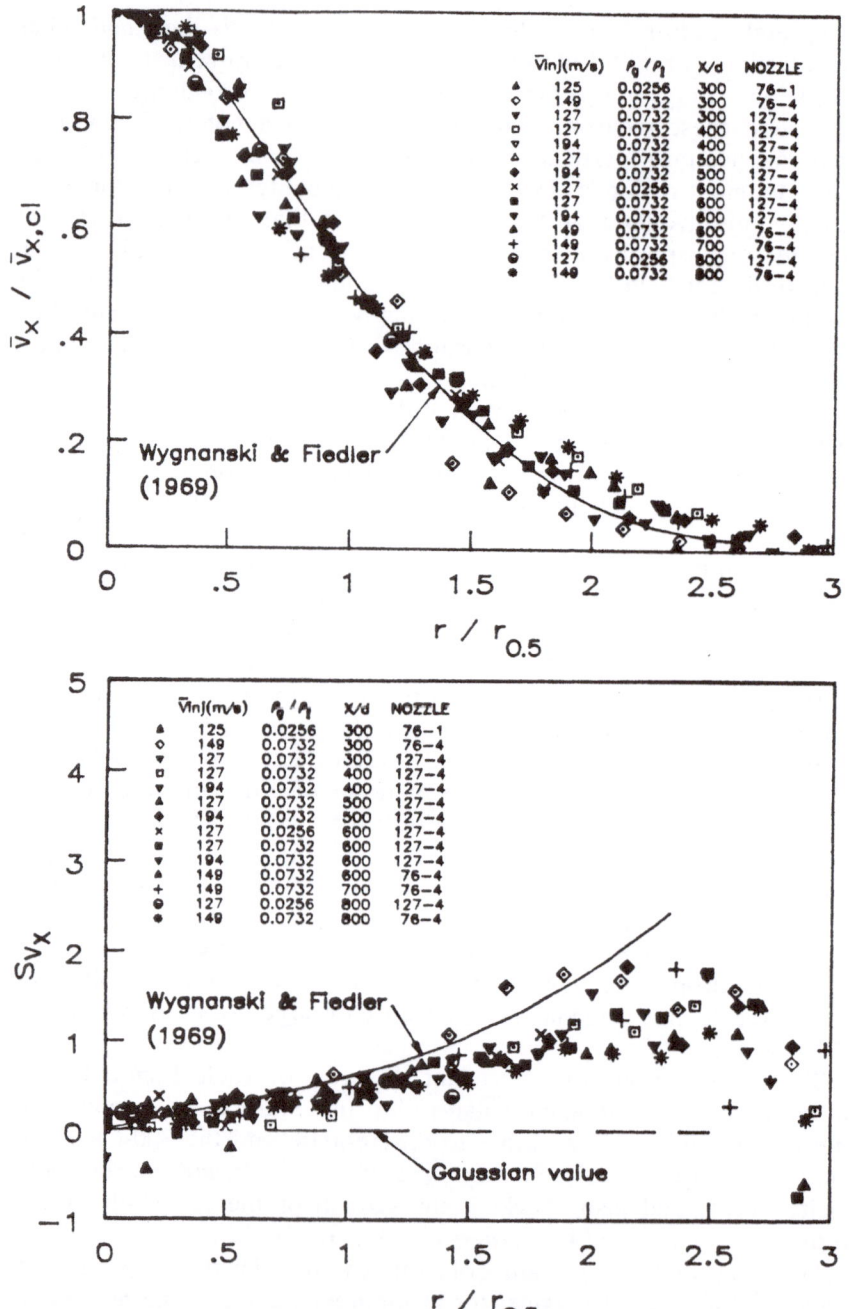

Fig. 3. Mean axial drop velocity and amplitude, skewness, and flatness of the fluctuation of the axial drop velocity for the atomizing sprays[15] of Table 1.

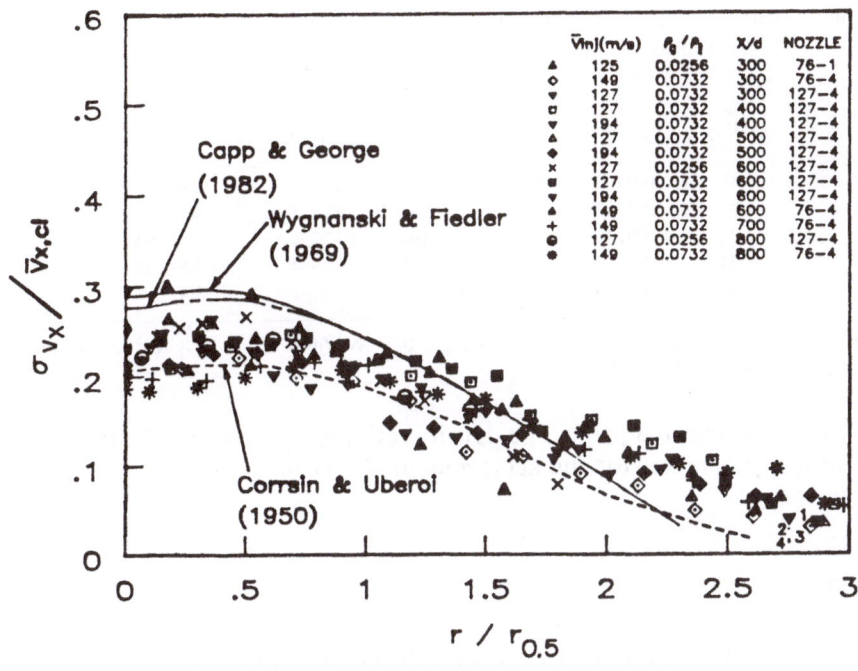

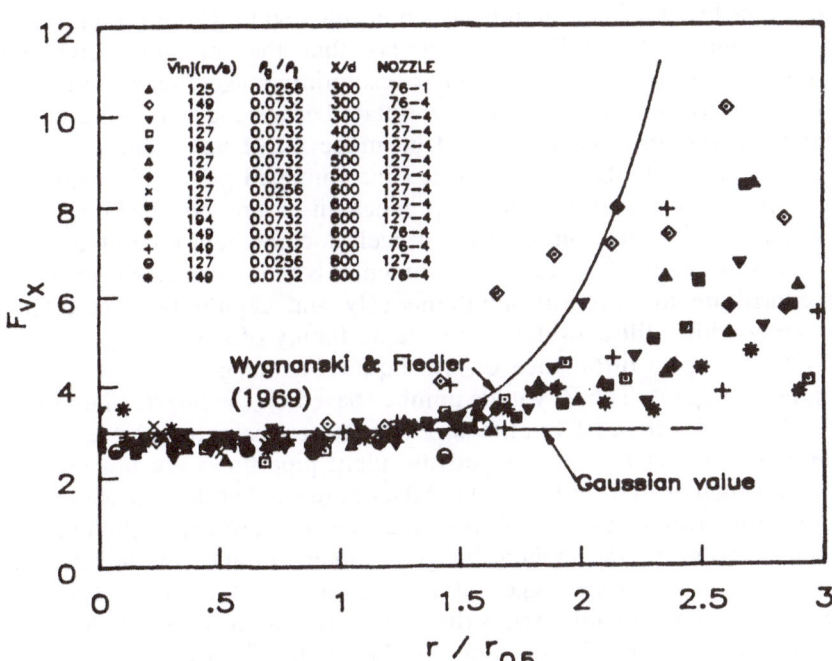

Table 1. Spray conditions for the drop velocity measurements of Figs. 3, 8, and 9 (after Wu et al.[15]) [a]

Series	P_g (MPa)	ρ_g/ρ_l	Δp (MPa)	$\bar{V}_{inj}$ (m/s)	Nozzle $d(\mu m) - L/d$	X/d
A	1.48	0.0256	11.0	127	127–4	600, 800
B	4.24	0.0732	11.0	127	127–4	300, 400, 500, 600
C	4.24	0.0732	26.2	194	127–4	400, 500, 600
D	4.24	0.0732	11.0	149	76–4	300, 600, 700, 800
E	1.48	0.0256	11.0	125	76–1	300

[a] Liquid: n-hexane, $\rho_l = 665$ kg/m³, $\mu_l = 3.2 \times 10^{-4}$ N·s/m², $\sigma_l = 1.84 \times 10^{-2}$ N/m. Gas: nitrogen. Room temperature.

increasing function of ρ_l/ρ_g. The density ratio, ρ_l/ρ_g, is seen to be the main additional parameter for the achievement and structure of the far field of these sprays.

3. THE STEADY NEAR FIELD

In considering the near field we will start from the outer part of the spray in the immediate vicinity of the nozzle exit. It is of interest to know what forces break up the liquid surface in this region. From photographs, such as Figures 1b and 2d, it appears that the jet starts diverging immediately at the nozzle exit. Higher-resolution images, such as Figure 6 which will be reconsidered later, show isolated drops and an opaque, highly irregular, diverging fluid that could be made up of any combination of drops, ligaments, blobs, and deformed intermingled gas–liquid continua.

Hypotheses about the breakup mechanism are not lacking,[1] but quantitative evidence from controlled, well-documented, unequivocal experiments has been very scant. One difficulty is that too many events that can contribute are present simultaneously and cannot be investigated separately while still considering the same family of sprays.

For example, turbulence of the liquid was suggested as the main destabilizing agent. The Reynolds number based on the nozzle diameter is generally greater than 10^4 but the nozzle length-to-diameter ratio is seldom greater than 6 and fully developed turbulent pipe flows are not present. Moreover increasing L/d from 10 to 80 leads to smaller divergence angles, i.e., to more stable jets.[19,20] Within the nozzle there are turbulent wall boundary layers and the sudden change of forces within them that occurs at the nozzle exit has been suggested as destabilizing, but even in the most carefully machined metal nozzles the surface roughness is still no less than 10 μm with diameters in the range of 70 μm to 300 μm. It is not clear whether the tall ridges and deep valleys formed by surface roughness trip the flow or trap it. In any case, radiusing the exit edge with various curvatures brought about no measurable change in the initial spray angle.[20] Cavitation is invariably present at the entrance of practical nozzles of this family and has been suggested as the main destabilizing agent. But cavitation-free nozzles have been found to give immediately diverging jets too.[20] Their angle was smaller, indicating greater stability, but the geometry of the cavitation-free nozzles was so different from the sharp-inlet, sharp-outlet, straight-wall geometry of standard nozzles that the entire nozzle flow field was also different.

Also suggested as possibly being responsible for the breakup are the rearrangement of the cross-section axial velocity profile, and liquid supply pressure oscillations. But none of these mechanisms *alone* was found adequate to explain the trends exhibited by the initial spray angle.[11,20] However, *any* of them could be contributing to the breakup process as explained below.

Since aerodynamic interaction is known to cause the breakup of jets at lower speeds, the suggestion that it may continue to do so at higher speeds too is a natural one. According to this view the length of jet surface over which unstable surface waves grow becomes shorter and shorter as the speed is increased, due to the faster growth rate of the unstable waves, until it becomes of the order of a few microns and is no longer detectable. Then the jet appears to diverge immediately at the nozzle exit. This first-order linear perturbation theory leaves a parameter unspecified that can be interpreted as the initial amplitude of the surface perturbations when the liquid first enters the gas. If this parameter is allowed to vary with nozzle geometry, as would be the case if different nozzle flows establish different initial perturbation levels through some combination of the previously mentioned processes, then a supplemented aerodynamic interaction mechanism results that seems to comply in a fitting way with a rather large set of experimental information. Castleman[21] was among the early supporters of this view. Ranz[22] produced a theoretical framework for it, extending the work of Taylor.[23] And Reitz[24,11] performed a comprehensive evaluation of most major proposals and sharpened the focus on it.

The supplemented aerodynamic theory leads to the prediction[22,24,25] of the initial spray angle, the initial average drop size, and the length of the intact core. The equations are particularly simple in the limit $\rho_l \, \mathrm{Re}_j^2 / \rho_g \, \mathrm{We}_j^2 > 1$, which is the one of practical interest in many cases of atomization. The angle is determined by combining the radial velocity of the fastest growing of the unstable surface waves with the axial injection velocity:

$$\tan \frac{\theta}{2} = \frac{1}{A} 4\pi \left(\frac{\rho_g}{\rho_l} \right)^{1/2} \frac{3^{1/2}}{6} \tag{15}$$

where A is a constant whose value depends on the nozzle geometry and must be determined experimentally. The initial average drop diameter is assumed to be proportional to the length of the most unstable wave:

$$\bar{d}_l = \frac{4\pi B\sigma}{\rho_g \bar{u}_{0,CL}^2} \frac{3}{2} \tag{16}$$

where B is a constant of order one and should be insensitive to the nozzle geometry (but dependent on the reference velocity that is chosen for the breakup process). And the length of the intact core is obtained by subtracting the mass of the drops from the intact liquid column as they are formed:

$$\frac{x_1}{d} = C\left(\frac{\rho_l}{\rho_g}\right)^{1/2} \tag{17}$$

where C is a proportionality constant.

The prediction that in the atomization regime the initial value of the spray angle depends almost exclusively on the density ratio and the nozzle geometry is surprising, considering the many parameters that could effect it, but is generally born out by measurement. Figure 4 shows that, for a

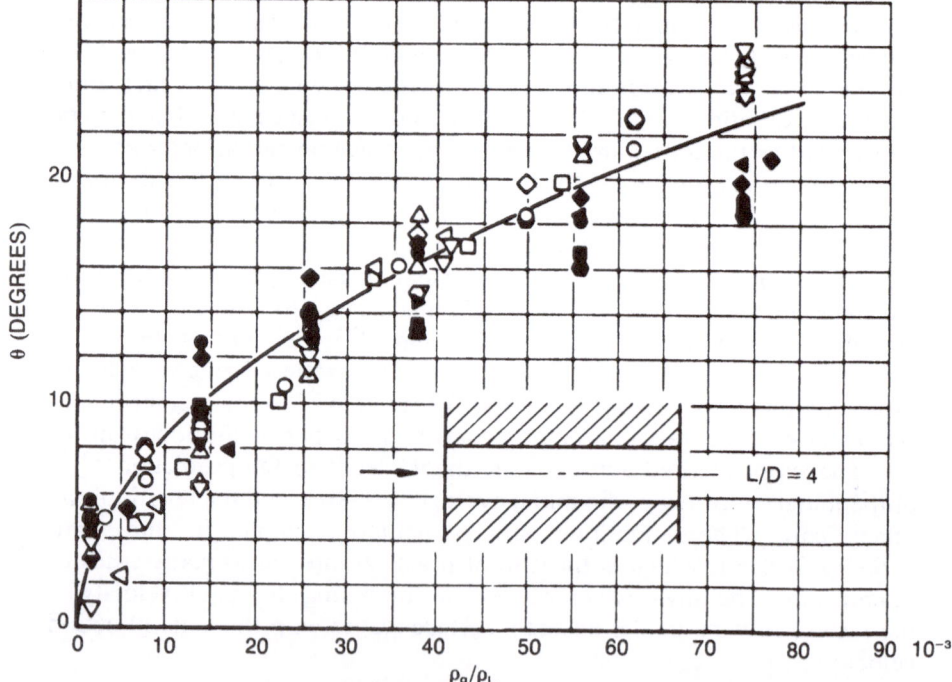

Fig. 4. Initial angle of atomizing jets versus density ratio with fixed nozzle geometry. Injections of glycerol–water, water, hexane, tetradecane into N_2, He, Xe, Ar at liquid pressures of 500–13,300 psi and $D = 254$, 343, and 610 μm. Room temperature.[20]

given nozzle geometry, large changes in liquid properties, injection velocity, and gas pressure bring about only very small systematic trends and that the density ratio dominates.[20] Figure 5 gives an example of the effect of nozzle geometry, all other parameters being the same.[20] In spite of the success there are limitations: the mechanism by which the nozzle geometry influences the angle is not known; although the angle is very reproducible in any given experiment, its value depends somewhat on definitions and measuring techniques; when a broad range of injection velocities is explored a mild trend is detected at lower density ratios that does not conform with the expected one.[20]

Recently[26,27] the diameter of drops was measured at the edge of atomizing sprays in the immediate vicinity of the nozzle exit using photographs such as that of Figure 6. The average diameter is given in Figure 7 for the conditions of Table 2. When compared with equation (16), the measured values exhibit the correct trends with respect to injection velocity, liquid properties, and nozzle geometry (it has little effect) but the incorrect one with respect to gas density. Moreover, the measured diameters are about a factor of 3 greater than expected. However, it would appear that a reasonable explanation for the disagreements exists.[27] The drops that could be and were measured are at the outer edge of the spray and not at their formation sites to which equation (16) applies. Due to the

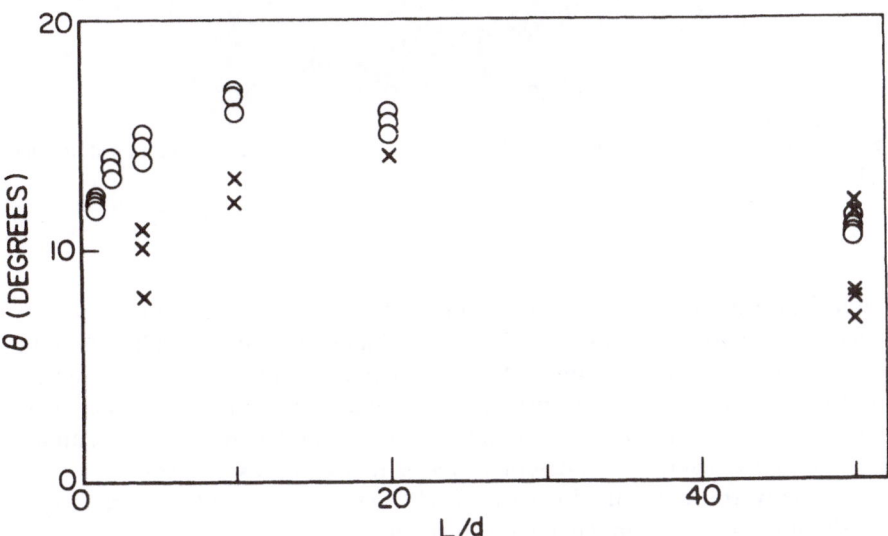

Fig. 5. Initial angle of atomizing jets versus nozzle length-to-diameter ratio. All other conditions constant: ○ after Wu et al.,[20] × after Hiroyasu et al.[19]

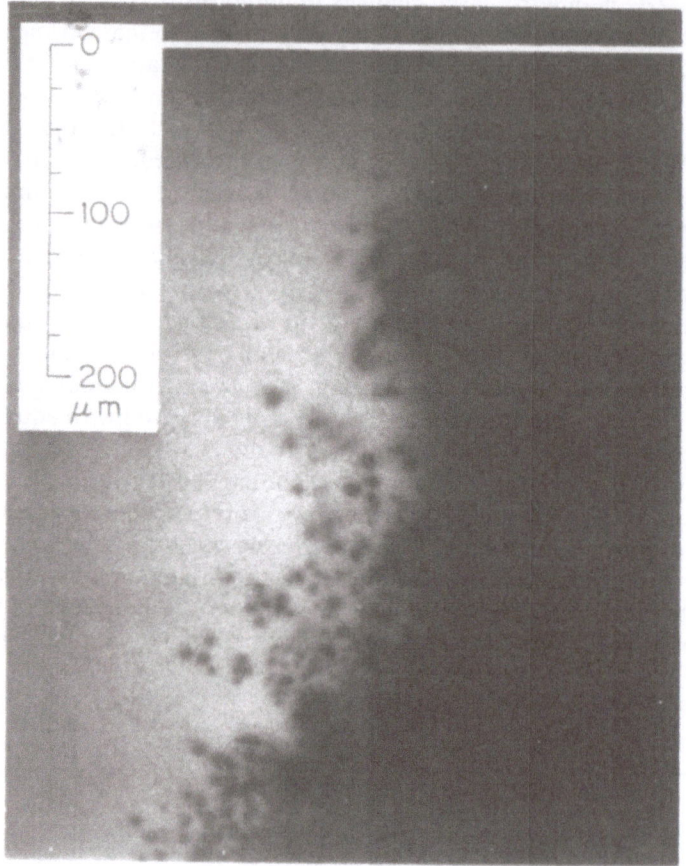

Fig. 6. Edge of spray and droplets in the immediate vicinity of the nozzle exit for an atomizing jet.[26,27]

high drop-number density in the region, it is unreasonable to expect that each drop clears the congested area and comes out without colliding with other drops. When collisions and coalescence are considered, with a model that will be mentioned presently, the discrepancies tend to disappear. It is true, however, that one set of measurements and computations cannot possibly be considered sufficient to close this complex subject.

Finally the data of Hiroyasu et al.[19] support the intact-core length prediction of equation (17) with respect to injection velocity and gas density. However the technique they used, based on measurement of the electrical resistance between the nozzle and a screen that could be moved

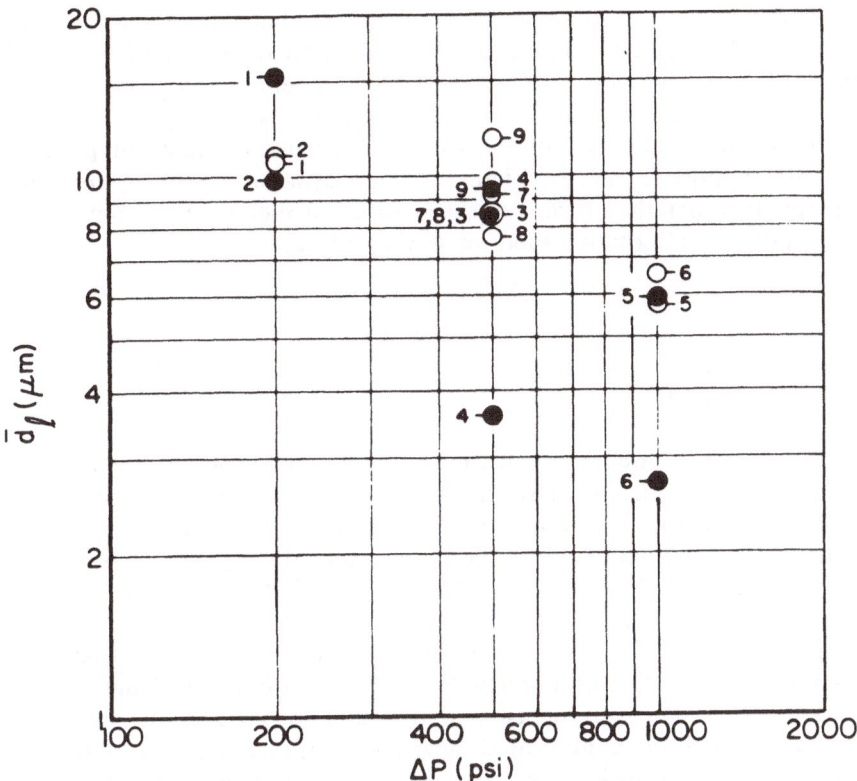

Fig. 7. Average diameter of drops in the immediate vicinity of the nozzle exit for the atomizing jets of Table 2: ○ measured; ● expected trends from the supplemented aerodynamic theory.[26,27]

Table 2. Spray conditions for the drop-size measurements of Figs. 6 and 7 (after Wu *et al.*[26,27])

Series	Nozzle $L-d$ (μm)	P_g (MPa)	ρ_g (kg/m³)	Liquid	Δp (MPa)	$\bar{V}_{inj}$ (m/s)	C_D	Re_j $\times 10^{-4}$	We_j $\times 10^{-4}$	No. of drops
1	335–4	1.48	17.0	n-C6H14	1.38	59.4	0.92	4.14	4.26	119
2	335–4	2.86	33.0	n-C6H14	1.38	52.4	0.81	3.65	3.32	109
3	335–4	1.48	17.0	n-C6H14	3.45	79.2	0.78	5.51	7.58	116
4	335–4	2.86	33.0	n-C6H14	3.45	92.8	0.91	6.46	11.85	111
5	335–4	1.48	17.0	n-C6H14	6.90	99.0	0.69	6.89	11.85	107
6	335–4	2.86	33.0	n-C6H14	6.90	111.0	0.77	7.73	14.89	117
7	335–10	1.48	17.0	n-C6H14	3.45	79.2	0.78	5.51	7.58	119
8	127–4	1.48	17.0	n-C6H14	3.45	78.1	0.77	2.06	2.79	114
9	335–4	1.48	17.0	n-C14H30	3.45	81.2	0.86	0.95	6.31	103

axially, detects any continuous liquid connection between the nozzle and the station of the screen. It gives an upper limit for the length of the intact core and no information about its radial dependence. Thus considerable uncertainty about the structure of the core persists.

Individually, any one of the quoted experiments may not provide adequate support for the supplemented aerodynamic theory of atomization. But together they form a rather consistent picture. By comparison, the support for alternative theories is very meager.

4. THE TRANSIENT

Because of its importance in Diesel engines, many measurements have been reported of the velocity of the tip of atomizing jets[28] and it is generally agreed that it is about 70% of the steady-state centerline velocity. But details are anything but clear, as indicated by the lack of consistency in the reported curve fits for the tip velocity. In fact, the transient of round jets has received very little detailed attention.

There seems to have been only one study of the transient of laminar incompressible jets[29] and it concludes with wrong time and length scales for it.[30] Recent measurements[31] of transient incompressible jets were limited in value by complicated nozzle conditions. Recent numerical studies[30] of the transient of laminar, turbulent, and spray jets gave reasonably complete information about laminar jets, because there are no uncertain physical parameters for such jets, but only indicative information about turbulent and spray jets, since there are too many uncertain parameters for them.

The diffusive nature of jets makes the identification of their most advanced position a matter of definition and a function of the experimental technique. However, it would appear that all jets scale up in time at least approximately. That a scaling exists is shown by pictures, such as those of Figure 1, and by the many correlations derived from them. That the scaling of sprays is complex, or just an approximation, is indicated by the small probability of the existence of exact similarity solutions of their constitutive equations.

Indeed if we reconsider Figure 1b and if we define the arrival of the spray as the time at which the centerline velocity first achieves a selected percent of its steady-state value, then we can identify different stages in the propagation that are controlled by different processes and therefore are likely to scale in different ways.

For distances up to several tens of nozzle diameters there is likely to be a compact liquid core that moves at the injection velocity. Its most forward surface, however, encounters the resistance of the ambient gas. A stagnation point can be conceived with gas forcing the liquid to flow radially away from it and then backward. Thus the observed tip velocity will be a function of the injection velocity, the density ratio, and the actual structure of the core.

Eventually the compact core disappears and ligaments and/or drops encounter the ambient gas first. Since drops and gases exchange forces differently than liquid columns and gases, the apparent tip velocity need not scale as it did earlier. At sufficient distance from the nozzle, most of the jet momentum is with the entrained gas and a gaseous head vortex moves into the gaseous environment and exchanges forces with it in the manner of incompressible jets, which is again different from that of drops and liquid columns. Thus the scaling is likely to be different at different stages of the tip propagation.

Reitz filmed the initial propagation of atomizing jets[8] but he was concerned with an overall description of the process and a broad set of conditions and his data are insufficient to determine accurate correlations for the initial tip speed. Far from the injector, the tip velocity is about 70% of the velocity given by equations (11) and (5) (with x_0 neglected with respect to x or estimated using the data of Wu *et al.*[15]). There is no dependable information for the scaling of the intermediate region and its merging with the other two.

5. COMPUTATIONS

A good test of the degree of our understanding of the structure of these sprays is the extent to which we can predict their details.

Navier–Stokes equations for two-phase flows are the appropriate conservation equations[32] but cannot be solved with adequate resolution for the entire flowfield either analytically or numerically.[33] Local averaging and the adoption of semiempirical equations to represent the effects of the neglected details are necessary. Even so, no attempt has been made to compute the flow within the nozzle to predict the initial perturbations required by the aerodynamic theory of breakup to account for the effects of the nozzle geometry.

Many difficulties remain even if one considers starting the computations at the nozzle exit from some arbitrary initial perturbations of the liquid–gas interface. Not predicted by the linear stability theory, and therefore unknown, are the size and size distribution of the unstable growths at breakup and the time between successive ruptures. Away from the nozzle exit plane, as the generating surface moves closer to the axis of the jet, there are questions as to what gas field is seen by the liquid interface. The velocity of the entrained gas is closer to that of the generating surface. Thus it would appear that the breakup process should become coupled to the structure of the two-phase flow field that exists between the presumed intact core and the unperturbed outer gas. As the relative velocity between liquid and gas decreases inside the jet, larger drops or ligaments or blobs should be formed, just as different breakup modes and larger drops are found when the injection velocity is decreased, as in going from Figure 2d to Figure 2a. Also, coalescence of the liquid fragments can be expected where formation occurs due to the locally large value of the liquid volume fraction. The net outcome may be the small difference between large formation and coalescence rates. Thus it is clear that equations (15)–(17) may provide some information about the outcome of the breakup, but in no way do they give all that is necessary.

Faced with so many unknowns, even the most advanced of current spray computations[34] are initiated at some distance from the nozzle from selected drop-size and velocity distributions obtained from Equations (15)–(17). When drop collisions and coalescence are considered, it is found that the far field results tend to become independent of the initial conditions because smaller drops exhibit a higher rate of coalescence and because, as previously discussed, the far field is dominated by the entrained gas. Thus the main conditions for the accurate computation of far-field mean quantities are knowledge and conservation of axial momentum and proper gas-phase turbulent diffusivity.[35] Even those minimal requirements are not met without some care. In general it is difficult to compute the injection momentum of an actual spray even when its mass flow rate is known, because the injection velocity profile is not known [C_m in equation (1) is not known even if C_q in equation (2) is known]. $k - \varepsilon$ models can be tuned to give the far-field diffusivity of incompressible jets if such diffusivity were known accurately. But this is not the case as shown, for example, by C_t varying within $\pm 10\%$ when different sets of experimental data are used.[15] Successful computations of fluctuations, such as those of the drop velocity, require that the model be accurate also at smaller scales. The comparisons[35] of Figures 8 and 9 show adequate reproduction of mean axial velocities but the amplitude of the drop axial velocity fluctuation is underestimated by a factor of two. Several reasons could be advanced for the specific disagreement of Figure 9 but, ultimately, current quantitative knowledge of the coupling between drop motion and turbulence gas eddies is incomplete.[14,36]

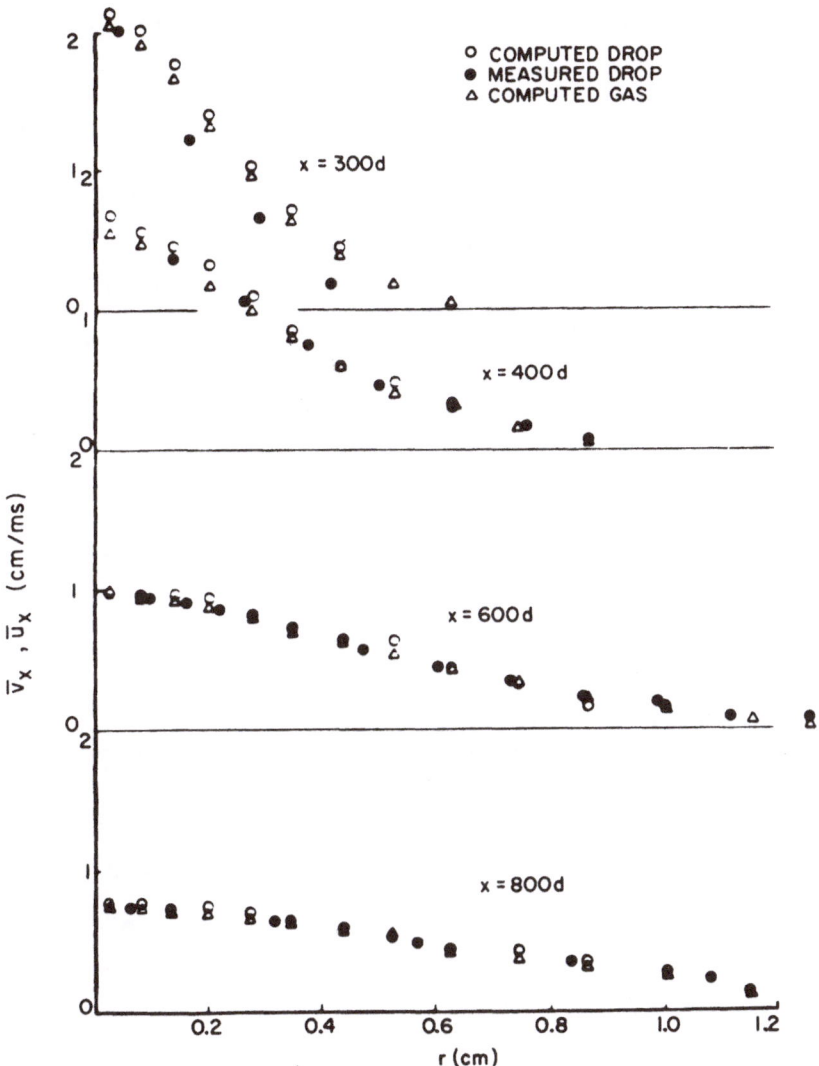

Fig. 8. Measured[15] mean axial drop velocity and computed[35] mean axial drop and gas velocity for Spray A of Table 1.

6. SUMMARY

What is known about the structure of nonvaporizing, nonreactive sprays from single-hole cylindrical nozzles can be summarized with the help of Figure 10.

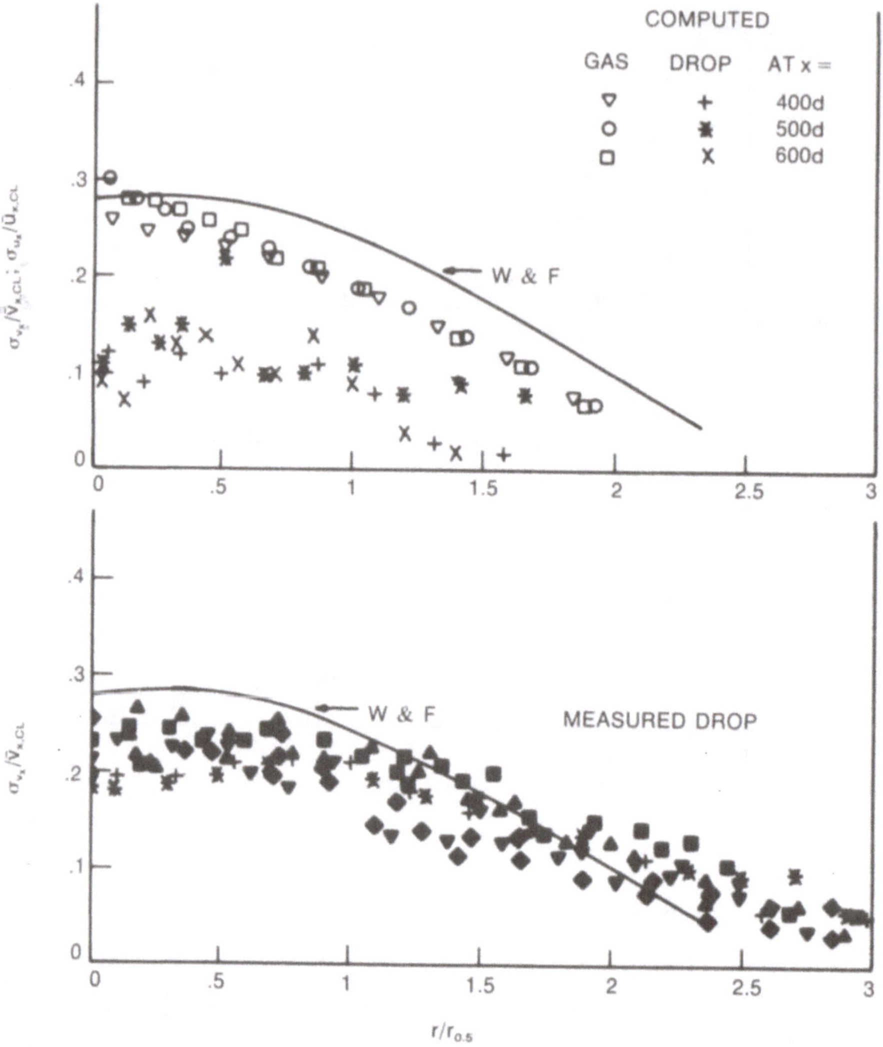

Fig. 9. Measured[15] amplitude of the fluctuation of the axial drop velocity and computed[35] amplitudes of the fluctuations of the axial drop and gas velocities for Spray C of Table 1.

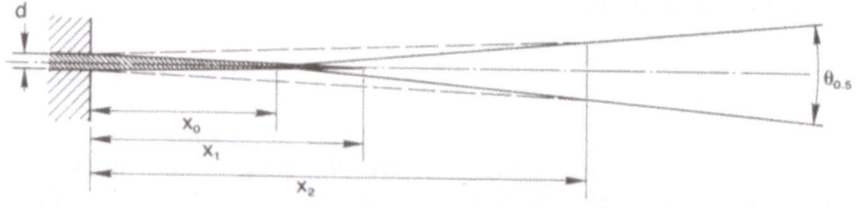

Fig. 10. Schematic structure of atomizing jets.

At sufficiently high injection velocity the jet is found to diverge immediately at the nozzle exit where fine drops are observed. The outer surface of the liquid is disrupted by the interaction with the ambient gas that leads to the rapid and selective growth of surface waves whose initial amplitudes are controlled by events that occur within the nozzle. This view allows one to predict the initial angle of the spray. The parallel prediction of the size of the drop thus formed also compares favorably with measurements, if collisions and coalescence of the drops after their formation are included.

The disruption of the liquid column eventually reaches the axis of the jet, i.e., is complete, because only isolated drops are found downstream. However the geometry, structure, and mode of breakup of the core of the jet are not known, but there are indications that the length of the intact core, x_1, may approach 100 nozzle diameters (versus 10, for incompressible jets).

While the internal breakup continues, gas from the environment is entrained rapidly. The entrained gas eventually becomes dominant and achieves the structure of fully developed incompressible jets. Within it, drops tend to reach equilibrium but selectively, since both drop and eddy sizes are distributed. The fully developed equilibrium distance, x_2, is several hundred nozzle diameters (versus several tens, for incompressible jets). The virtual origin of this fully developed far field, x_0, is of the order of tens of nozzle diameters (versus order of nozzle diameters, for incompressible jets).

The precise relationship between x_0, x_1, and x_2 and shape of the corresponding boundaries is not known because the structure of the transition region is not known. But the three lengths increase with increasing liquid–gas density ratio.

The propagation velocity of the tip, or head, of these sprays is a fraction of their steady-state centerline velocity. In the fully developed equilibrium region this fraction is about 70% but in the development region it is not known. Since the steady-state centerline velocity scales in an established way in the fully developed equilibrium region, so does the tip velocity. Experimental information suggests that the steady-state centerline velocity and the tip velocity scale also in the development region, at least approximately. But these scaling functions are determined by the structure of the development region and therefore are not known precisely.

ACKNOWLEDGMENTS. Support for this review was provided by the U.S. Army Research Office (DAAG 21-81-K-0135) and the U.S. Department of Energy (DE-AC-04-81AL16338).

NOMENCLATURE

A, B, C, C_m, C_q, C_t	dimensionless constants of equations (15)–(17), (1)–(3), respectively
C_D	nozzle discharge coefficient
d	nozzle diameter
d_l	drop diameter
D_t	turbulent diffusivity
F_{v_x}	flatness of the fluctuation of the axial component of the drop velocity
L	nozzle length
M	specific momentum flow rate
p	pressure
Q	volumetric flow rate
$r_{0.5}$	jet half-radius (half the width at half the depth)
Re_j	liquid jet Reynolds number, $\rho_l \bar{V}_{\text{inj}} d / \mu_l$
S_{v_x}	skewness of the fluctuation of the axial component of the drop velocity
$\mathbf{u}$	gas velocity vector of components u_x, u_r
$\bar{V}_{\text{inj}}$	mass mean injection velocity
$\mathbf{v}$	drop velocity vector of component v_x, v_r
W_j	jet Weber number, $\rho_l \bar{V}_{\text{inj}}^2 d / \sigma$
x	axial distance from the nozzle exit
x_0	virtual origin
x_1	end of intact liquid core
x_2	beginning of fully developed entrained gas jet
r	radial distance
Δp	injection pressure difference
θ	spray angle
$\theta_{0.5}$	angle corresponding to $r_{0.5}$
μ_g	gas viscosity
μ_l	liquid viscosity
σ	surface tension
σ_{u_x}	standard deviation of the fluctuation of the axial component of the gas velocity
σ_{v_x}	standard deviation of the fluctuation of the axial component of the drop velocity
ρ_g	gas density
ρ_l	liquid density
τ	characteristic time

Superscripts

'	fluctuation
*	applicable to far field of sprays

Subscripts

CL	on the centerline, $r = 0$
e	entrained
0	at the nozzle exit, $x = 0$

REFERENCES

1. E. Giffen and A. Muraszew, *The Atomization of Liquid Fuels*, John Wiley and Sons, New York (1953).
2. W. R. Marshal, Jr., *Atomization and Spray Drying*, AIChE Monogr. Ser. **50** (2)(1954).
3. N. Dombrowski and G. Mundy, *Spray Drying*, Biochemical and Biological Engineering Science, **22**, Academic Press, New York (1968).
4. H. Schlichting, *Boundary-Layer Theory*, McGraw-Hill Book Co., New York (1968).
5. J. O. Hinze, *Turbulence*, McGraw-Hill Book Co., New York (1975).
6. G. N. Abramovich, *The Theory of Turbulent Jets*, The MIT Press, Cambridge, Massachusetts (1963).
7. W. Rizk, Experimental studies of the mixing processes and flow configurations in two-cycle engine scavenging, *Proc. Inst. Mech. Eng.* **172**, 417–437 (1958).
8. R. D. Reitz and F. V. Bracco, Ultra-high-speed filming of atomizing jets, *Phys. Fluids* **22**, 1054–1064 (1979).
9. V. G. Levich, *Physicochemical Hydrodynamics*, Prentice-Hall, Englewood Cliffs, New Jersey (1962).
10. R. E. Phinney, The breakup of a turbulent liquid jet in a gaseous atmosphere, *J. Fluid Mech.* **60**, 689–701 (1973).
11. R. D. Reitz and F. V. Bracco, Mechanism of atomization of a liquid jet, *Phys. Fluids* **25**, 1730–1742 (1982).
12. M. W. Thring and M. P. Newby, Combustion length of enclosed turbulent jet flames, in: *Fourth Symp. (Int.) on Combust.*, pp. 789–796, Williams and Wilkins Co., Baltimore, Maryland (1953).
13. G. Kleinstein, Mixing in turbulent axially symmetric free jets, *J. Spacecraft* **1**, 403–408 (1964).
14. G. M. Faeth, Evaporation and combustion of sprays, *Prog. Energy Combust. Sci.* **9**, 1–76 (1983).
15. K.-J. Wu, A. Coghe, D. A. Santavicca, and F. V. Bracco, LDV measurements of drop velocity in Diesel-type sprays, *AIAA J.* (to appear).
16. I. Wygnanski and H. Fiedler, Some measurements in the self preserving jet, *J. Fluid Mech.* **38**, 577–612 (1969).
17. S. Corrsin and M. S. Uberoi, Further Experiments on the Flow and Heat Transfer in a Heated Turbulent Air Jet, NACA TR998 (1950).
18. S. P. Capp and W. K. George, Jr., Measurements in an axisymmetric jet using a two-color LDA and burst process, paper presented at Int. Symp. on Appl. of LDA to Fluid Mech., Lisbon, Portugal (1982).
19. H. Hiroyasu, M. Shimizu, and M. Arai, The breakup of high speed jet in a high pressure gaseous atmosphere, *ICLASS-82*, Madison, Wisconsin (1982).

20. K.-J. Wu, C.-C. Su, R. L. Steinberger, D. A. Santavicca, and F. V. Bracco, Measurements of the spray angle of atomizing jets, *J. Fluids Eng.* **105**, 406–413 (1983).
21. R. A. Castleman, Jr., The Mechanism of Atomization Accompanying Solid Injection, NACA Report No. 440 (1932).
22. W. E. Ranz, Some experiments on orifice sprays, *Can. J. Chem. Eng.* **36**, 175–181 (1958).
23. G. K. Batchelor (ed.), *Collected Works of G. I. Taylor*, Cambridge University Press, Cambridge, Mass. (1958).
24. R. D. Reitz, Atomization and Other Breakup Regimes of a Liquid Jet, Ph.D. Thesis No. 1375, Department of Mechanical and Aerospace Engineering, Princeton University, Princeton, New Jersey (1978).
25. R. D. Reitz and F. V. Bracco, On the Dependence of the Spray Angle and Other Spray Parameters on Nozzle Design and Operating Conditions, SAE Paper No. 790494 (1979).
26. K.-J. Wu, Atomizing Round Jets, Ph.D. Thesis No. 1612, Department of Mechanical and Aerospace Engineering, Princeton University, Princeton, New Jersey (1983).
27. K.-J. Wu, R. D. Reitz, and F. V. Bracco, Drop Sizes of Atomizing Jets (to appear).
28. N. Hay and P. L. Jones, Comparison of the Various Correlations for Spray Penetration, SAE Paper No. 720776 (1972).
29. S. Abramovich and A. Solan, The initial development of a submerged laminar round jet, *J. Fluid Mech.* **59**, 791–801 (1973).
30. T.-W. Kuo and F. V. Bracco, On the Scaling of Transient Laminar, Turbulent, and Spray Jets, SAE Paper No. 820038 (1982).
31. P. O. Witze, The Impulsively Started Incompressible Turbulent Jet, SANDIA 80–8617 (1980).
32. F. A. Williams, *Combustion Theory*, Addison-Wesley Publ. Co., Reading, Massachusetts (1965).
33. F. V. Bracco, Introducing a new generation of more detailed and informative combustion models, *SAE Trans.* **84**, 3317–3340 (1975).
34. P. J. O'Rourke and F. V. Bracco, Modelling of drop interactions in thick sprays and a comparison with experiments, *I Mech E Conference Publications* 1980–9, 101–116 (1980).
35. L. Martinelli, R. D. Reitz, and F. V. Bracco, Comparisons of computed and measured dense spray jets, paper presented at 9th Int. Coll. on Dynamics of Explosions and Reactive Systems, Poitiers, France (1983). Proc. to appear in *AIAA Progr. Astronaut. Aeronaut. Ser.*
36. S. E. Elghobashi and T. W. Abou-Arab, A two-equation turbulence model for two-phase flows, *Phys. Fluids* **26**, 931–938 (1983).

11

Linear Model of Convective Heat Transfer in a Spray

W. A. SIRIGNANO

ABSTRACT. The relative motion between dropets in a spray and the ambient gas is represented in an idealized fashion (Oseen-type approximation) that allows for a linear analysis of gas-phase unsteadiness and transient heating of the droplets.

One-dimensional, unsteady solutions for the gas temperature and droplet temperature are obtained. In addition to the general solution, asymptotic solutions for large liquid thermal inertia and small liquid thermal inertia are obtained and consistency is demonstrated. The solutions are expressed as combinations of exponential terms and convergent Taylor series; the solutions are piecewise analytic and a domain of independence of boundary values is obtained. In general, two characteristic times (and their associated space scales) appear in the solutions, namely a gas residence time and a droplet heating time.

1. INTRODUCTION

There has been substantial interest in recent years[1-8] in the behavior of vaporizing sprays. Many combustion scientists have been especially interested in the problem of a vaporizing spray in a high-temperature environment. Some interesting theoretical studies[1,2,6-8] have addressed this key problem.

This author is contributing elsewhere a review[9] of the theory of spray combustion and will not present here a detailed literature review. It should be understood, however, that the mathematical problem is unusually

W. A. SIRIGNANO • Department of Mechanical Engineering, Carnegie-Mellon University, Pittsburgh, PA 15213, U.S.A.

challenging in that: (i) the equations and the differential operations are nonlinear, (ii) a complex, coupled system of partial differential equations describes the phenomena, (iii) multitime scales occur (resulting in multi-space scales), and (iv) some of the differential operators are parabolic (elliptic in space) and others are hyperbolic. It is therefore usual to rely on large-scale computational methods; nonetheless they must be employed wisely on account of the multiscales and potential errors associated with the use of separate grids and separate numerical schemes for the parabolic and hyperbolic subsystems. Although this author has been involved in the development and use of numerical methods for spray calculations,[10-12] it does seem appropriate to have a parallel, analytical approach that is less costly but still retains most of the interesting physics.

In this paper, an attempt has been made to initiate such an approach. The development herein, like other first approaches, is modest and does not contain all of the interesting physics. A one-dimensional, unsteady representation is given of a spray moving through a hot gas. Transient heating of the spherical droplets is considered but vaporization is not considered in this paper. The analysis can be assumed to apply during the early portion of the droplets' lifetimes when significant heating but little vaporization occurs.[13] The differential operator will be linearized by an Oseen-type approximation. Two interesting time scales arise in this model: the heating time for the droplets and the residence time for an element of gas in the cloud of droplets.

For the present, the model disregards gas-phase diffusion, change in droplet size with time and mass transfer, droplet drag and momentum exchange, radiation, and higher-space dimension. Also, the droplet temperature is assumed uniform throughout the droplet at any instant of time; finite thermal conductivity and internal circulation for the liquid are not considered. There is a basis for confidence that these important effects can eventually be included in a more complete analytical representation via a quasi-linearization or iterative approach.

Section 2 contains the physical description of the model, the statement of the key assumptions, and the general mathematical formulation. In Sections 3 and 4, two asymptotic cases are solved. In each case, only one of the two characteristic times has an impact on the solution. The general solution is presented in Section 5 with conclusions stated in Section 6.

2. PHYSICAL DESCRIPTION OF THE MODEL

Consider a large cloud or spray of droplets (or solid spherical particles for that matter) moving through a gas. For our purposes, the gas is hot and the droplets are cool although the opposite situation could be analyzed in identical fashion. The average distance between droplet centers is b and the average droplet diameter is d; these averages are determined at each point of the spray by considering a set of droplets in the neighborhood of that point. These averages will be taken as constant with time and position in the spray.

In reality, the average velocity of the droplets and the average distance between droplets would vary due to a variation in initial droplet velocities and the effect of droplet drag. Our assumption of constancy of the droplet velocity results in a tremendous simplification. First, a Galilean transformation is made so that the droplets become stationary and the gas flows through the spray. Now the gas velocity field is constant. We note that the gas velocity at any point is taken to be the average value in a neighborhood of that point. As long as the total droplet volume is small compared to the total gas volume in that neighborhood, we may consider the average velocity as roughly equal to the local free-stream velocity a few diameters away from the droplet. The statement of constant gas velocity allows us to bypass the gas continuity equation, the gas momentum equation, and the droplet momentum equation. Furthermore, it enhances the linearization of the energy equation.

The standard averaging principles for the variables will be utilized in order to reduce a three-dimensional to a one-dimensional description. The one-dimensional, gas-phase energy equation may be written as

$$\rho c_\mathrm{p} \frac{\partial T}{\partial t} + \rho U c_\mathrm{p} \frac{\partial T}{\partial x} - \kappa \frac{\partial^2 T}{\partial x^2} = -nh(T - T_\mathrm{s})S \tag{1}$$

where $S = \pi d^2$ is the surface area of the average droplet, $n = b^{-3}$ is the number of droplets per unit volume, and h is a heat-transfer coefficient for the average droplet. For sufficiently small values of b/d, the Nusselt number $\mathrm{Nu} = hd/\kappa$ will depend upon that parameter.[14,15] Also, the Nusselt number depends upon the Reynolds number $\mathrm{Re} = Ud/\nu$.

The droplet energy equation may be written as

$$mc \frac{\partial T_\mathrm{s}}{\partial t} = hS(T - T_\mathrm{s}) \tag{2}$$

where m is the mass of the average droplet and c is the liquid specific heat. We may rewrite this equation in the form

$$\frac{\partial T_\mathrm{s}}{\partial \tau} = T - T_\mathrm{s} \tag{3}$$

where

$$\tau = \frac{hS}{mc} t$$

This representation involves an idealization in which the liquid temperature is assumed to be uniform throughout the droplet, essentially an infinite-conductivity limit. An extension to the case of spatial variations in the liquid temperature can be obtained by employing integral equations following eigenfunction expansions to develop the kernel. Both the case of spherically symmetric internal heating[10,16] and heating with internal circulation[17-20] have been studied by this approach. The linearity of the equations would be retained but now equation (2) would be replaced by an integrodifferential equation. Therefore, the approach is not believed to be inherently limited to that key approximation about uniform liquid temperature.

For the present analysis, the diffusion term is neglected so that only convective transport remains and we have

$$\frac{1}{k}\frac{\partial T}{\partial \tau} + \frac{\partial T}{\partial z} = T_s - T \tag{4}$$

where

$$\lambda \equiv \frac{nhS}{\rho U c_p}, \qquad z \equiv \lambda x, \qquad k \equiv \frac{\lambda U m c}{hS} = \frac{nmc}{\rho c_p}$$

Now z, τ, and k are nondimensional quantities, and k may be viewed as the ratio of liquid thermal inertia to gas thermal inertia, i.e., for large k, the gas temperature would change much more rapidly than the liquid temperature.

The initial conditions are

$$T_s = T_0 \qquad \text{at } t = 0 \ (\tau = 0); \quad x \geq 0 \ (z \geq 0) \tag{5}$$

$$T = T_1 \qquad \text{at } t = 0 \ (\tau = 0); \quad x \geq 0 \ (z \geq 0) \tag{6}$$

$$T = T_1 \qquad \text{at } x = 0 \ (z = 0); \quad t \geq 0 \ (\tau \geq 0) \tag{7}$$

This implies that the droplets suddenly appear with temperature T_0 in a gas at temperature T_1. After the initial instant, heat is transferred from the gas to the liquid droplet causing T to decrease with increasing τ and z and T_s to increase with increasing τ. It is, of course, an idealization to assume that the droplets appear in the gas instantaneously. It would take some time for them to occur at any position and some heat exchange would result during that period.

The problem now presents two characteristic times: a droplet heating time, mc/hS, and a residence time, $(\lambda U)^{-1} = \rho c_p/nhs$. We note that λ^{-1} is a characteristic length for the cooling of the gas, namely if T_s were constant, if $T = T_1$ at $z = 0$, and if T were steady, then $T - T_s = (T_1 - T_s)e^{-\lambda x}$. In our general problem, of course, both T_s and T will be unsteady and strongly coupled; however, the two time scales and the length scales will be seen in the solution.

It is convenient to represent T and T_s in the nondimensional form

$$\theta = \frac{T_I - T}{T_I - T_0}$$

$$\beta = \frac{T_s - T_I}{T_I - T_0}$$

Equations (3)–(7) then become

$$\frac{\partial \beta}{\partial \tau} = -(\beta + \theta) \tag{8}$$

$$\frac{1}{k} \frac{\partial \theta}{\partial \tau} + \frac{\partial \theta}{\partial z} = -(\beta + \theta) \tag{9}$$

$$\beta(0, z) = 1 \tag{10}$$

$$\theta(0, z) = 0 \tag{11}$$

$$\theta(\tau, 0) = 0 \tag{12}$$

The differential operations are linear and hyperbolic with constant coefficients. The coupling of the two equations leads to the more interesting mathematics.

A similar set of equations could be obtained to govern the vaporization and mass transport of dilute species in the liquid; θ would have a linear relation with gas-phase concentration of the dilute species while β would have a linear relation with liquid-phase concentration of the dilute species. The temperature field would be assumed to be uniform in that case. The above system of equations could also approximate convective heat transfer for a flow over an array of parallel cylinders. In that case S becomes the perimeter length of the cylinder cross-section, $n = b^{-2}$ where b is the average distance between cylinder centers, and h represents the heat transfer coefficient for an individual cylinder.

In Section 5, we shall discuss the general solution to the system of equations (8)–(12), but first we shall discuss some special cases.

3. LIMIT OF CONSTANT DROPLET TEMPERATURE

As mentioned in the Introduction, there are two characteristic times in the problem. In this section, we shall examine the case where the droplet heating time becomes infinite. In a physical sense, this means that the product of droplet mass and liquid specific heat becomes very large so that the temperature change in the droplet becomes insignificant although the gas temperature is decreasing. In a mathematical sense, τ becomes uniformly zero for finite time t; however, $k\tau$ remains finite and positive for finite and positive t.

The formal procedure to obtain the correct limit would be to set $k\tau \equiv y$, $\varepsilon \equiv k^{-1}$, and $\tau = \varepsilon y$. Then in the limit as ε tends to zero, equations (8) and (10) yield, after a simple zero-order solution, $\beta(y, z) = -1$. Now substitution into the system of equations (9), (11), and (12) yields

$$\frac{\partial \theta}{\partial y} + \frac{\partial \theta}{\partial z} + \theta = 1 \tag{13}$$

$$\theta(0, z) = 0 \tag{14}$$

$$\theta(y, 0) = 0 \tag{15}$$

Note that $\theta = 1$ is a particular solution and $\theta = e^{-z}$ and $\theta = e^{-y}$ are homogeneous solutions to equation (13). On physical grounds, one would expect a wave-like character to the general solution as well. In other words, information should be propagated along the gas element path $x - Ut = $ constant or equivalently, $z - y = $ constant. Also, we would expect after a long time to reach the steady-state solution given by $\theta = 1 - e^{-z}$.

Let us consider the Heaviside function defined by

$$H(z - y) = 0 \quad \text{for } z - y < 0$$
$$H(z - y) = 1 \quad \text{for } z - y > 0$$

$z = y$ may be viewed as the dividing line between the gases that were present in the cloud of droplets at the initial time $(z - y > 0)$ and those gases that entered the cloud after the initial time $(z - y < 0)$. The line $z = y$ is therefore an interesting wavefront.

The solution to the system (13)–(15) is

$$\theta = 1 + (H - 1)e^{-z} - He^{-y} \tag{16}$$

For $z - y > 0$ the above solution becomes

$$\theta = 1 - e^{-y} \tag{17}$$

while for $z - y < 0$ the solution is

$$\theta = 1 - e^{-z} \tag{18}$$

It is noteworthy that the expected steady solution has been obtained for large y without applying that condition as a constraint on the solution. In fact, the steady-state solution holds for all $y > z$. Also, it is interesting to see that the solution is continuous at $z = y$. Derivatives, however, are discontinuous at $z = y$. The sum of the two first derivatives is continuous across the line $z = y$ so that no discontinuity occurs in the sum of terms on the left-hand side of equation (13).

Equation (17) shows that the gas present in the cloud at the initial time cools uniformly in space. Its temperature is affected by the initial gas temperature but is not influenced by the temperature at the boundary $z = 0$. On the other hand, the temperature of the gas that flows into the cloud after the initial time depends on the boundary temperature but not on the initial temperature. It is noteworthy that if the boundary temperature and the initial temperature were taken to be different values, a discontinuity would exist along the line $z = y$ although the magnitude of the discontinuity would decrease exponentially with time. That is, if the initial value in equations (14) were $\theta_0 \neq 0$, the solution changes in that the last term of (16) is replaced by

$$- H(1 - \theta_0) e^{-y}$$

Calculated results for gas temperature are shown in Figure 1. It is seen that the solution is continuous but has two branches that meet precisely at $z = y$, as discussed above. The solution for $z > y$ is independent of z, while the solution for $z < y$ is steady. The peak value of θ is increasing with time, implying that the temperature is decreasing. The gas temperature approaches T_0, the fixed droplet temperature, asymptotically with increasing z and increasing y.

If the second-derivative term in equation (1) had been retained, we would expect that it should have an interesting effect in the vicinity of the wave front $z = y$ since gas-phase diffusion would smooth the solution there. In our own case, however, the solution behind the wave front only depends on the conditions upstream, not on downstream conditions or prior conditions. Even without consideration of gas-phase diffusion, this conclusion will be modified when the effects of droplet thermal inertia are taken into account. The temperature history of the droplet will influence the gas temperature in that case.

The limiting case opposite to the one considered in this section would involve $\varepsilon \to \infty$ or $k \to 0$. Then, the characteristic length and the characteristic residence time would go to infinity while the heating time remains finite. It can readily be shown that $\theta = 0$ and $\beta = - e^{-\tau}$ are the solutions here of the very simple ordinary differential equations that appear in this limit. A somewhat more interesting limiting situation will be examined in the next section.

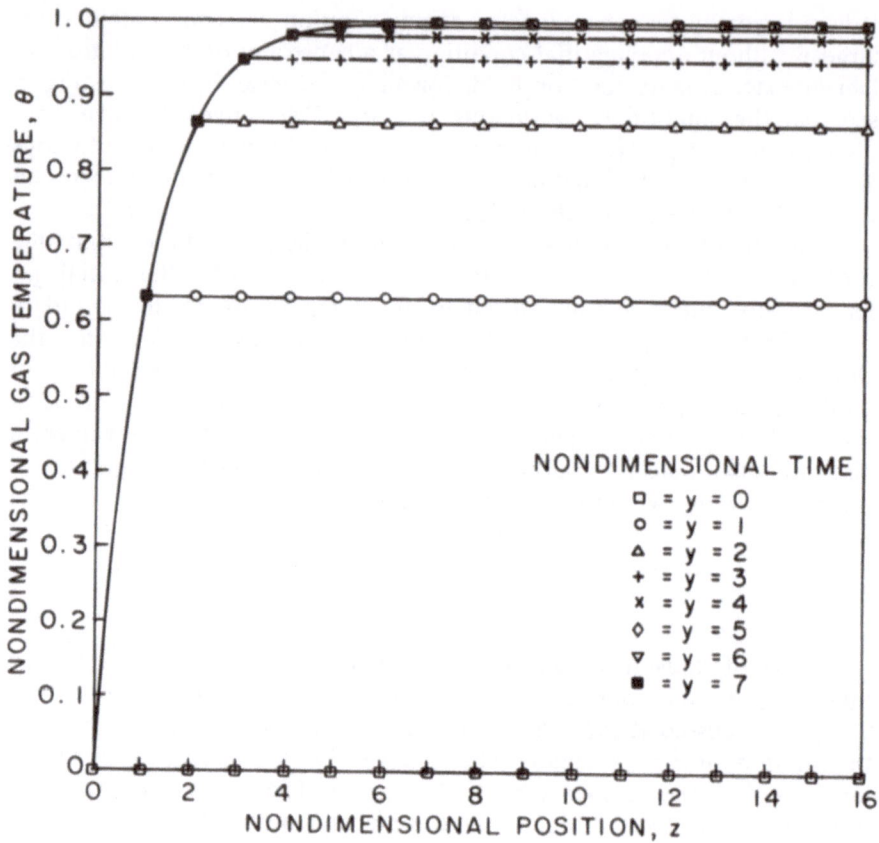

Fig. 1. Constant droplet temperature case: gas temperature behavior.

4. LIMIT OF QUASI-STEADY GAS-PHASE BEHAVIOR

Let us now consider a limit where a quasi-steady gas phase has been established but unsteady heating effects can occur for the droplets. Naturally, since the droplet and gas temperature are coupled we expect that both will vary with time. The limiting form of the equations may be obtained by letting $k \to \infty$ in equations (8)–(12) in which case

$$\frac{\partial \theta}{\partial z} + \theta = -\beta \qquad (19)$$

$$\frac{\partial \beta}{\partial \tau} = -(\beta + \theta) \qquad (20)$$

$$\theta(\tau, 0) = 0 \tag{21}$$

$$\beta(0, z) = -1 \tag{22}$$

This limit relates to the limit of the previous section in the following sense: we are now considering a limit where $\tau/\varepsilon = y \to \infty$ but $\tau = \varepsilon y$ is finite. The limit study of Section 3 dealt with a relaxation process to a quasi-steady gas-phase solution in which the droplet temperature remains fixed. In this section, after a longer time, the droplet temperature begins to increase while the gas temperature adjusts in a quasi-steady manner. Since a quasi-steady solution for θ is sought here, it is not necessary to apply initial conditions on θ. Essentially, the gas-phase solution after a long time becomes independent of the initial gas temperature but dependent on the boundary temperature. We should note, in this case, that the gas temperature will depend on the initial droplet temperature and the history of the droplet temperature.

Functions $\bar{\beta}(z)$ and $\bar{\theta}(z)$ are defined as the Laplace transforms (over time) of β and θ. Transformation of equation (20) yields

$$s\bar{\beta} = -(\bar{\beta} + \bar{\theta}) + \beta(0, z)$$
$$= -(\bar{\beta} + \bar{\theta}) - 1$$

or

$$\bar{\beta} = -\frac{1 + \bar{\theta}}{s + 1} \tag{23}$$

Substitution of equation (23) into the transformation of (19) gives the result

$$\frac{d\bar{\theta}}{dz} + \bar{\theta} = +\frac{1}{s+1}\bar{\theta} + \frac{1}{s+1} \tag{24}$$

Transformation of equation (21) yields the boundary condition

$$\bar{\theta}(0) = 0 \tag{25}$$

Formal integration of equation (24) subject to condition (25) can be carried out assuming that the right-hand side is given. The result is

$$\bar{\theta} = \frac{1}{s+1} \int_0^z e^{z'-z}\bar{\theta}dz' + \frac{1}{s+1}(1 - e^{-z}) \tag{26}$$

Now, $e^{-\tau}$ is the inverse transform of $(s+1)^{-1}$ so that, through use of the convolution theorem,

$$\theta(\tau, z) = \int_0^z \int_0^\tau e^{z'-z} e^{-(\tau-\tau')} \theta(\tau', z') d\tau' dz' + e^{-\tau}(1 - e^{-z}) \tag{27}$$

It can be seen that, in the limit $\tau \to 0$, equation (27) yields the result given by equation (18), as expected based upon the previous discussion of the relationship between the limiting cases of this section and of Section 3.

It is useful to define

$$w(\tau, z) \equiv e^z e^\tau \theta(\tau, z) \tag{28}$$

Then equation (27) may be written as

$$w(\tau, z) = \int_0^z \int_0^\tau w(\tau', z') d\tau' dz' + (e^z - 1) \tag{29}$$

The author has solved equation (29) by successive substitutions starting with the initial guess

$$w_0 = e^z - 1$$

and substituting this into the integrand to obtain

$$w_1 = e^z - 1 + \tau(e^z - 1 - z)$$

The general result is

$$w_n = e^z - 1 + \tau(e^z - 1 - z) + \frac{\tau^2}{2!}\left(e^z - 1 - z - \frac{z^2}{2!}\right) + \cdots$$

$$+ \frac{\tau^n}{n!}\left(e^z - 1 - z - \cdots - \frac{z^n}{n!}\right)$$

Term-by-term comparison with the Taylor-series expansion for the exponential function leads us to the conclusion that we have a convergent series. In fact, we can substitute the Taylor-series expansion for e^z and subtract terms to obtain

$$w(\tau, z) = \sum_{m=0}^\infty \left(\sum_{n=m+1}^\infty \frac{z^n}{n!}\right) \frac{\tau^m}{m!} \tag{30}$$

or, equivalently from equation (28),

$$\theta(\tau, z) = e^{-\tau} e^{-z} \sum_{m=0}^\infty \left(\sum_{n=m+1}^\infty \frac{z^n}{n!}\right) \frac{\tau^m}{m!} \tag{31}$$

One may wish to take advantage of numerical integration in order to solve equation (29) directly. Two approaches are possible, although neither has been pursued by the author. In one approach, successive substitution could be utilized with numerical quadrature of the integral in each step. Then a convergence criterion would be applied to the iteration process. Another approach would be to transform equation (29) to a partial differential equation with prescribed constraints and to solve that equation by a finite-difference scheme. In particular, equation (29) can be readily shown to be equivalent to

$$\frac{\partial^2 w}{\partial z \partial \tau} - w = 0, \qquad \text{for } \tau \geqslant 0, \quad z \geqslant 0 \tag{32}$$

with

$$w(0, z) = e^z - 1 \tag{33}$$

and

$$w(\tau, 0) = 0 \tag{34}$$

Equation (27) exhibits several interesting effects that are eventually reflected via the solution given by equation (31). First, the portion of the solution resulting from the "initial" condition [which is the final asymptotic behavior obtained in the previous section and is reflected through the last term in equation (27)] decays exponentially with time. Second, through the integral term, earlier values of θ and upstream values of θ affect the value of θ at a given position z and time τ. The weighting factor of the influence decreases exponentially with the distance (in both space and time) between the positions of the influenced value and the influencing value. In particular, the difference between the quantities $z + \tau$ at the two positions is critical in the weighting factor. This integral term is a consequence of the time-varying droplet temperature. Since an element of gas flows over many droplets, the gas temperature is affected by the upstream conditions at prior times.

Now let us solve equations (20) and (22) [or equivalently, equation (23)]. Either by solving the first-order ordinary differential equation or by an inverse transformation, we obtain

$$\beta = -e^{-\tau}\left[1 + \int_0^\tau e^{\tau'}\theta(\tau', z)d\tau'\right] \tag{35}$$

From (31), we see that

$$\int_0^\tau e^{\tau'}\theta(\tau', z)d\tau' = e^{-z}\sum_{m=0}^{\infty}\left(\sum_{n=m+1}^{\infty}\frac{z^n}{n!}\right)\int_0^\tau \frac{(\tau')^m}{m!}d\tau'$$

$$= e^{-z}\sum_{m=0}^{\infty}\left(\sum_{n=m+1}^{\infty}\frac{z^n}{n!}\right)\frac{\tau^{m+1}}{(m+1)!}$$

Therefore,

$$\beta(\tau, z) = -e^{-\tau} - e^{-\tau}e^{-z} \sum_{m=0}^{\infty} \left(\sum_{n=m+1}^{\infty} \frac{z^n}{n!} \right) \frac{\tau^{m+1}}{(m+1)!} \qquad (36)$$

It is obvious from equation (35) that the droplet temperature at a given position z depends on the history of the gas temperature at that same position with an exponential weighting factor that favors the more recent history.

Since, for positive z, we know by comparison of the Taylor series for e^z with equation (30) that

$$e^z > \sum_{n=m+1}^{\infty} \frac{z^n}{n!}$$

it follows that

$$w(\tau, z) < e^z e^\tau$$

and

$$\theta(\tau, z) < 1$$

In similar fashion, it can be seen that equation (36) implies $\beta(\tau, z) > -1$.

Figure 2 shows that θ is monotonically increasing with position z and monotonically decreasing with time. It follows that gas temperature decreases with z and increases with time. From Figure 3, it is seen that β (and therefore droplet temperature) decreases with z and increases with time. This indicates that the gas loses heat to the droplet as it flows downstream. Meanwhile, the droplet temperature increases with time due to the heating. As a net result, the temperature difference and the heat loss rate decrease with time; this implies that, with time, the spatial variation in gas temperature will decrease.

It is suspected from a physical viewpoint that for large τ the droplets and the gas tend toward the same temperature $T = T_s = T_l$ or $\theta = \beta = 0$. The physical argument is as follows: The energy required to bring the droplet temperature from T_0 to the final temperature is a bounded value (for a given finite volume), while the energy removed from the gas is given by a time integral of the product of mass flux through the volume and gas specific-enthalpy difference across the volume. Since the mass flux remains finite for an infinite time, this second energy integral can only remain finite and balance the first energy integral if the enthalpy difference goes to zero. This means that after a long time the gas temperature does not change in flowing through the volume. This argument will be somewhat modified when vaporization occurs, since the droplet temperature becomes limited by the boiling point.

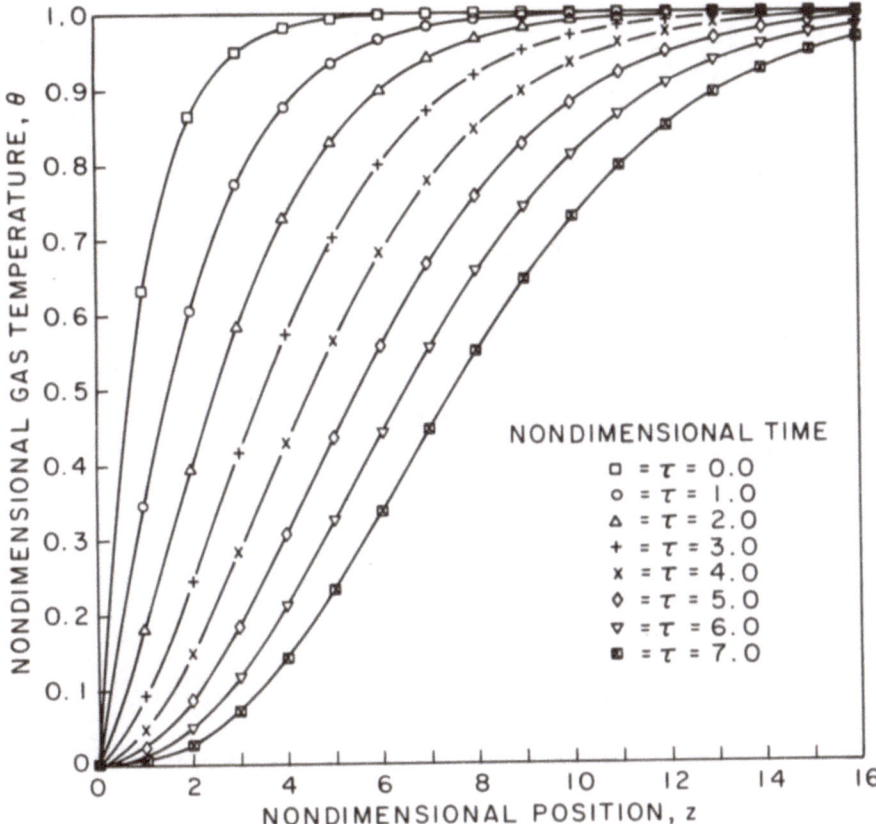

Fig. 2. Quasi-steady gas-phase case: gas temperature behavior.

5. GENERAL SOLUTION

Let us now examine the solution to the system of equations (8)–(12). Both characteristic times will remain in this analysis and no inherent assumption is made *a priori* about the relative magnitudes of the two times. The Laplace transformation of the system yields

$$\frac{d\bar{\theta}}{dz} + \left(\frac{s}{k} + 1\right)\bar{\theta} = \frac{1}{s+1}\,\bar{\theta} + \frac{1}{s+1} \tag{37}$$

$$\bar{\theta}(0) = 0 \tag{38}$$

We note that identical equations to equation (23) and therefore to equation

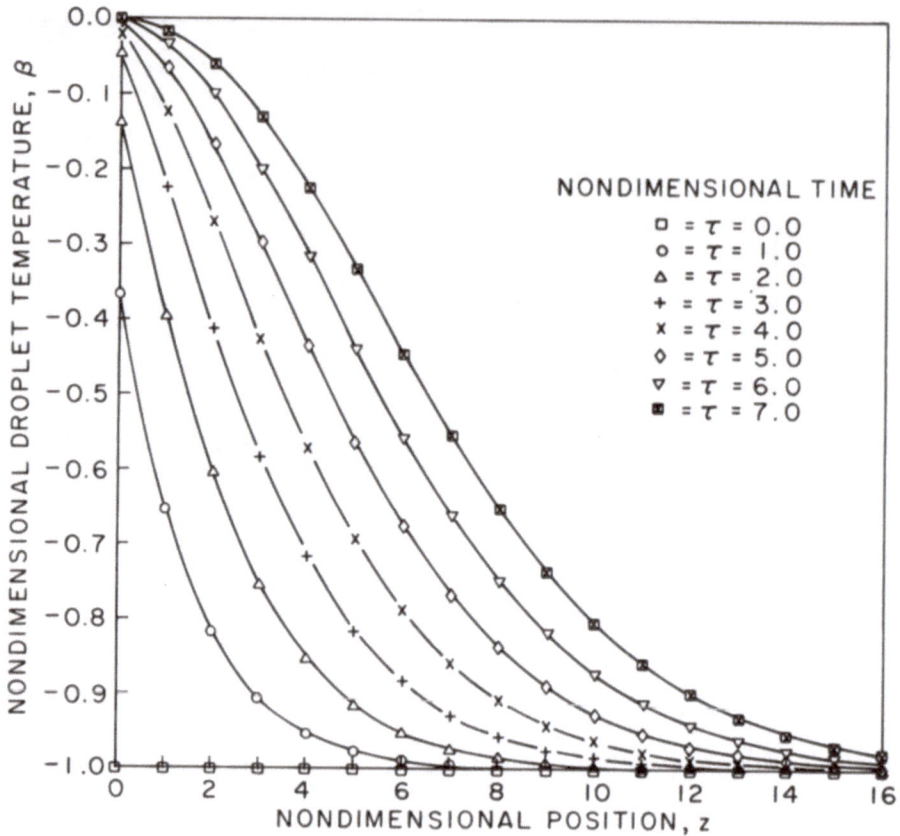

Fig. 3. Quasi-steady gas-phase case: droplet temperature behavior.

(35) would also be obtained via the transformation, algebraic manipulation, and inverse transformation.

The right-hand side of equation (37) will be treated formally as a known quantity in solving a first-order linear ordinary differential equation. The result is

$$\bar{\theta} = \frac{1}{s+1} \int_0^z \bar{\theta} e^{-s(z-z')/k} e^{-(z-z')} dz' + \frac{1}{s+1} \frac{k}{k+s} (1 - e^{-z} e^{-sz/k}) \quad (39)$$

Note that the inverse transform of $[1/(s+1)] e^{-s(z-z')/k}$ is given by

$$\int_0^\tau e^{-\tau'} \delta \left[\frac{z'-z}{k} + (\tau - \tau') \right] d\tau' = e^{-\tau} e^{(z-z')/k} H\left(\tau - \frac{z}{k} + \frac{z'}{k} \right)$$

where δ is the Dirac delta function. The algebraic sum of the terms in the exponents on the right-hand side must be negative or zero. In any domain where they seem to be a positive quantity, the exponential function is replaced by zero as indicated by the Heaviside function. Now the inverse transform of

$$\bar{\theta}\frac{1}{s+1}e^{-s(z-z')/k}$$

is

$$\int_0^\tau e^{-(\tau-\tau')}e^{(z-z')/k}\theta(\tau',z')[H(\tau-\tau'-z/k+z'/k)]d\tau'$$

where $\tau'-z'/k \leqslant \tau-z/k$ is understood. Otherwise the kernel is zero. The resulting integral equation is

$$\theta=\int_0^z\int_0^\tau e^{-(z-z')}e^{(z-z')/k}e^{-(\tau-\tau')}\theta(\tau',z')[H(\tau-\tau'-z/k+z'/k)]d\tau'dz'$$

$$+\frac{k}{1-k}(e^{-z}e^{z/k}-1)e^{-\tau}H(\tau-z/k)$$

$$+\frac{k}{1-k}(e^{-k\tau}-e^{-\tau})[1-H(\tau-z/k)] \tag{40}$$

Now define

$$\tilde{w}(\tau,z)\equiv e^z e^{-z/k}e^\tau\theta(\tau,z) \tag{41}$$

Then equation (40) becomes

$$\tilde{w}(\tau,z)=\int_0^z\int_0^\tau \tilde{w}(\tau',z')H(\tau-\tau'-z/k+z'/k)d\tau'dz'$$

$$+\frac{k}{1-k}(1-e^z e^{-z/k})H(\tau-z/k)$$

$$+\frac{k}{1-k}e^z e^{-z/k}(e^{(1-k)\tau}-1)[1-H(\tau-z/k)] \tag{42}$$

The forcing function is continuous but not smooth at $z=k\tau$. In the limit $k\to\infty$, equations (37), (39), (40), and (42) tend respectively to equations (24), (26), (27), and (29). We cannot reduce equation (42) to a form equivalent to equations (32), (33), and (34) since the forcing function is time-dependent (in a portion of the domain).

For $\tau-z/k<0$, the solution at a position τ, z is independent of the boundary values at $z=0$ since for such times the gas elements that were at $z=0$ at time $\tau=0$ have not yet reached the position z. The solution at such a position τ, z depends only on the initial values along the line $\tau = 0$ between the values of $z - k$ (the lower bound) and z (the upper bound). Since θ and β are given as constants along this line and since the distance from $z = 0$ is not relevant (for $\tau - z/k < 0$), we can intuitively expect a solution in this domain to be independent of z. The solution may be found by seeking a z-independent solution to equations (41) and (42) or, more readily, by solving equations (8)–(11), which reduce to a second-order system of ordinary differential equations in this zone. The solutions are

$$\theta = \frac{k}{k+1}(1 - e^{-(1+k)\tau}) \tag{43}$$

and

$$\beta = -\frac{k}{k+1} - \frac{1}{k+1} e^{-(1+k)\tau} \tag{44}$$

for $\tau \geq 0$ and $\tau - z/k \leq 0$.

It is seen that both characteristic times appear in this solution. In the limit $k \to \infty$, $\beta \to -1$ and θ goes to the limit given by equation (17). We note that $z - y \geq 0$ is the same statement as $\tau - z/k \leq 0$.

The integral equation (42) can be expressed in the domain $\tau - z/k \geq 0$ as follows:

$$\tilde{w}(\tau, z) = \int_0^z \int_{z'/k}^{\tau - z/k + z'/k} \tilde{w}(\tau', z') d\tau' dz' + \tilde{w}(z/k, z) \tag{45}$$

From equations (43) and (41), we know that

$$\tilde{w}(z/k, z) = \frac{k}{k+1}(e^z - e^{-z/k}) \tag{46}$$

The series solution may be generated by successive substitution using the forcing function as initial guess. The results are

$$\tilde{w} = \frac{k}{k+1} \sum_{m=0}^{\infty} \left(\sum_{n=m}^{\infty} \frac{z^n}{n!} \right) \frac{(\tau - z/k)^m}{m!} - \frac{k}{k+1} \sum_{m=0}^{\infty} \left(\sum_{n=m}^{\infty} \frac{(-z/k)^n}{n!} \right) \frac{(z - k\tau)^m}{m!} \tag{47}$$

and, from equation (41),

$$\theta = \frac{k}{k+1} e^{-z} e^{-(\tau - z/k)} \left\{ \sum_{m=0}^{\infty} \left(\sum_{n=m}^{\infty} \frac{z^n}{n!} \right) \frac{(\tau - z/k)^m}{m!} \right.$$
$$\left. - \sum_{m=0}^{\infty} \left(\sum_{n=m}^{\infty} \frac{(-z/k)^n}{n!} \right) \frac{(z - k\tau)^m}{m!} \right\} \tag{48}$$

for $\tau - z/k \geq 0$. Again, two time scales (namely τ and $k\tau$) as well as two space scales (z and z/k) are displayed in the solution. Equations (30) and (31) may be recovered from (47) and (48) in the limit $k \to \infty$. We note that constant values of $\tau - z/k$ (or $z - k\tau$) are gas-particle paths so that a Lagrangian time is seen to be important. The distances from the boundary, z and z/k, also appear independently of the Lagrangian time. On physical grounds, one would expect that for large τ, $T \to T_1$ or equivalently $\theta \to 0$. In the present case, however, the asymptotic behavior has not yet been carefully analyzed.

In comparing the kernels of equations (27) and (40), we can see the effect of gas-phase unsteadiness. The time required for a gas particle to move from z' to z is given by $(z - z')/k$. The influence of the temperature at τ', z' on the temperature at a point τ, z is greater because of this movement of the gas in a finite time. On the other hand, on account of this residence-time effect, there are values of $\theta(\tau', z')$ at certain $\tau' < \tau$ and $z' < z$ that have no influence on the solution of θ at particular τ and z values. That is, if $\tau' - z'/k > \tau - z/k$, the solution at τ' and z' will not count in the integral of equation (40). For the quasi-steady solution where residence time is zero, that solution would have an influence on the solution at the point τ and z.

Equations (35), (43), and (48) may be combined to yield the solution for β in the range $\tau - z/k \geq 0$, $z \geq 0$. The result is

$$\beta = -\frac{k}{k+1}\left[e^{-(\tau - z/k)} + \frac{1}{k}e^{-(z+\tau)}\right]$$

$$-\frac{ke^{-z}}{k+1}e^{-(\tau - z/k)}\left\{\sum_{m=0}^{\infty}\left(\sum_{n=m}^{\infty}\frac{z^n}{n!}\right)\frac{(\tau - z/k)^{m+1}}{(m+1)!}\right.$$

$$\left.+\frac{1}{k}\sum_{m=0}^{\infty}\left(\sum_{n=m}^{\infty}\frac{(-z/k)^n}{n!}\right)\frac{(z - k\tau)^{m+1}}{(m+1)!}\right\} \quad (49)$$

Again, two time scales and two space scales are seen in the solution for β, as in equation (48). The droplet temperature is affected by the gas temperature in its vicinity over the previous time, as indicated by equation (35). The gas in that vicinity at a given instant has a temperature that is history-dependent, so that the dependence on the Lagrangian time appears. Since the gas-energy equation is coupled to the droplet-energy equation, the same Lagrangian time appears in equation (48) for θ.

The convergence of the series in equations (47), (48), and (49) can be demonstrated by comparison with Taylor-series expansions for exponential functions that provide upper bounds. The exercise in Section 4 may be considered as an example. In the limit of $\varepsilon = k^{-1} \to 0$, $\tau = \varepsilon y \to 0$. Equation (48) yields the result of equation (18) and $\beta \to -1$. In the limit $k \to 0$,

$\tau - z/k$ is always negative for finite τ and z. Equations (47), (48), and (49) will therefore not apply. Equations (43) and (44) yield the limits $\theta = 0$ and $\beta = -e^{-\tau}$, which agree with the result stated at the end of Section 3.

Results for various values of k are shown in Figures 4–9. Temperature θ is displayed for $k = 10.0$, 1.00, and 0.10 in Figures 4, 6, and 8, respectively, while β is given for those k values in Figures 5, 7, and 9. Figures 2 and 3 essentially represent the case where $k = \infty$. We recall that a decrease in k results from a decrease in liquid thermal inertia or from an increase in gas-phase thermal inertia. The same qualitative trends are seen for all values of k, namely the gas temperature increases with time and decreases with position z while the droplet temperature increases with time and decreases with position z. Again, the gas loses heat to the droplets

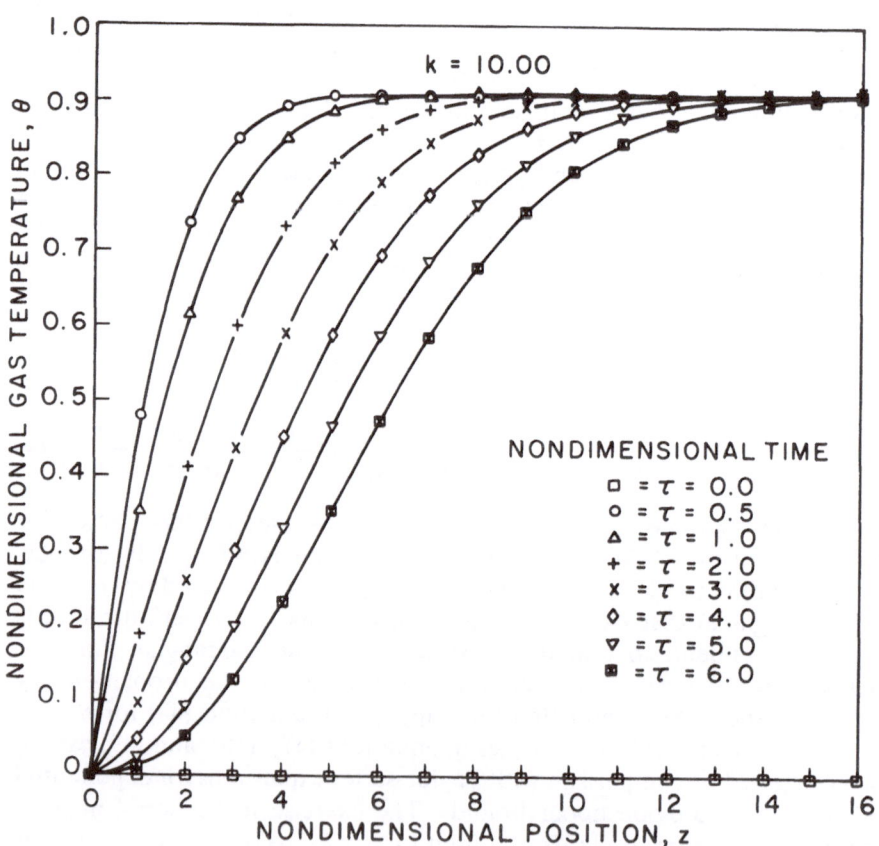

Fig. 4. General case: gas temperature behavior, $k = 10.00$.

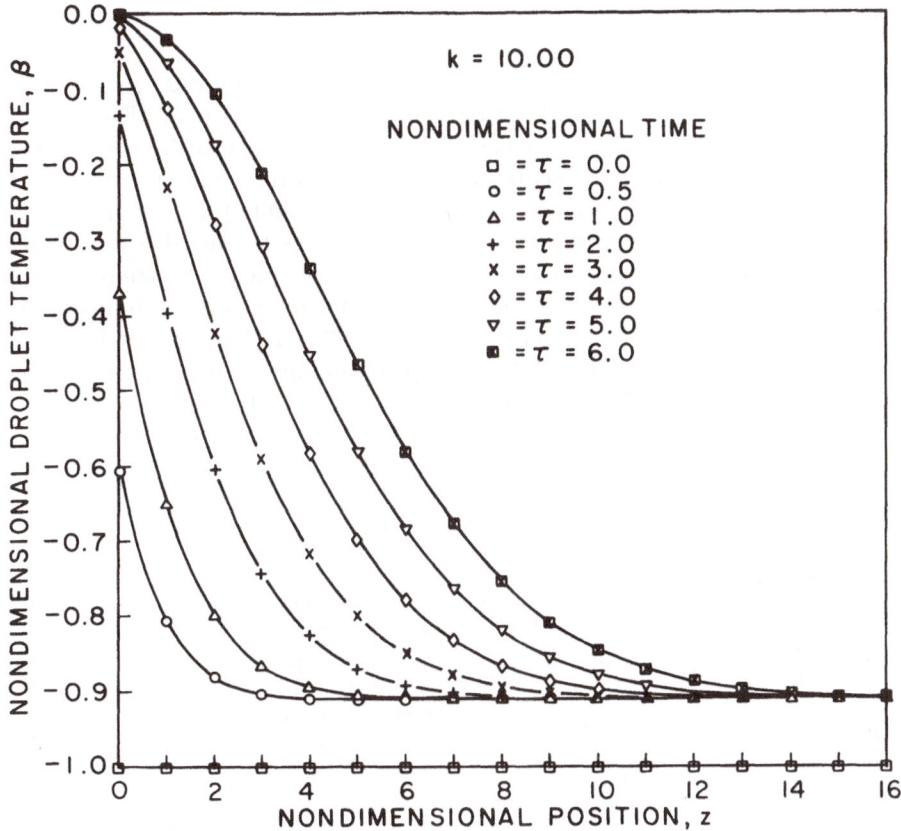

Fig. 5. General case: droplet temperature behavior, $k = 10.00$.

so causing the droplets to heat, thereby decreasing the temperature difference and the heat loss with time.

It should also be noted that $\theta = -\beta = $ constant is an asymptotic solution to equations (8) and (9), or to equations (19) and (20), for large z. The figures show that the value of the constant increases with increasing k, going to unity as k becomes infinite. Hence the larger the ratio of liquid thermal inertia to gas thermal inertia, the closer the asymptotic temperature is to the initial droplet temperature.

For finite z, the physical argument given at the end of Section 4 should still describe the long-time behavior, i.e., after a long time both the gas temperature and the droplet temperature at given position z should tend to the initial gas temperature.

6. CONCLUDING REMARKS

By introducing some simplifying physical assumptions, a tractable linear system of partial differential equations has been obtained. The effects of two different time scales and associated space scales are clearly discerned, so that insight has been obtained into the complexities of the unsteady problem. A region of domination by initial values has been identified as distinct from the region where both initial values and boundary values are important. The solutions are piecewise analytic and continuous. (A discontinuity along $z = y$ is expected when the initial value of $z = 0$ and the boundary value of $\tau = 0$ are different.) The dependence of the gas and droplet temperatures on upstream and prior temperatures is nicely demonstrated by an integral-equation representation.

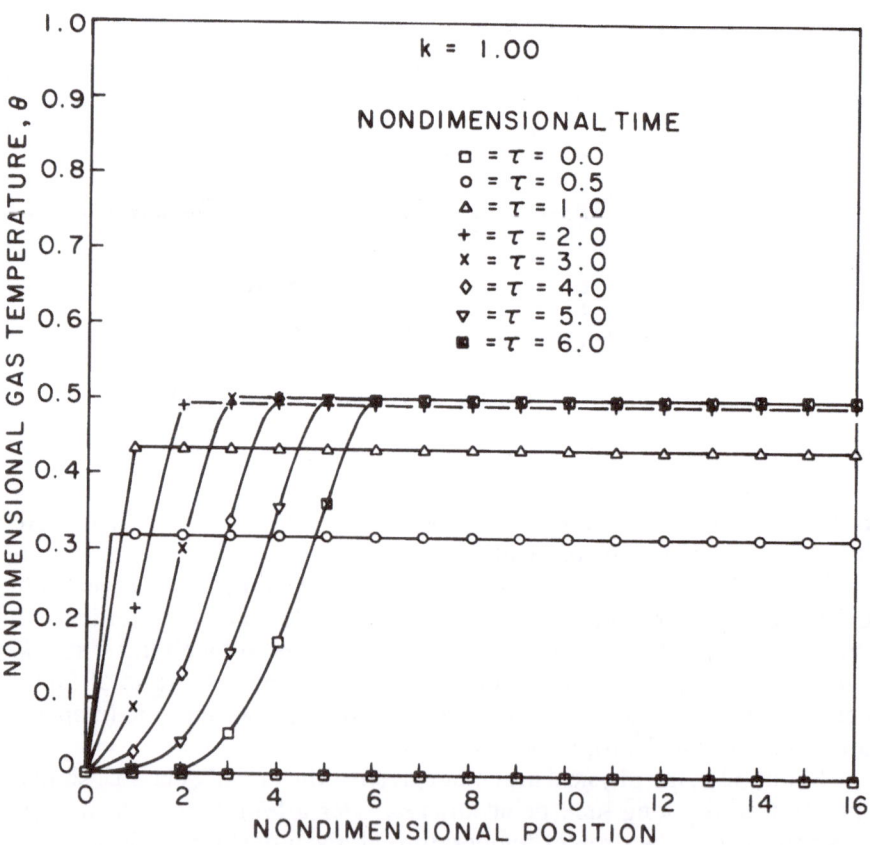

Fig. 6. General case: gas temperature behavior, $k = 1.00$.

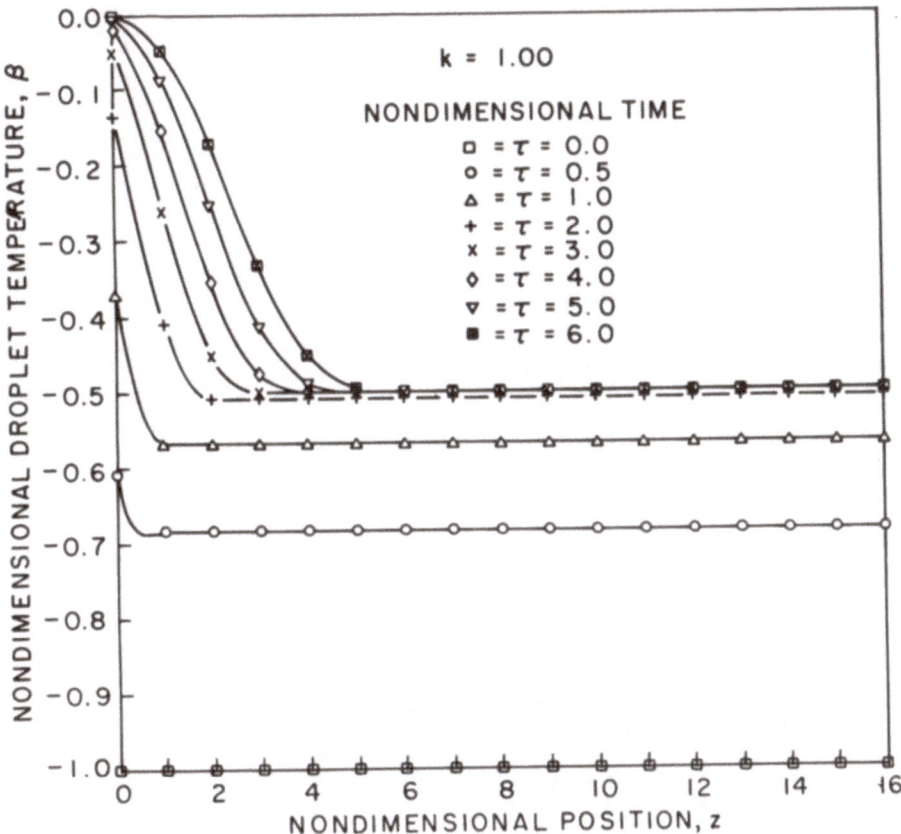

Fig. 7. General case: droplet temperature behavior, $k = 1.00$.

The formulation is promising as regards the addition of other physical complexities while retaining the linearity of the system. Potential examples are gas-phase diffusion, higher space-dimension, spatial temperature gradients and internal circulation in the droplets, and coupled heat and mass transfer. The added physics should introduce new characteristic time and space scales and may modify the present scales. The linear models should provide useful insights and guidelines for the numerical analyst attempting to solve the nonlinear problem. Possibly, the linear models can also provide a basis for testing some numerical schemes. Certain nonlinear physical effects may be analyzed by use of an iterative approach with the present type of formulation.

The general solution agrees with the asymptotic solutions. In particular, asymptotes have been obtained for (i) large liquid thermal inertia

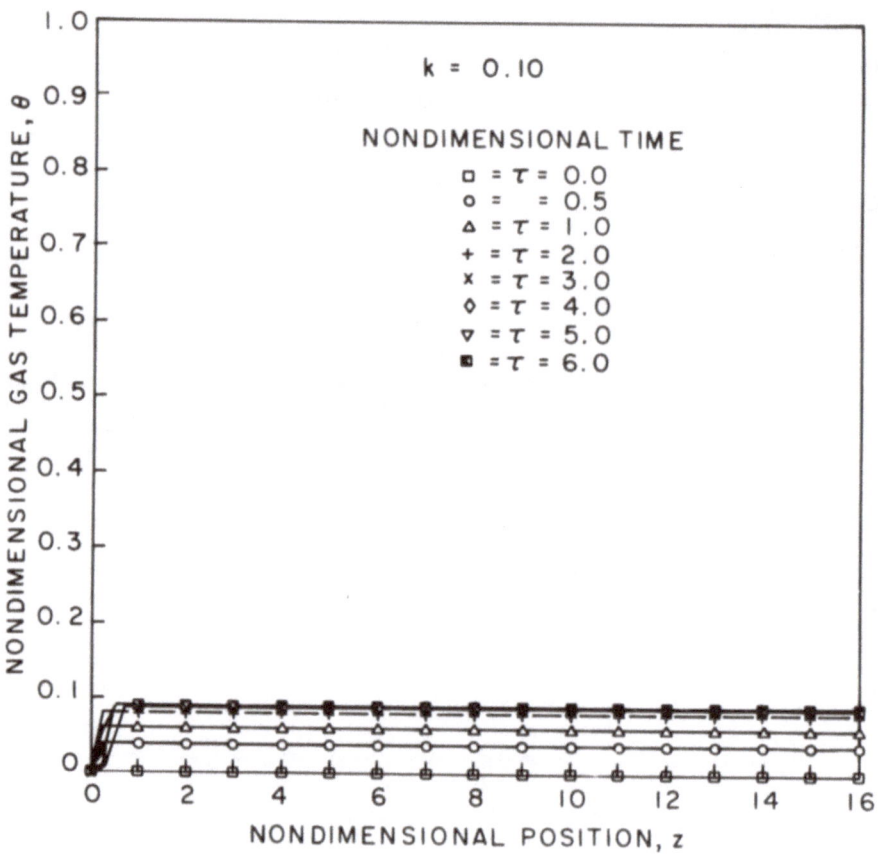

Fig. 8. General case: gas temperature behavior, $k = 0.10$.

$(k = \infty)$ and short time ($\tau = 0$, y finite), (ii) small liquid thermal inertia ($k = 0$), and (iii) large liquid thermal inertia ($k = \infty$) and long time (τ finite, y infinite). More importantly, the asymptotic solutions have been especially useful in clarifying the problem and its general solution.

ACKNOWLEDGMENT. The author greatly appreciates this opportunity to participate in the Festschrift for Professor Luigi Crocco, whose outstanding examples as scholar, teacher, gentleman, and friend have had a profound and lasting positive impact on many of our lives. The assistance of Mr. Paul Beatty in performing the calculations is appreciated.

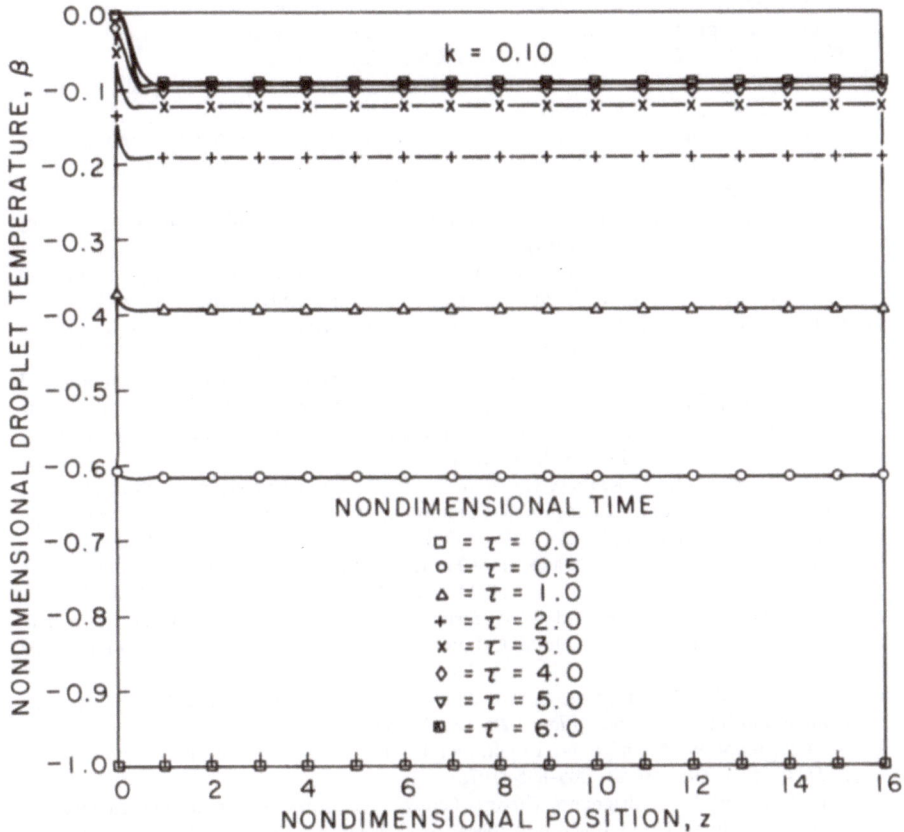

Fig. 9. General case: droplet temperature behavior, $k = 0.10$.

REFERENCES

1. T. Suzuki and H. H. Chiu, Multidroplet combustion of liquid propellants, in: *Proc. Int. Symp. Space Technol. Sci. 9th*, pp. 145–154 (1971).
2. H. H. Chiu, H. Y. Kim, and E. J. Croke, Internal group combustion of liquid droplets, *Nineteenth Symp. (Int.) on Combust.*, The Combustion Institute, Pittsburgh, PA (1983) (in press).
3. N. A. Chigier, Instrumentation techniques for studying heterogeneous combustion, *Prog. Energy Combust. Sci.* **3**, 175–189 (1977).
4. N. A. Chigier, *Energy, Combustion and Environment*, McGraw-Hill Book Co., New York (1981).
5. J. J. Sangiovanni and L. G. Dodge, Observations of flame structure in the combustion of monodispersed droplet streams, in: *Seventeenth Symp. (Int.) on Combust.*, pp. 455–465, The Combustion Institute, Pittsburgh, PA (1979).

6. M. Labowsky and D. E. Rosner, "Group" combustion of droplets in fuel clouds: I Quasi-steady predictions, in: *Evaporation-Combustion of Fuels, Advances in Chemistry Series 166* (J. T. Zung, ed.), pp. 63–79, ACS (1978).

7. M. Labowsky, The calculation of the burning rates of interacting fuel droplets, *Combust. Sci. Technol.* **22**, 217–226 (1980).

8. S. M. Correa and M. Sichel, The group combustion of a spherical cloud of monodisperse fuel droplets, in: *Nineteenth Symp. (Int.) on Combust.*, The Combustion Institute, Pittsburgh, PA (1983) (in press).

9. W. A. Sirignano, Fuel droplets vaporization and spray combustion, *Prog. Energy Combust. Sci.* (to appear in 1984).

10. B. Seth, S. K. Aggarwal, and W. A. Sirignano, Flame propagation through an air–fuel spray mixture with transient droplet vaporization, *Combust. Flame* **39**, 149–168 (1980).

11. S. K. Aggarwal, G. J. Fix, D. N. Lee, and W. A. Sirignano, Numerical optimization studies of axisymmetric unsteady spray, *J. Comput. Phys.* **50** (1) (1983).

12. S. K. Aggarwal, G. J. Fix, D. N. Lee, and W. A. Sirignano, Numerical computation of fuel–air mixing in a two-phase axi-symmetric coaxial free jet flow, in: *Advances in Computer Methods for Partial Differential Equations — IV* (R. Vichnevetsky and R. S. Stepleman, eds.), pp. 317–323, IMACS (1981).

13. W. A. Sirignano and C. K. Law, Transient heating and liquid-phase mass diffusion in fuel droplet vaporization, in: *Evaporation-Combustion of Fuels, Advances in Chemistry Series 166* (J. T. Zung, ed.), pp. 3–26, ACS (1978).

14. R. Tal (Thau) and W. A. Sirignano, Heat Transfer in Sphere Assemblages at Intermediate Reynolds Number: A Cylindrical Cell Model, ASME Preprint 81-WA/HT-44 (1981).

15. R. Tal (Thau), D. N. Lee, and W. A. Sirignano, Hydrodynamics and Heat Transfer in Sphere Assemblages: Multisphere Cylindrical Cell Models, AIAA Preprint 82-0302 (1982).

16. C. K. Law and W. A. Sirignano, Unsteady droplet combustion with droplet heating — II: Conduction limit, *Combust. Flame* **28**, 175–186 (1977).

17. S. Prakash and W. A. Sirignano, Liquid fuel droplet heating with internal circulation, *Int. J. Heat Mass Transfer* **21**, 885–895 (1978).

18. S. Prakash and W. A. Sirignano, Theory of convective droplet vaporization with unsteady heat transfer in the circulating liquid phase, *Int. J. Heat Mass Transfer* **23**, 253–268 (1980).

19. P. Lara-Urbaneja and W. A. Sirignano, Theory of transient multicomponent droplet vaporization in a convective field, in: *Eighteenth Symp. (Int.) on Combust.*, pp. 1365–1374, The Combustion Institute, Pittsburgh, PA (1981).

20. A. Tong and W. A. Sirignano, Analytical solution for diffusion and circulation in a vaporizing droplet, in: *Nineteenth Symp. (Int.) on Combust.*, pp. 1007–1020, The Combustion Institute, Pittsburgh, PA (1983).

V
COMBUSTION

12

Experiments on Solid Propellants Combustion: Experimental Apparatus and First Results

C. Casci

ABSTRACT. In solid propellant rocket combustion optical measurements at a fixed point of the gas phase thermal profile are very important, since they can be used to check the heterogeneous combustion theory briefly described in the first section of this paper.

An apparatus keeping the burning surface of the propellant sample fixed relative to the combustion chamber was realized by means of a stepping motor lifting the cylindrical propellant with a velocity equal to the burning-surface regression rate. The position of the burning surface is sensed by means of a laser system. A series of experimental runs were performed on an AP-based solid propellant at subatmospheric pressure. Under those conditions combustion of the propellant occurs with self-sustained burning oscillations. The signals of the feeding-motor velocity and of the light emitted by the gas phase at a distance of 5 mm from the burning surface were recorded. The system is capable of keeping the burning-surface position fixed, relative to the laboratory, within an error of a few tenths of a millimeter. The results show that the frequency of the burning oscillations and the mean burning rate agree with the data obtained with other experimental methods.

1. INTRODUCTION

Experimental work at ISREP on solid rocket propellant combustion is directed toward the measuring of quantities and parameters that character-

C. CASCI • Dipartimento di Energetica — Politecnico di Milano, Piazza Leonardo da Vinci, 32-20133 Milano, Italy.

ize the heterogeneous deflagration wave under steady or dynamic conditions. In both cases measurements of relevant quantities in the gas phase (like temperature and velocity) are very important. Until now this type of measurement has been performed with the propellant burning surface moving, relative to the laboratory reference frame, with a velocity equal to the regression (or burning) rate. The volume probed therefore moves along the gas profile at the same velocity, while leaving the burning surface.

In many instances it is especially interesting to maintain a fixed distance between burning surface and volume probed. In fact, this allows acquisition of data at a particular point over an extended period of time, a significant capability, for instance, when studying combustion transients.

Future experiments with an apparatus capable of performing this task will yield velocity and temperature data in the gas phase of a burning solid rocket propellant. They can be used to check the theoretical analysis developed by the heterogeneous combustion group at Dipartimento di Energetica of Politecnico di Milano.

2. THEORETICAL MODEL

Analysis of heterogeneous combustion is based on the mathematical description of a combustion wave propagating inside the condensed (solid) phase, and using the classic Fourier's equation. The system is assumed one-dimensional and, by means of a galilean transform, the origin of the reference frame is fixed to the burning surface. A sketch of the system is given in Figure 1. For a complete treatment of this subject see De Luca.[1]

The energy equation of the condensed phase, with the appropriate initial and boundary conditions in its simplest nondimensional form, is as follows (see nomenclature):

$$(\partial\theta/\partial\tau) + R(\partial\theta/\partial X) = \partial^2\theta/\partial X^2$$

$$\theta(X, \tau = 0) = \text{assigned function}$$

$$\theta(X \rightarrow -\infty, \tau) = \theta_{-\infty} \tag{1}$$

$$K_c(\partial\theta/\partial X)_{c,s} = K_g(\partial\theta/\partial X)_{g,s} + RH_s - \dot{q}_1(\theta_s)$$

To analyze the burning stability, an integral method, similar to one used by Von Karman and Polhausen,[2] was developed by De Luca.[3,4] A finite temperature disturbance

$$u(X, \tau) = \theta(X, \tau) - \bar{\theta}(X)$$

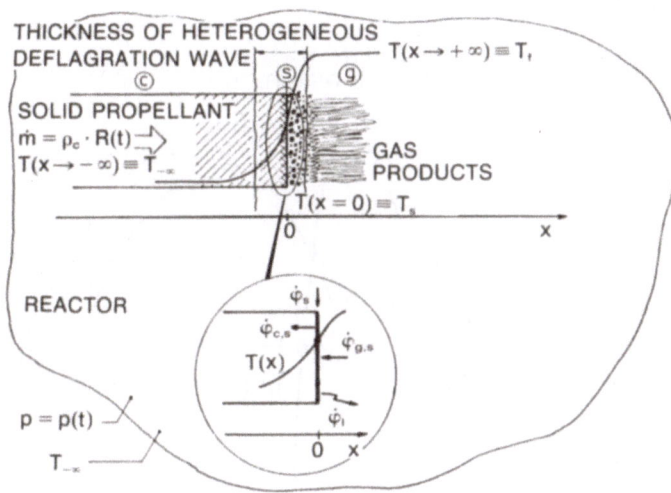

Fig. 1. Physical system schematic.

is introduced in equation (1) in order to have a penetration-type solution. After some mathematical manipulations an ordinary differential equation of the type[4]

$$d\theta_s/d\tau = f(\theta_s - \bar{\theta}_s)$$

is obtained, where $f(\theta_s - \bar{\theta}_s)$ is the so-called "Static Restoring Function." Its physical meaning is similar to the stiffness of a spring-mass oscillator, after a finite-size disturbance has moved it from its equilibrium position. A plot of a static restoring function is shown in Figure 2. From the zeroes of this function a bifurcation diagram can be obtained (see Figure 3), from which predictions about burning stability are possible. In the particular case of self-sustaining burning oscillations, the range of pressures where they occur is predicted. From this stability analysis the pressure deflagration limit and the dynamic extinction limit can also be predicted.[5,6]

Combustion simulations can be obtained from the numerical integration of equation (1). The numerical code developed by the heterogeneous combustion group in Milano can simulate different kinds of combustion situations, like pressure transients, shown in Figure 4, or the combustion oscillations of Figure 5. By means of this code a complete description is possible of the behavior of relevant quantities like burning rate, surface temperature, flame temperature, and gas velocity; checking of this mathematical model, however, requires a comparison between computed and measured quantities. In this frame, the experimental apparatus presented here can be extremely useful.

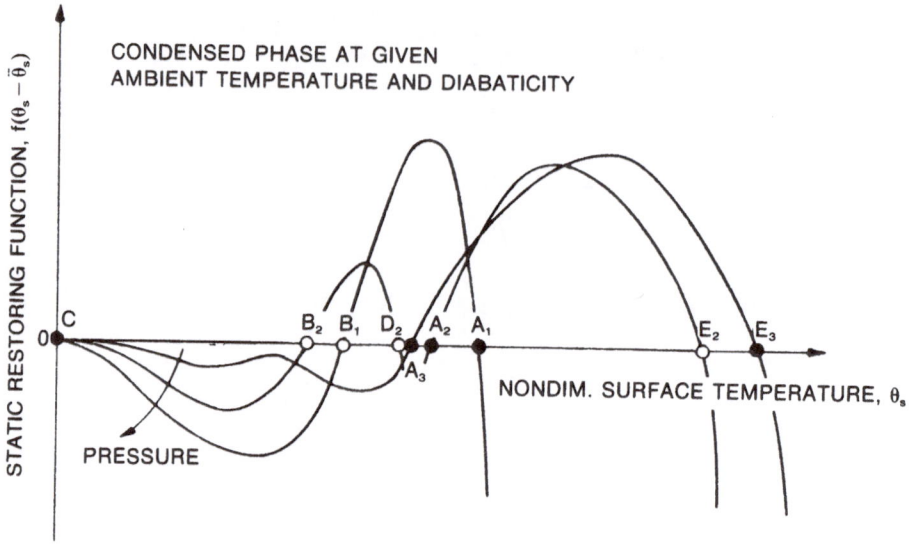

Fig. 2. State restoring function vs. burning-surface temperature plot.

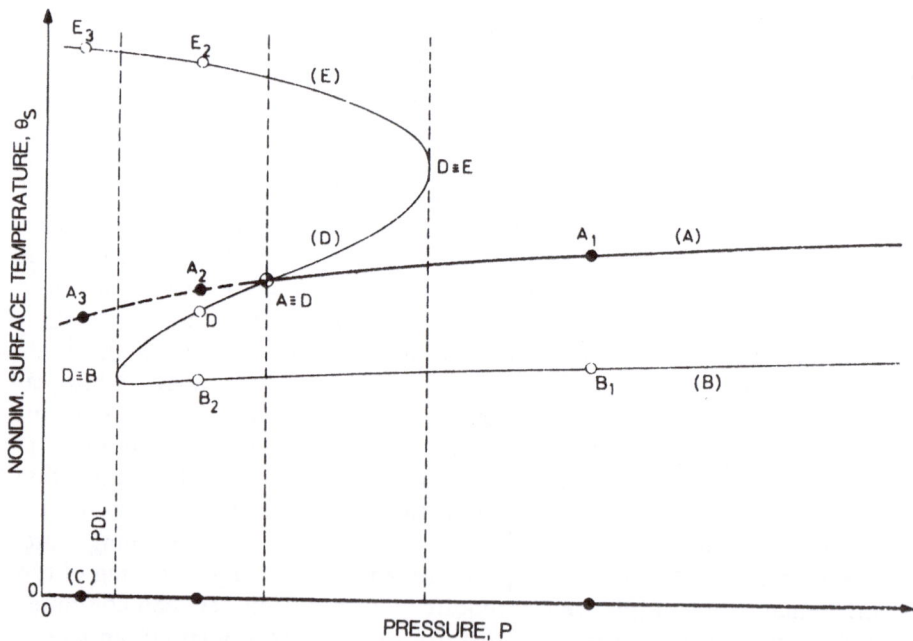

Fig. 3. Bifurcation diagram: burning-surface temperature vs. pressure.

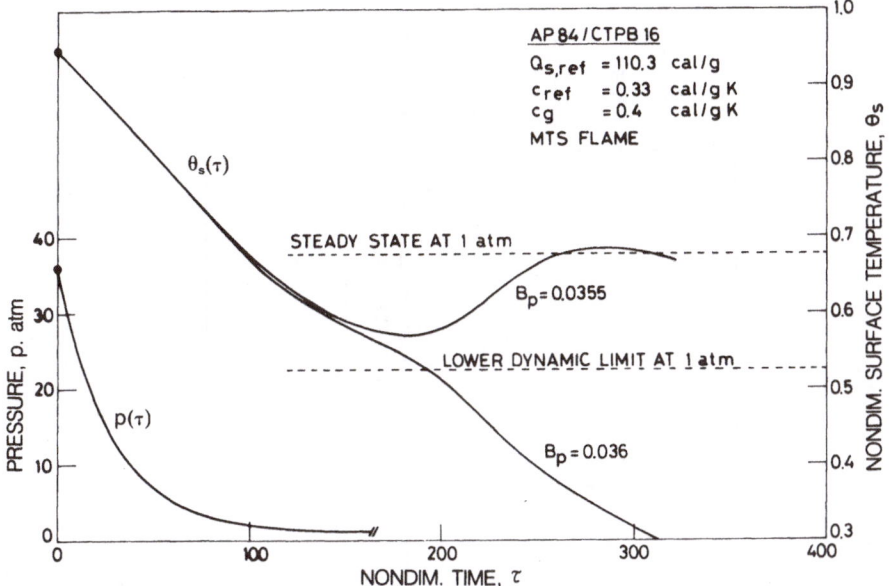

Fig. 4. Numerical simulation of a depressurization.

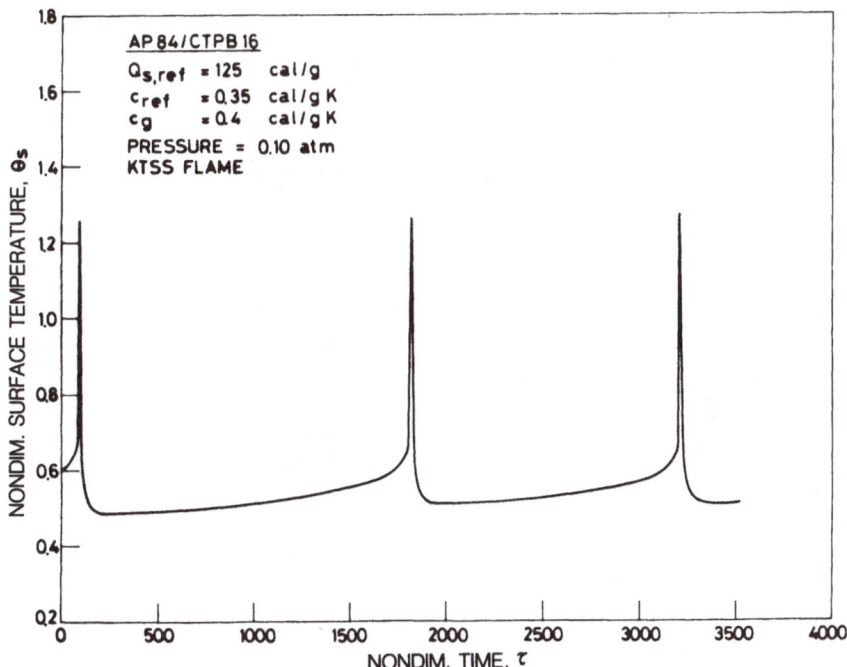

Fig. 5. Numerical simulation of self-sustaining burning oscillations.

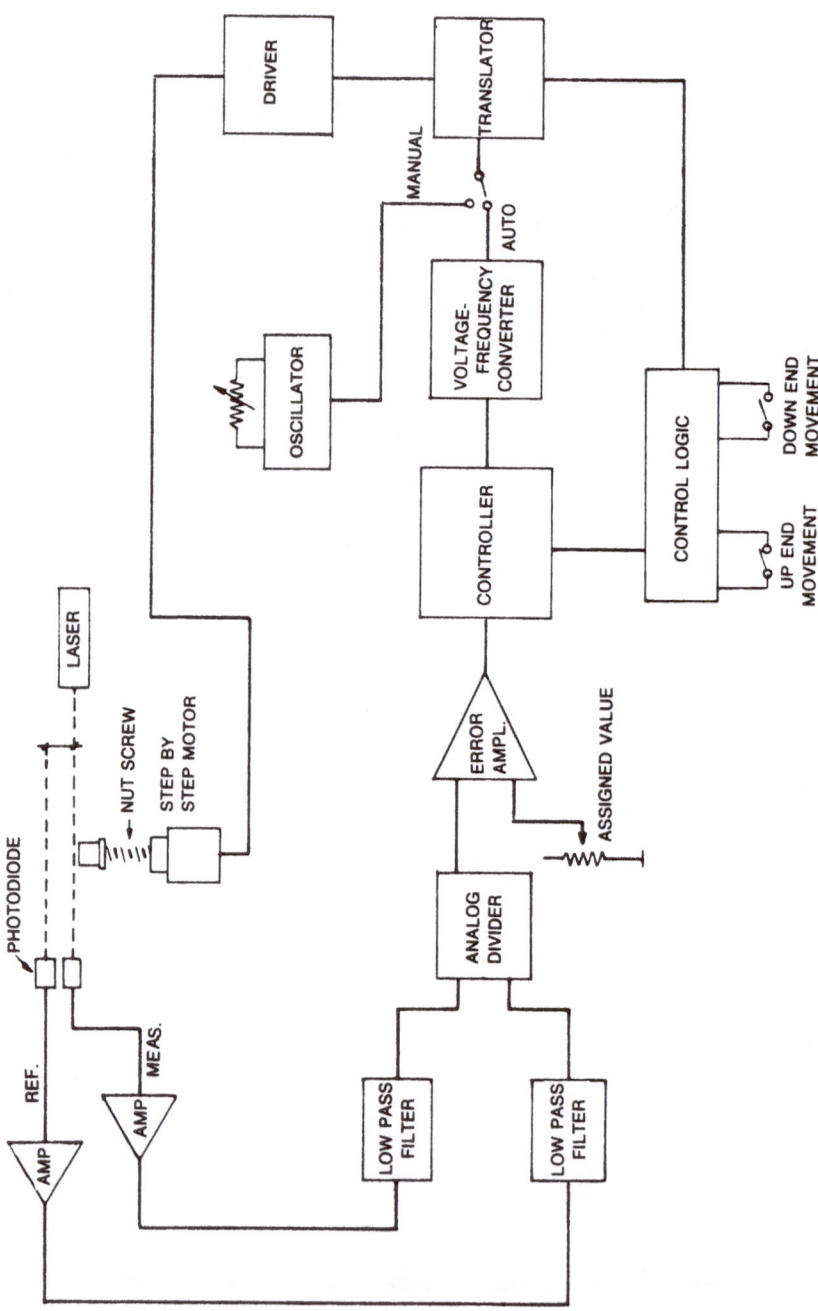

Fig. 6. Operating scheme of the position control system.

3. THE EXPERIMENTAL APPARATUS FOR CONTROL OF THE BURNING SURFACE POSITION

This apparatus keeps the volume probed by the diagnostic system at a constant distance from the burning surface. Many types of optical diagnostics can be used simultaneously; for example, the gas velocity can be measured by means of a LDV system, and gas temperature by means of a line-reversal method. High-speed cinematography can also be useful.

The burning-surface position control system consists of a servo-controlled stepping motor, moving the propellant sample during combustion at the consumption speed of the sample itself. The apparatus, whose scheme is shown in Figure 6, consists of four basic parts:

1. A laser system for determination of the burning surface position.
2. A stepping motor driving a lifting system for the propellant sample.
3. An electronic feedback system controlling the motor.
4. A combustion chamber containing the burning propellant sample.

3.1. Position Sensor

The position of the propellant burning surface is sensed by means of a laser system. The beam of a 15-mW He–Ne laser is first expanded by means of a beam expander, and successively made cylindrical again by means of a pinhole and condenser lens. Through a two-hole diaphragm the expanded beam splits into two parallel beams. The larger beam (measuring beam) has a diameter of about 3 mm and crosses the chamber at right angles to the axis of the propellant strand, grazing its burning surface. Under ideal tracking conditions, therefore, it is partially blocked by the edge of the burning surface. The second beam (reference beam) has an offset of about 10 mm with respect to the strand side, and crosses the chamber below the propellant-surface level. In this way its intensity is practically independent of the flame and affected, at most, only by the possible presence of smoke, or by window-fogging. Past the combustion chamber the two beams pass through a narrow-band interferential filter (10-nm width) centered on the laser wavelength, in order to cut off the wide-spectrum luminous emission of the propellant flame. After passing through the interferential filter each beam reaches a phototransistor. The burning-surface position is given by the ratio of the measuring-beam signal to the reference-beam signal.

3.2. Mechanical Lifting System and Combustion Chamber

The combustion chamber is shown in Figure 7. It was designed to accomodate the stepping motor and the lifting system part, together with their electrical connections.

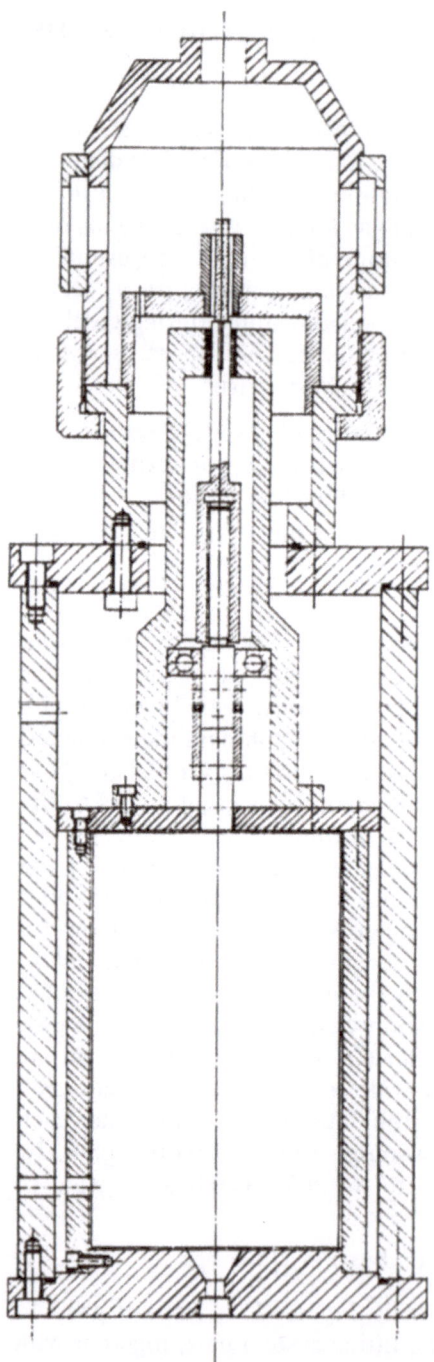

Fig. 7. Combustion chamber.

To keep the chamber and the lifting motor system clean of smoke and combustion products during test runs, inert gas is injected at the bottom of the chamber. Generally, a nitrogen flux is used in order to have a nonreacting atmosphere during the experiments. The lifting system consists of a feed piston lifted by a stepping motor by means of a precision nut–screw assembly having a 1.50-mm pitch. The excursion length of the lifter is limited by two microswitches and is at present 30 mm. The stepping motor has 200 steps per revolution and can reach its maximum speed of 1800 rpm in about 400 milliseconds.

The testing section, where the propellant sample actually burns, is situated at the top of the combustion chamber. It has, at present, four optical windows, two of which are used by the beams of the position sensor. The propellant sample is fastened to its holder on the surface of the feed piston. In the combustion chamber head are the pressure-transducer housing, the ignition system, and connections for probes, such as thermocouples. The purging gas flows from the bottom through the combustion chamber, and is exhausted through an exhaust port at the top of the chamber head, connected to a vacuum pump.

3.3. Electronic System

The experimental setup, including the position control system, is described in Figure 6. The reference beam and the measuring beam, after crossing the combustion chamber, are picked up by two photodiodes with amplifiers operating in photocurrent mode, and converted to electrical signals. As the amplifiers' gain is high, the use of low-pass filters was necessary in order to avoid high-frequency noise due principally to the photodiodes but also to the amplifiers as well.

It is seen from Figure 6 that the filter output constitutes the input for the analog divider that will be described later. The divider made low-pass filters much more necessary, since the noise components of the input signals are uncorrelated with each other, and therefore so large as to compromise the overall working of the equipment.

Neglecting the noise effect, the analog-divider output is proportional to the ratio V_{MEAS}/V_{REF}, and hence to the position of the propellant sample with respect to the measuring beam. The reason for using an analog divider is due to the presence of smoke and fog condensing on the windows, which could reduce the beam intensity received by the photodiodes compromising the measurements. In fact, it is not possible to distinguish between the attenuation produced by smoke and that produced by the blocking effect of the propellant sample edge.

The use of the reference beam minimizes this error source, since two laser beams are so close to each other that it is possible to assume the same attenuation for both.

The analog-divider output constitutes one of the inputs of the tracking error amplifier, the other one being a reference voltage corresponding to a preassigned position of the propellant sample. The output of the error amplifier is a voltage proportional to the difference of the two inputs and is therefore the positional error that must be nulled by the stepping motor.

The successive block in the system is the actual regulator and is composed of elements the function of which is to stabilize the control system and to minimize the response time while allowing a correct turning of the motor. The output voltage of this block is converted into a frequency that is sent to the translator. The translator is a module converting the input pulses coming from the V/F counter and the direction signals into sequenced phase-control signals to operate a bifilar stepping motor as a power driver. The function of the "control logic" block is to create the direction signal and to stop the motor whenever the end travel microswitches are activated, in order to avoid equipment damage. The system also allows manual feeding of the motor to set a correct initial positioning for the propellant sample and for testing and calibration purposes.

4. EXPERIMENTAL PROCEDURE AND RESULTS

A first series of tests was performed to check the apparatus behavior during the combustion of a double-base propellant sample. Ignition during these tests occurred when the propellant strand was in the initial position for the controlling system. The ignition burst caused some trouble to the system; nevertheless runs were performed at up to 7 atm with an uncatalyzed double-base propellant. The presence of a transparent so-called "dark zone" in this propellant flame made the position sensor work easier. A second series of test runs were performed during the combustion of an ammonium-perchlorate composite propellant. The pressure range in these runs was in the subatmospheric field. The choice of this pressure field was partially due to the difficulties arising in the system in its present configuration when studying a propellant having a very luminous flame anchored to the burning surface; the actual studies being carried on at ISREP also had a great influence on this choice. In fact, particular attention is devoted to low-pressure combustion phenomena and specifically to self-sustaining burning oscillations near the pressure-deflagration limit.

The results obtained in these runs will be compared with analog results recently obtained at ISREP by means of different experiment methods. In these runs ignition occurred when the propellant-sample surface was about

10 mm higher than the equilibrium position set for the controlling apparatus. After ignition the propellant burns while the motor is still, until the measuring beam of the position sensor is blocked. Then the position control system is triggered and the ratio between measuring-beam and reference-beam signals reaches the set value that defines the equilibrium position for the burning surface. From that moment on the system reacts to keep the burning-surface position fixed relative to the combustion chamber.

During these runs the propellant-sample image was focused, by means of a 150-mm focal lens, on the plane of a diaphragm with a 1-mm-diameter hole. The hole was positioned to collect the light of the flame at 5 mm over the burning surface. The light admitted by the diaphragm was collected by a phototransistor.

Motor velocity and luminosity signals were recorded by means of a Nicolet digital recorder having 4096 sample points on two channels. Figure 8 is a plot of a run at a pressure of 0.118 atm, corresponding to theoretically predicted burning oscillations. The lower trace represents the motor feeding velocity, the upper trace is the flame luminosity collected from a phototransistor. It is noticeable how the stepping motor velocity varies

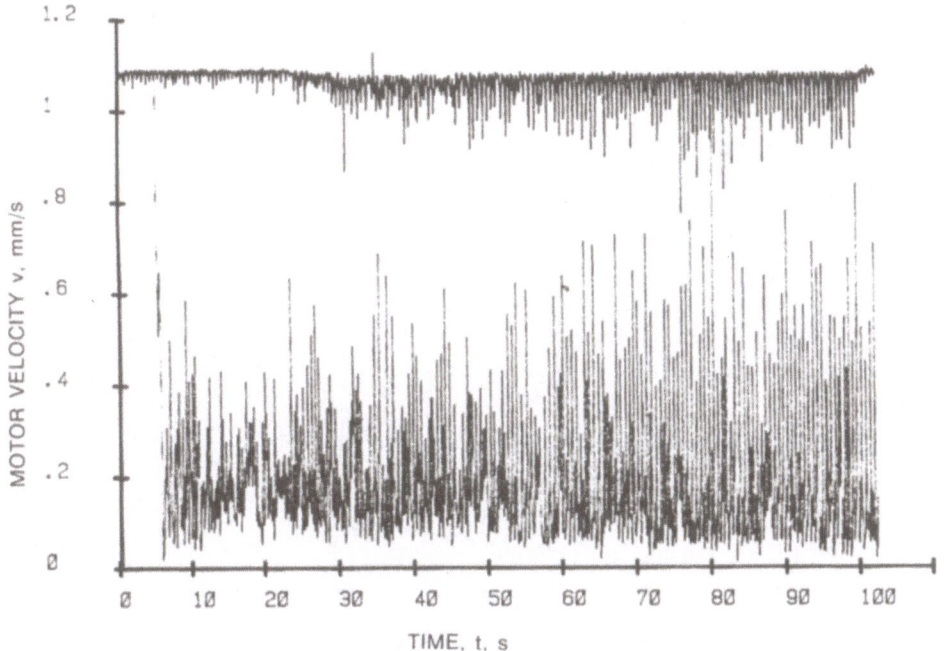

Fig. 8. Feeding velocity and luminosity traces during an experimental run.

greatly in order to keep the burning surface fixed to the reference position. The position of the burning surface during combustion was recorded also by means of cinematography. Analysis of the film confirmed the effective stability of the burning-surface position within the combustion chamber. Expanding the time scale of the same plot as in Figure 9 shows clearly flame oscillations (upper trace) that are very regular in frequency. These flame oscillations are accompanied by motor velocity oscillations with a delay of about a quarter-wave. Figure 10 shows steady-state burning-rate data vs. pressure, where the burning rate was defined as the motor mean velocity during each run. These data are compared with experimental results obtained from the same propellant by means of a laser method for measuring the steady-state burning rate at low pressure. The laser data are represented by their best fitting line. As shown, good agreement exists between the two methods of measurement.

However, difficulties arise in *instantaneous* measurements of burning rate, when no quantitative connection can be found between motor instantaneous velocity and burning rate. Frequency data of flame oscillations vs. pressure obtained in these runs are plotted in Figure 11. These

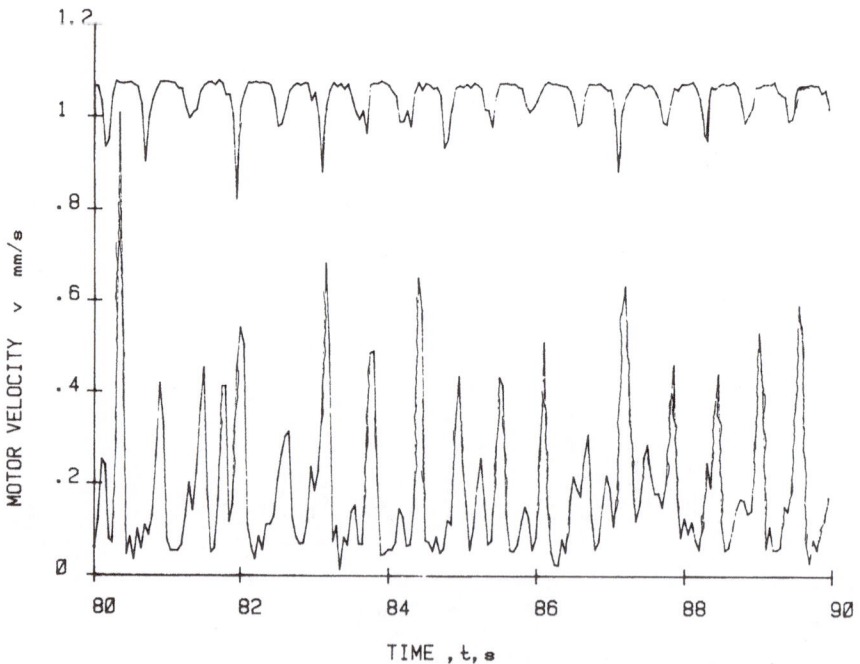

Fig. 9. Feeding velocity and luminosity traces (expanded scale).

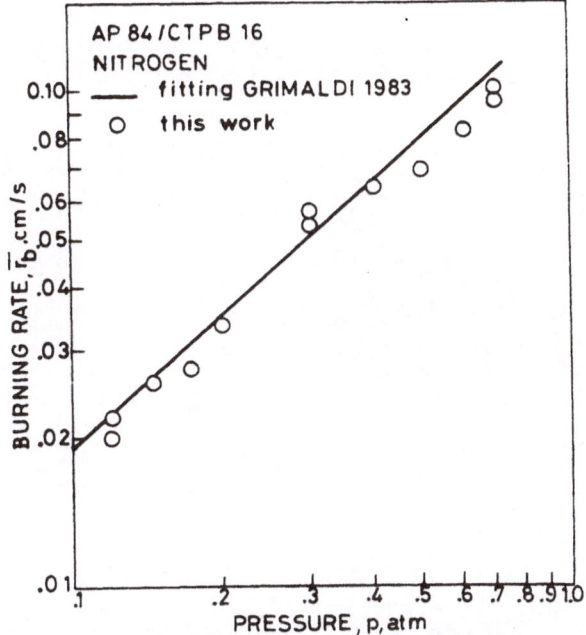

Fig. 10. Steady-state burning rate vs. pressure plot.

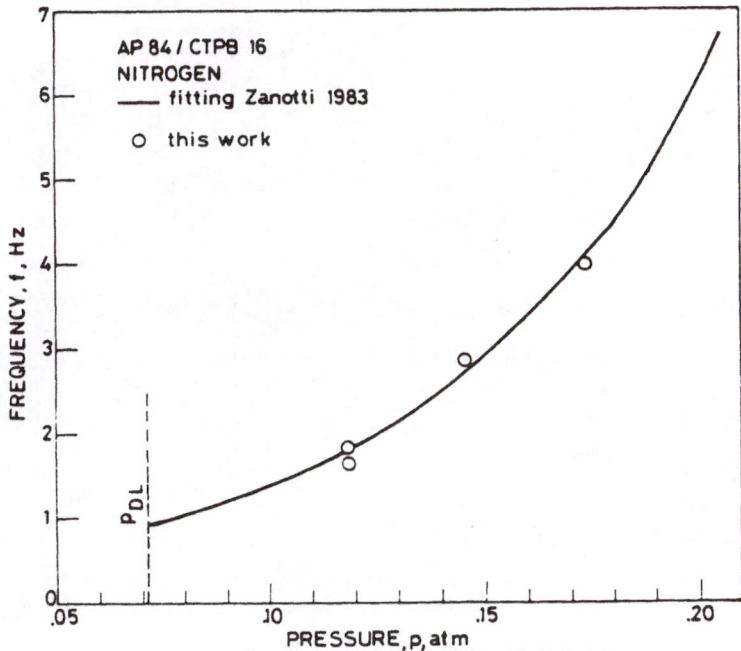

Fig. 11. Oscillation frequency vs. pressure plot.

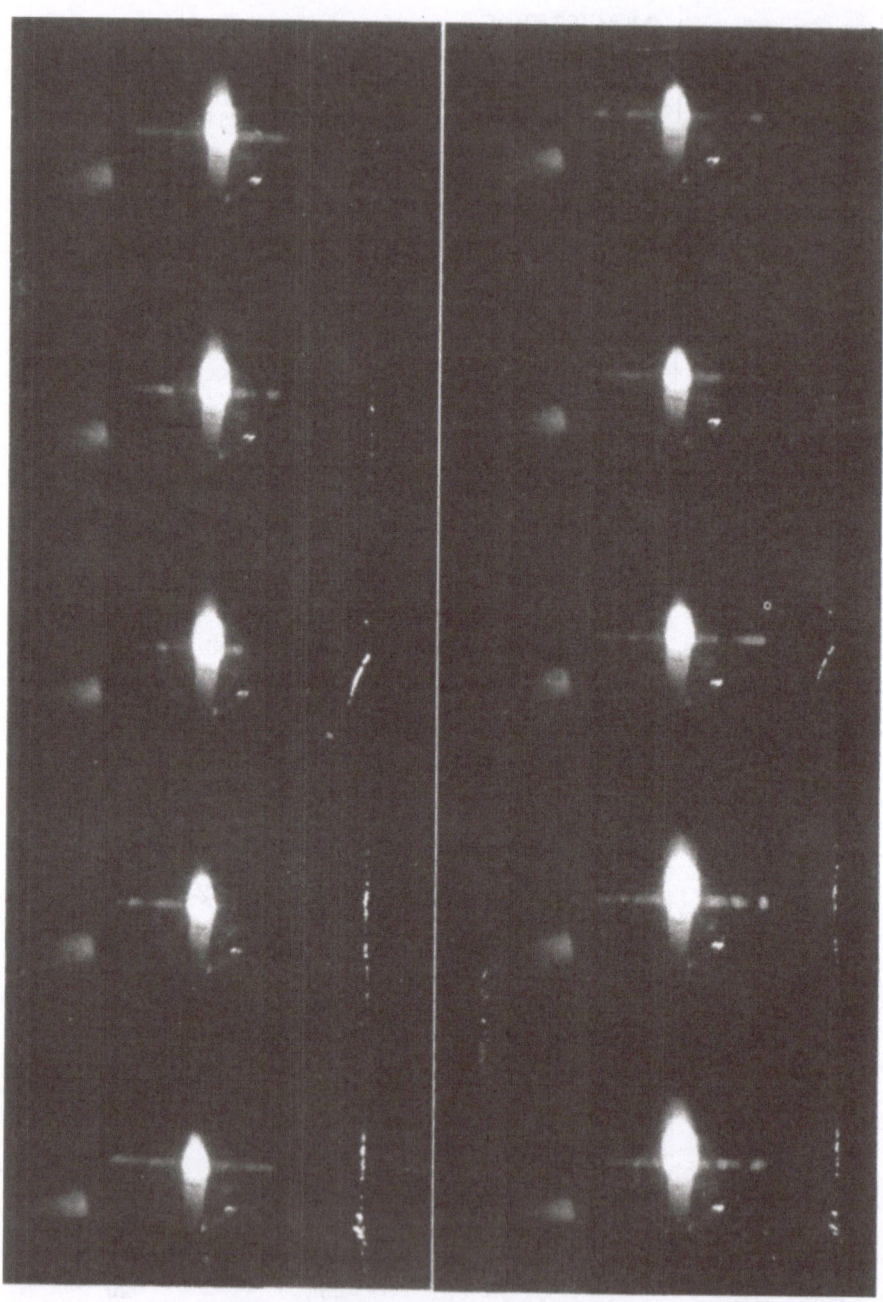

Fig. 12. Sequence of photograms showing a burning oscillation period.

data are compared with the curve fit of the results obtained with the same propellant in a different combustion chamber having a different internal volume. The agreement between the two series of results demonstrates that burning oscillation frequency is a characteristic of the propellant at given constant conditions and does not depend on acoustic phenomena. In fact no pressure variation was observed during the flame oscillation bursts.

In conclusion, as observed by the first results, the apparatus in its actual configuration is a good tool for gas-phase measurements, with some limitations due to the propellant type and pressure. The apparatus limits depend mainly on the burning-surface position-sensing system, in which direction efforts will be made to improve the apparatus response.

ACKNOWLEDGMENT. The author wishes to thank the research staff of the Solid Propellant Group in Milan and in particular R. Dondè, G. Riva, A. Volpi, and T. Ferrari for their assistance in preparing this article.

NOMENCLATURE

c	specific heat, cal/g K
d	layer thickness, cm
H	Q/Q_{ref}, nondimensional heat release
k	thermal conductivity, cal/cm s k
$\dot{m}$	mass flow rate, g/cm^2 s
p	pressure, atm
p_{ref}	68 atm, reference pressure, atm
P	p/p_{ref}, nondimensional pressure
$\dot{q}_{c,s}$	$\dot{\varphi}_{c,s}/\dot{\varphi}_{ref}$, nondimensional heat conducted away in condensed phase
$\dot{q}_{g,s}$	$\dot{\varphi}_{g,s}/\dot{\varphi}_{ref}$, nondimensional heat feedback from gas phase
$\dot{q}_l$	$\dot{\varphi}_l/\dot{\varphi}_{ref}$, nondimensional heat loss from burning surface
Q	heat release, cal/g
Q_{ref}	$C_{c,ref}(T_{s,ref} - T_{ref})$, reference heat release, cal/g
Q_s	net surface heat release (positive if exothermic), cal/g
r_b	burning rate, cm/s
$r_{b,ref}$	$r_b(p_{ref})$, reference burning rate, cm/s
R	$r_b/r_{b,ref}$, nondimensional burning rate
$\mathcal{R}$	universal gas constant, cal/mol K
t	time, s
T	temperature, K
T_{ref}	300 K, reference temperature, K
$T_{s,ref}$	$T_s(p_{ref})$, reference surface temperature, K
u	nondimensional finite-size disturbance of temperature (see text)

u_x	nondimensional finite-size disturbance of thermal gradient
v	motor feeding velocity
x	space variable, cm
X	$x/(\alpha_{c,ref}/r_{b,ref})$, nondimensional space variable

Greek Symbols

α	thermal diffusivity, cm^2/s
δ	$d/(\alpha_{c,ref}/r_{b,ref})$, nondimensional layer thickness
θ	$(T - T_{ref})/(T_{s,ref} - T_{ref})$, nondimensional temperature
ρ	density, g/cm^3
τ	$t/(\alpha_{c,ref}/r_{b,ref}^2)$, nondimensional time
$\dot{\varphi}_{c,s}$	heat conducted away in condensed phase, cal/cm^2 s
$\dot{\varphi}_{g,s}$	heat feedback from gas phase, cal/cm^2 s
$\dot{\varphi}_l$	heat loss from burning surface, cal/cm^2 s
$\dot{\varphi}_{ref}$	$\rho_c C_{c,ref} r_{b,ref}(T_{s,ref} - T_{ref})$, reference heat flux, cal/cm^2 s

Abbreviations

AP	ammonium perchlorate (NH$_4$ClO$_4$)
DB	double base
KTSS	Krier–T'ien–Sirignano–Summerfield
MTS	Merkle–Turk–Summerfield
PBAA	polybutadiene acrylic acid

Subscripts and Superscripts

b	burning
c	condensed phase
f	flame
g	gas phase
l	loss
s	surface
c, s	surface from condensed phase side
g, s	surface from gas phase side
th	thermal
ref	reference
–	steady-state or average value
$-\infty$	far upstream
$+\infty$	far downstream

REFERENCES

1. L. De Luca, Theoretical studies on heterogeneous deflagration waves: I. A partial differential equation formulation of the problem, *Meccanica* **13**, 16–27 (1978).
2. T. R. Goodman, Application of integral methods to transient nonlinear heat transfer, in: *Advances in Heat Transfer*, Vol. 1, pp. 51–122, Academic Press, New York (1964).
3. L. De Luca, Solid Propellant Ignition and Other Unsteady Combustion Pehnomena Induced by Radiation, Ph.D. Thesis, Department of Aerospace and Mechanical Sciences, Princeton University, AMS Report 1192-T (November, 1976).
4. L. De Luca, G. Riva, and C. Bruno, Bifurcation in Heterogeneous Combustion, NATO Advanced Study Institute on Non-Equilibrium Cooperative Phenomena in Physics and Related Fields, Madrid, Spain, 1–11 August, 1983, Proceedings (in press).
5. C. Bruno, G. Riva, C. Zanotti, R. Dondè, C. Grimaldi, and L. De Luca, Experimental and theoretical burning of solid rocket propellants near pressure deflagration limit, XXXIV International Astronautical Federation Congress, Budapest, Hungary, 10–15 October, 1983, IAF Paper 83–367.
6. R. Dondé, G. Riva, and L. De Luca, Experimental and theoretical extinction of solid rocket propellants by fast depressurization, XXXIII International Astronautical Federation Congress, Paris, France, 27 September–2 October, 1982, IAF Paper 82-361, to appear in *Astronaut. Acta.*

13

*L**-Combustion Instability in Solid Propellant Rocket Combustion

S. I. CHENG

ABSTRACT. An analysis of the *L**-combustion instability in solid propellant rockets is formulated to include (1) secondary or residual combustion in the rocket chamber and (2) the change of the mean chamber pressure. The aim was to explore if these factors might remedy the failure of the many transient heat-transfer theories with quasi-steady gaseous-phase reactions in predicting any *L**-combustion instability. It became clear soon how and why the response functions derived from such quasi-steady gas-phase reaction theories must fail regardless of the aforementioned remedies.

The time-lag formulation, which emphasizes the transient gaseous-phase reactions, is then examined in the same context. It encounters no such difficulties. If some form of Arrhenius transient reaction-rate law for the gaseous burning zone is postulated, the magnitude of the interaction index is more than adequate to explain the occurrence of *L**-instability in AP motors. The apparently anomalous behavior of *L**-combustion stability in rockets with nitrocellulose or nitramine propellants is a natural consequence of the time conditions of such instability. The interaction index and some mean time lag of a propellant can be determined from experimental data with sufficient details.

1. INTRODUCTION

Analytic estimates of the burning-rate response function of a solid propellant subjected to chamber pressure oscillations have been very

S. I. CHENG • Princeton University, Princeton, NJ 08540, U.S.A. and (Consultant) Princeton Combustion Research Laboratories, Inc., Princeton, NJ 08540, U.S.A.

extensively developed since 1958.[1-15] These formulations are all (except part of the paper by Hart and McClure[4]) based on the transient heat-transfer model in the solid phase while the gaseous (and/or solid phase) reactive processes are treated with some form of "quasi-steady" approximation, including the Zeldovich formulation. The burning-rate response function determined in such a manner naturally depends largely on solid-phase properties regardless of the various details. This is a significant departure from what was thought to be intuitively obvious prior to 1960,[14,15] that chemical reactions in gaseous phase are more sensitive to pressure oscillations than those in solid phase. This last view was suggested by the observation that with the same Boltzmann factor in the Arrhenius form of a gross reaction-rate law, a gaseous-phase reaction of second order will respond to pressure oscillations with an exponent of almost two, while the zeroth-order solid-phase reaction would be essentially nonresponsive. However, this was recognized as not assured since combustion reactions may not possess an overall kinetic-rate expression especially in the transient state. Whether the solid-phase or the gaseous-phase processes might be treated with the quasi-steady approximation in the modeling of the burning-rate response function of a solid propellant, could be decided only indirectly through comparison of theoretical predictions with reliable experimental data.

The burning-rate response function for the analysis of combustion instability in solid propellant rockets is, strictly speaking, not a "propellant property."[16] It is only convenient to suppose that such is the case, under some fixed, simple fluid-dynamics situation. Thus both the pressure-interaction index and the time lags were postulated as propellant properties.[14] Likewise, the newer, solid-phase heat-transfer theories[1-12] compute the burning-rate response functions without reference to the detailed chamber conditions. When and if any fluid-dynamic factor(s) in the chamber should prove important (such as velocity coupling), it could be incorporated as additional correction factor(s).[14] It was therefore proposed[16] that the time record of chamber pressure in some standard rockets of simple configuration be obtained and analyzed for different solid propellants to test if it is approximately a propellant property to be modeled. The time-lag theory[14] contains two parameters, a pressure interaction index m, and a mean time lag $\bar{\tau}$. This theory is phenomenological, but can be related to basic physicochemical constants[4,16] once the kinetic details are specified. The concept of sensitive time lag was introduced by Crocco[19] for analyzing combustion instability in liquid-propellant rockets under more complicated physical circumstances.

The "newer" solid-phase heat-transfer theories[1-12] bypass the above phenomenological stage and choose to relate the burning-rate response function directly to those basic constants of the physicochemical processes

according to some postulated model(s). Some form of quasi-steady approximation is generally adopted in treating the reactive regions of the
gaseous-flame and solid-pyrolysis zones so that no kinetics of chemical
reactions need to be specified. The burning-rate response function of these
models can be put into a universal form in terms of two gross parameters
A, $B^{(12)}$ containing the many physicochemical constants. We shall show
why such solid-phase heat-transfer theories fail to explain $L*$-instability,
and how the gas-phase theory in the simple form of sensitive time lag
succeeds in doing so.

2. FORMULATION OF $L*$-INSTABILITY ANALYSIS

$L*$-instability refers to unstable oscillations in solid-propellant rocket
combustion chambers at such low frequencies that the oscillatory chamber
pressure is spatially uniform at all times. The same oscillatory pressure
generates oscillations of the burning rate and the nozzle damping through
the supercritical outflow. The mass-conservation consideration alone governs the dynamics of the oscillatory system:

$$\frac{\partial \rho_g}{\partial t} + \text{div}\,(\rho_g \bar{q}_g) = 0 \qquad (1)$$

where ρ_g is the gas density and $\bar{q}_g$ is the gas velocity. Both are functions of
$(\bar{x}, t)$. Unlike the situation in liquid rocket combustion chambers, there is
no gas generation in the combustion chamber of a solid motor so that the
source term on the right-hand side of equation (1) vanishes. Integrating
equation (1) over the varying combustion-chamber volume $\bar{V}(t)$, with the
help of the divergence theorem, we obtain

$$\frac{\partial}{\partial t} \int_{V(t)} \rho_g dV + \int_{A_b} \rho_{gb} r_b(t) dA_b = \dot{M}_b - \dot{M}_e \qquad (1a)$$

where $\dot{M}_b$ is the gas generation rate on the burning surface A_b, i.e.,
$\int_{A_b} \rho_{sb} r_b(t) dA_b$, and $\dot{M}_e$ is the gas mass outflow rate at the chamber exit into
the nozzle. The integral $\int_{A_b} \rho_{gb} r_b(t) dA_b$ is the mass of gas filling the
incremental chamber volume due to the receding solid-propellant surface.
Here $r_b(t)$ is the linear burning rate of the solid propellant. It is convenient
to write

$$\dot{m}_b = \int_{A_b} [1 - (\rho_g/\rho_s)_b] \cdot \rho_{sb} \cdot r_b(t) dA_b$$

and let $\dot{M}_e \to \dot{m}_e$, so that the integrated mass-balance equation (1a) stands as

$$\dot{m}_b(t) - \dot{m}_e(t) = \frac{\partial}{\partial t} \int_{V(t)} \rho(t, \bar{x}) dV \tag{2}$$

where the subscript g for gas is dropped from $\rho_g(t, \bar{x})$.

The right-hand side of equation (2) is the time rate of incremental mass in the combustion chamber. Let the steady-state mass flow rate be $\bar{m} = \bar{m}_b = \bar{m}_e$ and define $\tilde{\mu}_b = \dot{m}_b(t)/\bar{m} - 1$; $\mu_e = \dot{m}_e(t)/\bar{m} - 1$. In terms of some average gas residence time in the combustion chamber, we introduce the reduced time by

$$\frac{dz}{dt} = \frac{\rho_e U_e A_e}{\rho_e V(t)} = \frac{\bar{m}_b}{\rho_e V(t)} = \frac{\bar{m}_e}{\rho_e V(t)}$$

For small oscillations of an ideal gas with negligible molecular-weight variations, equation (2) becomes

$$\tilde{\mu}_b(t) - \mu_e(t) = \frac{\partial}{\partial z} \int \frac{\bar{\rho}(x)}{\bar{\rho}_e} \cdot \left(\frac{\rho'}{\rho} \right) dV$$

$$= \frac{\partial \phi}{\partial z} \int \frac{\bar{T}_e}{\bar{T}(x)} \left\{ 1 - \left[\frac{\partial \theta(t, \bar{x})}{\partial z} \Big/ \frac{\partial \phi(t)}{\partial z} \right] \right\} d\bar{x} \tag{3}$$

where $\tilde{\mu}_b$ is the fractional variation of the instantaneous mass burning rate of the propellant from its steady-state value, constant over the entire burning propellant surface anywhere in the chamber. The "tilde" designates specifically that it is the local fractional variation of surface pyrolysis rate, generating the "intermediate" partially burned gas to become "completely" burned gas somewhere in the rocket chamber through some exothermic reactions.

In Figure 1a, a steady-state temperature distribution is sketched for AP-type solid propellants. The propellant surface temperature is substantially lower than the adiabatic flame temperature T_f. This primary flame is thin and the temperature gradient large. A small fractional variation of this large energy release rate can be significant.

When the gas pressure oscillates the chemical kinetics of heat generation and the heat-transfer processes are simultaneously modulated, through different mechanisms. Any departure from the exact balance between large contributions will cause significant $\partial T/\partial t$. The quasi-steady approximation of evaluating $\partial T/\partial t$ neglects all the unknown, detailed basic processes. Such a sweeping assumption of a dominant factor can, of course, harbor unpleasant consequences.

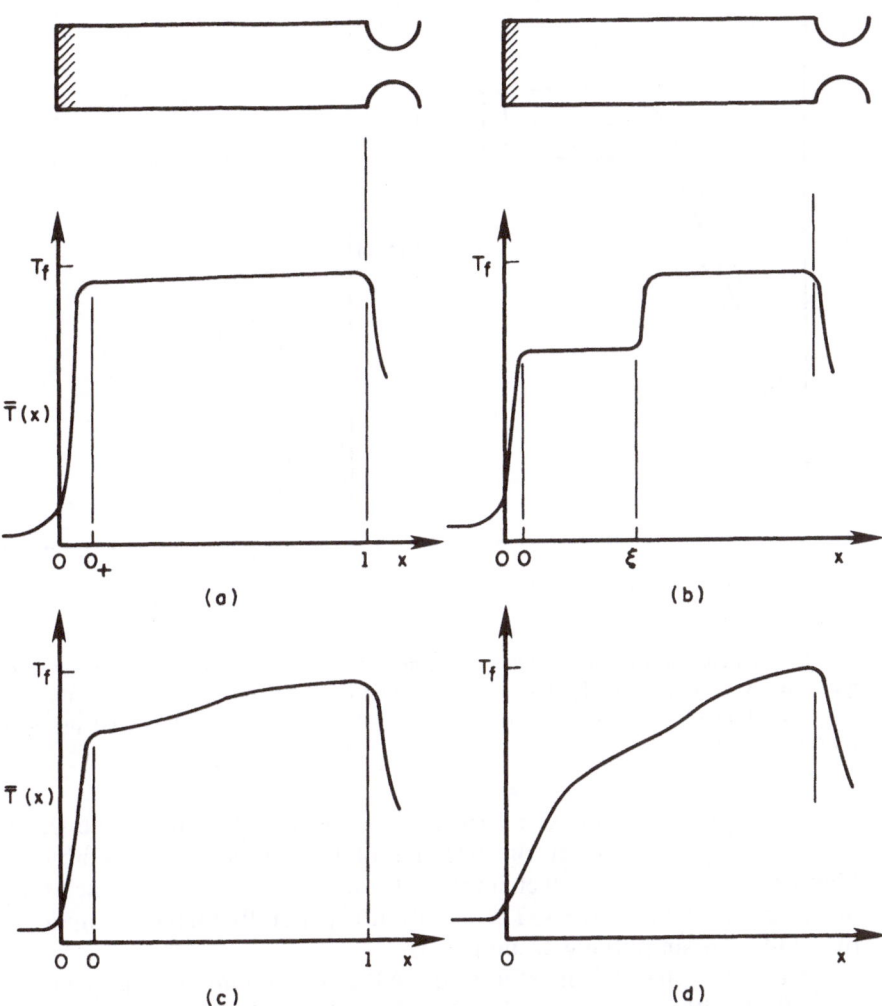

Fig. 1. Schematic diagrams of mean chamber temperature (or combustion) distribution along axis of symmetry. (a) Concentrated combustion at motor end. (b) With a residual or secondary combustion concentrated at ξ. (c, d) With distributed secondary combustion.

The essential ingredient of this quasi-steady approximation is that the temporal variation of some dimensionless property is independent of the local time rate $(\partial/\partial t)$ of this property. Thus when the chamber gas temperature $T_c(t)$ and the solid surface temperature $T_s(t)$ fluctuate (Figure 2) in response to chamber pressure oscillation $\phi(t)$, the dimensionless temperature profile $(T - T_s)/(T_c - T_s)$ is universal and remains the same as

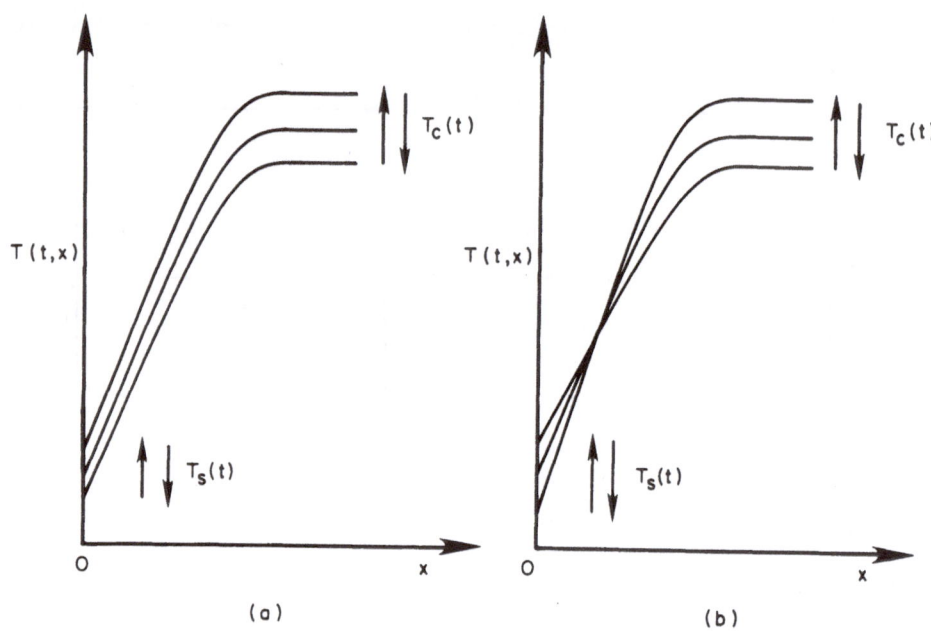

Fig. 2. Schematic diagram of instantaneous temperature profiles across a flame region under quasi-steady approximation $[T(t, x) - T_s(t)]/[T_c(t) - T_s(t)] = [\bar{T}(x) - \bar{T}_s]/\bar{T}_c - \bar{T}_s]$ with temperatures $T_c(t)$ and $T_s(t)$ down- and upstream of the flame region oscillating (a) in phase and (b) out of phase.

that in the steady state at all times. A sceptic will point to the drastic difference in the response mechanisms of heat transfer and combustion kinetics. An enthusiast will claim that at "sufficiently" low frequencies, this must be essentially correct. The situation is generally further complicated by additional simplifying assumptions.

Consider the L^*-instability of AP-type (ammonium perchlorate) propellant with a steady-state temperature distribution in the combustion chamber in Figure 1a. By excluding the surface-flame region, a new chamber volume extending from 0_+ to 1 (Figure 1a) will have a uniform temperature of the "burned" gas, T_c, slightly less than the adiabatic flame temperature T_f of the propellant under the prevailing chamber pressure. The volume integral in the mass-accumulation term on the right-hand side of equation (3) can be simply evaluated as $(1/\gamma)\partial\phi/\partial z$.

The fractional variation of the mass outflux μ_e in equation (3) is proportional to the fractional chamber-pressure variation ϕ with a response function $R_e = \mu_e/\phi = (1 + A_e)/\gamma$ where the nozzle admittance function A_e has been calculated[19] and its limiting value $(\gamma - 1)/2 + i\beta\omega$ as $\omega \to 0$ is adopted in what follows.

With the modified chamber volume, the fractional burned-gas-generation rate is $\mu_b(t)$ at $x = 0_+$, not the $\bar{\mu}_b(t)$ at $x = 0$ for the relatively cool, intermediate gases from solid surface pyrolysis. Thus equation (3) becomes

$$\mu_b(z) - \mu_e(z) = \frac{1}{\gamma} \frac{\partial \phi}{\partial z} \qquad (4)$$

The evaluation of the $\mu_b(z)$ will have to include the complex gaseous-phase interactions in the large-gradient region, in addition to the interactions in the thin solid region of surface pyrolysis.

The gas-phase interaction was emphasized both in Chengs' time-lag formulation[14] and in Hart and McClure's (4) kinetic model, while some form of quasi-steady approximation was adopted in treating the solid-phase interaction. On the other hand, the newer theories employed the quasi-steady approximation in both the gaseous- and the solid-phase reactive regions. The temperature oscillation in the solid phase provides the necessary phase lag in varied forms, leading to many different burning-rate response functions $R_b = \mu_b(t)/\phi(t)$ in the literature.[1-12] They have been conveniently summarized by Culick[12] in some general form in terms of two parameters A and B (vs. m and $\bar{\tau}$ in the time-lag theory) containing all the physicochemical constants.

The integrated mass-conservation equation (4), after dividing by $\phi(t)$, becomes

$$R_b - \frac{1}{\gamma}(1 + A_e) = \frac{1}{\gamma} \frac{\partial \ln \phi}{\partial z} \qquad (5)$$

which stands alone as a complex equation determining the logarithmic increment of ϕ as a function of the dimensionless time z. With $\phi(t) = \exp[(\alpha + i\omega)z]$, ω may be identified as the angular frequency and α the amplification coefficient of a specific Fourier component of the chamber pressure fluctuation. The amplitude of this component of oscillation increases (or decreases) with time if $\alpha > 0$ (or < 0). When $\alpha = 0$, it defines the stability boundary. The fastest amplifying component (largest $\alpha > 0$) is usually the most prominent component in the pressure record of physically observed unstable oscillations.

3. GENERAL PROPERTIES OF L*-INSTABILITY

Equation (5) is the characteristic equation, governing the oscillatory behavior of the linear system (small disturbances). This equation is nonlinear, where both the burning-rate response function $R_b = R_b^r + iR_b^i$

and the nozzle admittance function $A_e = A_e^r + iA_e^i$ are complex functions of the characteristic or eigenvalue $(\alpha + i\omega)$. For stability investigations $|\alpha|$ may be taken as a small quantity (which need not correspond to experimental situations) while the dimensionless frequency ω should be $O(1)$, with $A_e^r = (\gamma - 1)/2$ and $A_e^i = \omega \cdot \beta$ where β is a small positive constant depending on nozzle geometry. Without specifying the complex function R_b, equation (5) determines the growth rate α of the disturbances of a specific frequency ω through the following real equations:

$$\gamma R_b^r = 1 + A_e^r + \alpha = \frac{\gamma + 1}{2} + \alpha$$
$$\gamma R_b^i = \omega + A_e^i = \omega(1 + \beta) \tag{6}$$

By definition, ω and α can be physically meaningful only when they are both *real*. Such real solutions are, however, not ensured from a nonlinear system of equations! Moreover, to demonstrate unstable L^*-oscillations both α and ω must be nonnegative in a solid-propellant rocket. Thus, the burning-rate response function R_b of an AP-type propellant must possess a sufficiently large magnitude and a sufficiently large phase lag if the system is ever unstable, i.e.,

$$R_b^r \geqslant (\gamma + 1)/2\gamma \quad [\sim O(1)] \tag{6a}$$

$$R_b^i \geqslant \omega/\gamma \quad [O(1)] \tag{6b}$$

Condition (6a) requires that the excitation by the burning-rate response must overcome the damping through the nozzle outflow.[16] In other words, when the chamber pressure increases, the burned gas generated from the burning surface must exceed the corresponding increase of outflow through the choked nozzle so as to produce excess mass accumulation and further pressure rise in the chamber for an unstable situation. This requirement is less stringent than that in the steady state because of the possible presence of "phase lag" between oscillations of the burning rate and the chamber pressure. Condition (6b) states explicitly the Rayleigh criterion of reinforcing oscillations under proper phase relationship (the singing flame). The L^*-oscillation is commonly referred to as low-frequency instability (as opposed to acoustic frequencies) in a solid-rocket combustion chamber, with the oscillation period comparable to the gas residence time in the combustion chamber, i.e., the dimensionless frequency $\omega \sim O(1)$. Thus R_b^i must be positive and of $O(1)$.

Under quasi-steady gas-phase combustion, the local responses are instantaneous, i.e., without introducing any phase lags. (This should be carefully distinguished from quasi-steady approximation for rate processes.) Time-dependent heat-transfer processes in solids will have to provide all the phase lag. Thus condition (6b) can be demanding. The fractional variation of chamber gas temperature θ_e is no greater than $[(\gamma - 1)/\gamma]\phi$, and that of the propellant surface temperature θ_s must be significantly less than θ_e under the postulated quasi-steady heat-conduction process. Even with the instantaneous, Arrhenius-type pyrolysis law, it is difficult to visualize how the $|R_b|$ (and/or R_b^r) could be significantly different from the pressure exponent n in the steady-state burning-rate law $\bar{m} \sim \bar{p}^n$, where n is necessarily less than unity. Hence there can also be difficulty in fulfilling condition (6a). Thus the phenomenological model of gas-phase interaction in terms of the "sensitive time lag" with quasi-steady solid-phase interaction was preferred earlier.[14]

The measured pressure oscillations in L^*-motors are often accompanied by significant variations of mean chamber pressure. To use these L^*-motor data for comparative studies, we modify the accumulation term in equations (3) as follows:

$$\tilde{\mu}_b(t) - \mu_e(t) = \frac{\partial \phi}{\partial z} \int_0^1 T_c/\bar{T}(x) \left(1 - \frac{\partial \theta}{\partial z} \bigg/ \frac{\partial \phi}{\partial z}\right) dV$$

$$+ \frac{\partial \ln \bar{p}}{\partial z} \cdot \int_0^1 T_c/\bar{T}(x)(\phi - \theta) dV \qquad (3')$$

$$R_b - \frac{1}{\gamma}(1 + A_e) = \frac{1}{\gamma} \frac{\partial \ln \phi}{\partial z} + \frac{1}{\gamma} \frac{\partial \ln \bar{p}}{\partial z} \qquad (5')$$

and

$$\gamma R_b^r = \frac{\gamma + 1}{2} + \alpha + \frac{\partial \ln \bar{p}}{\partial z}$$

$$\gamma R_b^i = \omega(1 + \beta) \qquad (6')$$

While $\partial \ln \bar{p}/\partial z$ and α can become comparable in the experimental data, they are generally small compared with the nozzle damping contribution $(\gamma + 1)/2$. The rapidly amplifying L^*-oscillations, often observed shortly after ignition with rapidly rising mean chamber pressure, generally require a somewhat larger magnitude of the real part of R_b than that indicated by condition (6a).

4. COMPARATIVE STUDY WITH EXPERIMENTAL DATA

Among the many calculations[1-12] of the burning-rate response function $R_b = \mu/\phi$ [or admittance function $(u'/\bar{u})/(p'/\bar{p})$] based on the quasi-steady model of gas reaction, the treatment of Culick[12] is the most complete and provides a general form of R_b:

$$R_b = nAB[(\lambda + A/\lambda) - (1 + A) + AB]^{-1} \qquad (7)$$

where n is the pressure exponent of the steady-state burning-rate law $\bar{r} = \bar{r}_0 \bar{p}^n$. The quantity $\lambda = \lambda_r + i\lambda_i$ is the complex dimensionless wave number of the temperature oscillation in the solid propellant with $\lambda(\lambda - 1) = ik\omega/\bar{r}^2$ expressed in terms of the thermal diffusivity k, the oscillation frequency ω, and the propellant burning rate $\bar{r}$; $A = (1 - T_p/\bar{T}_s)E_s/R\bar{T}_s$ is purely a solid-propellant property in terms of its undisturbed propellant temperature T_p, surface temperature $\bar{T}_s$ with activation energy E_s for surface pyrolysis. The parameter B is again primarily a solid-phase property but can be made to vary slightly with gaseous-phase properties by incorporating various details. The magnitude of B does not change significantly within the quasi-steady-state approximation. The various models differ largely in the details of how B is evaluated from the integrated energy equation across the thin gaseous flame region with different postulated steady-state processes or temperature profiles.

When the surface-pyrolysis rate is assumed to be pressure sensitive with an exponent n_s (much like the interaction index m in the time-lag theory), an additive term would appear on the numerator but generally insignificant compared with nAB.

The response function R_b given by equation (7) is complex only because λ is complex; all other parameters are real and essentially determined by the solid-propellant properties. Its form is primarily determined by the quasi-steady-state assumption (Culick used the word "quasi-static"). The following observations were made in his analysis: (1) intuitively important parameters in the gaseous phase are conspicuously missing from the final results, (2) the qualitative effects of many solid-phase parameters appear disturbing, (3) sample computations show that the magnitude of the response function appears too low in the physical range of interest. Even so, it was hoped that there would have been "a great degree of freedom afforded in applying the results to experiment." This hope was, however, not realized. Thus condensed phase reaction was incorporated. Different gaseous-phase flame theories for steady state were introduced into the quasi-steady formulation, with many more gaseous-phase parameters.[12] Thus B was made to vary with gaseous-phase properties. The peak magnitude of the response function R_b could eventually be raised adequately but "certain aspects remain qualitatively unacceptable." This exercise was exhaustive but frustrating, as to prompt the statement that "it would be surprising if a result containing only two parameters should work."

Beckstead and Culick[13] finally reported that, despite the large number of degrees of freedom in adjusting the values of many physical constants for calculating the parameters A and B in the response function (7), it could not possibly give any unstable oscillatory solution of the characteristic equation (5) corresponding to the data of the A-13 propellant (76% AP of 90 μm size and 24% PBAN binder with epoxy curative). This is the reference propellant designated by ICRPE, having a steady-state burning rate of 0.1 in./s at 100 psi and 0.3 in./s at 1000 psi, and $n \simeq 0.5$. The A-13 data fell in the nonphysical range of parameter values $A < 0$.

It has been pointed out[16] that in the low-frequency limit $\omega \to 0$, the burning-rate response function of a solid propellant must become n, the pressure exponent in the steady-state burning (and pyrolysis) rate law. This is truly a quasi-steady requirement. Culick correctly observed and discarded those expressions of response functions that failed to meet this requirement from his comparative study. He then noticed that additional parameters and physical processes are ineffective in modifying the response function in any significant way. This quasi-steady requirement of $R_b \to n$ as $\omega \to 0$ and the quasi-steady approximation for the gaseous-phase reactions established the basic form and behavior of the response function (7) that contradicts the observed L^*-instability.

The failure of (7) to rationalize the observed L^*-instability will now be shown analytically. The physical constants for the data under consideration are such that $16\Omega^2 \gg 1$ (see equation (9) with $\Omega = k\omega/\bar{r}^2$ in Beckstead and Culick's paper[13]). This alone suggests that the underlying physical mechanism is dubious. Asymptotically, we have

$$\lambda = [\tfrac{1}{2} + (\Omega/2)^{1/2}] + i(\Omega/2)^{1/2} \tag{8}$$

and

$$R_b^r \simeq nAB\{(\Omega/2)^{1/2} - 1 - A[1 - (1/2\Omega)^{1/2} + B]\}/D^2$$

$$R_b^i \simeq -nAB(\Omega/2)^{1/2}(1 - A/2\Omega)D^2$$

and

$$D^2 = \{[(\Omega/2)^{1/2} - 1] - A[1 - (1/2\Omega)^{1/2} + B]\}^2 + \frac{\Omega}{2}\left(1 - \frac{A}{2\Omega}\right)^2$$

Clearly, R_b^i will ordinarily be negative, contradicting condition (6b); R_b^i could be positive if A (likewise B) should be assigned any nonphysical negative values. Alternatively R_b^i could also be positive if A should possess some large positive values with $A > 2\Omega \gg 1$. Such large values of A could be achieved only with physically unreasonable values of $E_s/R\bar{T}_s$; and in any case, with $B \simeq 1$, the magnitude of R_b^r will be unable to satisfy condition (6a). The failure is dramatic and the reason is fundamental.

The time-lag theory was proposed over a quarter of a century ago[14] based on transient gas-phase interaction that provides a simple response function:

$$R_b = (S + n/2) - (S - n/2)\exp(i\omega\bar{\tau}) \tag{9}$$

It contains two parameters, the interaction index $S = m - n/2$ and the mean time lag $\bar{\tau}$ with n fixed as the pressure exponent of the steady-state burning rate and m the pressure exponent of some global, transient rate processes in gas reactions. We shall show that the magnitudes of m and $\bar{\tau}$ required to produce L^*-instability are physically reasonable.

With the response function (9), the characteristic equation (6) or (6′) becomes

$$\gamma m - \gamma(m - n)\cos\omega\bar{\gamma} = \frac{\gamma + 1}{2} + \alpha + \frac{\partial \ln\bar{p}}{\partial z} \tag{10a}$$

$$\gamma(m - n)\sin\omega\bar{\tau} = \omega(1 + \beta) \tag{10b}$$

Unstable L^*-oscillations can be obtained when

(i) $(2j + 1)\pi + \sin^{-1}\dfrac{\omega(1+\beta)}{\gamma(m-n)} \leqslant \omega\bar{\tau} \leqslant 2(j+1)\pi - \sin^{-1}\dfrac{\omega(1+\beta)}{\gamma(m-n)}$

(ii) $\gamma^2(m-n)^2 = \left(\gamma m - \dfrac{\gamma+1}{2} - \alpha - \dfrac{\partial}{\partial z}\ln\bar{p}\right)^2 + \omega^2(1+\beta)^2$

The first condition (i) states that the total time lag must be in the proper range of values for each integral value of $j = 0, 1, 2, 3, \ldots$, so that $\omega\bar{\tau}$ will lie in the proper sector of the lower half plane. If ω should be too small, unstable oscillations would require very large time lags compared to residence time and thus are unlikely. With $\omega = O(1)$ unstable oscillations of period comparable to residence time can occur if the total time lag $\bar{\tau}$ is $O(1)$. At sufficiently large time lags containing one or more oscillation periods, the parameter j may take successively larger integral values. The second condition (ii) means that the pressure-interaction index m of the gaseous-phase reactions must be sufficiently large to give any real solutions of ω.

If the gaseous reaction rate is postulated to be of the Arrhenius form

$$G \sim p_g^2 T_g^{-1/2}\exp\left(\frac{E_g}{RT_g}\right)$$

for second-order reactions with frequency factor $\sim T_g^{-1/2}$, the interaction index m, as evaluated from equation (3.1) on p. 87 of Cheng,[16] is given by

$$m = \frac{d(\ln G)}{d(\ln p)} = \frac{3\bar{\gamma}+1}{2\bar{\gamma}} + \frac{E_g}{RT_g}\frac{\bar{\gamma}-1}{\bar{\gamma}}$$

where $\bar{\gamma}$ will be the specific-heat ratio $\gamma = c_p/c_v$ if the oscillation in the gaseous reactive region is isentropic. For a typical value of $E_g \sim 40$ kcal for AP propellant with $T_g \sim 2500$ K, the values of the pressure interaction index m will be 2.9 to 3.3 for $\bar{\gamma} \sim 1.2$ to 1.3. The corresponding values of m for first-order gaseous reactions are 2.0 to 2.5. All these values are well above requirement (ii) for the magnitude of m. Thus for AP-type propellants the only signficant requirement is that of time lag (i), i.e., $\omega\bar{\tau}$ must be at least between π and 2π or $\bar{\tau} = O(1)$ (order of residence time) to produce unstable oscillations.

When both m and $\bar{\tau}$ are given for a propellant, there can be zero, one, or more oscillation frequencies, primarily determined by (i) or (10b). Each of these oscillations with frequency ω will amplify or decay in gross terms according to whether

$$\alpha + \frac{\partial}{\partial z}\ln\bar{p} > 0 \quad \text{or} \quad < 0$$

as is determined from equation (10a). There may be more than one unstable component. The fastest amplifying component is the most easily observed in the experimental situation. This component corresponds to the smallest ω compatible with the time condition (i) for the given time lag $\bar{\tau}$. The larger the dimensionless mean time lag $\bar{\tau}$, the smaller the minimum ω value and the larger the amplification rate of the disturbance. Since the frequencies ω of all admissible oscillatory solutions must satisfy $\omega \leqslant \omega^* = \gamma(m-n)/(1+\beta)$, the time lag of a given propellant must exceed some minimum value $(\tau_{min})_j > (2j+1)\,\pi/\omega^*$ to exhibit any unstable oscillation in this j segment. If the time lag of a propellant under test should have $\tau < (\tau_{min})_j$ there will not be any unstable oscillations corresponding to the set of parametric values for that particular choice of j. The time condition may, however, be satisfied at one or more smaller integral values of j. The most prominent unstable oscillations observed in a given experiment will again be the fastest amplifying oscillation among those provided by the solutions in all the lower segments. If the mean time lag $\bar{\tau}$ is too small even for $j = 0$, there simply will not be any unstable oscillatory components and the system is "stable." The situation is complicated by the nonlinear nature of the characteristic equation (10) that defines the multivalued stability boundary of such linear oscillations. It is important to understand the complex situation to interpret properly the experimental results. They cannot be analyzed quantitatively without pertinent experimental details.

Schöyer[20] reported that for ANP-2608 propellant (82% AP with 12.5% polyurethane), 30 out of 60 test runs showed oscillatory combustion with L^* from 0.1 to 1 m, in the frequency range 75 to 300 Hz. The residence time ranges from 0.2 to 2 m-s. If it is postulated that the total time lag consists essentially of the transit time of the reactive gas leaving the solid-propellant burning surface at a velocity of 10 cm/s, the "thickness" of the gaseous flame over the burning surface is in the range 0.02 to 0.2 mm, which is comparable to experimentally observed values. Thus the time-lag formulation (10 a, b) is compatible with Schöyer's data. Details of the A-13 propellant data which Culick[12,13] referred to are said to be not as complete as Schöyer's.[21]

Figure 3 is taken from Figure 21 of Schöyer[20] where the dimensionless frequency $\Omega = \omega^* k / \bar{r}^2$ of all observed unstable L^*-oscillations for all the chamber pressures tested was shown to lie between two neighboring values. This Ω was defined in the heat-transfer theories and can be converted to the present variable $\omega = \omega^*(c_0^* \Gamma / L^*) = \Omega(\bar{r}^2/k)(c_0^* \Gamma / L^*)$.

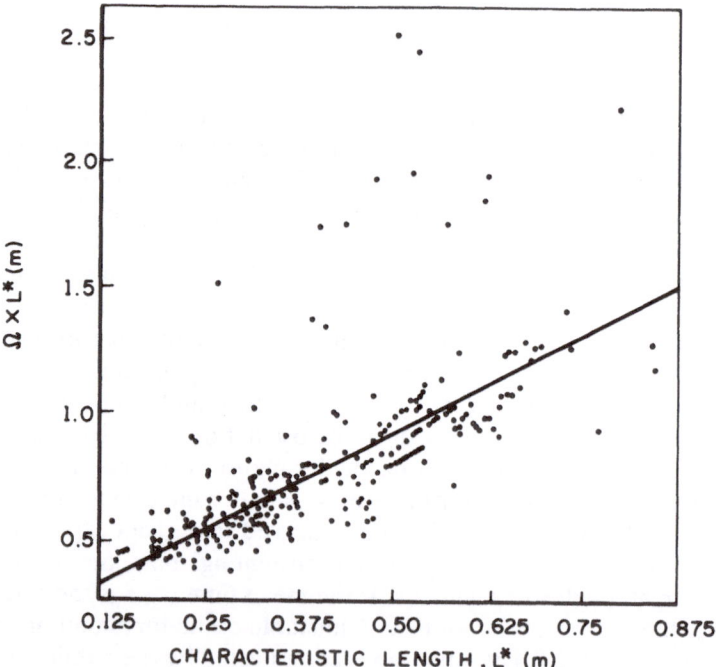

Fig. 3. Observed unstable L^*-oscillations of ANP-2608 propellant, ΩL^* vs. L^* with dimensionless frequency $\Omega = Fk/\bar{r}^2$ defined according to heat-transfer theories (after Schöyer,[20] Fig. 21). Dimensionless frequency ω in time-lag theory is $\omega = [\Omega \cdot (\bar{r}^2/k)]/(L^*/c_0^* \Gamma)$.

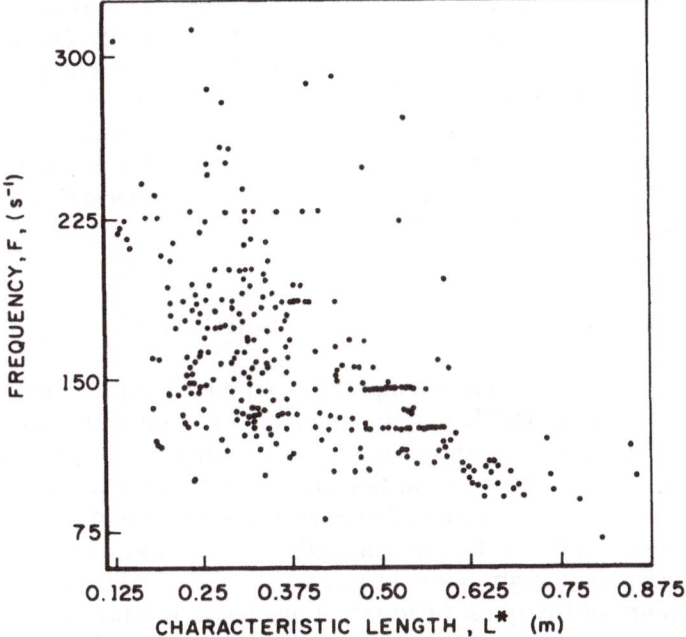

Fig. 4. Observed unstable *L**-oscillations of ANP-2608 propellant. Dimensional frequency *F* (in Hz) vs. *L** (after Schöyer,[20] Fig. 19).

This ω is thus expected to lie in between a corresponding range of values, as is suggested by conditions (i) and equations (10).

Figure 4 is taken from Figure 19 of Schöyer[20] displaying the dimensional unstable frequencies *F* (in hertz or cycles per second) for the range of *L** tested. Since

$$\omega = 2\pi F \cdot t_{\text{res}} = 2\pi F \frac{L^*}{c_0^* \Gamma}$$

the bounds of ω suggest that

$$c_0^*/\bar{\tau} > FL^* > \tfrac{1}{2} c_0^*/\bar{\tau}$$

i.e., all the observed unstable oscillations should lie between two hyperbola-like curves in the *F* vs. *L** plane. Figure 4 clearly demonstrates that this is so. Since both c_0^* and $\bar{\tau}$ vary with the mean chamber pressure corresponding to different *L**-values, it is not possible to analyze the result in any quantitative manner.

These observations are encouraging in that they seem to indicate that the time-lag formulation will not encounter the difficulty faced by the solid-phase heat-transfer theories with quasi-steady gaseous-phase processes. The interaction index m and the associated time lag $\bar{\tau}$ can indeed be determined for each experimental point. We would then be in a position to judge whether and how these two parameters m and $\bar{\tau}$ might be interpreted as the properties of this AP propellant, and to decide intelligently on the direction of further study.

5. RESIDUAL AND SECONDARY COMBUSTION

The steady-state chamber temperature (combustion) distribution of a solid rocket with a double-based propellant or a nitramine propellant displays the characteristics shown in Figure 1b and/or c, i.e., with substantial residual or secondary combustion. It was observed experimentally[20] that the L^*-instability behavior of motors with such propellants are quite different from that with AP propellants (Figure 5). It was suggested[21] that residual combustion might cause the significant change in stability behavior and the failure of the quasi-steady-state gas-phase reaction theories.

Let us consider the case (Figure 1b) where there is a steep combustion front at some station ξ with $0_1 < \xi < 1$. In this steep-gradient region there

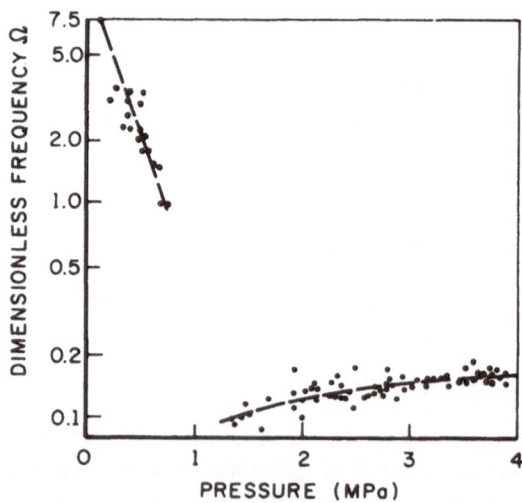

Fig. 5. Observed unstable L^*-oscillations of ARP propellant. Dimensional frequency F (in Hz) vs. mean chamber pressure $\bar{p}$ (after Schöyer,[20] Fig. 9).

will be nontrivial contributions from

$$\frac{\partial \theta}{\partial z} \bigg/ \frac{\partial \phi}{\partial z} = \frac{\partial \theta}{\partial \phi}$$

that cannot be "buried" in the burning-surface response function. If one should adopt the quasi-steady approximation, so that the nontrivial contribution of transient combustion in the steep-gradient region may be dropped, equation (3) would then stand as

$$\mu_b(t) - \mu_e(t) = \frac{1}{\gamma} \frac{\partial \phi}{\partial z} \left\{ \int_0^1 [T_e / \bar{T}(x)] dx \right\} \tag{11}$$

where $\bar{T}(x)$ can be quite arbitrary as illustrated. By comparing equation (11) with equation (4) for AP propellant without residual combustion, we note that the coefficient $1/\gamma$ in the accumulation term is modified by the multiplying factor f, given by

$$f = \int_0^1 [T_e / \bar{T}(x)] dx \tag{12}$$

which is always greater than unity. This multiplying factor is larger if the low-temperature region $[\bar{T}(x) < T_e]$ occupies a larger portion of the combustion chamber volume. The characteristic equation (6) becomes

$$\gamma R_b^r = (1 + A_e^r) + f \cdot \alpha = (\gamma + 1)/2 + f \cdot \alpha \tag{13a}$$

$$\gamma R_b^i = f\omega + A_e^i = \omega(f + \beta) \tag{13b}$$

With the secondary flame separated from the primary-flame zone over the solid burning surface (Figure 1b), R_b will not change with T_1 replacing T_e since parameters A and B equation (7) do not explicitly contain the burned-gas temperature. This is the consequence of the previously mentioned limit that R_b must reduce to n as $\omega \to 0$. If the secondary flame should be close to the primary flame, or should be well distributed as in Figure 1c, the effect is equivalent to a change in the steady-state temperature profile in the primary flame, so as to possess some form of a kink or a plateau. Within the quasi-steady-state approximation and the requirement of $\lim_{\omega \to 0} R_b = n$, the magnitudes of parameters A and B, and the response function R_b will remain essentially unchanged. Thus the characteristic equation is altered [comparing equations (6) and (11)] only by the presence of f. With f larger than unity, equations (13) would suggest larger positive values of R_b^r and R_b^i required to produce any L^*-instability if there is residual (or distributed) combustion in the chamber. Consequently, so long as the quasi-steady approximation is uniformly applied to the gaseous reactive regions, the presence of a residual combustion cannot remedy the frustrating situation of how the solid-phase heat-transfer theories fail to predict any L^*-combustion instability.

With the time-lag formulation the presence of a significant secondary flame will introduce considerable complication, because the kinetics of the secondary flame is probably quite different from that in the primary flame. Thus a new pair (or more) of phenomenological constants for the interaction index and the mean time lag will have to be introduced for a meaningful detailed treatment. Suppose the nontrivial contribution of the gaseous reaction in the primary flame to the response function is identified as

$$m_1[1 - \exp(i\omega\bar{\tau}_1)]$$

referring to the pressure oscillation at the entrance to the modified combustion chamber $x = 0_+$ at the local time. The intermediate gas generated from the primary flame occupies the fraction ξ of the chamber volume. Here m_1 is the pressure-interaction index in this primary gaseous flame with flame temperature T_1. To the same approximation, the nontrivial contribution in the secondary flame will be, referring to its local time at $x = \xi$,

$$(1 - \xi)m_2[1 - \exp(i\omega\bar{\tau}_2)]$$

where $(1 - \xi)$ is the factor introduced by the reduced chamber volume affected by the secondary flame; m_2 is the pressure-interaction index of the secondary flame. Let $\xi L/\bar{u}$ be the transit time for the intermediate gas at temperature T_1 to travel from the primary flame at 0_+ to the secondary flame at ξ, with the mean velocity $\bar{u}$ of the gas between the two stations. This $\xi L/\bar{u}$ is much larger than the reaction time $\bar{\tau}_2$.

Let us now consider the response of the gas system in the entire combustion chamber with reference to the pressure oscillation acting on the secondary flame at any instant t. The response of the primary flame is $\tau_2 + \xi L/\bar{u}$ earlier. Thus the overall burning rate response function is

$$\begin{aligned}
R_b = &(1 - \xi)m_2[1 - \exp(i\omega\bar{\tau}_2)] \\
&+ m_1[1 - \exp(i\omega\bar{\tau}_1)] \cdot \exp[i\omega(\bar{\tau}_2 + \xi L/\bar{u})] \\
&+ n \exp[i\omega(\bar{\tau}_1 + \bar{\tau}_2 + \xi L/\bar{u})]
\end{aligned} \tag{14}$$

This response function R_b can be used in equation (5') or (6') to determine the admissible oscillating solutions with frequency ω and gross amplification factor $\alpha' = \alpha + \partial \ln \bar{p} / \partial z$ for a specific set of parameter values. While the solution is tedious, it is straighforward as for the simpler case of equations (10a, b). In view of the fact that the transit time lag $\xi L / \bar{u}$ is significantly larger than the sensitive part of the time lags $\bar{\tau}_1$ and $\bar{\tau}_2$, the qualitative nature of the solution cannot be much different except for the presence of a very large effective time lag $\sim (\bar{\tau}_1 + \bar{\tau}_2 + \xi L / \bar{u})$ and the slow variation of some mean interaction index m of the overall system. As it was estimated previously that both m_2 and m_1 are probably sufficiently large, its slow variation around some mean value is of little consequence qualitatively. Thus the major effect of secondary combustion lies in the large effective mean time lag.

It was concluded from equation (10) that the significant requirement for the occurrence of L^*-unstable oscillations is the time condition. Unstable oscillations are not so "prominent" for AP-type propellants, most likely because the very thin flame gives rise to modest values of time lag compared to residence time. The presence of some secondary combustion can appreciably increase this mean time lag, and therefore will promote L^*-instability of AP solid rocket motors. For nitrocellulose or nitramine propellant, secondary combustion is significant. Not only is the mean-time-lag condition much more likely to be fulfilled at the lowest segment $j = 0$, it may be satisfied also at some nonzero values of j. Such motors are therefore liable to abrupt changes in the observed behavior of L^*-instability when the most unstable or the fastest amplifying oscillation changes from one segment to another with different j-values. Such abrupt changes have been reported by Schöyer in connection with his data of ARP propellant of 50% nitrocellulose and 37% glycerol trinitrate as detailed below.

Schöyer[20] presented in his Figures 9 and 10 the data of ARP-propellant unstable oscillations with dimensional frequency F (in Hz) and his dimensionless frequency Ω plotted against the mean chamber pressure $\bar{p}$. His figures are reproduced here as Figures 5 and 6. There is a startling "discontinuity" in unstable oscillation frequencies as the mean chamber pressure increases from low values to 4 MPa. The discontinuity in the frequencies of unstable L^*-oscillations occurs in the vicinity of $\bar{p} \simeq 1$ MPa. There is also a sudden change in the slope of the graph above and below this pressure. Such an abrupt change in the frequency of unstable oscillations with the increase in chamber pressure is suggested by the above time-lag formulation. At very low pressures, less than 1 MPa, the flame spreads almost over the entire chamber with very large mean time lag $\bar{\tau}$. Thus unstable oscillations in some segments with $j \geq 1$ can result. The effective mean time lag $\bar{\tau}$ decreases rapidly as the flame shortens (smaller effective ξ) with increasing chamber pressure. The slight decrease of the

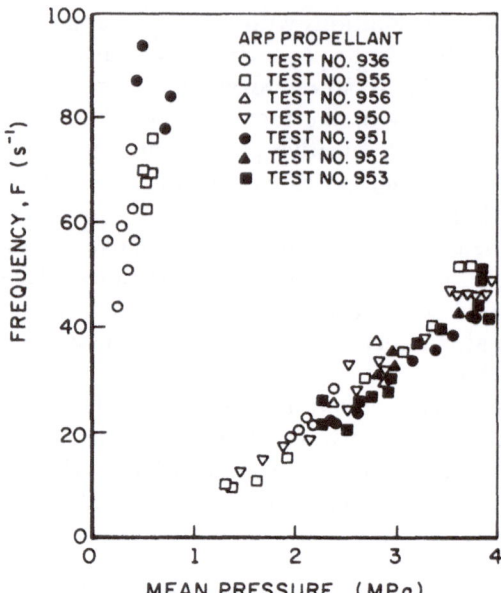

Fig. 6. Observed unstable L^*-oscillations of ARP propellant. Dimensionless frequency $\Omega = Fk/\bar{r}^2$ according to heat-transfer theories vs. mean chamber pressure $\bar{p}$ (after Schöyer,[20] Fig. 10). Dimensionless frequency ω in time-lag theory is $\omega - [\Omega \cdot (\bar{r}^2/k)]/(L^*/c_0^*\Gamma)$.

residence time t_{res} with increasing chamber temperature is insignificant. With decreasing $\bar{\tau}$, the oscillation frequencies F or ω increase in the segment. When the chamber pressure reaches about 1 MPa, the minimum value of $\bar{\tau}$ in this interval, namely $(\tau_{min})_j$, was reached. Further increase of chamber pressure beyond 1 MPa led to a discontinuous jump of the observed unstable frequency to much lower values, probably corresponding to $j = 0$. While remaining in this segment $j = 0$, both F and ω increase with further increase of chamber pressure accompanied by the reduction of $\bar{\tau}$.

It may be noted in Figure 5 that the dimensionless frequency $\Omega = \omega^* k/\bar{r}^2$, adopted by Schöyer following the heat-transfer theories, appears to decrease slightly in the lower pressure segment. This is due to the rapid increase of the mean burning rate $\bar{r} \sim \bar{p}^n$ under the rising chamber pressure $\bar{p}$. In terms of the present dimensionless frequency variable $\omega = \Omega(\bar{r}^2/k)(c_0^*\Gamma/L^*)$, this ω will consistently increase with increasing chamber pressure while in the segment. There is no change in stability behavior of the ARP propellant in the two segments. There is no need for any evidence of significant changes in the interaction index m of the ARP propellant at low and high pressures to explain the observed phenomena. There is only the obvious drastic change of the time lag from the very extended combustion zone at low pressures to the relatively well-defined flame zone at higher pressures. The details of the chemical kinetics of the combustion reactions at low and high pressure can of course be quite different, but the pressure dependence of the global rate processes need not be. The discontinuous jump and the qualitative variation of F and ω with increasing chamber pressure are correctly predicted with the time-lag formulation. The magnitudes of both m and $\bar{\tau}$ appear to be of correct order for explaining the observations of the unstable phenomena. It appears that the interaction index m and the mean time lag $\bar{\tau}$ of a propellant system can be determined when pertinent details of the experimental data are made available.

6. CONCLUDING REMARKS

This paper has reviewed and explained how and why the many solid-phase heat-transfer models with quasi-steady gaseous-phase reactions fail in predicting the L*-combustion instability in solid-propellant rockets. The large number of adjustable constants introduced into the response function as calculated by these theories cannot help. Neither can the presence of secondary combustion. The quasi-steady assumption must not be used to treat the gaseous flame.

One of the fundamentally transient formulations is the quarter-century-old "time-lag theory" which encounters none of the difficulties of the theories of L*, as well as the acoustic modes of combustion instability in solid rockets. It is shown that the presence of secondary combustion increases the mean time lag, and thus can render a marginally stable motor unstable, and *vice versa*. Extensive secondary combustion, as is observed in motors with double-base propellants, can cause L*-instabilities at very low frequencies, and can display sudden change in oscillation frequency as chamber pressure rises.

In view of the experimental results and the theoretical developments given above, it appears beneficial to examine the time-lag theory closely in the light of the available experimental data. If the phenomenological constants could be designated in some form as propellant properties, there is no fundamental difficulty in the contruction of some reasonable microscopic model to calculate such phenomenological constants, so long as one is willing to postulate or specify the detailed kinetics. If not, some other phenomenological theories should preferably be developed and tested prior to engaging in the analytic developments of comprehensive models.

REFERENCES

1. L. Green, Jr., Some properties of solid propellant burning, *Jet Propulsion* **28**, 6 (1958).
2. W. Nachbar and L. Green, Jr., Analysis of a simplified model of solid propellant resonant burning, *J. Aero/Space Sci.* **26**, 8 (1959).
3. R. Akiba and M. Tanno, Low frequency instability in solid propellant rocket motors, in: First Symposium on Rockets and Astronautics, Tokyo Proceedings (1959).
4. W. R. Hart and F. T. McClure, Combustion instability: Acoustic interaction with a burning propellant surface, *J. Chem. Phys.* **30**, 9 (1959).
5. J. F. Bird, L. Harr, R. W. Hart, and F. T. McClure, Effect of solid propellant compressibility on combustion instability, *J. Chem. Phys.* **32**, 5 (1960).
6. M. R. Denison and E. Baum, A simplified model of unstable burning in solid propellants, *ARS J.* **31**, 8 (1961).
7. R. W. Waesche and M. Summerfield, Solid Propellant Combustion Instability: Oscillatory Burning of Solid Rocket Propellants, Princeton Univ. AMS Report 751, p. 8, Princeton, New Jersey (1965).
8. R. W. Hart, R. A. Farrel, and R. H. Cantrell, Theoretical study of a solid propellant having a heterogeneous surface reaction. I. Acoustic response, low and intermediate frequencies, *Combust. Flame* **10**, 12 (1966).
9. F. E. C. Culick, Calculation of admittance function for a burning surface, *Astronaut. Acta* **13**, 3 (1967).
10. H. Krier, J. S. T'ien, W. A. Sirignano, and M. Summerfield, Nonsteady burning phenomena of solid propellants: Theory and experiment, *AIAA J.* **6**, 2 (1968).
11. G. A. Marxman and C. E. Wooldridge, Effects of surface reactions on the solid propellant response function, *AIAA J.* **6**, 3 (1968).
12. F. E. C. Culick, A review of calculations for unsteady burning of a solid propellant, *AIAA J.* **6**, 12 (1968).
13. M. W. Beckstead and F. E. C. Culick, A comparison of analysis and experiment for solid propellant combustion instability, *AIAA J.* **9**, 1 (1971).
14. S. I. Cheng, High frequency combustion instability in solid propellant rockets, *Jet Propulsion*, January–April, Parts I and II (1954).
15. F. K. Moore and S. H. Maslen, Transverse Oscillations in a Cylindrical Combustion Chamber, NACA TN 3152 (1954).
16. S. I. Cheng, Unstable combustion in solid propellant rocket motors, in: *Eight Symp. (Int.) on Combust.*, pp. 81–96, Williams and Wilkins, Baltimore (1962).
17. T-Burner Manual, Chemical Propulsion Information Agency, CPIA Publ. 191, Silver Springs, Maryland (1969).
18. Experimental Studies on the Oscillatory Combustion of Solid Propellants, Aerothermo Chemistry Division, Naval Weapons Center Report NWC TP 4393, China Lake, California (1969).
19. L. Crocco and S. I. Cheng, Theory of Combustion Instability in Liquid Propellant Rocket Motors, AGARDograph No. 8, Chapter 1 and Appendix 2, Butterworths, London (1956).
20. H. F. R. Schöyer, Low Frequency Oscillatory Combustion, Experiments and Results Agard Conf. Proc. CP259, p. 25, London (1979).
21. M. Summerfield, private communication.

14

Theory of Idealized Coal Devolatilization

P. A. LIBBY

ABSTRACT. The aerothermochemical interaction between a devolatilizing coal particle and its ambient is studied by means of an idealized model consisting of two solid-phase species, which thermally decompose to yield a single gas-phase species. Devolatilization is described by a classical linear theory involving both equilibrium and kinetic characteristics of each solid-phase species. The temperature history of the particle is treated in two different ways reflecting different experimental and applied situations. One simulates the direct heating of a particle, e.g., on a wire grid, so that the temperature history is prescribed. The second relates to injection of a cold particle into a hot ambient either under experimental conditions or in an entrained flow reactor. In the second situation both a frozen gas phase corresponding to devolatilization in inert ambients and equilibrium chemistry in the gas phase as described by a flame-sheet model are discussed. It is concluded that, despite the high degree of idealization involved in the analysis, a large number of thermochemical parameters, some with uncertain values, enter the description of particle behavior. It is difficult to determine kinetic data from direct heating experiments. The thermal response of a cold particle injected into a hot ambient is sensitive to the kinetics of devolatilization and to the heats of pyrolysis.

1. INTRODUCTION

The motivation for our previous studies[1-3] on the burning of carbon particles resides in the relevance of the results to the combustion of

P. A. LIBBY • Department of Applied Mechanics and Engineering Sciences, University of California San Diego, La Jolla, CA 92093, U.S.A. and Systems Science and Software, P.O. Box 1620, La Jolla, CA 92038, U.S.A.

pulverized coal. One aspect of coal behavior not accounted for in treatments of carbon particles is devolatilization, the thermal decomposition of various coal constituents prior to significant oxidation of the carbon. During this process the particle can lose up to 40–50% of its original mass in the form of gases, which interact fluid mechanically and thermochemically with the ambient surrounding the particle. It is our purpose to examine this interaction in terms of a highly idealized devolatilizing particle.

It is clear from the literature related to devolatilization (cf., e.g., the review of Anthony and Howard[4]) that complex processes, difficult to study even experimentally, are involved; a consequence is that considerable controversy surrounds even basic features determining the aerothermochemical interaction of the particles and their ambient. With respect to such difficulties Anthony and Howard note: "... the equipment itself often significantly influences the results." Thus we find important differences in the implications derivable from experiments involving the heating of particles spread on a grid (cf., e.g., Suuberg, et al.[5]) versus those involving heating in a gas stream (Kobayashi et al.[6] and McLean et al.[7]).

A further indication of basic features that remain beclouded is provided by the magnitude and sign of the heats of pyrolysis, the energy released or absorbed by a species in the course of devolatilization. The combustion of a coal particle may be studied in terms of its sudden immersion in a hot oxidizing ambient; in the energy balance describing the response of such a particle the energy of pyrolysis would be expected to play an important role, and uncertainty as to its value diminishes confidence in the resulting predictions.

Finally, relative to the complexity of the situation we note the variety of gases that are devolatilized; water, carbon dioxide, carbon monoxide, various hydrocarbons, and hydrogen in that order are the main constituents as a particle is heated. Each species must in general be considered to be produced at various rates from various species and molecular complexes in the solid state. The different thermochemical characteristics of the resulting gas-phase species cause them to interact differently with gases in the surrounding ambient and with one another, thereby altering the thermal response of the particle.

Given this situation it seems worthwhile to study a simplified model of a particle undergoing devolatilization in order to expose basic features of the resulting interaction of the particle with its ambient and to prepare thereby for a more complete treatment. Such a simplified model may in fact suggest focuses for additional research. Accordingly, we consider the devolatilization of a spherical particle of fixed radius consisting of residual, nondevolatilizing material and two solid-phase species that devolatilize to yield a single gas-phase species. The hot ambient is treated differently in two separate cases; one simulates devolatilization in an inert atmosphere as in the experiments of Kobayashi et al.[6] so that no gas-phase reactions are taken into account. In the second case we assume that the surroundings consist of oxygen and an inert diluent, such as nitrogen, and that a flame sheet which may be contiguous with, or removed from, the particle yields a single product of reaction. Oxidation of the residual material and its attack by the devolatilizing gas are neglected; although such processes may be significant at temperatures above 1200 K, we defer their consideration to a future study. Even with these rather severe idealizations a large number of parameters, presumably with varying degrees of importance, enter the analysis.

We find in the literature no closely related study. Many of the references related to devolatilization, such as those cited in Anthony and Howard,[4] contain simplified descriptions of particle behavior suitable for correlating experimental results but not concerned with the fluid-mechanical and thermochemical interactions of interest to us. Other references, such as Ubhayakar et al.,[8] describe the global effects of devolatilization on flows involving that process; of necessity the detailed interactions between a single particle and its ambient are treated casually in such analyses. Finally detailed treatments of the dynamics of pore formation and of the processes within such pores are available (cf., e.g., Simons[9,10]). In our study we assume these details of the internal behavior are adequately described by the identification of an equilibrium state and the kinetics of devolatilization in order to focus on the interactions external to the particle.

2. ANALYSIS

We consider a spherical particle in a hot ambient. The initial composition of the particle is determined by values of the partial densities ρ_{si}, $i = 1, 2$ and by the residual density ρ_r of the permanent, nondevolatilizing material. As the particle is heated the densities ρ_{si} decrease with time to yield a single devolatilizing gas with a mass fraction Y_D.

2.1. History of the Partial Densities Within the Solid

We assume the following general kinetic form for the partial densities:

$$\frac{d\rho_{si}}{dt} = k_{fi}(\rho_{si}^{(e)} - \rho_{si}) \tag{1}$$

where k_{fi} is a rate of devolatilization in s^{-1} and $\rho_{si}^{(e)}$ is the equilibrium partial density of the ith species, a given function of the particle temperature T_p.

Suppose we nondimensionalize the partial densities with the initial particle density ρ_0, define $R_{si} \equiv \rho_{si}/\rho_0$ and take the temperature dependence of R_{si} in the form

$$R_{si}^{(e)} = \tfrac{1}{2} R_{si}(0)\{1 - \text{erf}[\beta_i(T_p - T_i^*)]\} \tag{2}$$

where $R_{si}(0)$ is the initial mass fraction of the jth species, and β_i and T_i^* are parameters selected to simulate the composition of the particle if held indefinitely at the temperature T_p. Clearly T_i^* is the temperature at which the density of the ith species is one-half its initial value. The parameter β_i can be determined from the temperature at which devolatilization of the ith species either begins or is completed under equilibrium conditions.

It is interesting to note that with appropriate values for β_i and T_i^* equation (2) represents quite accurately the data on yield of devolatilized species from both lignite and bituminous coal given by Suuberg et al.[5] These data correspond to various rates of particle heating, up to 10^4 K/s, but in fact within the experimental scatter it is not possible to discern any effect of heating rate on yield. This result implies that in equation (1) k_{fi} is sufficiently large so that $\rho_{si} \cong \rho_{si}^{(e)}$; an analysis based on this approximation and on equation (2) can be developed and yields results that are consistent with Suuberg et al.[5] We discuss later the simulation of experiments along these lines in more detail.

Returning to develop more of our analysis, we introduce a nondimensional time $\tau = t\eta_\infty/\rho_0 r_p^2$ where η_∞ is the viscosity coefficient in the surrounding ambient and r_p is the radius of the particle, taken as constant. Thus we have from equation (1)

$$\frac{dR_{si}}{d\tau} = \bar{k}_{fi}(R_{si}^{(e)} - R_{si}) \tag{3}$$

where $\bar{k}_{fi} = k_{fi}\rho_0 r_p^2/\eta_\infty$ is a nondimensional reaction-rate parameter. Consistent with the exploratory nature of this study we represent $\bar{k}_{fi}$ in the form

$$\bar{k}_{fi} = A_i \exp(-T_{ai}/T_p) = \exp\{(T_{ai}/T_{ci})[1 - (T_{ai}/T_p)]\}$$

where $A_i = B_i\rho_0 r_p^2/\eta_\infty$ involves a preexponential factor B_i, where $\ln A_i = T_{ai}/T_{ci}$ defines the temperature T_{ci}, and T_{ai} is the activation temperature associated with devolatilization of the ith species. The form of the reaction-rate parameters given by equation (4) is especially convenient in the practically interesting case of $T_{ai}/T_{ci} \gg 1$; then T_{ci} plays the role of a critical temperature. For $T_p < T_{ci}$ we see that $\bar{k}_{fi} \ll 1$ and R_{si} can lag significantly behind $R_{si}^{(e)}$. For $T_p > T_{ci}$ on the contrary $\bar{k}_{fi} \gg 1$ and $R_{si} \cong$

$R_{si}^{(e)}$. Typical experimental data suggest that $T_{ai}/T_{ci} > 10$ with values as high as 35.

The experimentally determined values of the activation temperatures T_{ai} are widely scattered, depending on the type of coal and on the experimental techniques used in their determination. It is suggested, however, that for low temperatures, i.e., less than 1000 K, a value of 2000 K is representative while at higher temperatures a value of 10^4 is similarly representative.* Consistent with these values would be critical temperatures of 600 K and 1200 K, respectively. We remark in this connection that an alternative to the present idealization of two devolatilizing species in the solid phase involving distinct kinetic parameters is a distribution of activation temperatures as proposed by Anthony and Howard.[4] The present treatment can be readily extended to incorporate a large number of devolatilizing species so as to describe more closely a continuous distribution and is chosen because its numerical analysis is more straightforward.

Several remarks regarding equation (3) and these reaction-rate parameters are appropriate. We shall show the effect of variations in the activation temperature on predicted particle behavior; in making these variations we shall also vary the associated critical temperature so that the preexponential factors A_i and the ratio T_{ai}/T_{ci} remain constant. We also note that if $T_p > T_{ci}$ such that $\tilde{k}_{fi} \gg 1$, then $R_{si}^{(e)} \cong R_{si}$ and equation (3) degrades to an algebraic equation; the numerical analysis must be capable of handling this change in the nature of the describing equations. Finally, we note that the analysis is readily extended to include many more than the two devolatilizing species considered here; permitting such species to yield additional gas-phase species and taking into account the possible gas-phase reactions that might result would, on the contrary, complicate the analysis.

It is useful to note that in the present analysis two sets of quantities describe the characteristics of each devolatilizing species; the quantities $R_{si}(0)$, T_i^* and β_i relate to the equilibrium properties of each species and are presumably determined in experiments involving low rates of heating. The second set of quantities, namely the temperatures T_{ai} and T_{ci}, describe the kinetics of devolatilization for the species in question and are determined from experiments involving high rates of heating. If the different functions of these two sets are kept in mind, it is readily possible to select values describing a wide variety of devolatilization characteristics. For example, a high value of T_i^* accompanying a low value of β_i relates to a species, which under equilibrium conditions devolatilizes over a broad range of temperatures. However, if that same species involves a high activation temperature, only moderate heating rates are needed to assure

---

* See Figure 5 of Kobayashi et al.[6]

that the species will not devolatilize to any significant extent until the
temperature reaches T_{ci}.

Equations (3) and (4) provide interesting insight into some experiments designed to provide data on devolatilization. Suppose that the
temperature T_p is a known function of time, e.g., a linear increase at·a
known rate to a peak temperature followed by a linear decrease back to
ambient temperature. In this case equation (3) becomes a single equation
for $R_{si}(\tau)$ with the following solution:

$$R_{si} = \exp\left(-\int_0^\tau \tilde{k}_{fi}d\tau'\right)\left[R_{si}^{(0)} + \int_0^\tau \exp\left(\int_0^{\tau'} \tilde{k}_{fi}d\tau''\right)\tilde{k}_{fi}R_{si}^{(e)}d\tau'\right] \quad (4)$$

Devolatilization from such experiments is frequently presented in terms of
"yield in percent of original mass"; this equals in the present analysis the
porosity

$$\phi = \sum_{i=1}^2 (R_{si}(0) - R_{si})$$

since we admit a single devolatilizating gas-phase species. Thus we can
make a direct connection between the present analysis and such experiments. The study of particle behavior in terms of equation (5) simulates
one type of devolatilization experiment and has the advantage that various
uncertain thermodynamic and fluid-mechanical parameters are absent.
However, the simulation of particle behavior under entrained flow conditions requires the thermal response of the particle in a hot ambient to be
calculated.

2.2. The Flux from the Surface and the Energy Balance

Under the assumption that the devolatilizing species yield a single
gas-phase species with a mass fraction Y_D, conservation of mass yields

$$-\tfrac{4}{3}\pi r_p^3 \sum_{j=1}^2 \frac{d\rho_{sj}}{dt} = 4\pi r_p^2 \sum_{j=1}^2 (\rho_D v_D)_{jp} = 4\pi r_p^2\left[(\rho v)Y_D - \eta_\infty \frac{\partial Y_D}{\partial r}\right]_p$$

which becomes upon nondimensionalization

$$-\sum_{j=1}^2 \frac{dR_{sj}}{d\tau} = 3\left(KY_{Dp} - \frac{\partial Y_D}{\partial \xi}\bigg|_{\xi=1}\right) = 3K \quad (5)$$

where $K = (\rho v)_p r_p/\eta_\infty$ is the same nondimensional mass loss parameter
used in our earlier studies, $\xi = r/r_p$ is a nondimensional radius variable,

and Y_{Dp} is the mass fraction of the devolatilizing species at the surface of the particle. Both portions of equation (5) are significant: that involving K alone permits calculation of the rate of mass loss by summing the ratio of mass loss of the two solid-phase species, while that involving the characteristics of the devolatilizing species at the surface of the particle enters the energy balance. The gradient at the surface $(\partial Y_D / \partial \xi)(\xi = 1)$ depends on whether the gas phase is chemically frozen or in chemical equilibrium and will be eliminated in subsequent developments.

As a preliminary to calculating the energy balance of the particle, some comments regarding the heat of pyrolysis are indicated. Suuberg *et al.*[5] consider the thermodynamic aspects of pyrolysis of coal and conclude that for lignite the process is "... endothermic over a range of lower to intermediate temperatures ..." but at "... higher temperatures ... may be thermally neutral or exothermic." For bituminous coal they find little endothermicity or exothermicity at any temperature. However, the calculations used to reach these conclusions are subject to some uncertainty and it is therefore reasonable to include in the present analysis a heat of pyrolysis for each of the devolatilizing species. We also note that Kobayashi *et al.*,[6] in calculations used to estimate particle temperatures in their gas flow, assume a zero heat of pyrolysis.

In our treatment of the relevant thermochemistry the heat of pyrolysis of the ith species enters as the difference between the enthalpy of that species in the solid and of the gas resulting from devolatilization. We follow our earlier practice and assume linear relations between the species enthalpy and the temperature; thus for the gaseous species we have

$$h_i = c_p T + \Delta_i$$

and for the solid phase

$$h_{si} = c_s T + \delta_i$$

where c_p and c_s are the coefficients of specific heat taken to be constants. We interpret the heat of pyrolysis of the ith species to be $\delta_i - \Delta_D$, where for the single devolatilizing species we replace Δ_i by Δ_D. If $(\delta_i - \Delta_D) > 0$ the process of devolatilization is exothermic, while if $(\delta_i - \Delta_D) < 0$ the process is endothermic.

On the basis of these considerations we have as a primitive form of the equation of energy balance

$$\tfrac{4}{3}\pi r_p^3 \frac{d}{dt}\left(c_s \rho_s T_p + \sum_{i=1}^{2} \rho_{si}\delta_i\right) = 4\pi r_p^2 \left(q_c - q_r - \sum_{i=1}^{2} \rho_i v_i h_i\right)_p \tag{6}$$

where $\rho_s = \Sigma\,\rho_{si} + \rho_r$ is the instantaneous density of the solid. By following the treatment that leads to equation (5) we can rewrite equation (6) in the more convenient form

$$-(1-\phi)\frac{d\theta}{d\tau} = 3[(K+H)\tilde{h}_p - H\tilde{h}_\infty + Q_r - K\theta] + \sum_{j=1}^{2} \tilde{\delta}_i \frac{dR_{si}}{d\tau} \qquad (7)$$

where $\theta = T_p/T_\infty$, $\phi = 1 - \rho_s/\rho_0$ is the instantaneous porosity, $H = H(K) \equiv Ke^{-K}/(1-e^{-K})$, and $Q_r = (q_r r_p/\eta_\infty c_s T_\infty)$ is a nondimensional radiation parameter; the tilde over δ_i, h_p, and h_∞ implies nondimensionalization by the product $c_s T_\infty$, and unit Prandtl number has been assumed.

The various terms on the right side of equation (7) can be identified as follows. The product $K\tilde{h}_p$ describes the convective transfer due to devolatilization. The term $H(\tilde{h}_\infty - \tilde{h}_p)$ accounts for conductive heating as modified by convection. Clearly Q_r relates to the influence of radiative transfer, while the $K\theta$ term arises from the factor $T_p(d\rho_s/dt)$ and describes the reduction in the heat capacity of the particle resulting from its mass loss. Finally, the last term on the right side contains another effect of mass loss.

2.3. Frozen Gas-Phase Chemistry

We are now able to consider separately the two limiting cases of no gas-phase reaction and of equilibrium chemistry in the gas phase. In this section we take up the former case, which is of practical interest either if devolatilizing takes place at sufficiently low pressures or if the reactions among devolatilized hydrocarbons may be neglected. If there is no gas-phase reaction, then the solution for elements [equation (7) of Libby and Blake[1]] applies to individual species and

$$\left.\frac{\partial Y_D}{\partial \xi}\right|_{\xi=1} = -HY_{Dp} \qquad (8)$$

In equation (8) we assume the devolatilizing species to be absent from the ambient, i.e., that $Y_{D\infty} = 0$. The only other species present in this case is a neutral diluent with mass fraction Y_N; thus $Y_{N\infty} = 1$ and $Y_{Np} = 1 - Y_{Dp}$ at the surface of the particle.

When equation (8) is substituted into equation (5), there result from equations (2), (5), and (7) five equations in the five unknowns: R_{si}, $i = 1, 2$; θ; K; and Y_{Dp}. Three initial conditions, $R_{si}(0)$, $i = 1, 2$, and $\theta(0)$ must be specified along with a variety of parameters related to the thermochemical characteristics of the particle and its ambient.

2.4. Equilibrium Composition

When a coal particle devolatilizes in a hot oxidizing ambient under suitably high pressures, the devolatilized gases can react with the oxygen, yield a gaseous product, and alter significantly the thermal response of the particle to its ambient. To examine this case we assume that the gas-phase reactions occur only at a flame sheet located at $\xi_f \geqq 1$ to be determined as part of the solution, and that we have the simple reaction $D + 0 \rightarrow P$. Thus we deal with four species with mass fractions Y_D, Y_0, Y_P and Y_N. The composition in the ambient is specified by a value of Y_0 so that $Y_N = 1 - Y_{0\infty}$.

Two cases must be considered; if the rate of mass loss is sufficiently high, the flame sheet will be displaced from the surface of the particle and $\xi_f > 1$. In this case we can identify two regions: exterior to the flame $Y_D \equiv 0$ while in the interior region, $1 \leqq \xi \leqq \xi_f$, $Y_0 \equiv 0$. In each region the absence of gas-phase reactions implies that the conservation equation for each species has a solution of the form

$$Y_i = \beta_i e^{-K/\xi} + C_i \tag{9}$$

where the subscript i denotes any one of the four species and where β_i and C_i are arbitrary. For species destroyed or created at the flame sheet separate pairs of values for these constants prevail in the two regions, while for $i \rightarrow N$ a single pair pertains for $\xi \geqq 1$. Application of primitive and obvious boundary conditions at $\xi = 1$, ξ_f and ∞ leads to

$$Y_D = \frac{Y_{Dp}(e^{-K/\xi_f} - e^{-K/\xi})}{e^{-K/\xi_f} - e^{-K}}, \qquad 1 < \xi \leqslant \xi_f$$

$$= 0, \qquad \xi \geqslant \xi_f$$

$$Y_0 = 0, \qquad 1 < \xi \leqslant \xi_f$$

$$= \frac{Y_{0\infty}(e^{-K/\xi} - e^{-K/\xi_f})}{1 - e^{-K/\xi_f}}, \qquad \xi \geqslant \xi_f \tag{10}$$

$$Y_P = \frac{(Y_{Pf} - Y_{Pp})e^{-K/\xi} + (Y_{Pp}e^{-K/\xi_f} - Y_{Pf}e^{-K})}{1 - e^{-K}}$$

$$= \frac{Y_{Pf}(1 - e^{-K/\xi})}{1 - e^{-K/\xi_f}}, \qquad \xi \geqslant \xi_f$$

$$Y_N = \frac{(Y_{N\infty} - Y_{Np})e^{-K/\xi} + (Y_{Np} - Y_{N\infty}e^{-K})}{1 - e^{-K}}$$

In these equations Y_{Dp} and Y_{Pp} depend on the mass loss parameter K and are calculated as part of the solution as functions of time. The flame-sheet location ξ_f and the product concentration at the sheet Y_{Pf} can be eliminated as follows. Since the oxidation of the devolatilizing species is assumed to be first order, we have the requirement that

$$w \frac{\partial Y_D}{\partial \xi}\bigg|_{\xi=\xi_f^-} = -\frac{\partial Y_0}{\partial \xi}\bigg|_{\xi=\xi_f^+} \tag{11}$$

where w is the ratio of the molecular weight of oxygen to that of the devolatilizating species.* Equation (11) leads to

$$\xi_f = -\frac{K}{\ln\{[1 + (Y_{0\infty}/w Y_{Dp})e^{-K}]/[1 + (Y_{0\infty}/w Y_{Dp})]\}} \tag{12}$$

We also have the obvious condition $Y_{Pf} = 1 - Y_{Nf}$ so that

$$Y_{Pf} = 1 - \frac{(Y_N - Y_{Np})e^{-K/\xi_f} + (Y_{Np} - Y_N e^{-K})}{1 - e^{-K}} \tag{13}$$

Equations (12) and (13) when substituted into equations (11) determine the entire composition surrounding the particle provided Y_{Dp}, Y_{Pp}, and K are known.

We now turn to conditions at the particle surface; clearly we have from the first of equations (10)

$$\left(\frac{\partial Y_D}{\partial \xi}\right)_{\xi=1} = -\frac{Y_{Dp}Ke^{-K}}{e^{-K/\xi_f} - e^{-K}} \tag{14}$$

so that equation (5) is now completed for the case of equilibrium in the gas phase. We must also require that there be no flux of product or diluent through the surface of the particle. Thus

$$KY_{Pp} - \frac{\partial Y_P}{\partial \xi}\bigg|_{\xi=1} = KY_{Pp} - H(Y_{Pf} - Y_{Pp}) = 0$$

$$KY_{Np} - \frac{\partial Y_N}{\partial \xi}\bigg|_{\xi=1} = KY_{Np} - H(Y_{N\infty} - Y_{Np}) = 0 \tag{15}$$

---

* Except for the case of hydrogen devolatilizing from the particle, it is sufficiently accurate for our purposes to take $w = 1$.

During the early phase of particle heat-up the rate of devolatilization is insufficient for all of the oxygen to be consumed by the devolatilizing species. In this case the flame sheet is on the surface of the particle and two distinct regions external to the particle do not prevail. This case can be analyzed as follows. If we calculate the flux of the devolatilizing species and the flux of oxygen into the exterior side of the flame sheet required to consume those devolatilizing molecules, we can determine the oxygen concentration on that exterior side, namely

$$Y_{0p} = (HY_{0\infty} - wK)/(K + H) \qquad (16)$$

The second of equations (15) continues to apply so that with $Y_{Dp} = 0$

$$Y_{Pp} = 1 - Y_{0p} - Y_{Np} \qquad (17)$$

and we have the complete composition at the surface of the particle. The critical vaue of K, namely K_{cr}, at which the flame lifts from the surface of the particle is given by equation (16) with $Y_{0p} = 0$, hence

$$K_{cr} = \ln(1 + Y_{0\infty}/w) \qquad (18)$$

Thus the case of a devolatilizing particle with equilibrium composition in the gas phase is now complete.

For the situations we study here the initial conditions are such that there is always an initial period of particle heat-up during which $K < K_{cr}$, so that initially equations (3), part of equation (5), equations (8), (16), and (17) determine R_{si}, $i = 1, 2$; θ; K; Y_{0p}; and Y_{Pp}. During an intermediate stage in the particle history $K > K_{cr}$, so equations (3), (5), (7), (10), and (15) constitute eight equations for the eight unknowns: R_{si}, $i = 1, 2$; θ; K; Y_{0p}; Y_{Pp}; ξ_f; and Y_{pf}. Finally, during the final period in which devolatilization is nearly complete, the mass loss again decreases so that $K < K_{cr}$ and the first set of equations applies again.

3. NUMERICAL RESULTS

In this section we discuss the results of applying the analysis of the previous section. Because of the large number of parameters to be examined, it is convenient to establish a prototypical set of values and to vary those considered to be of greatest significance. In Table 1 we specify such a set, values chosen to be representative of those of interest in both laboratory and entrained flows. The large number of parameters entering the description of even the highly idealized model of devolatilization considered here is to be noted.

Table 1. Parameters defining prototypical conditions

c_p	$= 0.3$ cal/g K	T_∞	$= 1500$ K
c_s	$= 0.5$ cal/g K	T_{a1}	$= 20(10^3)$ K
Δ_D	$= -1(10^3)$ cal/g	T_{a2}	$= 10(20^3)$ K
Δ_P	$= -2(10^3)$ cal/g	T_{c1}	$= 600$ K
$\delta 1$	$= -1(10^3)$ cal/g	T_{c2}	$= 1.2(10^3)$ K
$\delta 2$	$= -1(10^3)$ cal/g	T_1^*	$= 800$ K
$R_{s1}(0) = R_{s2}(0) = 0.2$		T_2^*	$= 1600$ K
w	$= 1$	β_1	$= 5 (10^{-3})$ (K)$^{-1}$
r_p	$= 50$ microns	β_2	$= 5 (10^{-3})$ (K)$^{-1}$
Y_0	$= 0.2$		

Several comments regarding the entries in Table 1 are indicated. The values of T_{ci} and T_i^* for the two species considered are taken to be characteristic of low-temperature and high-temperature pyrolysis; thus the corresponding β_i's imply that devolatilization under equilibrium conditions is complete at 1200 K and 2000 K, respectively. The values of T_{ci} are selected so that devolatilization essentially begins at 600 K and 1200 K, respectively, as noted earlier.

Because of the simplified chemistry employed, the pressure does not enter our calculations. In general we include radiation in the energy balance of the particle with an emissivity of unity and with an effective radiation temperature equal to T_∞; we do so because in several experiments the particles enter a chamber with walls roughly equal to the hot-gas temperature. We present one calculation with radiation suppressed to examine its effect on particle behavior.

3.1. Particle Temperature Specified

Our first results relate to experiments in which the temperature history of the particle is imposed, e.g., by heating the particles on a grid (cf. Suuberg et al.[5]). In this case the history of the partial densities is given by equation (4) and the results of interest are the histories of the porosity $\phi(\tau)$. We are concerned with the influence of the kinetic parameters T_{ai} and T_{ci} on the latter histories and thus by implication with the deduction of such parameters from direct heating experiments.

We assume the particle is heated at a constant rate from room temperature to a peak temperature and then instantaneously cooled back to room temperature. This history represents a convenient idealization of the various histories used in the laboratory and is adequate for our considerations. The results are presented in terms of ϕ versus peak temperature for various rates of heating.

The first case we consider is shown in Figure 1a; prototypical conditions given in Table 1 are assumed throughout. We see that the separation between the two sets of temperatures characterizing the two devolatilizing species is such that kinetic effects on the porosity are likewise separated. The plateau that arises because devolatilization of the low-temperature species is nearly complete before that process begins for the second species effectively isolates the two histories. It should be noted that despite such isolation considerable accuracy is apparently called for if kinetic information is to be deduced from experiments of this sort.

The influence of kinetic parameters is more evident in the results shown in Figure 1b. In this case we retain all the quantities shown in Table 1 with the subscript 1 but alter those of the second species, so that the two

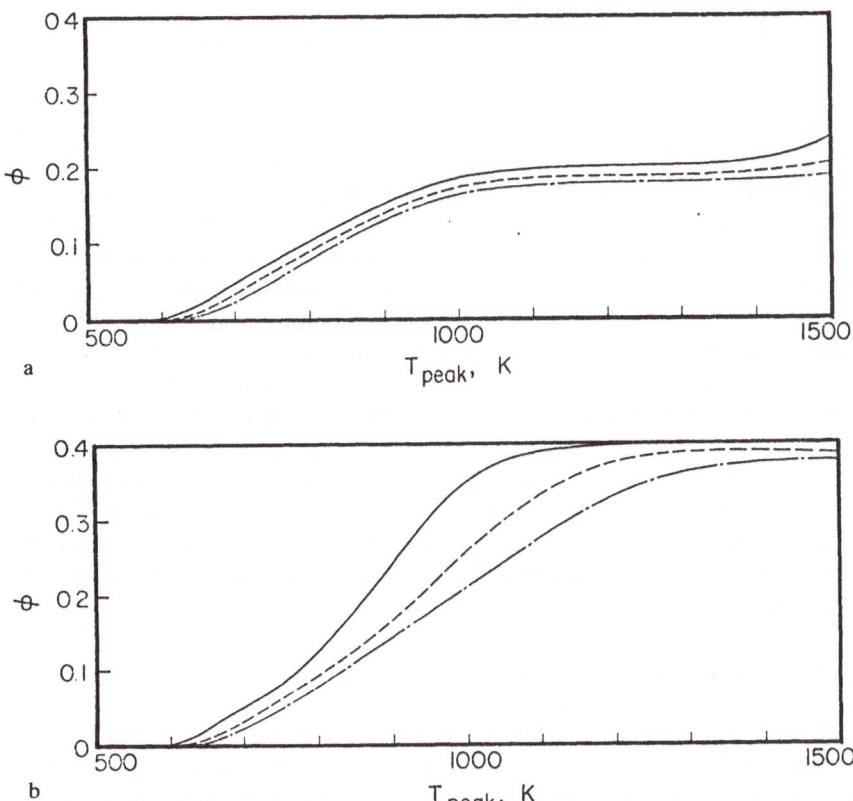

Fig. 1. History of particle porosity with prescribed temperature histories: ——— $dT_p/dt = 1(10^3)$ K/s, ------ $dT_p/dt = 5(10^3)$ K/s ——·—— $dT_p/dt = 10(10^3)$ K/s. (a) Prototypical conditions (see Table 1). (b) Characteristics of species 1 as in Table 1, $T_2^* = T_{c2} = 800$ K, $\beta_2 = 5(10^{-3})$ (K)$^{-1}$.

sets of characterizing temperatures are not as separate as in the previous example. In particular we take $T_2^* = T_{c2} = 800$ K while leaving unaltered β_2, so that under equilibrium conditions devolatilization is complete at 1200 K. We see from Figure 1b that there is in this case no plateau and the two devolatilizing processes overlap. In this case the possibility of deducing kinetic data would appear to increase.

In actual experiments such as those of Suuberg et al.[5] yields of various individual gas-phase species are determined; such yields resemble those indicated by Figure 1a. Our results seem to suggest that the determination of kinetic data for individual species may be difficult; however, direct heating experiments with low rates of heating are needed for the determination of the equilibrium composition appearing in the present analysis via the functions $R_{si}^{(e)}$.

3.2. Sensitivity to Kinetic and Thermochemical Parameters

The remainder of our numerical results relate to thermal-response calculations. We have studied the sensitivity of the thermal response of cold particles suddenly immersed into hot inert ambients to the various kinetic and thermochemical parameters entering the present theory. In the interests of brevity we simply summarize the results of these studies. Our most significant conclusion is that the response of particles under these circumstances is highly sensitive to the heats of pyrolysis; for instance, with only one devolatilizing species a slight exothermicity $(\delta_1 - \Delta_0) = 200$ cal/g causes K to increase without apparent limit. An analysis shows that the conditions necessary for $dK/d\tau = 0$, and thus that a maximum in K prevails, are sensitive to $d\theta/d\tau$. Accordingly, the detailed heat balance and the heats of pyrolysis play important roles in particle response.

We conclude from these results that in devolatilization studies involving the sudden immersion of a particle into a hot gas stream the kinetic characteristics of the devolatilization process may be difficult to determine with accuracy. Careful monitoring of particle temperature would appear to be important.

3.3. Particle Behavior for Frozen Flow

Figures 2a through c show the results of calculations for frozen flow. Consider first particle behavior under prototypical conditions. We see that our idealization of the characteristics of the particle into a low- and a high-temperature species results in a rate of mass loss in terms of $K(\tau)$ with distinct features. Species 1 with its critical temperature $T_{c1} = 600$ K is devolatilized as the particle heats rapidly with the consequence that the first peak in the value for K is relatively large. Species 1 is essentially

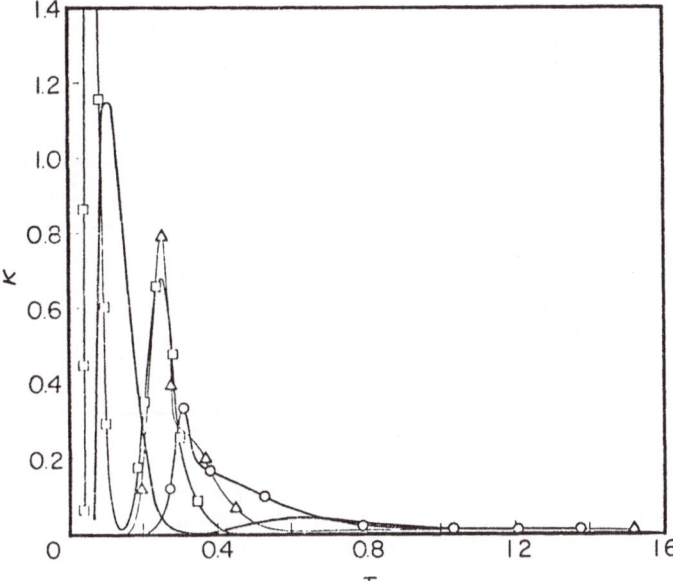

Fig. 2a.

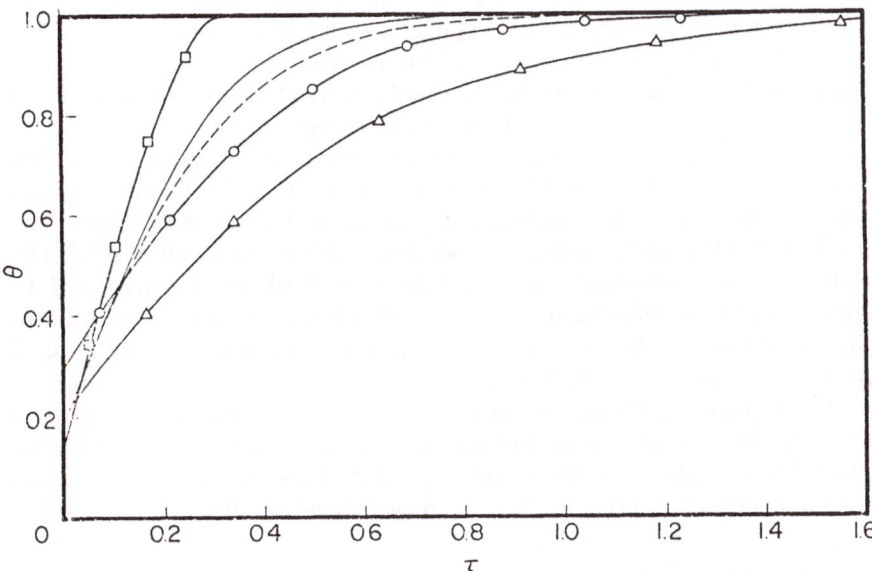

Fig. 2b.

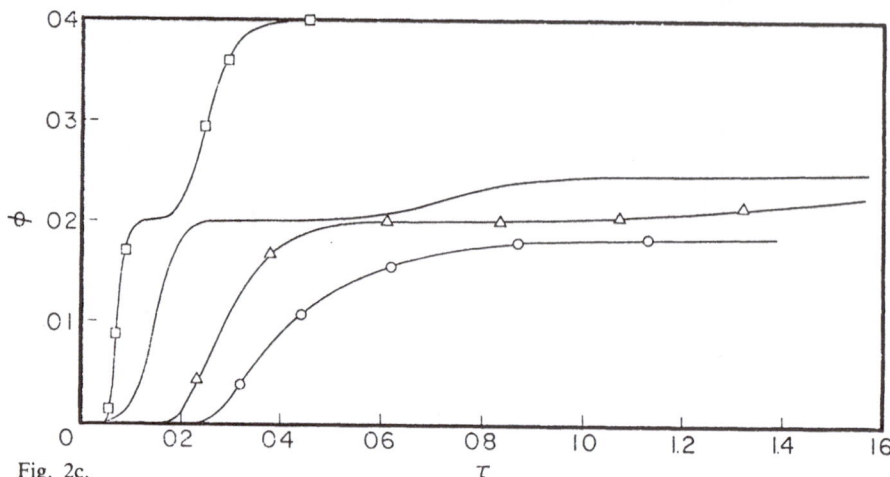

Fig. 2c.

Fig. 2. Particle behavior for frozen flow: ———— inert particle, ------ prototypical conditions, —— □ —— $T_\infty = 2000$ K, —— ○ —— $T_\infty = 1000$ K, —— △ —— prototypical
conditions with $Q_r = 0$. (a) History of mass-loss parameter. (b) Temperature history.
(c) History of porosity.

completely driven off before species 2 begins to be pyrolyzed to any extent.
Species 2 with its higher critical temperature of $T_{c2} = 1200$ K devolatilizes
as the rate of particle heating is reduced and is thus given off at a lower rate
and over an extended period of time.* Accordingly, different time scales
prevail for the two devolatilizing species. These scales are clearly seen in
Figure 2c, which gives the history of the porosity $\phi(\tau)$; a distinct plateau
corresponding to the complete devolatilization of species 1 prior to the
onset of the devolatilization of species 2 is seen.

If we compare the particle behavior under these prototypical conditions with that under variations thereof, we see that T_∞, influencing as it
does the rate of particle heating, has a profound effect on particle behavior;
for $T_\infty = 2000$ K all devolatilizing species are driven off when $\tau \cong 0.5$. This
is the only case considered where complete devolatilization is predicted. Of
course, if the calculations had been continued indefinitely, we would find
that in due course the particles would always be completely devolatilized,
provided T_∞ is greater than T_{ci}, $i = 1, 2$.

Although our idealized particle is not studied with the purpose of
carrying out comparison with experiment, it is interesting to compare the
times for complete devolatilization predicted by our calculations with
related experimental results. For our prototypical conditions t (ms) $= 62\ \tau$.

---

* The bimodal distribution of the mass loss parameter shown in Fig. 2a is reminiscent of
  "Instantaneous surface flux density of volatiles" given by Suuberg et al. (Ref. 5) from an
  entirely different analysis developed to model their experimental results.

Thus for $T_\infty = 2000$ K we predict devolatilization to be complete in 30 ms. This can be compared with a value of 40 ms given by Kobayashi *et al.*[6] for lignite particles with $30 \lesssim r_p \lesssim 40$ microns and of 30 ms for bituminous particles in the same size range. The smaller particle size associated with the experimental results reduces the predicted time to completion via its effect on r_p and increases that time due to diminished importance of radiative heating.

In an experiment similar to that of Kobayashi *et al.*,[6] McLean *et al.*[7] study devolatilization in oxidizing gas streams and find completion times of 5 ms for particles of 30 micron radius. If we simply alter the relation $t = t(\tau)$ to reflect the smaller particle, we predict for frozen flow a time of 11 ms. However, we show later that for equilibrium composition the temperature history of the particle is altered so that the completion time is reduced to one-half its frozen-flow value. Thus good agreement is found with the results of McLean *et al.*[8]

We note that other features of the data of Kobayashi *et al.*[6] are not in accord with our calculations, presumably because of the prototypical parameters we have chosen. For example, we must increase the fraction of devolatilizing species so that $[R_{s1}(0) + R_{s2}(0)] = 0.6$ and must make $T_{c1} > 1000$ K in order to predict negligible weight loss when $T_\infty = 1000$ K. Further comparison along these lines can be carried out but does not appear indicated.

Consider further the results shown in Figures 2a through c. We see that with radiative transfer set to zero, the rates of devolatilization and particle heating are slowed, changes expected on physical grounds. Although not shown, the results for $\delta_1 = \delta_2 = \Delta_D = -2(10^3)$ cal/g do not differ significantly from those corresponding to prototypical conditions; it would appear that the heats of pyrolysis are more significant than the enthalpy levels.

These results support our earlier findings on the close dependence of devolatilizing behavior on particle temperature. Thus T_∞, radiative transfer, and the heats of devolatilization are significant factors in the devolatilization process. Moreover, the rapid heating a particle experiences when first immersed in a hot gas stream results in rapid devolatilization of species given off at low temperatures in contrast to the relatively slow heating as $T_p \sim T_\infty$, which tends to stretch out completion for high-temperature devolatilizers.

3.4. Particle Behavior for Equilibrium in the Gas Phase

We now turn to the results of calculations with equilibrium composition in the gas phase. For this case the parameters Δ_P and Y_0 must be assumed in order to complete definition of the prototypical conditions; representative values as shown in Table 1 are therefore adopted.

Results for prototypical conditions are shown in Figures 3a through d. Comparison of the first three of these figures with their counterparts for frozen composition in the gas phase indicates that the added heating of the particle due to exothermicity in the gas phase accelerates devolatilization. To emphasize this point consider the following comparison. When $\tau \cong 0.75$, $\phi \cong 0.4$ for the equilibrium case but only 0.23 for frozen composition

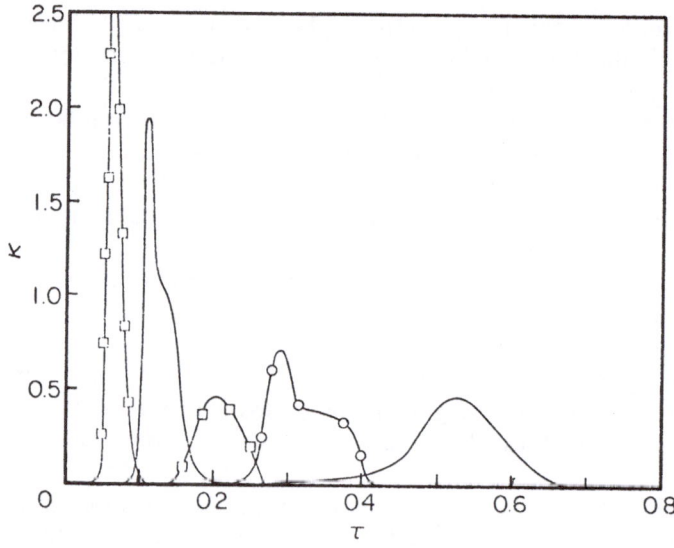

Fig. 3a.

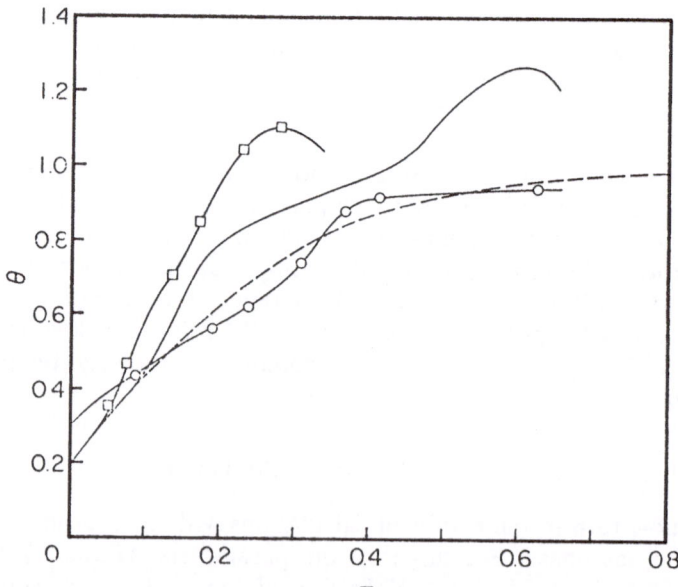

Fig. 3b.

in the gas phase with all other conditions identical in the two cases. Again we have evidence of the sensitivity of the devolatilization process to particle temperature.

With respect to the histories of the mass loss parameter $K(\tau)$ shown in Figure 3a, we see the same bimodal distribution noted earlier. Again the species devolatilizing at low temperature is completely driven off before

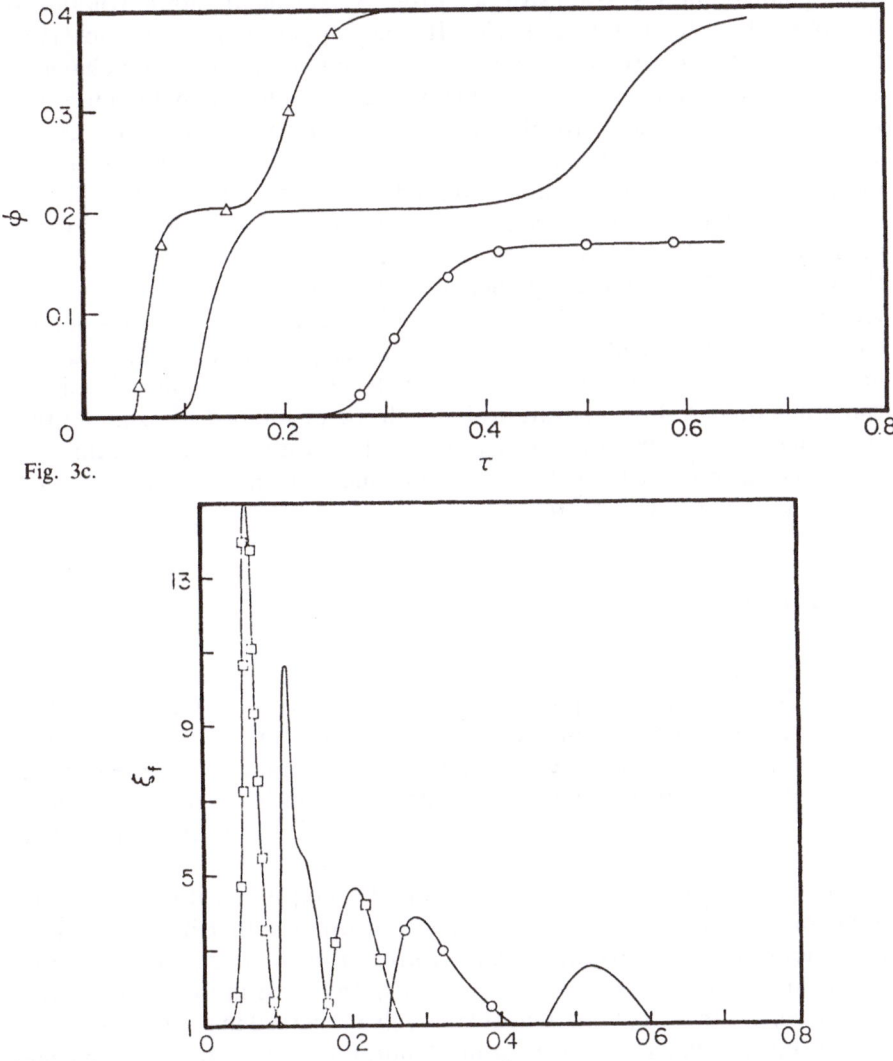

Fig. 3c.

Fig. 3d.

Fig. 3. Particle behavior for equilibrium flow (legend as for Figure 2). (a) History of mass-loss parameter. (b) Temperature history. (c) History of porosity. (d) History of flame-sheet position.

the species devolatilizing at high temperature begins to volatilize. The position of the flame sheet as a function of time exhibits a related behavior, as shown in Figure 3d. The flame moves abruptly from the surface of the particle and under these prototypical conditions reaches a position nearly ten radii from the particle. Just as abruptly the flame returns to the surface of the particle when the low-temperature species is exhausted. A second, more gradual and less extensive departure is associated with devolatilization of the high-temperature species. It would appear that careful observation should permit assessment of these predictions of flame-sheet behavior.

Two other cases are shown in Figures 3a through d; all parameters are held as in Table 1 except for the ambient temperature T_∞. For the case of $T_\infty = 2000$ K we see little qualitative change from the results previously discussed, but the more rapid heating reduces the time to completion and causes the somewhat larger excursions of the flame sheet from the surface of the particle. The case of a low-temperature ambient, $T_\infty = 1000$ K, on the contrary, does involve significantly different behavior; only the low-temperature species is devolatilized within the time period considered. Thus the results resemble those for frozen gas phase, i.e., the additional heating associated with exothermicity in the gas phase is insufficient to cause augmented devolatilization. It would be of interest to examine the combinations of ambient temperature and thermochemical parameters that might lead to such augmentation, but such an examination is beyond the scope of the present study.

4. CONCLUDING REMARKS

A theoretical investigation of the devolatilization of an idealized coal particle is carried out. Despite the idealization reference is made to related experimental techniques and results. It is found that even for the highly idealized particle considered, a large number of parameters related to the kinetic and equilibrium behavior of devolatilizing species and to their thermochemistry must be specified in order to make a prediction of particle behavior.

It is generally concluded that devolatilization is sensitive to particle temperature; the implication from this conclusion is that it is difficult to estimate accurately activation temperatures of devolatilization unless the temperature history of the particle is carefully monitored. Under the circumstances that a cold particle is suddenly immersed in a hot gas stream, circumstances which prevail in some laboratory experiments and in entrained flow reactors, the ambient temperature, radiative transfer, and heats of devolatilization are factors influencing particle temperature and therefore devolatilization.

Further studies along these lines should include additional species in the solid phase and additional gas-phase species. Although devolatilization is generally a low-temperature process, the temperatures considered here and in related laboratory experiments are sufficiently high so that heterogeneous reactions leading to char attack can occur and can influence high-temperature devolatilization.

ACKNOWLEDGMENTS. This paper is an abbreviated version of a report by the same title written while the author was a consultant to Systems Science and Software on work supported by the United States Department of Energy. Helpful discussions with Dr. Thomas R. Blake and Dr. Gary Schneyer are gratefully acknowledged. The relations among the element mass fractions, species mass fractions, and enthalpy used here are a generalization of the well-known relations due to L. G. Crocco.

REFERENCES

1. P. A. Libby and T. R. Blake, Theoretical study of burning carbon particles, *Combust. Flame* **36**, 139–169 (1979).
2. P. A. Libby and T. R. Blake, Burning carbon particles in the presence of water vapor, *Combust. Flame* **41**, 123–147 (1981).
3. P. A. Libby, Ignition, combustion and extinction of carbon particles, *Combust. Flame* **38**, 285–300 (1980).
4. D. B. Anthony and J. B. Howard, Coal devolatilization and hydrogenation, *AIChE J.* **22**, 625–656 (1976).
5. E. M. Suuberg, W. A. Peters, and J. B. Howard, Product compositions and formation kinetics in rapid pyrolysis of pulverized coal — implications for combustion, in: *Seventeenth Symp. (Int.) on Combust.*, pp. 117–143, The Combustion Institute, Pittsburgh, PA (1979).
6. H. Kobayashi, J. B. Howard, and A. E. Sarofim, Coal devolatilization at high temperatures, in: *Sixteenth Symp. (Int.) on Combust.*, pp. 411–425, The Combustion Institute, Pittsburgh, PA (1977).
7. W. J. McLean, D. R. Hardesty, and J. H. Pohl, Direct observations of pulverized coal particles in a combustion environment, *Eighteenth Symp. (Int.) on Combust.*, pp. 1239–1248, The Combustion Institute, Pittsburgh, PA (1980).
8. S. K. Ubhayakar, D. B. Stickler, C. W. von Rosenberg, Jr., and E. E. Gannon, Rapid devolatilization of pulverized coal in hot combustion gases, *Seventeenth Symp. (Int.) on Combust.*, The Combustion Institute, Pittsburgh, PA (1977).
9. G. A. Simons and M. L. Finson, The structure of coal char: Part I. Pore branching, *Combust. Sci. Technol.* **19**, 217–225 (1979).
10. G. A. Simons, The structure of coal char: Part II. Pore combination, *Combst. Sci. Technol.* **19**, 227 (1979).

15

An Analytical Solution of Diffusion-Controlled Gasification of Carbon Particles in the Presence of Steam, Oxygen, and Nitrogen*

S. S. PENNER, M. R. BRAMBLEY, AND M. Y. BAHADORI

ABSTRACT. Analytical solutions are derived for diffusion-controlled gasification of carbon particles. The solutions apply to arbitrary gas mixtures containing steam, oxygen, or nitrogen. The solution is developed in stages, beginning with isothermal and isobaric carbon gasification. Solutions are derived for constant values of the product of density and diffusion coefficient.

A general procedure is described for decoupling the species- and energy-conservation equations for problems with radial symmetry. The temperature profiles are obtained by iteration.

Particle lifetimes, mass-loss rates, particle radii, and species concentrations are in fair agreement with numerical data for finite chemical-reaction rates for the special cases for which numerical solutions are available.

1. INTRODUCTION

Several coal-gasification models have been concocted. None of these represents a realistic description of the burning mechanisms of coals. In particular, none includes a chemical model for the operationally important formation of clinkers during gasification. The pyrolysis and surface reac-

---

* A preliminary version of this study was presented by one of us (SSP) during 1976 at Punta Ala, Italy, at a Conference of the Engineering Societies of Tuscany.

S. S. PENNER, M. R. BRAMBLEY, AND M. Y. BAHADORI • University of California, San Diego, La Jolla, CA 92093, U.S.A. Present address for M. R. Brambley: School of Engineering and Applied Science, Washington University, St. Louis, MO 63130, U.S.A.

tions are grossly simplified and chemical species such as atoms and free radicals are notably absent, even from the most elaborate models for which only numerical solutions are obtainable.

In view of the extreme complexity of an adequate model for coal gasification, it is worthwhile to search for manageable descriptions that contain essential features of the surface and gas-phase reactions and may be verified by comparisons with limited available diagnostic information.

We first describe isothermal and isobaric carbon gasification determined by diffusion processes (Section 2). In Section 3, we obtain the solution for the species-conservation equations for constant mean values of the parameter ρD (which represents the product of gas density and diffusion coefficient and actually varies as T^m with $0.5 \leq m \leq 1$). Next, we describe a general procedure for decoupling the species- and energy-conservation equations for species-independent diffusion coefficients and specific heats (Section 4) and then present an iterative procedure for the determination of steady-state temperature profiles and time-dependent particle temperature (Section 5). Particle lifetimes corresponding to the model in Section 3 are obtained in Section 6 and comparisons are made with numerical results that include finite chemical-reaction rates. We show mass-fraction profiles and temperature profiles in Section 7 and make comparisons of mass-loss rates, particle radii, and species concentrations at the particle surface with results derived from a numerical model. In Section 8, comparisons are made with the burning rates obtained by using the Shvab–Zeldovich procedure for particle combustion, which involves only a single, one-step chemical reaction in the gas phase.

2. ISOTHERMAL AND ISOBARIC CARBON GASIFICATION DETERMINED BY DIFFUSION PROCESSES INVOLVING OXYGEN AND STEAM

We consider the special case in which a flame surface exists at $r = R$ and the concentrations of the product species, H_2O and CO_2, become vanishingly small as the surface is approached. At the flame surface, H_2 and CO are rapidly oxidized to H_2O and CO_2, respectively, while O_2 is prevented from entering the spherical shell between r_d and R.

The diffusion equation is

$$4\pi D \frac{d}{dr}\left(r^2 \frac{dc}{dr}\right) = 0 \qquad \text{for } r_d < r < R \text{ and } R < r < \infty \qquad (1)$$

for constant D since there are no chemical sources or sinks in these regions. Thus,

$$r^2(dc/dr) = \text{constant} \tag{2}$$

for each chemical species. If the local Reynolds and Grashof numbers are less than unity, forced and natural convection are unimportant. Integrating equation (2) for each species in both regions and using appropriate conservation relations for atomic species or applicable stoichiometric relations, we find that

$$R/r_d = 2 + (H_2O)_\infty/(O_2)_\infty \tag{3}$$

which reduces to $R = 2r_d$ as $(H_2O)_\infty \to 0$ and thus yields the result of Avedesian and Davidson[1] for gasification of carbon particles with oxygen alone.

2.1. Overall Gasification Rate and Particle Burning Times

The overall gasification rate must be such that all of the CO_2 diffusing from the flame front is supplied by reactions of carbon with oxygen and steam. The analysis leading to equation (3) shows that

$$(CO_2) = (O_2)_\infty R/r \tag{4}$$

The rate of mass loss (in g/s) from the carbon particle is

$$\dot{m}_C = -12 \times 4\pi D R^2 \left[\frac{d(CO_2)}{dr} \right]_{r=R^+} \tag{5}$$

if D is the diffusion coefficient in cm^2/s. Using equation (3) for R, as well as equations (4) and (5), we obtain

$$\dot{m}_C = 1.51 \times 10^2 D (O_2)_\infty \left[2 + \frac{(H_2O)_\infty}{(O_2)_\infty} \right] r_d \tag{6}$$

which shows that the rate of mass loss is proportional to the particle radius and depends on both the ambient oxygen and steam concentrations, as well as on the constant assumed value of the diffusion coefficient.

For carbon burning in pure oxygen, it follows from equation (6) that

$$-4\pi r_d \rho_p dr_d = 1.51 \times 10^2 D (O_2)_\infty dt$$

where ρ_p represents the particle density. If τ is the burning time at which the particle radius has shrunk to the extinction radius r_{ext},

$$\tau = \frac{2\pi\rho_p}{1.51 \times 10^2 D (O_2)_\infty} (r_{d,0})^2 [1 - (r_{ext}/r_{d,0})^2] \tag{7}$$

where $r_{d,0}$ is the initial value of the particle radius. According to equation (7), τ varies linearly as $r_{d,0}^2$ and inversely as $(O_2)_\infty$ for $r_{ext}/r_{d,0} \ll 1$. For a carbon sphere with diameter $2r_{d,0} = 10^{-2}$ cm burning in pure oxygen with $(O_2)_\infty = 1.30 \times 10^{-3}$ g/cm^3 at 1 atm and 300 K, $\rho_p = 1.5$ g/cm^3, and $D \simeq 0.608$ cm^2/s (diffusion of O_2 in a CO_2–O_2 mixture at 650 K), $\tau = 63 \times 10^{-3}$ s. This value of τ is in good agreement with both laser-ignition[2] and other[3] experiments. The scaling of τ with $(r_{d,0})^2$ and $(O_2)_\infty^{-1}$ also agrees quite well with the available experimental data. Replacement of the arbitrarily chosen effective temperature of 650 K by a proper value requires solution of the coupled energy- and species-conservation equations.

In a gasifier, we expect to encounter very many particles and flamelets, the cooperative effects of which will be a nearly constant value of a high reactor temperature throughout the gasifier, which is determined by high levels of radiative and turbulent heat exchange. Under these conditions, the isothermal and isobaric model may well constitute an acceptable first approximation.

3. ANALYTICAL SOLUTIONS FOR CONSTANT ρD

For steady, spherically symmetric flow, the conservation equation for species K is[4]

$$\dot{m}\frac{dY_K}{dr} = \frac{d}{dr}\left(4\pi r^2 \rho D \frac{dY_K}{dr}\right) \tag{8}$$

We require solutions of equation (8) for each species subject to the boundary conditions shown in Figure 1. For $\alpha = \dot{m}/4\pi\rho D = $ constant and

$$f(r) = 4\pi r^2 \rho D / \dot{m} \tag{9}$$

it may be shown that

$$Y_{H_2} = f(R)\left(\frac{dY_{H_2}}{dr}\right)_{R^-}\left\{\exp\left[-\alpha\left(\frac{1}{r}-\frac{1}{R}\right)\right]-1\right\} \quad \text{for } r_d \le r \le R \tag{10}$$

$$Y_{CO} = f(R)\left(\frac{dY_{CO}}{dr}\right)_{R^-}\left\{\exp\left[-\alpha\left(\frac{1}{r}-\frac{1}{R}\right)\right]-1\right\} \quad \text{for } r_d \le r \le R \tag{11}$$

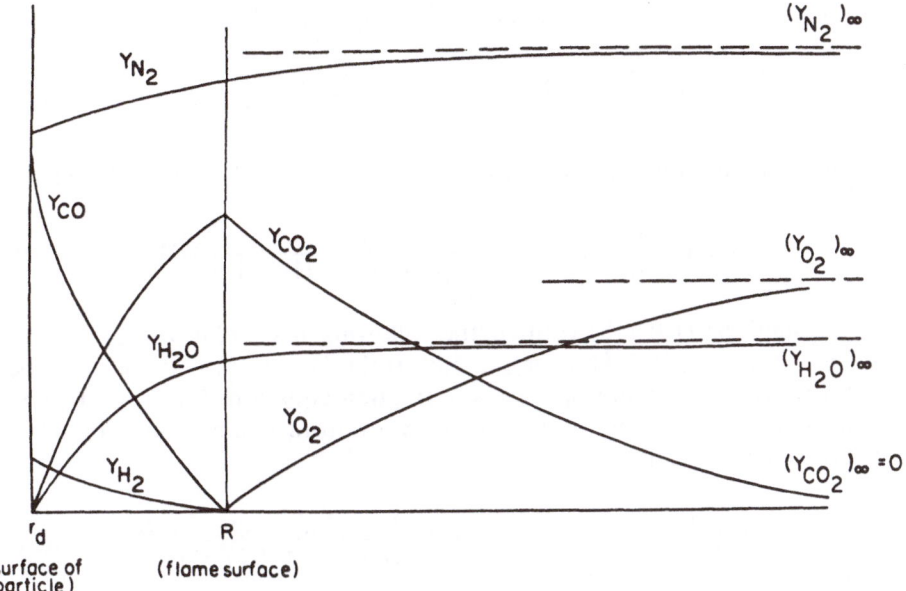

Fig. 1. Mass fraction profiles with appropriate boundary conditions are shown for gasification of a carbon particle by oxygen and steam in the presence of inert nitrogen. The flame surface is located at $r = R$, where O_2 is completely consumed; rapid surface reactions reduce the concentrations of CO_2 and H_2O to zero at $r = r_d$. Mass fraction profiles that are consistent with our analytical solutions are shown in Figure 5.

$$Y_{H_2O} = \begin{cases} f(r_d)\left(\dfrac{dY_{H_2O}}{dr}\right)_{r_d^+} \left\{\exp\left[\alpha\left(\dfrac{1}{r_d}-\dfrac{1}{r}\right)\right]-1\right\} & \text{for } r_d \leqslant r \leqslant R \quad (12) \\[2em] 1-(Y_{CO_2})_R\left[\dfrac{1-\exp(-\alpha/r)}{1-\exp(-\alpha/R)}\right] \\[1em] \quad -(Y_{O_2})_\infty\left\{\dfrac{\exp\alpha[(1/R)-(1/r)]-1}{\exp(\alpha/R)-1}\right\} \\[1em] \quad -(Y_{N_2})_\infty\exp(-\alpha/r) & \text{for } R \leqslant r < \infty \quad (13) \end{cases}$$

$$Y_{CO_2} = \begin{cases} f(r_d)\left(\dfrac{dY_{CO_2}}{dr}\right)_{r_d^+} \left\{\exp\left[\alpha\left(\dfrac{1}{r_d}-\dfrac{1}{r}\right)\right]-1\right\} & \text{for } r_d \leqslant r \leqslant R \quad (14) \\[2em] (Y_{CO_2})_R\left[\dfrac{1-\exp(-\alpha/r)}{1-\exp(-\alpha/R)}\right] & \text{for } R \leqslant r < \infty \quad (15) \end{cases}$$

$$Y_{O_2} = (Y_{O_2})_\infty\left\{\dfrac{\exp\alpha[(1/R)-(1/r)]-1}{\exp(\alpha/R)-1}\right\} \qquad\qquad \text{for } R \leqslant r < \infty \quad (16)$$

Continuity for the mass fraction of CO_2 at $r = R$ shows that

$$f(r_d) \left(\frac{dY_{CO_2}}{dr} \right)_{r_d^+} = \frac{(Y_{CO_2})_R}{\{\exp \alpha [(1/r_d) - (1/R)] - 1\}} \tag{17}$$

similarly, continuity for H_2O at $r = R$ and equations (12) and (13) lead to

$$f(r_d) \left(\frac{dY_{H_2O}}{dr} \right)_{r_d^+} = \frac{1 - (Y_{CO_2})_R - (Y_{N_2})_\infty \exp(-\alpha/R)}{\{\exp \alpha [(1/r_d) - (1/R)] - 1\}} \tag{18}$$

Equations (10)–(18) contain the following five unknown quantities: α, R, $(Y_{CO_2})_R$, $f(R)(dY_{H_2}/dr)_{R^-}$, and $f(R)(dY_{CO}/dr)_{R^-}$; these may be determined from atomic species conservation equations for H, O, C, and stoichiometric relations at r_d^+ and at R or from combinations of these expressions. We find

$$f(R) \left(\frac{dY_{H_2}}{dr} \right)_{R^-} = -\frac{1}{9} \left\{ \frac{1 - (Y_{CO_2})_R - (Y_{N_2})_\infty \exp(-\alpha/R)}{\exp \alpha [(1/r_d) - (1/R)] - 1} \right\} \tag{19}$$

$$f(R) \left(\frac{dY_{CO}}{dr} \right)_{R^-} = \frac{\frac{28}{99}(Y_{CO_2})_R - \frac{14}{9}[1 - (Y_{N_2})_\infty \exp(-\alpha/R)]}{\exp \alpha [(1/r_d) - (1/R)] - 1} \tag{20}$$

$$f(R) \left(\frac{dY_{CO}}{dr} \right)_{R^-} = -\frac{7}{11} \left\{ \frac{(Y_{CO_2})_R}{\exp \alpha [(1/r_d) - (1/R)] - 1} \right\} - \frac{7}{3} \tag{21}$$

$$(Y_{CO_2})_R = \frac{\exp(\alpha/R) - 1 + (Y_{O_2})_\infty}{\exp(\alpha/R)} \tag{22}$$

$$(Y_{CO_2})_R = \frac{11}{3}[1 - \exp(-\alpha/R)] \tag{23}$$

$$(Y_{CO_2})_R = \{\frac{11}{2}[\exp(\alpha/r) - 1] - \frac{11}{16}(Y_{O_2})_\infty\} \exp(-\alpha/R) \tag{24}$$

$$(Y_{CO_2})_R = \frac{22}{13}[1 - (Y_{N_2})_\infty \exp(-\alpha/R)] - \frac{33}{13} \exp\left[\alpha\left(\frac{1}{r_d} - \frac{1}{R}\right)\right] - 1 \tag{25}$$

Equations (22)–(25) provide four relations for $(Y_{CO_2})_R$ that may be equated to obtain

$$\frac{R}{r_d} = \frac{\ln[1 + \frac{2}{3}(Y_{H_2O})_\infty + \frac{3}{4}(Y_{O_2})_\infty]}{\ln[1 + \frac{3}{8}(Y_{O_2})_\infty]} \tag{26}$$

$$\alpha = r_d \ln[\frac{5}{3} - \frac{2}{3}(Y_{N_2})_\infty + \frac{1}{12}(Y_{O_2})_\infty] \tag{27}$$

$$\dot{m} = 4\pi\rho D r_d \ln\left[1 + \tfrac{2}{3}(Y_{H_2O})_\infty + \tfrac{3}{4}(Y_{O_2})_\infty\right] \tag{28}$$

Equations (26) and (28), for constant ρD, may be compared, respectively, with equations (3) and (6) for isothermal and isobaric gasification of carbon particles with constant D. In the limit as $(H_2O)_\infty \to 0$ and $(Y_{O_2})_\infty = 1$, $R = 2r_d$ for isothermal and isobaric gasification and $R = 1.76r_d$ for constant ρD. Similarly, for $(H_2O)_\infty = (O_2)_\infty$, $R = 3r_d$ for isothermal, isobaric gasification and $R = 2.52\,r_d$ for constant ρD with $(Y_{N_2})_\infty = 0$. Thus, the locations of the flame sheet for isothermal and isobaric gasification and for constant ρD agree within 19% for $0 \le (H_2O)_\infty/(O_2)_\infty \le 1$ with $(Y_{N_2})_\infty = 0$. The difference between the locations of the flame sheets for the two models increases somewhat as the ratio $(H_2O)_\infty/(O_2)_\infty$ increases, with approximately a 27% difference for $(H_2O)_\infty/(O_2)_\infty = 10$. The relations for mass-loss rate in equations (6) and (28) show the same dependence on diffusion coefficient (D) and particle radius (r_d), viz. $\dot{m} \propto D r_d$.

3.1. Utilization of Stoichiometry Conditions at the Flame and Particle Surfaces

The present formulation of the problem leads to solutions satisfying the surface- and flame-front boundary conditions. These boundary conditions may be compatible with one or more models for the surface- and flame-front reaction processes. Our solution satisfies the following overall surface reaction:

$$C + nH_2O + (1-n)CO_2 \to (2-n)CO + nH_2 \tag{29}$$

where the value of n depends on the ambient-gas concentrations at infinity. The preceding process is compatible, for example, with each of the following surface-reaction mechanisms:

$$C + H_2O \to CO + H_2 \tag{30}$$

$$(n-1)CO + (n-1)H_2O \to (n-1)CO_2 + (n-1)H_2$$

$$C + CO_2 \to 2CO, \quad mC + mH_2O \to mCO + mH_2, \quad m = n/(1-n) \tag{31}$$

The stoichiometric relation at R corresponding to the postulated overall surface reaction is

$$(2-n)CO + nH_2 + O_2 \to (2-n)CO_2 + nH_2O \tag{32}$$

Stoichiometric ratios may be determined by using the specified stoichiometric conditions at r_d^+ and R; for example, the molar flow-rate ratio of H_2O to C at r_d^+ must equal n, viz., $(\dot{m}_{H_2O})_{r_d^+} = -(18/12)n\dot{m}$.

4. SOLUTIONS OF STEADY, SPHERICALLY SYMMETRIC, LAMINAR FLOW PROBLEMS WITH CONSTANT $\bar{D}$ AND $\bar{c}_p$

Before combining the solutions of the species- and energy-conservation equations, we find it convenient to derive a generally applicable expression relating mass flow and heat conduction for a spherically symmetric flow problem.

The energy conservation equation is[4]

$$4\pi r^2 \rho v \frac{d}{dr} \sum_K \varepsilon_K h_K = 4\pi r^2 \rho v \frac{d}{dr} \sum_K Y_K h_K \left(1 - \frac{D_K}{v Y_K} \frac{dY_K}{dr}\right)$$
$$= \frac{d}{dr}\left(4\pi r^2 \lambda \frac{dT}{dr}\right) \tag{33}$$

where r = radial distance, ρ = density, v = mass-weighted average velocity, ε_K = mass flux fraction carried by species K, h_K = specific enthalpy of species K, Y_K = mass fraction of species K, λ = thermal conductivity, T = temperature. The product $4\pi r^2 \rho v$ is the total mass flow rate and represents a constant of the motion. For steady, spherically symmetric flow, the species conservation relation [equation (8)] becomes

$$4\pi r^2 \rho v \frac{dY_K}{dr} = \frac{d}{dr}\left(4\pi r^2 \rho D_K \frac{dY_K}{dr}\right) \tag{34}$$

We rewrite the left-hand side of equation (33) as

$$4\pi r^2 \rho v \frac{d}{dr} \sum_K \varepsilon_K h_K = \sum_K \left[4\pi r^2 \rho v \left(Y_K \frac{dh_K}{dr} + h_K \frac{dY_K}{dr}\right)\right.$$
$$\left. - h_K \frac{d}{dr}\left(4\pi r^2 \rho D_K \frac{dY_K}{dr}\right) - 4\pi r^2 \rho D_K \frac{dY_K}{dr} \frac{dh_K}{dr}\right] \tag{35}$$

and, after replacing $4\pi r^2 \rho v(dY_K/dr)$ in the second term of equation (35) by using equation (34),

$$4\pi r^2 \rho v \frac{d}{dr} \sum_K \varepsilon_K h_K = \sum_K 4\pi r^2 \rho v \frac{dh_K}{dr}\left(Y_K - \frac{D_K}{v} \frac{dY_K}{dr}\right) \tag{36}$$

Also,

$$h_K = h_K^0 + \bar{c}_p(T - T_0) \tag{37}$$

where h_K^0 is the standard heat of formation of species K at the reference temperature T_0 and we are assuming a constant and species-invariant value for the specific heat at constant pressure for all of the species K. Next, we replace the D_K by $\bar{D}$. It then follows that

$$\sum_K \frac{dh_K}{dr} Y_K = \bar{c}_p \frac{dT}{dr}$$

$$\sum_K \frac{dh_K}{dr} D_K \frac{dY_K}{dr} = \bar{D}\bar{c}_p \frac{dT}{dr} \sum_K \frac{dY_K}{dr} = 0 \tag{38}$$

so that equation (36) becomes

$$4\pi r^2 \rho v \frac{d}{dr} \sum_K \varepsilon_K h_K = 4\pi r^2 \rho v \bar{c}_p \frac{dT}{dr} \tag{39}$$

and equation (33) reduces to

$$\dot{m}\bar{c}_p \frac{dT}{dr} = \frac{d}{dr} \left(4\pi r^2 \lambda \frac{dT}{dr} \right), \quad r_d < r < R \text{ and } R < r < \infty \tag{40}$$

4.1. Solution of the Energy Equation for Constant Values of $\lambda/\bar{c}_p$

In equation (40), we write $a = \dot{m}\bar{c}_p/4\pi\lambda$, viz.,

$$\frac{dT}{dr} = \frac{1}{a} \frac{d}{dr} \left(r^2 \frac{dT}{dr} \right) \tag{41}$$

With the boundary conditions $T = T^*$ at $r = R$ and $\lim_{r \to \infty} T = T_\infty$,

$$T = T_\infty + (T^* - T_\infty) \left[\frac{1 - \exp(-a/r)}{1 - \exp(-a/R)} \right] \quad \text{for } R \leq r < \infty \tag{42}$$

Similarly, with $T = T_d$ at $r = r_d$ and $T = T^*$ at $r = R$,

$$T = T^* - (T^* - T_d) \left\{ \frac{1 - \exp - a[(1/r) - (1/R)]}{1 - \exp - a[(1/r_d) - (1/R)]} \right\} \quad \text{for } r_d \leq r \leq R \tag{43}$$

Furthermore,

$$\lim_{r \to \infty} \left[-\left(\frac{dT}{dr} \right) \right] = 0, \qquad \lim_{r \to \infty} \left[-r^2 \left(\frac{dT}{dr} \right) \right] = \frac{a(T^* - T_\infty)}{1 - \exp(-a/R)}$$

$$-\left(\frac{dT}{dr} \right)_{R^+} = (T^* - T_\infty)\frac{a}{R^2} \left[\frac{1}{\exp(a/R) - 1} \right]$$

$$\left(\frac{dT}{dr} \right)_{R^-} = (T^* - T_d)\frac{a}{R^2} \left\{ \frac{1}{1 - \exp - a[(1/r_d) - (1/R)]} \right\}$$ \hfill (44)

$$\left(\frac{dT}{dr} \right)_{r_d^+} = (T^* - T_d)\frac{a}{r_d^2} \left\{ \frac{1}{\exp a[(1/r_d) - (1/R)] - 1} \right\}$$

It is interesting to note that $-(dT/dr)_{R^+} = (dT/dr)_{R^-}$ if

$$T_d = T^* - (T^* - T_\infty) \left\{ \frac{1 - \exp - a[(1/r_d) - (1/R)]}{\exp(a/R) - 1} \right\} \qquad (45)$$

4.2. Determination of $T^*(T_d)$ at $r = R$

The temperature at the flame boundary ($r = R$) is determined by the condition that, at steady state, the heat released by reaction and the heat losses by conduction must be equal. We have postulated the following stoichiometric reactions:

$$C + (1 - n)CO_2 + nH_2O \rightarrow nH_2 + (2 - n)CO \qquad \text{at } r = r_d \qquad (46)$$

$$nH_2 + (2 - n)CO + O_2 \rightarrow (2 - n)CO_2 + nH_2O \qquad \text{at } r = R \qquad (47)$$

The surface area of the flame front is $4\pi R^2$. Hence, the mass-flow rate at the flame surface is $\dot{m}/4\pi R^2$ g/cm²-s, where $\dot{m}$ is given by equation (28). The rate of heat conduction from the flame surface is $4\pi\lambda R^2[(-dT/dr)_{R^+} + (dT/dr)_{R^-}]$ cal/s if λ is expressed in cal/cm²-s-(K/cm) and the temperature gradients are given in K/cm.

The rate of heat release at the flame surface is the heat of reaction $\Delta H_R(T^*)$ for the process defined by equation (47), which must be evaluated at T^*. Hence

$$4\pi R^2 \rho D \left[\left(\frac{dY_{CO}}{dr} \right)_{R^-} \left(\frac{-\Delta H_3}{28} \right) + \left(\frac{dY_{H_2}}{dr} \right)_{R^-} \left(\frac{-\Delta H_4}{2} \right) \right]$$

$$= 4\pi\lambda R^2 \left[\left(-\frac{dT}{dr} \right)_{R^+} + \left(\frac{dT}{dr} \right)_{R^-} \right] \qquad (48)$$

where $-\Delta H_3 =$ heat released per mole of CO oxidized, $-\Delta H_4 =$ heat released per mole of H_2 consumed, and we have used a constant (species-independent) product ρD. In view of equations (44), equation (48) is one of two needed relations for T^* and T_d. Although we have neglected minor species in the present simplified analysis, it is appropriate to evaluate T^* by performing a full equilibrium calculation.

5. THE TIME-DEPENDENT TEMPERATURE WITHIN THE UNIFORMLY HEATED SOLID

At steady state, heat absorption by the endothermic surface-gasification processes* and heat conduction to the interior of the solid must be compensated for by heat conduction from the flame surface if we again consider isolated particles and assume that the total radiant heat transfer to the surface is negligibly small. Thus,

$$4\pi r_d^2 \left[\left(\rho D_{CO_2} \frac{dY_{CO_2}}{dr} \right)_{r_d^+} \left(\frac{\Delta H_1}{44} \right) + \left(\rho D_{H_2O} \frac{dY_{H_2O}}{dr} \right)_{r_d^+} \left(\frac{\Delta H_2}{18} \right) \right]$$

$$+ 4\pi r_d^2 \lambda_d \left(\frac{dT}{dr} \right)_{r_d^-} = 4\pi r_d^2 \lambda_{r_d^+} \left(\frac{dT}{dr} \right)_{r_d^+} \tag{49}$$

where we use a sign convention such that $\Delta H_1 =$ heat absorption per mole of CO_2 consumed for the CO_2–C reaction and $\Delta H_2 =$ heat absorption per mole of H_2O consumed in the steam–carbon reaction; also λ_d represents the (assumed constant) thermal conductivity within the solid. It follows that

$$4\pi r_d^2 \lambda_d \left(\frac{dT}{dr} \right)_{r_d^-} = 4\pi r_d^2 \left[\lambda_{r_d^+} \left(\frac{dT}{dr} \right)_{r_d^+} - \left(\rho D_{CO_2} \frac{dY_{CO_2}}{dr} \right)_{r_d^+} \left(\frac{\Delta H_1}{44} \right) \right.$$

$$\left. - \left(\rho D_{H_2O} \frac{dY_{H_2O}}{dr} \right)_{r_d^+} \left(\frac{\Delta H_2}{18} \right) \right] \tag{50}$$

determines the rate of heat conduction to the solid if we use equation (44) for $(dT/dr)_{r_d^+}$. Introduction of previously derived relations for the species mass fraction profiles and use of a constant, species-independent value $D = D_{H_2O} = D_{CO_2}$ may be shown to lead to

$$\lambda_d \left(\frac{dT}{dr} \right)_{r_d^-} = \lambda_{r_d^+} \left(\frac{dT}{dr} \right)_{r_d^+} - \frac{3}{2} \frac{\rho D}{r_d} \frac{\ln\left[\frac{5}{3} - \frac{2}{3}(Y_{N_2})_\infty + \frac{1}{12}(Y_{O_2})_\infty \right]}{[1 - (Y_{N_2})_\infty - \frac{7}{16}(Y_{O_2})_\infty]}$$

$$\times \left\{ [1 - (Y_{N_2})_\infty - (Y_{O_2})_\infty] \frac{\Delta H_2}{18} + \frac{11}{8}(Y_{O_2})_\infty \frac{\Delta H_1}{44} \right\} \tag{51}$$

---

*We neglect surface-pyrolysis processes in the specified combustion model.

We may now formulate the time-dependent heat-transfer problem within the solid particle. The simplest assumption is that the temperature is uniform within the particle (i.e., thermal accommodation occurs instantaneously in the solid). In this case, we may use equation (51) to determine the energy transfer by heat conduction to the carbon sphere, i.e., $4\pi r_d^2\lambda_d(dT/dr)_{r_d}$, as a function of T_d. The time-dependent temperature of the carbon sphere is then given by

$$d\bar{T}_d/dt = (3\lambda_d/r_d\rho_d c_d)(dT/dr)_{r_d} \tag{52}$$

where c_d represents the specific heat of the solid carbon. This formulation has been used previously[5,6] to determine $\bar{T}_d$. Equations (51) and (52) may be solved to find $\bar{T}_d$ as a function of time.

6. CALCULATION OF THE RATE OF DECREASE OF PARTICLE RADIUS WITH TIME; PARTICLE LIFETIMES

The rate of decrease of particle radius with time is conveniently obtained by combining equation (28) with the mass-conservation relation

$$\dot{m} = -\frac{d}{dt}(\tfrac{4}{3}\pi r_d^3\rho_d) = -4\pi r_d^2\rho_d(dr_d/dt) \tag{53}$$

from which it follows that

$$r_d/r_{d,0} = \left\{1 - 2\frac{\rho D}{\rho_d r_{d,0}^2}\ln\left[1 + \tfrac{2}{3}(Y_{H_2O})_\infty + \tfrac{3}{4}(Y_{O_2})_\infty]t\right\}^{1/2} \tag{54}$$

where $r_{d,0}$ is the initial particle radius. The particle lifetime for complete consumption is

$$t_l = r_{d,0}^2/\{2(\rho D/\rho_d)\ln[1 + \tfrac{2}{3}(Y_{H_2O})_\infty + \tfrac{3}{4}(Y_{O_2})_\infty]\} \tag{55}$$

It should be noted that our model becomes physically equivalent to that of Libby and Blake[5,6] only in the limiting case when diffusion processes are rate-controlling.

6.1. Particle Lifetimes

Libby and Blake[5,6] define a dimensionless time $\tau = (\eta t/\rho_d r_{d,0}^2)$, where $\eta = \eta_\infty$ is the viscosity coefficient for the ambient conditions and the Schmidt number (Sc) has been set equal to unity for all species (i.e.,

$\rho D = \eta$). Thus, the particle lifetime for the Libby–Blake model is

$$t_{l,L-B} = \tau_{l,L-B} \rho_d r_{d,0}^2 / \eta \tag{56}$$

where $\tau_{l,L-B}$ may be read from plots of their results. Taking the ratio of equation (55) to equation (56), we find that

$$\frac{t_l}{t_{l,L-B}} = \{2\tau_{l,L-B} \ln[1 + \tfrac{2}{3}(Y_{H_2O})_\infty + \tfrac{3}{4}(Y_{O_2})_\infty]\}^{-1} \tag{57}$$

which, for $T_\infty = 1500$ and 2000 K and the ranges of mass fractions at infinity considered by Libby and Blake, has values from 0.39 for $T_\infty = 1500$ K to 0.90 for $T_\infty = 2000$ K.

7. COMPARISONS WITH THE RESULTS OF NUMERICAL CALCULATIONS

We have compared the results obtained for our diffusion-flame model with the numerical results of Libby and Blake (for which we use the subscript L–B) for carbon oxidation controlled by chemical reaction rates. Our dimensionless mass-loss parameter α/r_d corresponds to the value of K of L–B, viz.,

$$\frac{\alpha}{r_d}\left(=\frac{\dot{m}}{4\pi\rho D r_d}\right) = K\left(=\frac{\dot{m}}{4\pi\eta_\infty r_d}\right) \tag{58}$$

for a Schmidt number of unity and η_∞ equal to the constant viscosity coefficient.

In view of the differences between a diffusion-controlled model and a model involving both diffusion and chemical reaction rates, the following conclusions are evident:

1. $(Y_{N_2})_{r_d} = (Y_{N_2})_{r_d,L-B}$ because the (Y_{N_2}) profile is independent of the model at low temperatures for which N_2 is inert;
2. $\lim_{T_d \text{ or } T_\infty \to \infty}(K)_{L-B} = K$ and $K_{L-B}/K \leqslant 1$;
3. $\lim_{T_d \text{ or } T_\infty \to \infty} t_{l,L-B} = t_l$ and $t_l/t_{l,L-B} \leqslant 1$;
4. $(Y_{CO} + Y_{H_2})_{r_d} \geqslant (Y_{CO} + Y_{H_2})_{r_d,L-B}$ because H_2O and CO_2 are assumed to be completely consumed by reaction with C at the particle surface in our diffusion-controlled model;
5. $\lim_{T_\infty \to \infty} T_{d,L-B} = T_d$, $T_d \leqslant T_{d,L-B}$, because we do not allow for direct exothermic surface oxidation of carbon, i.e., the process $C + \tfrac{1}{2}O_2 \to CO$ cannot occur.

S. S. Penner, M. R. Brambley, and M. Y. Bahadori

Figures 2–4 illustrate the stated conclusions. In these figures, we have plotted (K_{L-B}/K) vs. $T_{d,L-B}$, $(r_d/r_{d,0})$ vs. time and $(Y_{CO} + Y_{H_2})_{r_{d,L-B}}/(Y_{CO} + Y_{H_2})_{r_d}$ vs. $T_{d,L-B}$. We also show mass-fraction and temperature profiles for our diffusion-controlled model in Figures 5 and 6.

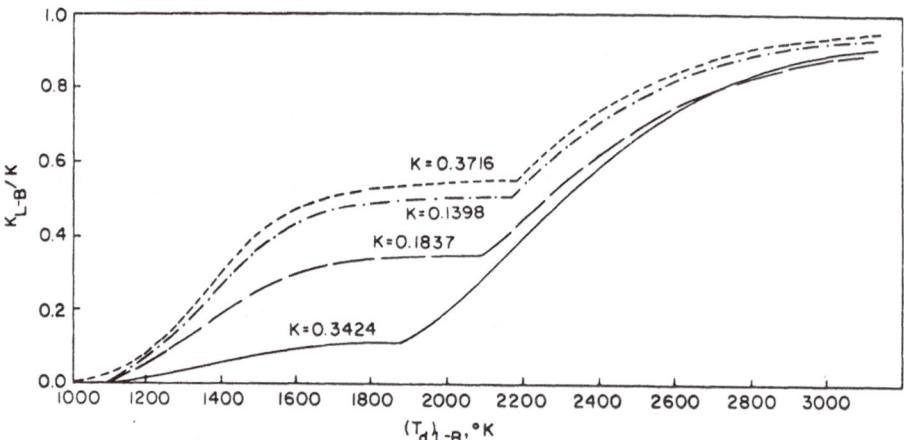

Fig. 2. The ratios of the dimensionless mass-loss rates, K_{L-B}/K, are shown as functions of the particle temperature, $T_{d,L-B}$, for $pr_d = 10^2$ atm-μm and equilibrium gas-phase chemistry in the Libby–Blake model and for the following ambient mass fractions: ——— for $(Y_{O_2})_\infty = 0.1$, $(Y_{N_2})_\infty = 0.4$, $(Y_{H_2O})_\infty = 0.5$; ---- for $(Y_{O_2})_\infty = 0.18$, $(Y_{N_2})_\infty = 0.72$, $(Y_{H_2O})_\infty = 0.10$; –·–·– for $(Y_{O_2})_\infty = 0.2$, $(Y_{N_2})_\infty = 0.8$, $(Y_{H_2O})_\infty = 0.0$; ------ for $(Y_{O_2})_\infty = 0.6$, $(Y_{N_2})_\infty = 0.4$, $(Y_{H_2O})_\infty = 0.0$.

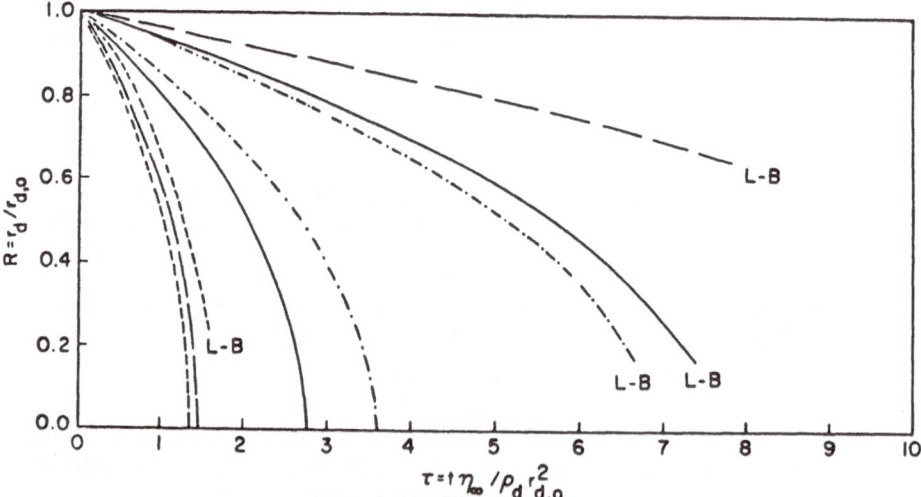

Fig. 3. The dimensionless particle radius, $r_d/r_{d,0}$, is shown as a function of nondimensional time τ: ——— for $(Y_{O_2})_\infty = 0.18$, $(Y_{N_2})_\infty = 0.72$, $(Y_{H_2O})_\infty = 0.10$; ---- for $(Y_{O_2})_\infty = 0.6$, $(Y_{N_2})_\infty = 0.4$, $(Y_{H_2O})_\infty = 0.0$; --- for $(Y_{O_2})_\infty = 0.1$, $(Y_{N_2})_\infty = 0.4$, $(Y_{H_2O})_\infty = 0.5$; –·–·–·– for $(Y_{O_2})_\infty = 0.2$, $(Y_{N_2})_\infty = 0.8$, $(Y_{H_2O})_\infty = 0.0$. The abbreviation L–B indicates the results of Libby and Blake.

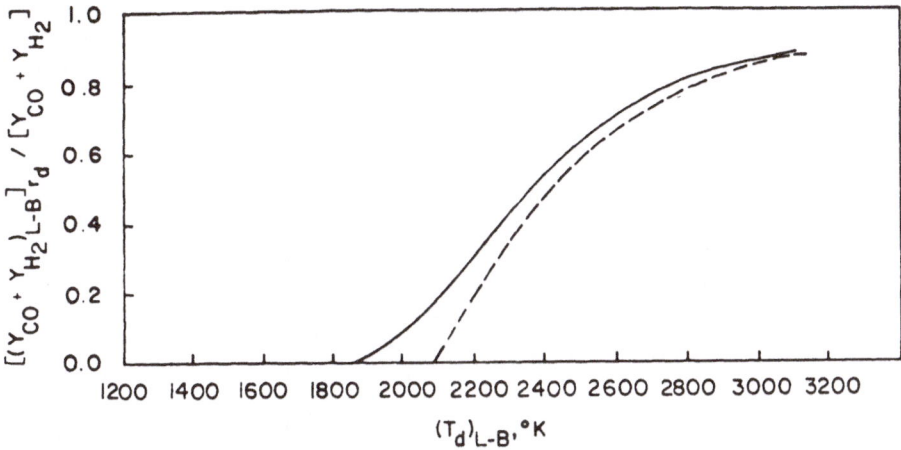

Fig. 4. The ratio $(Y_{CO} + Y_{H_2})_{r_d, L-B}/(Y_{CO} + Y_{H_2})_{r_d}$ is shown as a function of the particle temperature, $T_{d,L-B}$, for the following ambient conditions: ——— for $(Y_{O_2})_\infty = 0.1$, $(Y_{H_2O})_\infty = 0.5$, $(Y_{N_2})_\infty = 0.4$; ---- for $(Y_{O_2})_\infty = 0.18$, $(Y_{H_2O})_\infty = 0.10$, $(Y_{N_2})_\infty = 0.72$.

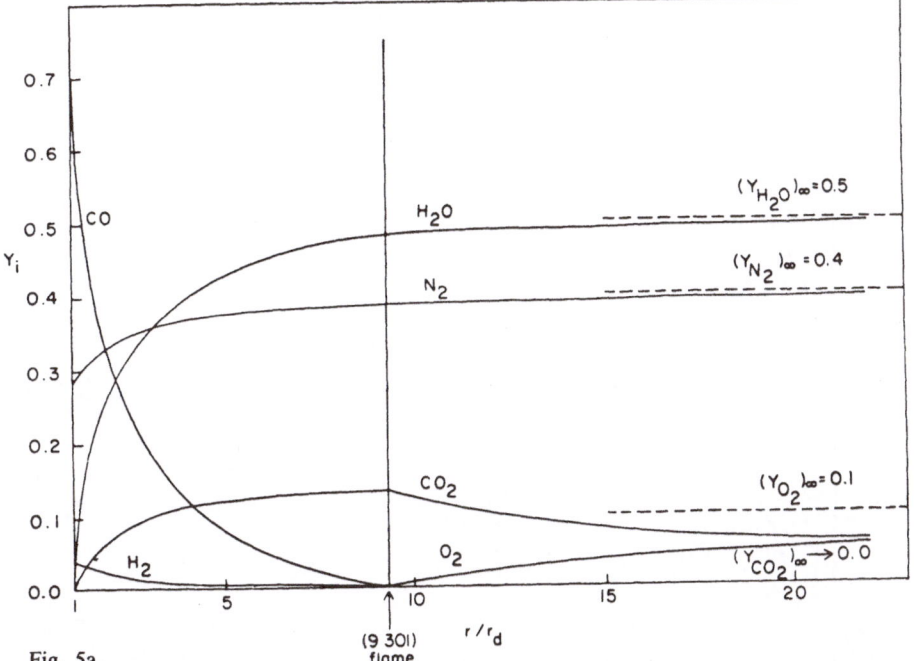

Fig. 5a.

Fig. 5. The mass fraction profiles in the gas-phase are shown for (a) $(Y_{O_2})_\infty = 0.1$, $(Y_{H_2O})_\infty = 0.5$, $(Y_{N_2})_\infty = 0.4$; (b) $(Y_{O_2})_\infty = 0.18$, $(Y_{H_2O})_\infty = 0.10$, $(Y_{N_2})_\infty = 0.72$; (c) $(Y_{O_2})_\infty = 0.6$, $(Y_{H_2O})_\infty = 0.0$, $(Y_{N_2})_\infty = 0.4$.

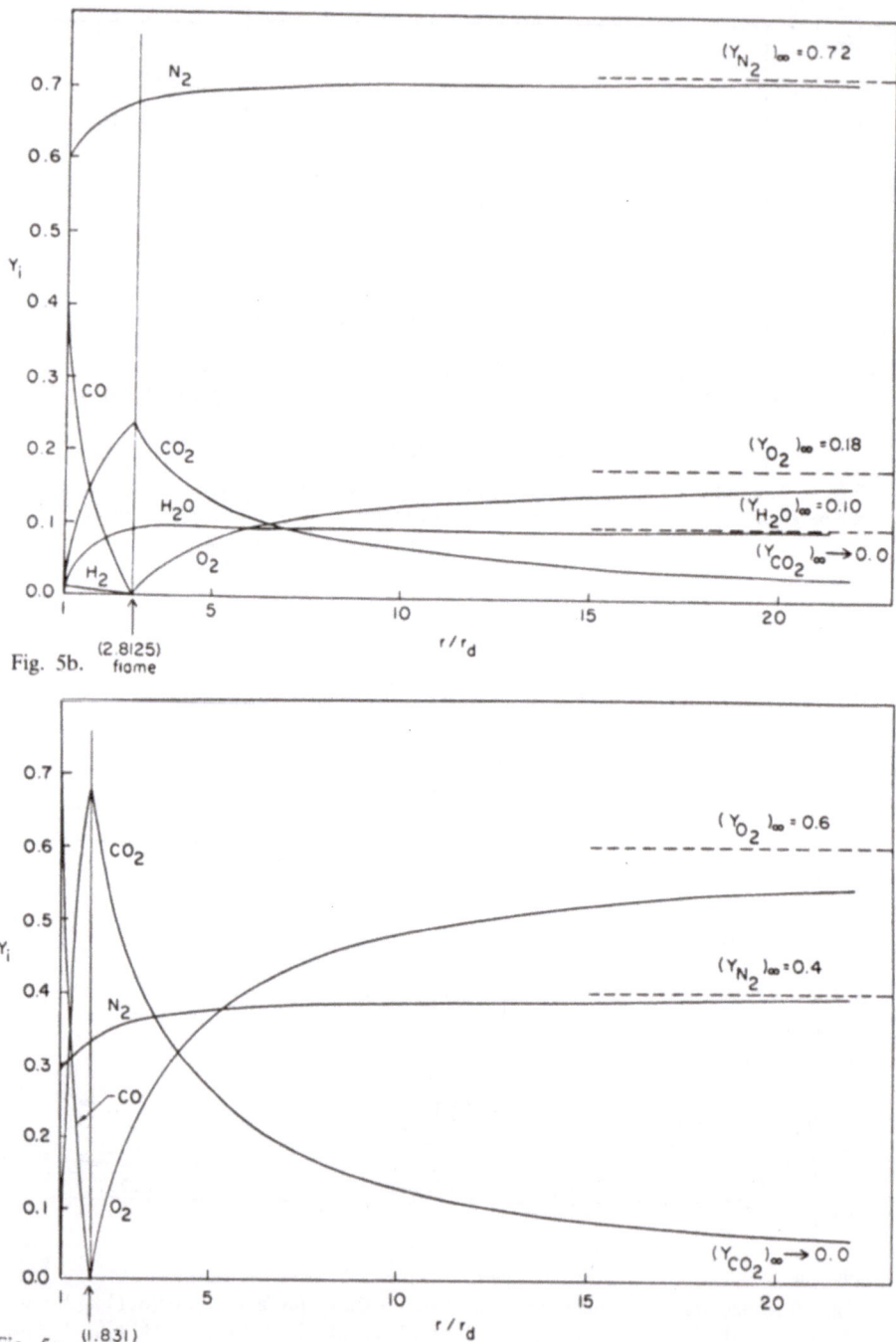

Fig. 5b.

Fig. 5c.

Equation (28) for $\dot{m}$ shows that the mass-loss rate increases with increasing mass fractions of the reactants $(Y_{H_2O})_\infty$ and $(Y_{O_2})_\infty$ and that oxygen is more effective than steam in gasifying carbon particles. These relationships are illustrated in Figure 3, where the particle radius is shown to decrease more rapidly for increasing values of the sum of $(Y_{O_2})_\infty$ and $(Y_{H_2O})_\infty$ in the diffusion-controlled model. The results of Libby and Blake show a stronger dependence of $\dot{m}$ on $(Y_{O_2})_\infty$ than on $(Y_{O_2})_\infty + (Y_{H_2O})_\infty$ because they allow for the direct surface oxidation of carbon. In the limit as $(Y_{O_2})_\infty$ approaches unity, the mass-loss rates and the particle-radius histories of Libby and Blake must approach the results of the diffusion flame model (compare the dotted curves in Figure 3) because of higher flame temperatures and thus greater oxidation rates in the gas phase. The species profiles of the L–B model should approach the profiles for the diffusion flame for $(Y_{O_2})_\infty \rightarrow 1$.

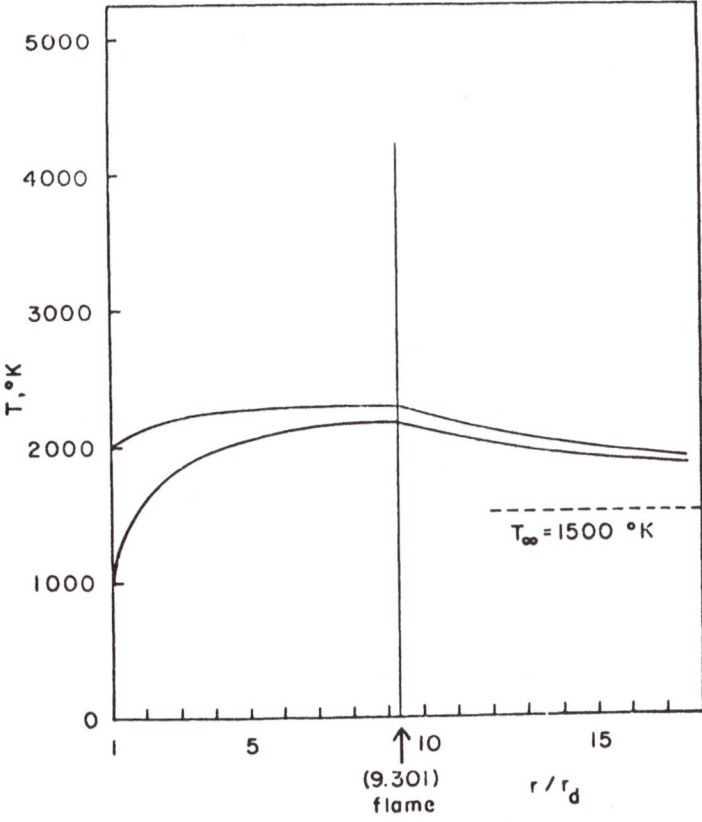

Fig. 6a.

Fig. 6. The gas-phase temperature, T, is shown as a function of the dimensionless radial distance, r/r_d, with T_d as a parameter for (a) $(Y_{O_2})_\infty = 0.1$, $(Y_{H_2O})_\infty = 0.5$, $(Y_{N_2})_\infty = 0.4$; (b) $(Y_{O_2})_\infty = 0.18$, $(Y_{H_2O})_\infty = 0.10$, $(Y_{N_2})_\infty = 0.72$; (c) $(Y_{O_2})_\infty = 0.6$, $(Y_{H_2O})_\infty = 0.0$, $(Y_{N_2})_\infty = 0.4$.

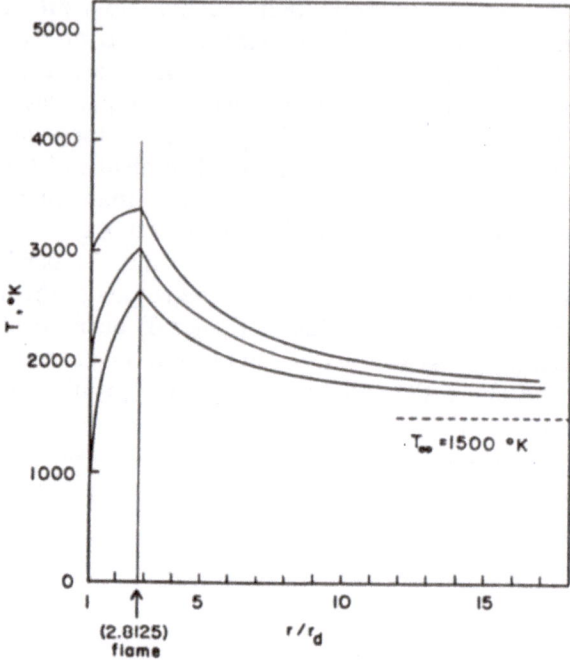

Fig. 6b.

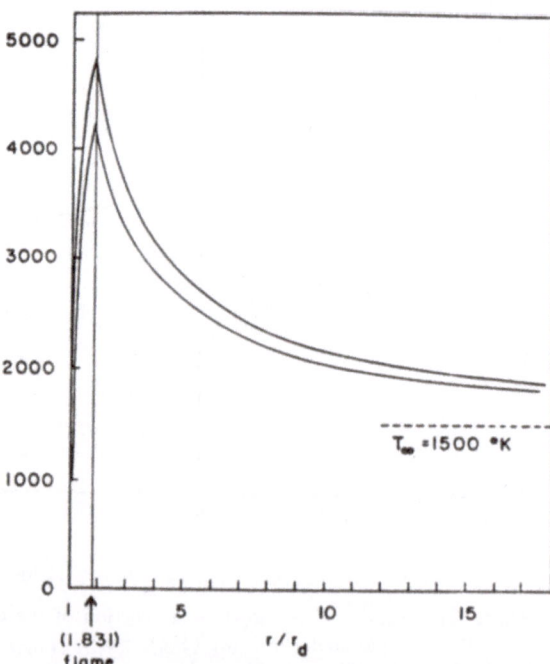

Fig. 6c.

Reference to Figures 5(a)–(c) shows that the flame front moves toward the particle surface as the ambient concentration of oxygen increases; it reaches a limiting value of $R = 1.76r_d$ for $(Y_{O_2})_\infty = 1$.

For sufficiently high flame temperatures (at $r = R$), dissociation occurs and leads to the formation of minor species (e.g., OH, O, H, C, N, etc.), thereby reducing the flame temperature well below T^* (to ~3000–4000 K). Heat is released as these minor species diffuse away from the flame front and recombine. As a result, the peak temperature T^* is reduced below the values shown in Figure 7 and the temperature profiles shown in

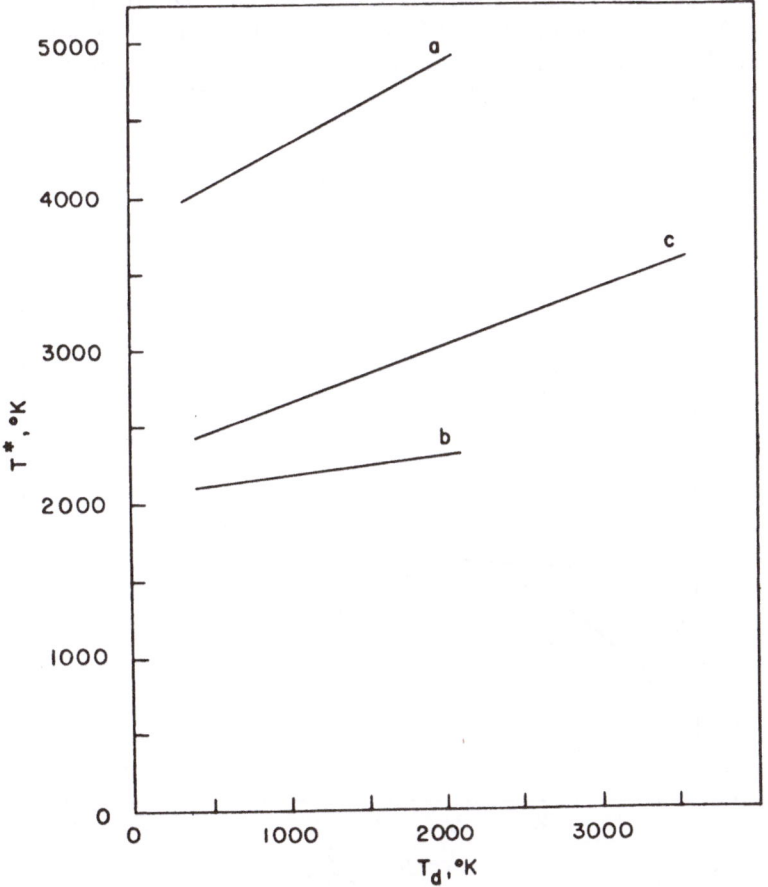

Fig. 7. Flame temperature, T^*, vs. particle temperature, T_d, for the following ambient conditions: (a) $(Y_{O_2})_\infty = 0.6$, $(Y_{H_2O})_\infty = 0.0$, $(Y_{N_2})_\infty = 0.4$; (b) $(Y_{O_2})_\infty = 0.1$, $(Y_{H_2O})_\infty = 0.5$, $(Y_{N_2})_\infty = 0.4$; (c) $(Y_{O_2})_\infty = 0.18$, $(Y_{H_2O})_\infty = 0.10$, $(Y_{N_2})_\infty = 0.72$. Dissociation will reduce the actual maximum values of T well below the values shown, particularly for $T^* \geqslant 3500$ K.

Figure 6 are flattened out appreciably. In order to determine temperature profiles properly, full equilibrium calculations that are inconsistent with our diffusion-flame approximation must be performed.

Dimensionless particle temperatures, T_d/T_∞, are shown as functions of the dimensionless time τ in Figure 8 for our diffusion-flame model and for the L–B model. The particle temperature approaches $T_{d,L-B}$ as τ increases. Detailed specifications for the two models (e.g., the reactions assumed to occur at the particle surface) are so different that particle-radius histories are not expected to agree.

The rate processes and chemical mechanisms for the gasification of carbon are not so well known that models including chemical reactions will necessarily be better approximations to experimental data than a diffusion model. At high temperatures (T_d or $T_\infty \to \infty$), N_2 can no longer be considered inert and may react according to any of several processes, including the following: $M + N_2 + O_2 \to 2NO + M$, $2NO + 2CO \to 2CO_2 + N_2$, $NO + CO \to CONO \to CN + O_2$.

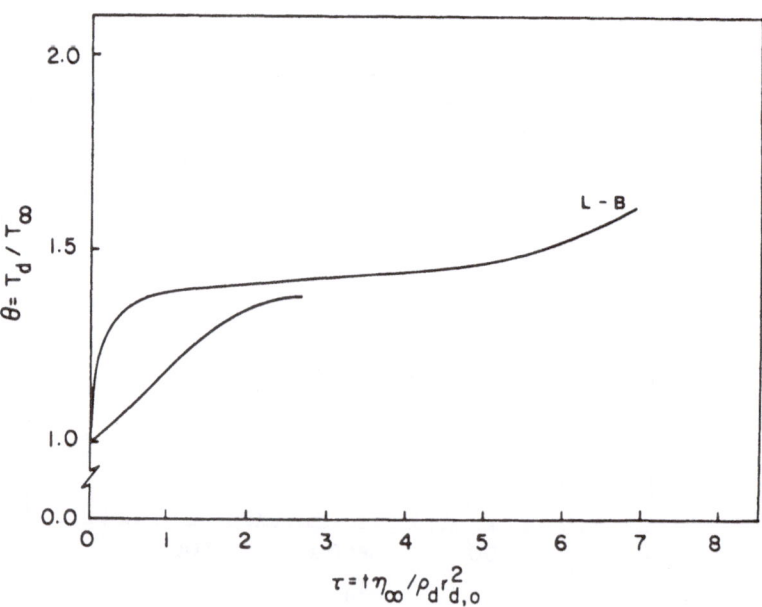

Fig. 8. Nondimensional particle temperature, T_d/T_∞, vs. nondimensional time, $\tau = t\eta_\infty/\rho_d r_{d,0}^2$, for $(Y_{O_2})_\infty = 0.18$, $(Y_{N_2})_\infty = 0.72$, $(Y_{H_2O})_\infty = 0.10$.

8. COMPARISONS WITH RESULTS DERIVED FROM APPLICATION OF THE SHVAB–ZELDOVICH PROCEDURE

In applications of the Shvab–Zeldovich procedure, the detailed gasification models are replaced by the overall reaction $C + O_2 \rightarrow CO_2$. Carbon sublimes at the particle surface before undergoing homogeneous oxidation. The dimensionless mass-loss rate K obtained by using the Shvab–Zeldovich procedure is[7]

$$K = \ln \left\{ 1 + \frac{\bar{c}_p T^*}{\Delta l_s} \left[\frac{(T_\infty - T_d)}{T^*} + \frac{(Y_{O_2})_\infty}{Y^*_{O_2}} \right] \right\} \tag{59}$$

where Δl_s is the specific heat of sublimation for carbon and $Y^*_{O_2}$ is the mass fraction of oxidizer entering into the overall chemical reaction. For T_∞ close to T_d, the second term in equation (59) is much smaller than unity and equation (59) becomes

$$K \simeq \frac{\bar{c}_p T^*}{\Delta l_s} \left[\frac{(T_\infty - T_d)}{T^*} + \frac{(Y_{O_2})_\infty}{Y^*_{O_2}} \right] \tag{60}$$

When dissociation is properly included in the determination of T^*, equation (60) yields values of K that are lower by factors of about 3–5 than the values obtained from our diffusion-flame model. This result is obtained because in the present model we have implicitly replaced endothermic surface reactions of carbon with CO_2 and H_2O (with heats of reaction of approximately 38 and 32 kcal/mol, respectively) by the highly endothermic (172 kcal/mol) sublimation of carbon. Reference to equation (60) shows that K varies inversely with the heat of sublimation of carbon; this parameter does not appear at all in the diffusion-flame model.

NOMENCLATURE

a	$\dot{m}\bar{c}_p/4\pi\lambda$, cm^{-1}
c	concentration, g/cm^3
$c_p(\bar{c}_p)$	specific heat (average) of the gas at constant pressure, $cal/g \cdot K$
c_d	specific heat of the carbon particle, $cal/g \cdot K$
$D_K(\bar{D})$	binary diffusion coefficient for species K (species-invariant), cm^2/s
$f(r)$	$4\pi r^2 \rho D/\dot{m}$, cm
h_K	specific enthalpy of species K, cal/g

h_K^0 — specific standard heat of formation of species K at the reference temperature T_0, cal/g

ΔH_1 — heat absorbed per mole of CO_2 consumed for the CO_2–C reaction, cal/mol

ΔH_2 — heat absorbed per mole of H_2O consumed in the H_2O–C reaction, cal/mol

$-\Delta H_3$ — heat released per mole of CO oxidized, cal/mol

$-\Delta H_4$ — heat released per mole of H_2 consumed, cal/mol

$\Delta H_R(T^*)$ — rate of heat release by reaction at the flame temperature, cal/s

K — dimensionless mass-loss rate $= \alpha/r$

K_{L-B} — dimensionless mass-loss rate used by Libby and Blake[5,6]

Δl_s — specific heat of sublimation for carbon, cal/g

$\dot{m}_C$ — rate of mass loss of a particle, g/s

$\dot{m}$ — total mass flow rate $= 4\pi r^2 \rho v$ (which is numerically equivalent to $\dot{m}_C$), g/s

m, n — stoichiometric coefficients

r — radial distance from the center of the particle, cm

r_d — particle radius, cm

$r_{d,0}$ — initial particle radius, cm

r_{ext} — extinction radius, cm

R — flame location, cm

Sc — Schmidt number $= \eta/\rho D$

T — temperature, K

T^* — temperature at the flame, K

T_∞ — ambient temperature, K

T_d (or $\bar{T}_d$) — uniform particle temperature, K

t — time, s

t_l — particle lifetime for complete consumption, s

v — mass-weighted average gas velocity, cm/s

Y_K — mass fraction of species K

$(Y_K)_\infty$ — mass fraction of species K in the ambient gases at $r \to \infty$

α — $\dot{m}/4\pi\rho D$, cm

ρ — gas density, g/cm^3

ρ_d (or ρ_p) — particle density, g/cm^3

τ — burning time (defined in Section 2.1), s

$\tau(\tau_l)$ — dimensionless time (lifetime) used by Libby and Blake[5,6] and defined in Section 6.1

λ — thermal conductivity, cal/cm·s·K

$\lambda_{r_d^+}$ — thermal conductivity of the gas mixture adjacent to the particle surface, cal/cm·s·K

λ_d — thermal conductivity of the carbon particle, cal/cm·s·K

ε_K mass flux fraction of species $K \equiv \rho Y_K (v + V_K)/\rho v$, where V_K is the diffusion velocity of species K

η viscosity coefficient, g/cm · s

η_∞ viscosity coefficient for ambient conditions, g/cm · s

REFERENCES

1. M. M. Avedesian and J. F. Davidson, Combustion of carbon particles in a fluidised bed, *Trans. Inst. Chem. Eng.* **51**, 121–131 (1973).
2. S. K. Ubhayakar and F. A. Williams, Burning and extinction of a laser-ignited carbon particle in quiescent mixtures of oxygen and nitrogen, *J. Electrochem. Soc.* **123**, 747–756 (1976).
3. M. F. R. Mulcahy and I. W. Smith, Kinetics of combustion of pulverized fuel: A review of theory and experiment, *Rev. Pure Appl. Chem.* **19**, 81–108 (1969).
4. S. S. Penner, *Chemistry Problems in Jet Propulsion*, pp. 280–291, Pergamon Press, London (1957).
5. P. A. Libby and T. R. Blake, Theoretical study of burning carbon particles, *Combust. Flame* **36**, 139–169 (1979).
6. P. A. Libby and T. R. Blake, Burning carbon particles in the presence of water vapor, *Combust. Flame* **41**, 123–147 (1981).
7. S. S. Penner and B. P. Mullins, *Explosions, Detonations, Flammability, and Ignition*, pp. 9–15 and 101–106, Pergamon Press, London (1959).

16

On the Dynamics of a Combustible Atmosphere

J. F. Clarke

ABSTRACT. The paper considers plane gas-dynamic waves of small amplitude that propagate through an exothermically reacting (combustible) atmosphere. The latter is spatially uniform but its temperature and pressure are both increasing with time. Any such problem needs several parameters for its description, but four have particular significance; they are the amplitude and frequency of the disturbance, a Reynolds number, based on the initial ambient frozen sound speed and the induction time for the atmosphere, and the activation energy for the chemical reaction as a multiple of the initial level of thermal energy; the inverse of this activation energy number is written as ε ($\ll 1$).

By identifying certain relationships between the parameters a number of physical situations can be modelled by rational approximate (parameter-perturbation) methods. In particular the following results are derived.

For simple waves of audible acoustic amplitude it is shown that the amplifying effect of the perturbed reacting atmosphere will exceed any damping due to the actions of viscosity and heat conduction; waves of low frequency will be most strongly amplified. Similar results apply to standing acoustic waves, with the fundamental mode being most strongly amplified. Combustible gas mixtures will therefore be naturally "noisy."

When the dimensionless amplitude and frequency of a simple wave becomes comparable with ε, nonlinear gas-dynamic effects become significant. The motion is governed by a modified form of the Burgers equation, and a rapidly amplifying temperature peak can propagate at the local frozen sound speed relative to the gas. In the neighborhood of the consequent imminent local ignition, parameter-perturbation analysis shows that the gas-dynamic character of the

J. F. CLARKE • Aerodynamics, Cranfield Institute of Technology, Bedford MK43 OAL, England.

process changes. Waves are created that propagate in both directions as each fluid element experiences its own ignition event in conditions that lie between the limits of combustion at constant pressure and constant volume; transport effects are negligible to a first approximation.

1. INTRODUCTION

In studies of the dynamics of chemically reacting gases the ambient atmosphere is usually assumed to be in a state of uniform equilibrium or, occasionally, of metastable equilibrium. The general task addressed in the following pages is to describe the progress of plane disturbances as they propagate through a reacting gas that is in a state of chemical disequilibrium, even before the advent of the imposed perturbations. Such an atmosphere must be evolving toward a state of ambient equilibrium, which it will achieve in the manner described by Semenov's model of an explosion, but our prime interest is not in this ambient explosive event so much as in the way in which ambient atmosphere and imposed perturbations interact with one another, especially during the induction period of the spatially uniform ambient event.

In order to make the problem as simple as possible, it is assumed that there is but one reacting species in the mixture which is undergoing an irreversible chemical reaction dominated by an Arrhenius-type of rate process. The equations of conservation then reduce to four (mass, momentum, energy, and species) for, essentially, the four dependent variables: pressure p, density ρ, gas velocity u, and nonequilibrium variable (or mass fraction of the reactant) q. Transport terms are retained in the equations.

Analysis of these relations relies heavily on the use of parameter perturbations. One such technique, called variously the characteristic parameter or the coordinate-straining method, was exploited to a first order by Crocco and Sirignano[1] in their study of a problem posed by interactions between shock waves and combustion in rocket motors. By postulating the existence of shock waves that reflect from end to end of the rocket chamber, it was possible to show that these waves could be substained by the presence of a localized region of combustion-energy release at one end of the duct. The general coordinate-straining method was subsequently taken by Crocco[2] to a high level of both accuracy and facility and used in conjunction with the method of multiple scales to analyze a variety of problems in nonlinear gas-dynamic oscillations and supersonic aerofoil theory.

In the present short paper there is space for only two illustrations of the use of techniques that are allied to the coordinate-stretching methods used so successfully by Crocco. The first example shows that acoustic-amplitude plane standing waves can be sustained or even amplified by their interaction with a spatially distributed combustion reaction. This fact has a bearing on the evolution of explosions in closed vessels, as shown by the second illustration, which examines behavior near the peak of an initially small-amplitude wave of overpressure. Consequent disturbances to the Arrhenius kinetics of the ambient combustion reaction give rise to the appearance of a local ignition center that travels with the wave and gives rise to the appearance of specifically combustion-generated gas-dynamic waves.

2. NONDIMENSIONAL EQUATIONS

Choose the set of dimensionless quantities, indicated by an overbar ($^-$), as follows:

$$p - p_0 = \rho_i a_{f0i}^2 \bar{p}, \qquad p_0 = \rho_i a_{f0i}^2 \bar{p}_0$$

$$\rho - \rho_i = \rho_i \bar{\rho}, \qquad u = a_{f0i} \bar{u}, \qquad a_f - a_{f0} = a_{f0i} \bar{a}_f$$

$$a_{f0} = a_{f0i} \bar{a}_{f0} = a_{f0i} (\gamma \bar{p}_0)^{1/2}, \qquad q - q_0 = q_{0i} \bar{q}, \qquad q_0 = q_{0i} \bar{q}_0 \qquad (2.1)$$

$$\eta = \eta_i (\bar{\eta}_0 + \bar{\eta}), \qquad \lambda = \lambda_i (\bar{\lambda}_0 + \bar{\lambda}), \qquad \mathcal{D} = \mathcal{D}_i (\bar{\mathcal{D}}_0 + \bar{\mathcal{D}})$$

$$x = \gamma t_1 a_{f0i} \bar{x}, \qquad t = \gamma t_1 \bar{t}$$

A subscript zero indicates a value in the undisturbed atmosphere, and thus a function of time only; a subscript i indicates an initial value. Apart from u, p, ρ, and q already defined, a_f is the frozen speed of sound (a_f^2 is equal to $\gamma p / \rho$), x and t are the space and time variables, and η, λ, and $\mathcal{D}$ are the shear viscosity, thermal conductivity, and mass diffusion coefficient, respectively; γ is the ratio of the frozen specific heats (assumed constant), and t_1 is the induction time in the ambient explosive atmosphere. It is useful to define the number

$$\varepsilon = p_{0i} / \rho_i E_A \qquad (2.2)$$

where E_A is the activation energy for the Arrhenius factor in the combustion reaction.

The dimensionless mass and momentum equations can now be written as

$$\bar{\rho}_{\bar{t}} + \bar{u}\bar{\rho}_{\bar{x}} + (1 + \bar{\rho})\bar{u}_{\bar{x}} = 0 \tag{2.3}$$

$$(1 + \bar{\rho})(\bar{u}_{\bar{t}} + \bar{u}\bar{u}_{\bar{x}}) + \bar{p}_{\bar{x}} = \frac{1}{\mathrm{Re}}\left[(\bar{\eta}_0 + \bar{\eta})\bar{u}_{\bar{x}}\right]_{\bar{x}} \tag{2.4}$$

while the energy equation can be manipulated into the forms

$$\{\bar{p}_{\bar{t}} + [\bar{u} \pm (\bar{a}_{\mathrm{t}0} + \bar{a}_{\mathrm{t}})]\bar{p}_{\bar{x}}\} \pm (1 + \bar{\rho})(\bar{a}_{\mathrm{t}0} + \bar{a}_{\mathrm{t}})\{\bar{u}_{\bar{t}} + [\bar{u} \pm (\bar{a}_{\mathrm{t}0} + \bar{a}_{\mathrm{t}})]\bar{u}_{\bar{x}}\}$$

$$= \varepsilon C\left\{(1 + \bar{\rho})\exp\left[\frac{1}{\gamma\varepsilon}\left(\frac{1}{\bar{p}_0} - \frac{1+\bar{p}}{\bar{p}_0+\bar{p}}\right)\right](1 + \bar{q}/\bar{q}_0) - 1\right\}$$

$$+ \frac{1}{\mathrm{Pr}\,\mathrm{Re}}\left[(\bar{\lambda}_0 + \bar{\lambda})\bar{\theta}_{\bar{x}}\right]_{\bar{x}} \tag{2.5$\pm$}$$

$$+ \frac{(\gamma - 1)}{\mathrm{Re}}\left[(\bar{\eta}_0 + \bar{\eta})\bar{u}\bar{u}_{\bar{x}}\right] \pm \frac{1}{\mathrm{Re}}(\bar{a}_{\mathrm{t}0} + \bar{a}_{\mathrm{t}})[(\bar{\eta}_0 + \bar{\eta})\bar{u}_{\bar{x}}]_{\bar{x}}$$

where $\bar{\theta}$ is the dimensionless temperature, equal to $\gamma(\bar{p}_0 + \bar{p})/(1 + \bar{\rho})$, and

$$C = C(\bar{t}; \varepsilon) \equiv \exp\left[\frac{1}{\varepsilon}\left(1 - \frac{1}{\gamma\bar{p}_0}\right)\right]\bar{q}_0 \tag{2.6}$$

In addition Re is a Reynolds number and Pr a Prandtl number, defined as follows:

$$\mathrm{Re} = \gamma t_1 a_{\mathrm{f}0i}^2 \rho_i/\tfrac{4}{3}\eta_i, \qquad \mathrm{Pr} = \tfrac{4}{3}\eta_i C_{\mathrm{pf}}/\lambda_i \tag{2.7}$$

The dimensionless species equation can be written as

$$\bar{q}_{\bar{t}} + \bar{u}\bar{q}_{\bar{x}} = -\frac{\varepsilon C}{\bar{Q}}\left\{\exp\left[\frac{1}{\gamma\varepsilon}\left(\frac{1}{\bar{p}_0} - \frac{1+\bar{p}}{\bar{p}_0+\bar{p}}\right)\right](1 + \bar{q}/\bar{q}_0) - 1\right\}$$

$$+ \frac{1}{\mathrm{Sc}\,\mathrm{Re}\,(1 + \bar{\rho})}\left[(1 + \bar{\rho})(\bar{\mathscr{D}}_0 + \bar{\mathscr{D}})\bar{q}_{\bar{x}}\right]_{\bar{x}} \tag{2.8}$$

where

$$\mathrm{Sc} = \tfrac{4}{3}\eta_i/\rho_i\mathscr{D}_i, \qquad \bar{Q} = (\gamma - 1)q_{0i}\hat{Q}a_{\mathrm{f}0i}^{-2}$$

and $\hat{Q}$ is the energy of formation of the fuel species.

Finally the dimensionless forms of the ambient-atmosphere equations are

$$\bar{p}_0 + \bar{Q}\bar{q}_0 = \bar{p}_{0m} = \bar{p}_0(0) + \bar{Q}$$

$$\frac{d\bar{p}_0}{d\bar{t}} = \varepsilon C \tag{2.9}$$

The initial condition for equations (2.9) is

$$\bar{p}_0(0) = 1/\gamma, \qquad \bar{q}_0(0) = 1 \tag{2.10}$$

Various parameters appear in the foregoing equations; ε and Re^{-1} may subsequently be assumed to have very small values but γ, Pr, Sc, and $\bar{Q}$ will always be of magnitude unity.

From equations (2.9) and (2.10) it can be seen that $\gamma\bar{Q}$ is the same as $-1 + \bar{p}_{0m}/\bar{p}_0(0)$ since $\gamma\bar{p}_0(0)$ is equal to one. The induction time for a first-order reaction can therefore be written as

$$\gamma t_1 = \gamma\varepsilon \exp(1/\varepsilon)/W\mathscr{P}(p_{0m}/p_{0i} - 1)$$

where $\mathscr{P}$ is a preexponential factor, assumed constant from now on.

The group of terms $\frac{4}{3}\eta_i/\rho_i a_{f0i}^2$ that appears in the expression for Re in (2.7) is of the order of a mean molecular collision interval, which is roughly of the same order of size as the preexponential factor $\mathscr{P}$. Accordingly

$$Re \approx t_1\mathscr{P} = \varepsilon \exp(1/\varepsilon)/W(p_{0m}/p_{0i} - 1)$$

Representative numbers for p_{0m}/p_{0i} and W are 5 and 30, respectively, and Table 1 lists some values for Re and t_1, the latter on the assumption that $\mathscr{P}$ is 10^9 Hz, for given activation-energy numbers ε.

3. ACOUSTIC DISTURBANCES AND STANDING WAVES

Suppose that disturbances experienced by the ambient atmosphere are such that

$$\bar{\psi} \sim M\sigma\psi^{(1)}(T, X), \qquad \psi = p, \rho, u, a_t, q \tag{3.1}$$

Table 1. Activation energy and Reynolds number[a]

$1/\varepsilon$	15	20	25	30	35	40	45
Re	1.5(3)	1.6(5)	2.4(7)	3.0(9)	3.8(11)	4.9(13)	6.5(15)
$Re^{-1/2}$	2.6(−2)	2.5(−3)	2.0(−4)	1.8(−5)	1.6(−6)	1.4(−7)	1.2(−8)
t_1, s	1.5(−6)	1.6(−4)	2.4(−2)	3.0(0)	3.8(2)	4.9(4)	6.5(6)
$-\varepsilon \ln Re^{-1/2}$	0.24	0.30	0.34	0.36	0.38	0.39	0.41

[a] The number in parentheses is the power of 10 by which the preceding number must be multiplied.

where

$$\sigma T = \bar{t}, \qquad \sigma X = \bar{x}$$

and M and σ are a pair of parameters such that ord $\sigma < 1$ and ord $M \leqslant 1$. It is clear that in using these assumed forms the intention is to examine disturbances whose typical time scale is very much less than the induction time and whose amplitude is also very small. Substituting equation (3.1) into equations (2.3)–(2.5), and (2.8), gives

$$\rho_T^{(1)} + u_X^{(1)} + O(M\sigma) = 0 \tag{3.2}$$

$$u_T^{(1)} + p_X^{(1)} + O(M\sigma) = \frac{\sigma}{\mathrm{Re}\,\sigma^2}[\bar{\eta}_0 u_{XX}^{(1)} + O(M\sigma)] \tag{3.3}$$

$$p_T^{(1)} + \bar{a}_{t0}^2 u_X^{(1)} + O(M\sigma) = \frac{\varepsilon C}{M}\left\{ \exp\left\{ \frac{\sigma M}{\varepsilon \gamma \bar{p}_0^2}[p^{(1)} - \bar{p}_0\rho^{(1)} + O(M\sigma)] \right\} \right.$$
$$\left. \times (1 + M\sigma\rho^{(1)})(1 + M\sigma q^{(1)}/\bar{q}_0) - 1 \right\}$$
$$+ \frac{\sigma}{\mathrm{Re}\,\sigma^2}\left[\frac{\bar{\lambda}_0}{\mathrm{Pr}}(\gamma p_{XX}^{(1)} - \gamma\bar{p}_0\rho_{XX}^{(1)}) + O(M\sigma) \right] \tag{3.4}$$

$$q_T^{(1)} + O(M\sigma) = -\frac{\varepsilon C}{\bar{Q}M}\left\{ \exp\left\{ \frac{\sigma M}{\varepsilon \gamma \bar{p}_0^2}[p^{(1)} - \bar{p}_0\rho^{(1)} + O(M\sigma)] \right\} \right.$$
$$\left. \times (1 + M\sigma q^{(1)}/\bar{q}_0) - 1 \right\}$$
$$+ \frac{\sigma \mathscr{D}_0}{\mathrm{ScRe}\,\sigma^2}[q_{XX}^{(1)} + O(M\sigma)] \tag{3.5}$$

Equations (3.2)–(3.5) are valid versions of the exact set provided that $C = C(\bar{t};\varepsilon)$ defined in equation (2.6) is $O(1)$ in the limit as σ, $M \to 0$. Since all the ambient atmospheric quantities, such as $\bar{p}_0$, are $O(1)$ it follows that either ε must be $O(1)$ or, if ε is $o(1)$, $\gamma\bar{p}_0$ is limited to be $1 + \varepsilon p_0^{(1)}(\bar{t})$ with $p_0^{(1)}(\bar{t}) = O(1)$. It will be assumed for the present that one or the other of these conditions is met.

Even so, final approximate forms of equations (2.3)–(2.5) cannot be written down until a decision is made about the parameter groupings that appear in association with the [ ] and { } terms in these equations. Table 1 shows that Re is always a large number.

Choosing

$$\sigma^2 \mathrm{Re} = R', \qquad \sigma = \mathrm{Re}^{-1/2}R'^{1/2}, \qquad R' = O(1) \tag{3.6}$$

makes the transport terms in equations (3.3)–(3.5) into small, $O(\sigma)$

corrections to the main convective movements of momentum, energy, and mass, provided that ord $M < 1$, and this is the situation that will be modeled here. If $M = 1$ some nonlinear terms are of the same order of magnitude as transport effects, with the implication that both sets of terms should either be neglected or retained. If $\sigma M / \varepsilon$ should be $o(1)$ as σ, $M \to 0$, the $\exp(\ )$ terms in $\{\ \}$ in equations (3.4) and (3.5) can be expanded in powers of $(\sigma M / \varepsilon)$; these perturbations to the exponential Arrhenius chemical reaction rate factor could be nominally of the same order of significance as changes in density and composition. It should be noted that $\bar{q}_0$ does not diminish significantly during the induction period and that $\bar{q}/\bar{q}_0$ will be treated as $O(\bar{q})$ here.

In these circumstances the right-hand side of equation (3.5) becomes of order σ and it can be seen that $q_T^{(1)}$ is therefore likewise of order σ. Provided that the time integral of the right-hand side of equation (3.5) does not result in any increase above this order of magnitude, it is clear that $q^{(1)}$ is not $O(1)$, as implied in (3.1), but $O(\sigma)$ [in other words $\bar{q}$ is $O(M\sigma^2)$]. In such circumstances the $q^{(1)}$ problem decouples from the $u^{(1)}$, $p^{(1)}$, $\rho^{(1)}$ problem, which then appears in the form

$$\rho_T^{(1)} + u_X^{(1)} = 0 \tag{3.7}$$

$$u_T^{(1)} + p_X^{(1)} = \frac{\sigma}{R'} [\bar{\eta}_0 u_{XX}^{(1)}] \tag{3.8}$$

$$p_T^{(1)} + \bar{a}_{f0}^2 u_X^{(1)} = \sigma C \left[\frac{1}{\gamma \bar{p}_0^2} p^{(1)} - \left(\frac{1}{\gamma \bar{p}_0} - \varepsilon \right) \rho^{(1)} \right] + \frac{\sigma}{R'} \left[\frac{\bar{\lambda}_0}{\Pr} (\gamma p_{XX}^{(1)} - \gamma \bar{p}_0 \rho_{XX}^{(1)}) \right] \tag{3.9}$$

Table 1 shows that when (3.6) is true, not only is σ very small, in the region of 10^{-3} or less for $\varepsilon \leqslant 1/20$, but $-\varepsilon \ln \sigma$ is roughly equal to $\frac{1}{3}$; thus ε is small like $-1/\ln \sigma$ as $\sigma \to 0$. With the restrictions described after equation (3.5), equations (3.7)–(3.9) make up a valid set of approximate governing equations, provided that M is $o(1)$ in the σ-limit. Signals audible to the human ear lie roughly in the intervals $M\sigma$ equal to 10^{-9} to 3×10^{-4}. Supposing that M is $\sigma^{1/2}$ for purposes of illustration, this implies that σ lies between 10^{-6} and 4×10^{-3}; comparing these numbers with Table 1, and remembering that the latter provides only a fairly rough general guide to the size of the various quantities listed therein, it can be seen that equations (3.7)–(3.9) describe disturbances of acoustic amplitudes during induction periods before ambient explosion. For short induction periods the frequencies of waves described by these equations are likely to be ultrasonic (remember that it is σ itself that governs frequency).

Equations (3.7)–(3.9) are linear but have nonconstant coefficients that are functions of the "long" time $\bar{t}$. Time rates of change have been chosen to take place on the "short" time scale T, so that evolution of the ambient atmosphere can be seen to involve only a relatively slow modulation of any fundamentally harmonic disturbance. Solutions to equations (3.7)–(3.9) could be sought through the method of multiple scales (e.g., Nayfeh,[3] Chapter 6) but it is quite straightforward in the present case to use an approach that is linked with the ideas of geometric acoustics and is described by Whitham[4] (§11.8). Thus the variables described as $\psi^{(1)}$ in (3.1) will be assumed to have the form

$$\psi^{(1)} = \exp\left(\frac{i}{\sigma}\,\Theta\right)(\psi_1 + \sigma\psi_2 + \cdots), \qquad \psi = u, p, \rho \qquad (3.10)$$

where Θ is a phase function that depends on the slow variables $\bar{t}$, $\bar{x}$, while ψ_1 and ψ_2 are amplitude functions that also depend upon $\bar{t}$, $\bar{x}$. It is required that Θ, ψ_1, ψ_2, etc., and all of their derivatives with respect to $\bar{t}$, $\bar{x}$, shall be $O(1)$ as $\sigma \to 0$. The high-frequency character of $\psi^{(1)}$ is guaranteed by the denominator σ in the exponential index. Substituting equation (3.10) into equations (3.7)–(3.9) and equating coefficients of $\sigma^0, \sigma^1, \ldots$ to zero leads to a set of linear algebraic equations for ψ_1, ψ_2, etc.

The equations for ψ_1 are consistent only if

$$\Theta_{\bar{t}}\,(\Theta_{\bar{t}}^2 - \bar{a}_{t0}^0\Theta_{\bar{x}}^2) = 0 \qquad (3.11)$$

The solution $\Theta_{\bar{t}} = 0$ requires both u_1, and p_1 to vanish but admits an arbitrary ρ_1 so that $\rho^{(1)} \simeq \rho_1 \exp[i\Theta(\bar{x})/\sigma]$. This form of static density perturbation will be assumed to be ruled out by the initial-value data in the present case and attention confined to the remaining pair of phase functions in the interests of brevity. In view of equation (3.11) the phase must be such that

$$\Theta = \Theta(\phi_\pm), \qquad (\partial\bar{x}/\partial\bar{t})_{\phi_\pm} = \pm\,\bar{a}_{t0} \qquad (3.12)$$

In view of (3.11) and $\Theta_{\bar{t}} \neq 0$ the right-hand sides of the ψ_2 equations cannot be independent. Ensuring their compatibility leads to a single equation, for u_1 say, for each phase variable $\phi_\pm$. Remembering that $\gamma\bar{p}_0$ is equal to $\bar{a}_{t0}^2$ these equations can be integrated to give

$$\bar{a}_{t0}^{1/2}u_1 = \exp\left[\frac{1}{2}\int^{\bar{t}} \mathscr{C}(\bar{t})d\bar{t}\right] A_\pm(\phi_\pm) \qquad (3.13)$$

where $A_\pm$ are as yet arbitrary functions of $\phi_\pm$ and

$$\mathcal{C}(\bar{t}) = C[(\gamma - 1)\bar{a}_{t0}^{-4} + \varepsilon\bar{a}_{t0}^{-2}] - (\Theta_{\bar{x}}^2/R')[\bar{\eta}_0 + \bar{\lambda}_0(\gamma - 1)/\text{Pr}] \quad (3.14)$$

Equation (3.13) gives a general solution for u_1, which can now be seen to suffer reductions in amplitude from the transport effects, summarized in the last part of $\mathcal{C}(\bar{t})$ containing $\Theta_{\bar{x}}^2$ (note that the last square-bracket term in equation (3.14) is the classical diffusivity of sound; see Lighthill[5]), and to undergo amplification from the effects of the perturbed ambient reaction [the C-dependent term in equation (3.14)]. A similar result for the combined effects of transport processes and combustion chemistry has been given before, but for waves traveling in only one direction[6]; indeed the possible amplifying effect of a combustible atmosphere on acoustic disturbances was identified some time ago, with Toong[7] taking a particularly significant step forward. The present results for high-frequency harmonic disturbances are useful because they generalize previous work somewhat, include transport effects, and present the results in analytical form, and, in particular, because they expose limitations of the formulation through use of the parameter-perturbation approach. In addition it is now possible to construct simple standing-wave solutions, as follows.

Setting $\phi_+ = \bar{x} = \phi_-$ when $\bar{t} = 0$, equation (3.12) gives

$$\bar{x} \mp \int_0^{\bar{t}} \bar{a}_{t0} d\bar{t} = \phi_\pm \quad (3.15)$$

Then equations (3.10) and (3.15) can be used to construct the solution

$$u^{(1)} \simeq \bar{a}_{t0}^{-1/2} \exp\left(\tfrac{1}{2}\int_0^{\bar{t}} \mathcal{C} d\bar{t}\right)\{A_+(\phi_+)\cos[\Theta(\phi_+)/\sigma] + A_-(\phi_-)\cos[\Theta(\phi_-)/\sigma]\} \quad (3.16)$$

where real parts of (3.10) have been selected and $A_\pm$ are now the values of the amplitude of u_1 when $\bar{t}$ is zero. Choosing constant values for A_+ and A_- it can be seen that if $u^{(1)}$ is zero when $\bar{t} = 0$, then A_- is equal to $-A_+$. If one now sets

$$\Theta(\phi_\pm) = k\left(\bar{x} \mp \int_0^{\bar{t}} \bar{a}_{t0} d\bar{t}\right) \quad (3.17)$$

where k is a constant, it can be seen that equation (3.16) becomes

$$u^{(1)} \simeq \bar{a}_{t0}^{-1/2} \exp\left(\tfrac{1}{2}\int_0^{\bar{t}} \mathcal{C} d\bar{t}\right) A_+ \sin\left(\frac{k\bar{x}}{\sigma}\right) \sin\left(\frac{k}{\sigma}\int_0^{\bar{t}} \bar{a}_{t0} d\bar{t}\right) \quad (3.18)$$

This solution represents a standing wave whose amplitude and frequency both change with time, between the planes $\bar{x} = 0$ and σL, provided that

$$kL = n\pi, \qquad n = 1, 2, 3, \ldots \tag{3.19}$$

Equation (3.17) shows that $\Theta_{\bar{x}}$ is equal to the wave number k and, as a consequence of (3.19), it can be seen from equation (3.14) that the fundamental wave $n = 1$ is most likely to be amplified by the effects of combustion chemistry. Of course if the standing wave exists in a tube whose diameter is substantially less than L, rather than in the simple $\bar{x}$-space bounded only by planes at $\bar{x} = 0$ and L, there will be a stronger influence of viscous drag of the side walls on damping of $u^{(1)}$ than is provided by the diffusivity term in equation (3.17). This situation, which will not be analyzed here, is clearly one of some practical importance for explosions in tubes.

First-order solutions for p_1 and ρ_1 follow readily; standing-wave solutions to go with equation (3.18) are constructed in the obvious way. No attempt is made here to analyze the ψ_2 terms, but it can be seen that they will introduce small phase shifts between $p^{(1)}$, $\rho^{(1)}$, and $u^{(1)}$ (the first-order, ψ_1-type, solutions make the $p^{(1)}$, $\rho^{(1)}$, and $u^{(1)}$ quantities all have the same phase).

The amount of possible amplification from the $\mathscr{C}$-factor in equations (3.15) and (3.16) can clearly not be large during the induction interval, since $\bar{t}$ is essentially less than, or at most equal to, unity there and the $C[\quad]$ term in equation (3.14) is also limited to about unit size. Evidently amplification begins to become large as ambient explosion approaches; since this is synonymous with very large increases in $C(\bar{t}; \varepsilon)$ the present approximations break down there and the waves become dispersive and unstable, with a rapidly changing relaxation time.[6] One way round this difficulty is via numerical solution of the set of formally linearized versions of equations (2.3)–(2.6) such as has been undertaken by Toong,[7] Garris et al.,[8] and Gilbert et al.[9] All of these and other works, by the authors of the references just quoted, deal with the case of fixed wavelength. That this is not always physically correct, in the case of boundary- as opposed to initial-value problems for example, can be seen from either the present work or from Clarke's paper.[6]

When σ increases, so that the initial frequency of the disturbance decreases, the transport terms diminish steadily in general importance [note from (3.6) that an increase in σ is like an increase in R' to begin with, and then consult equation (3.14) for $\mathscr{C}$]. Disturbance amplitudes are kept small by making M smaller as σ is allowed to grow (so that $M\sigma$ is fixed, for example). When σ is so large that only a few oscillations occur within the induction interval the slowly-varying wave idea, in the form of equation

(3.10) here, begins to fail; coefficients of perturbation, $\psi^{(1)}$, quantities in equations (3.7)–(3.9), now begin to alter appreciably within one cycle of the motion and it may be necessary to resort to numerical integrations like those mentioned above. The appearance of shock waves in the essentially nonlinear situation ord $\sigma < 1$, ord $M = 1$ and related matters have been discussed by Clarke[10,11] using analytical methods.

4. BEHAVIOR AT A POINT OF LOCAL EXPLOSION

Assume that the disturbances are such that

$$\bar{\psi} = \sigma\psi^{(1)}(\beta, T), \qquad \psi = u, p, \rho, a_{\mathrm{f}}; \qquad \bar{q} = \sigma^2 q^{(1)}(\beta, T)$$

where   (4.1)

$$T = \bar{t}, \qquad \sigma(\partial\xi/\partial\bar{t})_\beta = 1 - \bar{u} - \bar{a}_{\mathrm{f}0} = \bar{a}_{\mathrm{f}}, \qquad \sigma\xi = \bar{t} - \bar{x}$$

Justification for the $O(\sigma^2)$ ordering of $\bar{q}$ follows an argument very like the one given in the previous section. Equations (4.1) describe a system of waves propagating along the positive $\bar{x}$ direction such that changes from wavelet to wavelet are more rapid than those experienced by any particular wavelet ("slowly-varying" waves). Choosing the wavelet parameter β to be such that $\sigma\beta$ is equal to T on the path along which gas velocity is specified (the "piston path") provides the following boundary condition:

$$\bar{u}(T/\sigma, T) = \sigma D'(T/\sigma), \qquad \xi = (T/\sigma) - \sigma D(T/\sigma), \qquad \sigma\beta = T \qquad (4.2)$$

The function D is given, and D' is its derivative with respect to its argument.

It will now be assumed that σ is so large that

$$\sigma = E\varepsilon, \qquad E = O(1) \qquad \text{as } \sigma \to 0 \qquad (4.3)$$

and that, as a consequence, modeling is to rely on the proposition that the activation energy of the reaction is large. This fact has some useful consequences as far as the ambient atmosphere behavior is concerned. In particular an asymptotic solution for $\bar{p}_0$ given by

$$1/\gamma\bar{p}_0 \sim 1 + \varepsilon \ln(\gamma\hat{T} - \gamma T) + \cdots, \qquad \gamma\hat{T} \sim 1 + \varepsilon(2 + n/\gamma\bar{Q}) + \cdots \qquad (4.4)$$

is good for all times $T < \hat{T}$ to within an exponentially [i.e. $\exp(-1/\varepsilon)$] small interval, the ambient frozen sound speed $\bar{a}_{\mathrm{f}0}$ behaves like

$$\bar{a}_{\mathrm{f}0} = (\gamma\bar{p}_0)^{1/2} \sim 1 - \tfrac{1}{2}\varepsilon \ln(\gamma\hat{T} - \gamma T) + \cdots \qquad (4.5)$$

and, while (4.4) remains a valid estimate of $\bar{p}_0$, $\bar{q}_0$ remains within $O(\varepsilon)$ of unity.

With all of the foregoing information equations (2.5+) and (2.8) reduce to the following forms under the limit $\varepsilon \to 0$ with β and T fixed:

$$2Eu_T^{(1)} = (\gamma\hat{T} - \gamma T)^{-1}\{\exp[(\gamma - 1)Eu^{(1)}] - 1\}$$
$$+ [\bar{\eta}_0 + \bar{\lambda}_0(\gamma - 1)\Pr^{-1}](\sigma^2 \operatorname{Re}\xi_\beta)^{-1}(u_\beta^{(1)}/\xi_\beta)_\beta \qquad (4.6)$$

$$\bar{Q}Eq_\beta^{(1)} = -\xi_\beta(\gamma\hat{T} - \gamma T)^{-1}\{\exp[(\gamma - 1)Eu^{(1)}] - 1\} \qquad (4.7)$$

since equations (2.3) and (2.4) show in the circumstances that

$$p^{(1)} = \rho^{(1)} = u^{(1)} \qquad (4.8)$$

Equation (4.6) has also been derived by Abousieff and Toong[12] using different methods.

One must always encounter a need to retain the diffusive term in equation (4.6) when $\xi_\beta \to 0$, as in the neighborhood of shock waves, regardless of the value of Re. However the present relatively large values of σ^2 Re will be invoked as good reasons for neglect of the diffusive terms in equation (4.6), and this will indeed be done in the work that follows, for which it can be confirmed that ξ_β is not too small. The solution of the shortened form of equation (4.6) that satisfies (4.2) is readily found to be

$$(\gamma - 1)Eu^{(1)}(\beta, T) = -\ln\left\{1 - [1 - e^{-(\gamma-1)ED'(\beta)}]\left(\frac{\hat{T} - \sigma\beta}{\hat{T} - T}\right)^a\right\}, \quad a = \frac{\gamma - 1}{2\gamma}$$
$$(4.9)$$

For short input pulses such that D and D' vanish for all $\beta > \beta_t = O(1)$, function (4.9) goes over to a form that is recognizable as the solution of the initial-value problem studied before, at least in some of its aspects, by Blythe[13] (for expansion waves) and Clarke[10,11] (for compression waves). One of several interesting features of the solution (4.9) is the existence of a local, as opposed to a global ambient atmospheric, ignition or induction time. In the short-pulse case this time is given to first order by T_I where

$$(1 - T_I/\hat{T})^a = \{1 - \exp[-(\gamma - 1)ED'(\beta)]\} \qquad (4.10)$$

since (4.9) shows that $u^{(1)}$ then becomes logarithmically infinite. The question of shock formation in such pulses has been discussed at some length by Clarke[11]; it suffices to say for the present that local explosion can take place, either before shocks appear or in parts of the pulse that are well removed (in $\bar{x}$) from shock positions. The earliest local ignition clearly occurs when $ED'(\beta)$ has its maximum positive value and it is interesting to list the values of this quantity that are associated with various reductions in T_i below the homogeneous value $\hat{T}$; this is done in Table 2 for $\gamma = 1.4$.

The boundary condition (4.2) shows that

$$\bar{u} = \sigma D'(\beta) = \varepsilon ED'(\beta)$$

at the piston face, so that $ED'(\beta)$ is proportional to the amplitude of the dimensionless "input" velocity, u/a_{f0i}. Since ε could well be in the range 1/20 to 1/40, as can be seen from Table 1, it is first of all clear that the notion of a small perturbation is being pushed rather close to its limits of validity if, in the present circumstances, one is modeling an event that leads to local explosion times significantly smaller than the ambient homogeneous value $\hat{T}$.

Certainly small perturbations as large as the ones envisaged in the foregoing paragraph have been modeled in other branches of aerodynamics by methods that are, in principle, exactly like those being used here. It must be remembered that the present perturbation scheme is not a *linearizing* process. Since one can anticipate a measure of qualitative exactitude from the methods employed here, it is worth proceeding for that end alone; experience, such as that described in Crocco's paper,[2] suggests that quantitative accuracy is achievable by the use of suitable second approximations, but they will not be pursued in the present paper.

Apart from anything else it is valuable to be able to use the short-pulse result (4.10), and its numerical illustration in Table 2, as an indication of the magnitude of those gas-dynamic disturbances that are capable of advancing the ignition time, locally, to an extent that is significant enough to involve some new physical phenomena when compared with the previous, smaller disturbance, case.

The general local-explosion criterion, namely the criterion for disturbances that are *not* of the short-pulse kind, follows from (4.9), i.e.,

$$(1 - T_i/\hat{T}) = (1 - \sigma\beta/\hat{T})\{1 - \exp[-(\gamma - 1)ED'(\beta)]\}^{1/a} \qquad (4.11)$$

Table 2. Local explosion times

$T_i/\hat{T}$	$\frac{1}{2}$	$\frac{2}{3}$	$\frac{3}{4}$	$\frac{4}{5}$	$\frac{5}{6}$	$\frac{6}{7}$	$\frac{7}{8}$	$\frac{8}{9}$
$ED'(\beta)$	5.9	4.8	4.3	4	3.7	3.5	3.4	3.3

If the signal launched by the piston is harmonic, for example, so that $D'(\beta)$ goes through a succession of maxima all of the same magnitude, equation (4.11) shows that it will be at the first of these maxima that the local explosive processes will be initiated, since the factor $(1 - \sigma\beta/\hat{T})$ diminishes monotonically as β grows from zero. It is not *a priori* obvious that this must be so, as successive maximum compressions are sent into an atmosphere whose chemical activity is growing.

It will be assumed from now on that the pulse is short, in the sense defined above, and that (4.9) for $u^{(1)}$ can therefore be adequately approximated by the relation

$$(\gamma - 1)Eu^{(1)}(\beta, T) = -\ln[1 - P(\beta)(1 - T/\hat{T})^{-a}] \qquad (4.12)$$

where

$$P(\beta) = \{1 - \exp[-(\gamma - 1)ED'(\beta)]\} \qquad (4.13)$$

Then

$$(\gamma - 1)Eu_T^{(1)} = [(\gamma - 1)/2\gamma\hat{T}(1 - T/\hat{T})]P(\beta)[(1 - T/\hat{T})^a - P(\beta)]^{-1} \qquad (4.14)$$

$$(\gamma - 1)Eu_\beta^{(1)} = (\gamma - 1)ED''(\beta)\exp[-(\gamma - 1)ED'(\beta)][(1 - T/\hat{T})^a - P(\beta)]^{-1} \qquad (4.15)$$

It can be seen from equation (4.14) that there are two fundamental reasons why $u_T^{(1)}$ will begin to grow larger than the strictly $O(1)$ magnitude that it must have for validity of the approximations that lead to its derivation. One is the factor $(1 - T/\hat{T})^{-1}$, which represents the direct influence of the onset of ambient explosion and which exactly parallels the breakdown behavior described by Clarke[6] for the case $E < \text{ord } 1$. It is noteworthy that this factor does *not* appear in (4.15) for $u_\beta^{(1)}$. The second and presently more interesting reason for the approaching misbehavior of $u_T^{(1)}$ is the onset of *local* explosion at times T_l given by

$$(1 - T_l/\hat{T})^a = P(\beta_m) \equiv P_m \qquad (4.16)$$

where β_m is the value of β that gives P its maximum positive value.

It is hypothesized from now on that such a β_m-value exists and is unique. More complicated cases must be considered separately and it is clearly wise to await resolution of the simplest problem first.

When local explosion is imminent it can now be seen from equation (4.15) that $u_\beta^{(1)}$ also becomes large, at least away from β_m itself, where

$$D''(\beta_m) = 0 = P'(\beta_m) \qquad (4.17)$$

For values of $T/\hat{T}$ near to $T_l/\hat{T}$, say

$$T/\hat{T} = T_l/\hat{T} - \Delta T, \qquad 0 < \Delta T \ll 1 \tag{4.18}$$

and β near to β_m, say

$$\beta = \beta_m + \Delta\beta, \qquad |\Delta\beta|/\beta_m \ll 1 \tag{4.19}$$

it can be shown after some analysis that $u_T^{(1)}$ is behaving uniformly like $1/\Delta T$, while $u_\beta^{(1)}$ values behave like $1/(\Delta T)^{1/2}$ within intervals $|\Delta\beta|$ that are proportional to $(\Delta T)^{1/2}$.

In order to continue the analysis into time and wavelet positions in the neighborhood of T_l and β_m it is evidently now necessary to abandon the representation $\bar{u} \sim \sigma u^{(1)}(\beta, T)$ and its associated limit process and to use instead a new asymptotic estimate of $\bar{u}$, based on new time and wavelet-position coordinates τ, B. In view of the foregoing analysis these are defined to be

$$T = T_l + \delta\tau, \qquad \beta = \beta_m + \delta^{1/2} B \tag{4.20}$$

where $\delta = \delta(\varepsilon) = o(1)$ is a scale factor that is to be determined.

It is also helpful to define the intermediate coordinates τ_i, B_i via

$$T = T_l + \delta_i \tau_i, \qquad \beta = \beta_m + \delta_i^{1/2} B_i \tag{4.21}$$

$$\delta_i \tau_i = \delta\tau, \qquad \delta_i^{1/2} B_i = \delta^{1/2} B$$

where $(\delta_i, \delta_i/\delta) \to (0, \infty)$ in the $\varepsilon \to 0$ limit. In terms of these variables (4.12) and (4.21) show that $\bar{u}$ is given by

$$\bar{u} \sim -\varepsilon(\gamma - 1)^{-1} \ln[1 - P(\beta_m + \delta_i^{1/2} B_i)(1 - T_l/\hat{T} - \delta_i \tau_i/\hat{T})^{-a}] \tag{4.22}$$

whence, using the intermediate limiting process ($\varepsilon \to 0$ with τ_i, B_i fixed), it follows that in the intermediate domain

$$\bar{u} \sim -\varepsilon(\gamma - 1)^{-1}\{\ln \delta_i + \ln(-\tau_i)$$
$$+ \ln[a(\hat{T} - T_l)^{-1} + \tfrac{1}{2} P_m'' P_m^{-1} B_i^2 \tau_i^{-1}] + O(\delta_i)\} \tag{4.23}$$

A suitable new series for $\bar{u}$ must therefore be

$$\bar{u} \sim -\varepsilon(\gamma - 1)^{-1} \ln \delta + \varepsilon u_e^{(1)}(\tau, B) \tag{4.24}$$

with

$$u_e^{(1)}(\tau \to -\infty, \, B^2\tau^{-1}\text{fixed})$$

$$\to -(\gamma-1)^{-1}\ln(-\tau)-(\gamma-1)^{-1}\ln\left\{\frac{a}{\hat{T}-T_l}+\tfrac{1}{2}\frac{P_m''}{P_m}\frac{B^2}{\tau}\right\} \qquad (4.25)$$

Observe that the matching that leads to relationships (4.24) and (4.25) is independent of the magnitude of δ so long as the latter is $o(1)$. This remark will be significant in what follows. In the present short-pulse situation it can be shown that when T and β are in the ranges implied by (4.20),

$$\xi_\beta \sim 1 - O(\delta^{1/2}\ln\delta) \qquad (4.26)$$

Consequently, *no shock waves will form in the neighborhood of local explosion*, since $\xi_\beta \neq 0$ in the time intervals defined by (4.20), given that there are no shock waves in the pulse anyway [cf remarks following equation (4.10)].

To find new asymptotic series for $\bar{p}$ and $\bar{\rho}$ it is necessary to look for the effect of the intermediate limiting process on the old solutions (4.8), namely $\dot{\bar{p}} \sim \bar{u} \sim \bar{\rho}$. Thus both $\bar{p}$ and $\bar{\rho}$ must behave like $\bar{u}$ in (4.23) in this limit, and suitable estimates for their new representations are evidently

$$\bar{\psi} + \varepsilon(\gamma-1)^{-1}\ln\delta \sim \varepsilon\psi_e^{(1)}(\tau, B), \qquad \psi = p, \rho \qquad (4.27)$$

In terms of the new variables the mass and momentum equations (2.3) and (2.4) become

$$\varepsilon^2 E\xi_\beta\rho_{e\tau}^{(1)}\delta^{-1} + \varepsilon\rho_{eB}^{(1)}\delta^{-1/2} - \varepsilon u_{eB}^{(1)}\delta^{-1/2} + \cdots = 0 \qquad (4.28)$$

$$\varepsilon^2 E\xi_\beta u_{e\tau}^{(1)}\delta^{-1} + \varepsilon u_{eB}^{(1)}\delta^{-1/2} - \varepsilon p_{eB}^{(1)}\delta^{-1/2} + \cdots = 0 \qquad (4.29)$$

where only the dominant piece of each separate left-hand side term is displayed. (Remember that $\sigma\,\mathrm{Re}$ is effectively infinite in this present analysis.)

Clearly a decision about the relative behavior of δ and ε will now influence the character of the new domain (τ, B).

If δ is such that $\varepsilon\delta^{-1/2}$ is $o(1)$ a detailed analysis[14] makes it clear that ε is the only reasonable choice for δ and the new $u_e^{(1)}$, etc., solutions can only match with the original solutions if the latter are extended from $O(\varepsilon)$ to $O(\varepsilon)$-plus-$O(\varepsilon^2)$ terms. The situation is reminiscent of the one encountered in spatially homogeneous explosions (e.g., Kassoy[15]; Dold,[16] Appendix F) and in the spontaneous ignition of initially unmixed reactants [Dold,[16] §4(iii)] and serves only to refine the estimate of local explosion time T_l past the $O(1)$ approximation of equation (4.10).

When δ is of order ε^2 the character of the field described by $u_e^{(1)}$, $p_e^{(1)}$, and $\rho_e^{(1)}$ changes appreciably. It is evidently correct to take

$$\delta = (\varepsilon E)^2 \tag{4.30}$$

and it is noted that a time rate of change at fixed β (and B) values, such as occurs on the left-hand side of Eq. (2.5+), is then such that

$$(\partial f/\partial T)_B = \varepsilon^{-2}(\partial f/\partial \tau)_B = O(\varepsilon^{-2}f)$$

since it is axiomatic that $\partial f/\partial \tau$ is of order f in the domain of validity of the variables τ and B. Thus equations (4.28) and (4.29) become

$$\xi_\beta \rho_{e\tau}^{(1)} + \rho_{eB}^{(1)} - u_{eB}^{(1)} = 0 \tag{4.31}$$

$$\xi_\beta u_{e\tau}^{(1)} + u_{eB}^{(1)} - p_{eB}^{(1)} = 0 \tag{4.32}$$

in the limit as $\varepsilon \to 0$ with τ, B fixed. From (4.4) and (4.27) the exponential terms in equation (2.5+) must now be written as

$$(\varepsilon\gamma)^{-1}[\bar{p}_0^{-1} - (1+\bar{\rho})(\bar{p}_0 + \bar{p})^{-1}] \sim \varepsilon^{-1}(\bar{p}\bar{p}_0^{-1} - \bar{\rho}) \sim -\ln\delta + \gamma p_e^{(1)} - \rho_e^{(1)}$$

and equation (2.5+) now gives

$$p_{e\tau}^{(1)} + u_{e\tau}^{(1)} = (\gamma\hat{T} - \gamma T_i)^{-1}\exp(\gamma p_e^{(1)} - \rho_e^{(1)}) \tag{4.33}$$

The character of the motion near local explosion has changed substantially. In the absence of $u_{e\tau}^{(1)}$ and $\rho_e^{(1)}$ from equation (4.33), that equation mimics the type of relation that describes a spatially homogeneous explosion near the ignition time. That there is therefore a strong element of homogeneous-explosion-like behavior in the present local-ignition phenomenon is clear, but it is equally clear that the true development of local pressures is now intimately linked with the way in which pressure waves with their associated $u_e^{(1)}$ and $\rho_e^{(1)}$ perturbations can carry away the locally rising pressures, due to explosion, at local acoustic speeds. The details of how this proceeds must await the solution of equations (4.31)–(4.33), with matching condition (4.25) for $u_e^{(1)}$ and equivalent requirements for $p_e^{(1)}$ and $\rho_e^{(1)}$.

In the meantime it is quite useful to convert the present equations in the independent variables τ and B into their equivalents in terms of an $\bar{x}$-like coordinate and the time τ. It can be shown with the aid of equations (4.5) and (4.24) that $(\partial f/\partial \tau)_B$ is the same as $(\partial f/\partial \tau)_{\bar{\xi}}$ to first-order accuracy, where $\bar{\xi}$ is equal to $\xi/\varepsilon E$. Subsequently defining $\bar{x}$ so that

$$T - \bar{x} = \varepsilon E\xi = (\varepsilon E)^2\bar{\xi} = \text{constant} + (\varepsilon E)^2(\tau - \bar{x}) \tag{4.34}$$

allows equations (4.32) and (4.33) to be transformed into

$$\rho_{e\tau}^{(1)} + u_{e\bar{x}}^{(1)} = 0 \tag{4.35}$$

$$u_{e\tau}^{(1)} + p_{e\bar{x}}^{(1)} = 0 \tag{4.36}$$

It then quickly follows that equation (4.33) is the same as

$$p_{e\tau}^{(1)} - \rho_{e\tau}^{(1)} = (\gamma\hat{T} - \gamma T_i)^{-1} \exp(\gamma p_e^{(1)} - \rho_e^{(1)}) \tag{4.37}$$

(The τ-derivatives in the last three equations are all taken at fixed $\bar{x}$.)

The model of physical behavior near the local explosive event, in the strict sense implied by (4.20) with δ equal to $(\varepsilon E)^2$, is perhaps more clearly revealed by equations (4.35)–(4.37) than by the equations in their original forms (4.31)–(4.33). The first two equations, (4.35) and (4.36), are the simple linearized continuity and momentum equations. As such they describe acoustic-like wave propagation once the (p, ρ), or specifically the $(p_e^{(1)}, \rho_e^{(1)})$, "compressibility" link has been ascertained. The latter must follow the form laid down for it by equations (4.37), which shows that the requisite (p, ρ) behavior is strongly governed by the rapidly-developing chemical events that take place near local explosion, and whose presence is manifested in equation (4.37) by the exponential term on its right-hand side.

It can be seen that equation (4.6), with $\mathrm{Re} = \infty$, and equation (4.7) combine to give

$$-\tfrac{1}{2}\bar{Q}q_{\beta}^{(1)} = u_T^{(1)}\xi_\beta \tag{4.38}$$

Assuming that the pulse is shock-free, so that its head is at $\beta = 0$,

$$-\tfrac{1}{2}\bar{Q}q^{(1)}(\beta, T) = \int_0^\beta u_T^{(1)}(T, \tilde{\beta})\xi_{\tilde{\beta}}d\tilde{\beta} \tag{4.39}$$

and it has already been stated that ξ_β is well-behaved in the present circumstances with magnitude equal to about one. In range $\delta^{1/2}$ of β_m, $u_T^{(1)}$ has a magnitude $O(1/\delta)$; there is therefore a contribution to $q^{(1)}$ in this neighborhood from the integral in equation (4.39) that is $O(\delta^{-1/2})$ or, in view of (4.30), $O(\varepsilon^{-1})$. It follows that the perturbation $\bar{q}$ to the nonequilibrium variable must be $O(\varepsilon)$, and *not* $O(\varepsilon^2)$, in the neighborhood of local explosion. The situation is reminiscent of the conditions found near breakdown of unidirectional wave propagation when $E = o(1)$,[6] and indicates that reactant consumption begins to assume a first-order role as local ignition develops.

Numerical solutions of equations (4.35)–(4.37) for a set of circumstances different from those demanded by the matching (initial) condition (4.25) show rather nicely how fluid elements begin to react at something between constant-volume and constant-pressure conditions, thereby generating pressure signals that propagate out into the field to affect the progress of reaction in neighboring elements (private communication from R. S. Cant). Similar general behavior may be anticipated in the present situation and it is hoped to report on the next stage of the work in due course.

REFERENCES

1. L. Crocco and W. A. Sirignano, A shock wave model of unstable rocket combustors, *AIAA J.* **2**, 1285–1296 (1964).
2. L. Crocco, Coordinate perturbation and multiple scale in gas dynamics, *Philos. Trans. R. Soc. London, Ser. A* **272**, 275–301 (1972).
3. A. H. Nayfeh, *Perturbation Methods*, John Wiley and Sons, New York (1973).
4. G. B. Whitham, *Linear and Nonlinear Waves*, John Wiley and Sons, New York (1974).
5. M. J. Lighthill, Viscosity effects in sound waves of finite amplitude, in: *Surveys in Mechanics* (G. K. Batchelor and R. M. Davies eds.), pp. 250–351, Cambridge University Press (1956).
6. J. F. Clarke, A generalised Burgers equation for plane waves in a combustible atmosphere, *Trans. Twenty-Seventh Conf. of Army Mathematicians*, pp. 551–562, U.S. Army ARO Report 82–1 (1982).
7. T.-Y. Toong, Chemical effects on sound propagation, *Combust. Flame* **18**, 207–216 (1972).
8. C. A. Garris, T.-Y. Toong, and J.-P. Patureau, Chemi-acoustic instability structure in irreversible reacting systems, *Acta Astronaut.* **2**, 981–997 (1975).
9. R. G. Gilbert, P. J. Ortoleva, and J. Ross, Nonequilibrium relaxation methods. Acoustic effects in transient chemical reactions, *J. Chem. Phys.* **58**, 3625–3633 (1973).
10. J. F. Clarke, Small amplitude disturbances in an exploding atmosphere, *J. Fluid Mech.* **89**, 343–355 (1978).
11. J. F. Clarke, On the evolution of compression pulses in an exploding atmosphere; initial behaviour, *J. Fluid Mech.* **94**, 195–208 (1979).
12. G. E. Abousieff and T.-Y. Toong, Non-linear wave-kinetic interactions in irreversibly reacting media, *J. Fluid Mech.* **103**, 1–31 (1981).
13. P. A. Blythe, Wave propagation and ignition in a combustible mixture, in: *Seventeenth Symp. (Int.) on Combust.*, pp. 909–916, The Combustion Institute, Pittsburgh, PA (1978).
14. J. F. Clarke, Plane waves in reacting gases. Pt. II, waves in explosive atmospheres. Sonderforschungsbereich 27, RWTH, Aachen, W. Germany; lecture notes. Also College of Aeronautics Memo 8106, Cranfield Institute of Technology (1981).
15. D. R. Kassoy, Extremely rapid transient phenomena in combustion, ignition and explosion, *SIAM-AMS Proc.* **10**, 61–75 (1976).
16. J. W. Dold, On the Autoignition and Combustion of a Finite Region of Gaseous Fuel, Ph.D. Thesis, Cranfield Institute of Technology (1979).

17

Physical and Chemical Kinetic Effects in Soot Formation

I. GLASSMAN, K. BREZINSKY, A. GOMEZ, AND F. TAKAHASHI

ABSTRACT. Evidence indicates that the fuel pyrolysis rate and the particular burn-up time are the controlling factors in soot formation in practical combustion systems. Under premixed flame conditions, the rate of increase with temperature of fuel pyrolysis to form the soot precursors is slower than the rate of increase of the oxidative attack on the pyrolysis products. Thus increasing the flame temperature decreases the tendency to soot. Chemical kinetic and sooting equivalence ratio data will be presented to support this postulate for aliphatic fuels and to show that previous lists categorizing the tendency of fuels to soot are misleading. New chemical kinetic evidence shows that aromatic fuels may not necessarily follow this temperature trend and that the great tendency of aromatic fuels to soot may be due not only to the initial fuel structure but also to the intermediates formed during the breakdown of the ring. Since there is no oxidative attack of the fuel in a diffusion flame, sooting tendency increases with increasing flame temperature. Smoke-point data as a function of temperature reveal the importance of initial fuel structure and fuel pyrolysis kinetics.

1. INTRODUCTION

Accumulated experimental evidence suggests the possibility that the rates and mechanisms of the pure and oxidative pyrolysis of the fuel may be the controlling factors in determining the amount of soot formed in practical combustion systems. Obviously the rate of particle oxidation is

I. GLASSMAN, K. BREZINSKY, A. GOMEZ, AND F. TAKAHASHI • Department of Mechanical and Aerospace Engineering, Princeton University, Princeton, NJ 08544, U.S.A.

345

another important element in controlling the soot emitted from such practical systems. The importance of fuel pyrolysis indicates the necessity, as well, of understanding the basic chemical-kinetics processes involved. The work reported here seeks to show certain interrelationships between flame studies of sooting tendencies and research on oxidation kinetics of fuels.

2. SOOT PROCESSES

Numerous reviews[1-8] detail the many complexities involved in determining the extent of soot particle emission from any combustion system. Many individual processes occur in sequential and overlapping steps: initial pure or oxidative pyrolysis of the fuel, creation of a precursor mono-element, subsequent formation of key precursors, nucleation reactions, absorption of other high-molecular-weight compounds on the condensed phase, dehydrogenation of the virgin soot particles, cyclization and dehydrogenation of the absorbed hydrocarbons, agglomeration, conglomeration, thermal ionization of the small particles, particle repulsion, and finally oxidation at every stage above and oxidation of the final aged particle.

A general consensus with regard to the complexities in this field has been expressed by Bittner and Howard,[5] who in a brief statement in their article comment: "A fact that has troubled researchers looking for a single unifying mechanism to explain soot formation (*although it is not likely that such a mechansim exists*) has been . . . ." That many investigators simply do not believe that soot formation is a unique process for all fuels burning in flames is succinctly stated by Cullis *et al.*,[9] . . . "that there is no single mechanism for the formation of carbon, and it is probable that many different radicals are involved in these reactions." From a global point of view, given the essential C/H ratio differences of virgin soot particles and the fact that there is a wide discrepancy in the tendency to soot among various fuels in diffusion flames, in the strictest sense one would have to agree with Cullis *et al.*[9] But these authors and others were seeking an understanding of the detailed and precise steps that constitute the entire soot-formation process without endeavoring to analyze whether there were controlling mechanisms in the various types of flames which would predict the quantitative yield of soot and give some insight into the relative tendency of various fuels to soot. It is rather remarkable, as Palmer and Cullis[1] point out, that: "With diffusion flames and premixed flames investigations have been made both of the properties of the carbon formed and of the extent of carbon formation under various conditions. In general, however, the properties of the carbons formed in flames are remarkably little affected by the type of flame, the nature of the fuel being burnt and the other conditions under which they are produced. Any complete and comprehensive theory of carbon formation must, of course, be able to account for this striking experimental finding." That the properties of the carbons formed are the same may be due to the fact that, as is now known,[10] the C/H ratio of virgin soot particles decreases as the particles pass through the hot burned gases of a combustor. It is possible, of course, that the similarities referred to by Palmer and Cullis[1] come about from the later pyrolytic dehydrogenation in the hot burned gases of a flame.

Influenced by these insights, sooting tendency in flames,[4] and some ongoing work on the pyrolysis and oxidation of hydrocarbons, we proposed earlier[4] that fuel pyrolysis and particle oxidation were the controlling steps in soot emission from practical combustion systems.

A key element that has intrigued investigators in the soot field is the rapidity with which soot particles form in flames (essentially of the order of 1 ms or less). An early soot model[6] was that the precursors are acetylenic in character, that these acetylenics polymerize and go through a cyclization process to form the virgin soot particle elements. This mechanism relies heavily on free radical reactions. However, as Calcote[8] points out, there appears to be some difficulty in accounting for the rapid production of soot by reasonable free radical reaction rates,[11] some problems in explaining the rearrangement of CH chains to produce cyclic aromatic hydrocarbons,[12] and some thermodynamic constraints on the growth of aromatics from radicals.[13] In order to circumvent these possible limitations to free radical mechanisms for soot formation, Calcote[8] has led a school of thought that ions in flames permit a series of much faster ion-neutral reactions that account for the rapid soot production under appropriate stoichiometric conditions. Homann[14] earlier questioned whether ions can be the complete story since he argued that the concentration of ions in the absence of a field is several orders of magnitude lower than the concentration of hydrocarbon radicals, and that, if ions were important, the number of particles in the flames of rich mixtures should decrease as well as the ion concentration, but that this is contrary to observed facts. There is some question whether Homann's second point was valid since recent results show ony a slight decay in ion concentration as one proceeds from stoichiometric to rich mixtures. Nor, in our point of view, does one have to accept the concept that all persisting hydrocarbon radicals be involved in the nucleation step. They could add to surface growth, be oxidized in later stages of the flame, or suffer other fates not contributing to soot process. There are still some open questions with respect to the role of ions in soot formation, especially in regard to tendencies in sooting diffusion flames.

Realizing that any soot precursor and its subsequently formed larger elements must be conjugated to withstand the high temperatures that exist in flames, it had been proposed earlier[4] that butadiene, diacetylene, methyl and vinyl acetylene could be important candidates for soot precursors. Since it is extremely difficult to form the polynuclear aromatic structure that makes up soot from phenyl radicals alone, it is very apparent, as Bittner and Howard[5] have pointed out, that aromatic fragments such as vinyl acetylene are necessary intermediates in soot formation, even from aromatic fuels. The attractiveness of butadiene and vinyl acetylene is that they undergo relatively rapid Diels–Alder reactions that make the formation of polynuclear species quite structurally feasible. Whether these reaction rates are of the order required for the rapid nucleation reactions may be open to question due to the small A factors in the rate term. Butadiene, and possibly vinyl acetylene, form strongly stabilized ions,[15] which themselves may have much faster Diels–Alder-type reactions.

Thus it appeared that fuels such as butadiene and vinyl acetylene and their Diels–Alder-type product, styrene, should be prolific sooters. Initial evidence was that this speculation was valid. Schalla and co-workers[16,17] earlier had studied the sooting tendency of various fuels in laminar diffusion flames. Their results are shown in Figure 1. Various important observations can be made from this figure. First, 1,3 butadiene appeared to be the most prolific sooter of any fuel tested, including aromatics. But, perhaps more important, for the olefin series, the C_4 compounds had the greatest tendency to soot. Interesting also is that the alkylated aromatics showed a greater tendency to soot than did benzene. Consideration will be given to these points later, particularly with regard to previous[18–21] and new experimentation performed at Princeton.

3. THE USE OF FLAMES IN SOOT ANALYSES

The basic literature for soot formation in flames has evolved in a somewhat historical manner. The earliest work developed procedures for measuring the tendency of fuels to soot and the effect of additives on this tendency. As chemical analytical equipment developed, work began on actual soot mechanisms to explain the general flame observations. Laser developments then permitted analysis of the particle-growth problem.[22–24] The pioneering work of Weinberg and Calcote on ion effects has finally taken hold and many recent research efforts are concentrating on the role of ions. As implied earlier, these efforts have been initiated in an attempt to elucidate further the soot nucleation process. Some of these studies[24] are also endeavoring to explain the interesting effect of certain metals (particularly the alkaline earths) on the general sooting tendency of a fuel.

The tendency of fuels to soot has been measured by various means: laminar premixed flames,[25] laminar diffusion flames,[26,27] turbulent diffu-

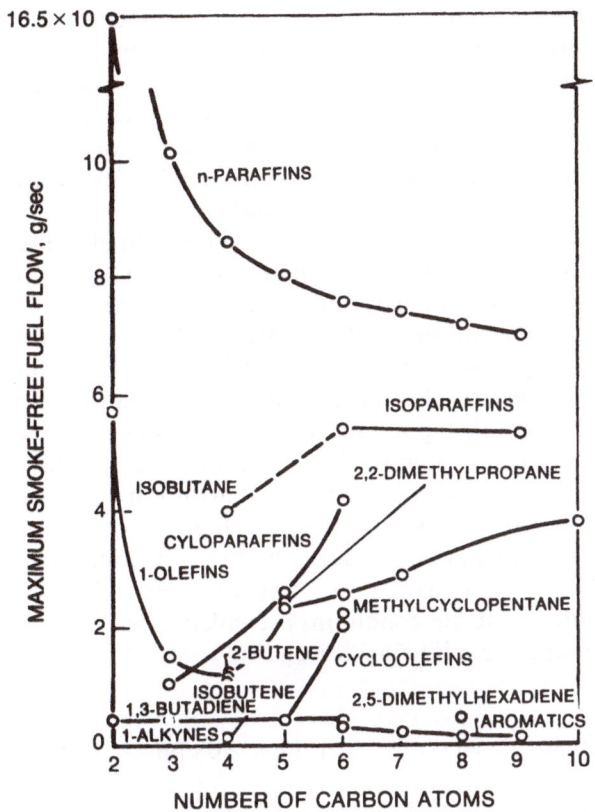

Fig. 1. Variation of maximum smoke-fire fuel flow with number of carbon atoms (after Schalla and Hibbard[17]).

sion flames,[28] and stirred reactors.[29,30] As discussed earlier, the amount of soot that will emerge from a particular combustion process is a result of a series of sequential and overlapping steps. It is well to note, however, that in laminar flame processes the spacial distribution of the steps is such that they are essentially noninterfering (nonoverlapping). Thus from a given laminar experiment one is more likely to determine which step is affected by a change in a given physical or chemical variable. In turbulent diffusion flames and stirred reactors the processes are mixed and although these types of experiment probably approach more realistic practical conditions, it is more difficult to compare results of various investigators because of the lack of detailed turbulence and degree-of-mixing measurements. The fact that stirred reactors give the same tendency to soot trends as do premixed laminar flames,[4,30] but somewhat different quantitative trends, points to the dominance of a step or steps that are not affected by back mixing. Indeed, the steps must be related to the fuel pyrolysis.

In premixed flames the tendency to soot is correlated with the equivalence ratio at which noticeable sooting just begins (luminous, continuum radiation). When the equivalence ratio is defined as the actual fuel–air ratio to the stoichiometric value, the smaller the equivalence ratio at the sooting point, the greater the tendency to soot.[25,31] In diffusion flames the tendency to soot is measured by the height of the fuel jet[26,27] or the mass flow of fuel[17] at which the luminous flame breaks open near its apex and emits a stream of particulates. The smaller the flame height at the breakthrough (called sooting height), or the smaller the mass flow rate, the greater the tendency to soot. Schug et al.[18] have shown that the flame height or the volumetric flow rate of the fuel is the more fundamental parameter. However, they point out that if all experiments are carried out at the same initial fuel temperature, the mass flow rate allows more convenient, overall correlations and essentially normalizes the results with respect to number of carbon atoms in the fuel.

The sooting tendency as measured by the techniques described above shows very different effects for fuel structure depending upon whether premixed or diffusion flame conditions prevailed. For premixed flames, the literature[25] reports the following descending tendency to soot according to fuel type:

aromatics > alcohols > paraffins > olefins > acetylene

For diffusion flames, the order reported[17] is

aromatics > acetylenes > olefins > paraffins > alcohols

Unfortunately, reporting such trends is misleading in understanding how fuel structure affects sooting tendency. The essential reason is that the flame temperature plays an extremely important role.

In this regard the early work of Milliken on premixed ethylene–air flames is particularly worth considering. Using a cooled flat flame burner Milliken[32] showed that, as the ethylene flame temperature was lowered, the flame emanated soot at a lower equivalence ratio, that is, it had a greater tendency to soot. Milliken proposed that it was the fate of the acetylene formed that determined the tendency to soot. He showed that although the rate of pyrolysis of acetylene increased with temperature, the rate of hydroxyl attack increased even faster with temperature. Thus the higher the premixed flame temperature the less the tendency to soot. Since there is no oxidative attack in diffusion flames, the higher the flame temperature, the greater must be the tendency to soot.

4. SOOTING TENDENCIES IN PREMIXED FLAMES

Although Milliken's experiments[32] were for only one fuel, his concept is applicable to all aliphatic fuels. In flow-reactor studies of hydrocarbon oxidation at Princeton,[33] it has been shown that all aliphatics break down (are pyrolyzed) to the three lower olefins: ethene, propene, and butylene. As one would expect, the most prominent species found is ethene. Figure 2 shows the product distribution in this flow reactor for the oxidation of octane. These results were from another study[34] and there were no fuel-rich runs made. However earlier work[33] with the rich ethene oxidation revealed that acetylene does form (Figure 3) as one would expect and as Milliken had hypothesized. Thus, if all the paraffins pyrolyze to ethene and then to acetylene, it is surprising (and misleading) to state that in premixed flames paraffins have a greater tendency to soot than the olefins,

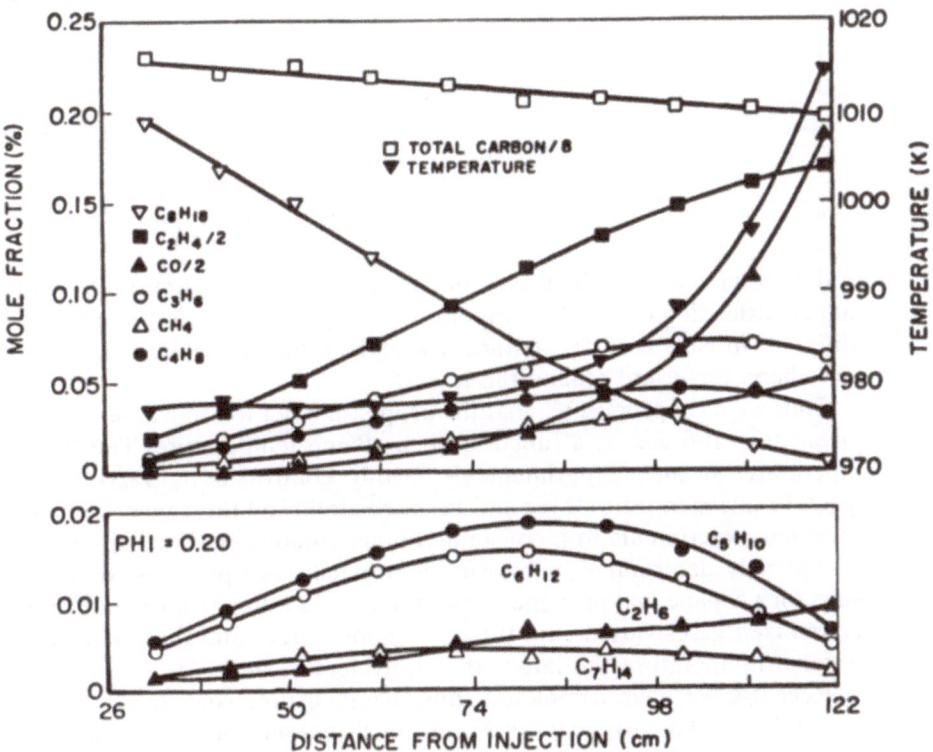

Fig. 2. Flow reactor specie concentration distribution for the oxidation of octane.

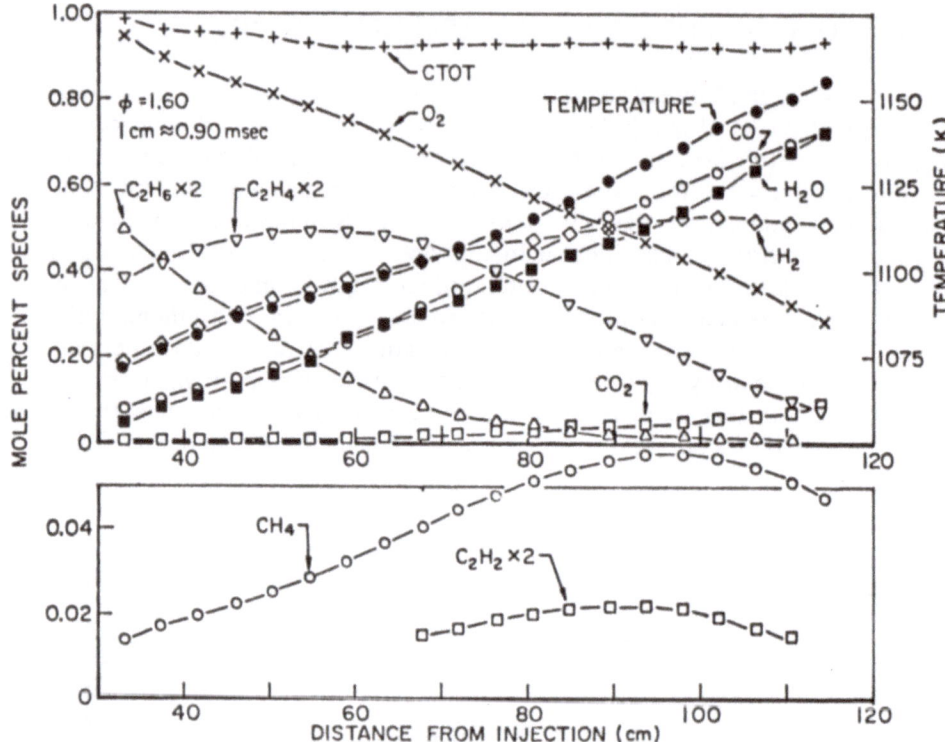

Fig. 3. Flow reactor specie concentration distribution for rich ethene oxidation.

and the acetylenes have the least tendency to soot. The reason this type of result is misleading is that the temperature at the sooting equivalence ratio is different in each case and naturally acetylene has the highest temperature, ethene next, and ethane the lowest.

Thus we have proposed[4] that the proper manner to report the sooting equivalence ratio was as a function of the flame temperature. The flame temperature in such experiments is readily controlled by varying the oxygen-to-nitrogen ratio. Dyer and Flower[35] followed this suggestion and performed experiments in a constant-volume combustion bomb in which they optically determined luminosity as a flame wave propagated. Their results for propane and propene showed that the sooting equivalence ratio for premixed gases varied with the flame temperature and that the higher the flame temperature the larger the equivalence ratio.

Recently we have completed some detailed premixed flame studies very similar to the early extensive work of Street an Thomas,[25] but with the important difference that the flame temperature for each fuel was varied. The experimental configuration was essentially the same as Street and Thomas: a long flow tube through which the premixed fuel, oxygen, and nitrogen mixture passed. The fuel–oxygen equivalence ratio was varied until flame luminosity was visually observed. This equivalence ratio (ϕ) was then reported as a function of the calculated adiabatic flame temperature (T_f) of the complete mixture. The results for 10 different fuels are shown in Figure 4 in which ϕ is plotted versus T_f. It is important to note that the ϕ plotted is the actual fuel–oxygen ratio divided by the stoichiometric fuel–oxygen ratio for all the carbon and hydrogen in the fuel being converted to carbon dioxide and water, respectively.

There are many interesting observations to be made by examining Figure 4. First note that the cross point on every fuel line represents the results for air. One thus observes that, for air, acetylene has the least tendency to soot, ethylene next, and then ethane, exactly the same as the

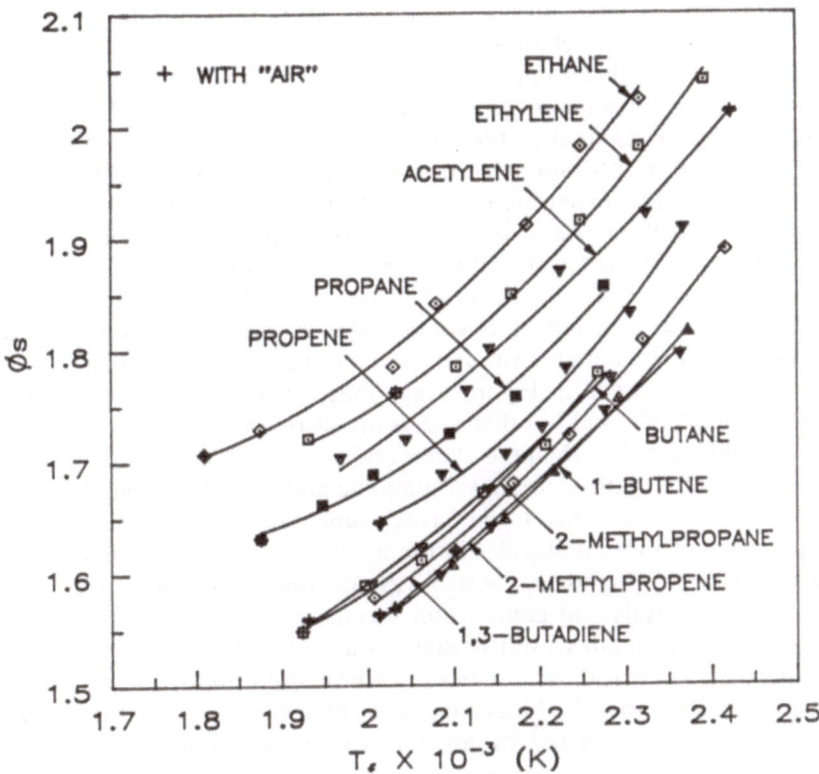

Fig. 4. Sooting equivalence ratios for various fuels in premixed flames as a function of temperature. Equivalence ratio based on carbon conversion to CO_2.

early trend reported.[25] Indeed all the cross data are in good agreement with the results of Street and Thomas.[25] However, notice that for a fixed temperature, acetylene has the smallest ϕ, for C_2 hydrocarbons. Thus as one would expect from the understanding of the kinetic mechanism, given a temperature, acetylene has the greatest tendency to soot of the C_2 hydrocarbons. Indeed Figure 4 shows as well that for the C_2, C_3, C_4 hydrocarbons the olefins have a greater tendency to soot than the paraffins. What is unusual about these results is that the C_2s appear to have a lesser tendency to soot that the C_3s and the C_4s. It would appear that the size of the fuel molecule plays a role and certainly this trend is inconsistent with oxidation kinetic results. Here again one must realize the fundamental kinetic realities of the process. All data obtained are in the very fuel-rich region; thus normalizing the data on the basis of an equivalence ratio where stoichiometric is defined as the carbon going to CO_2 is unrealistic. Kinetically in most cases the H must be abstracted by an oxygen species and it is somewhat realistic to assume that each two hydrogens will consume one oxygen atom. A more realistic normalization procedure would be to use a stoichiometric relation in which the carbon is converted to carbon monoxide and the hydrogen to water.[4] Figure 5 is a replot of the data in Figure 4 in a manner in which the equivalence ratio is based on C to CO and H to H_2O; that is, the term in the ordinate of Figure 5 written as $[C + (H/2)]$ is the stoichiometric oxygen requirement on this new basis, and the O is the actual experimental value at the sooting point. Simple analysis will show that the normal equivalence ratio (Figure 4) written in the same manner would be $[2C + (H/2)]/O$. It is kinetically unrealistic[4,6] to plot the data such that the ordinate would be simply $[C/O]$. In premixed flames the hydrogen in the fuel consumes some of the oxygen and it is for this reason that experimental data for air mixtures do not correlate[4,6] with $(C/O) = 1$.

The data trends in Figure 5 are much more appealing. The trends among the C_2, C_3, and C_4s themselves are as one would expect. In general the paraffins soot less than the olefins and acetylene. More importantly, the data indicate that the C_4 olefins, butadiene and acetylene have the greatest tendency to soot. This trend gives some credence to the postulated importance of these molecules in the nucleation process discussed earlier. Comparison of Figures 4 and 5 reveals the importance of the basic oxidative pyrolysis and combustion mechanisms of the fuel. The importance of the ordinates used is that, to a first order, they normalize the results with respect to the number of carbon atoms in the fuel. However, it is also apparent that the kinetic mechanism also prevails. For paraffins the hydrogens are abstracted by an oxidative specie. For the multibonded compounds, the oxidative species can add directly to the multiple bond[36,37] and thus not all the hydrogens must be abstracted initially by an oxidizing specie. Thus, in reality, due to kinetics there is no totally correct standard

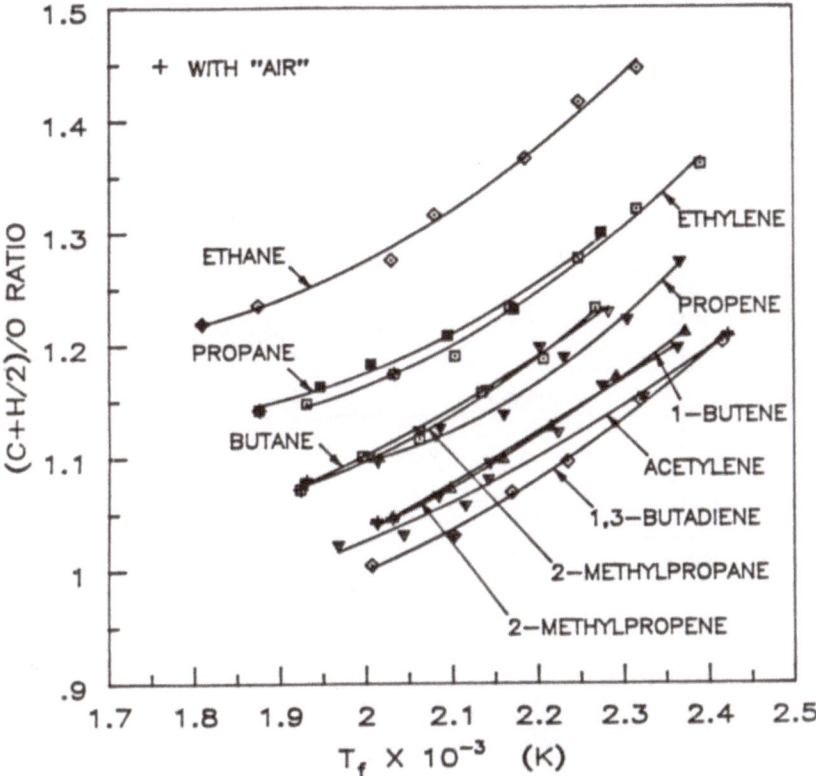

Fig. 5. Sooting equivalence ratios for various fuels in premixed flames as a function of temperature. Equivalence ratio based on carbon conversion to CO.

normalization procedure based on an equivalence ratio for the data presented in Figures 4 and 5.

The major point above is that the combustion temperature may be one of the most significant parameters in sooting premixed systems and that the higher the temperature the less the tendency to soot. The conclusion, however, holds only for aliphatic fuels. The aromatics have adiabatic flame temperatures close to the olefins,[4] yet they have a far greater propensity to soot than do all aliphatics. It is very apparent then that their basic mechanisms leading to the soot-nucleation process must be very much different than that of the aliphatics. Recent results on the Princeton flow reactor confirm this reasoning.[38,39]

Plotted in Figures 6 and 7 are flow reactor results for benzene and toluene. A detailed discussion of these results is reported elsewhere.[38,39] The important aspects represented in these figures and other data for ethyl benzene[38] are that the alkylated aromatics proceed through a series of

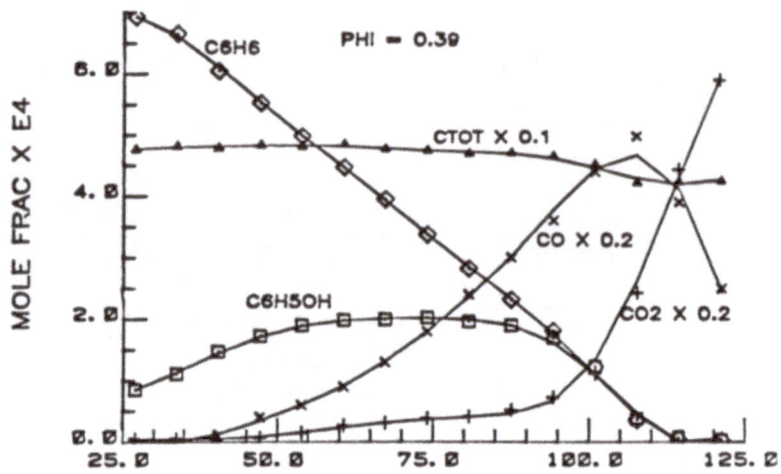

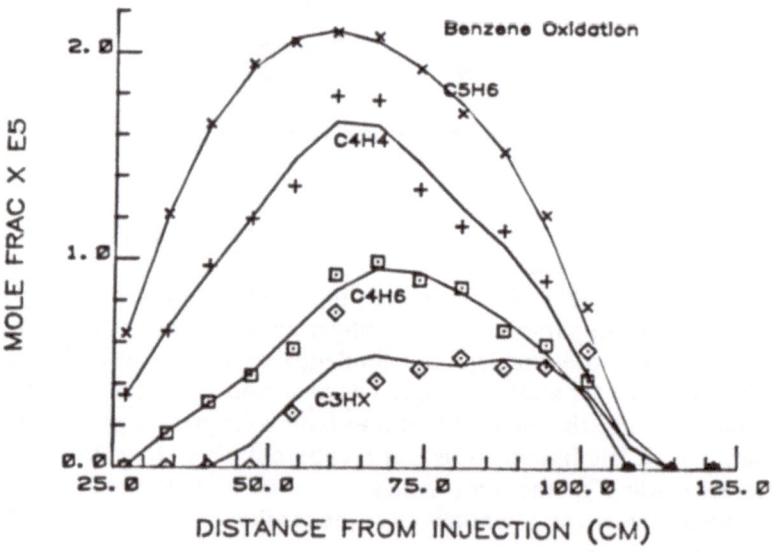

Fig. 6. Flow reactor specie concentration distribution for the oxidation of benzene.

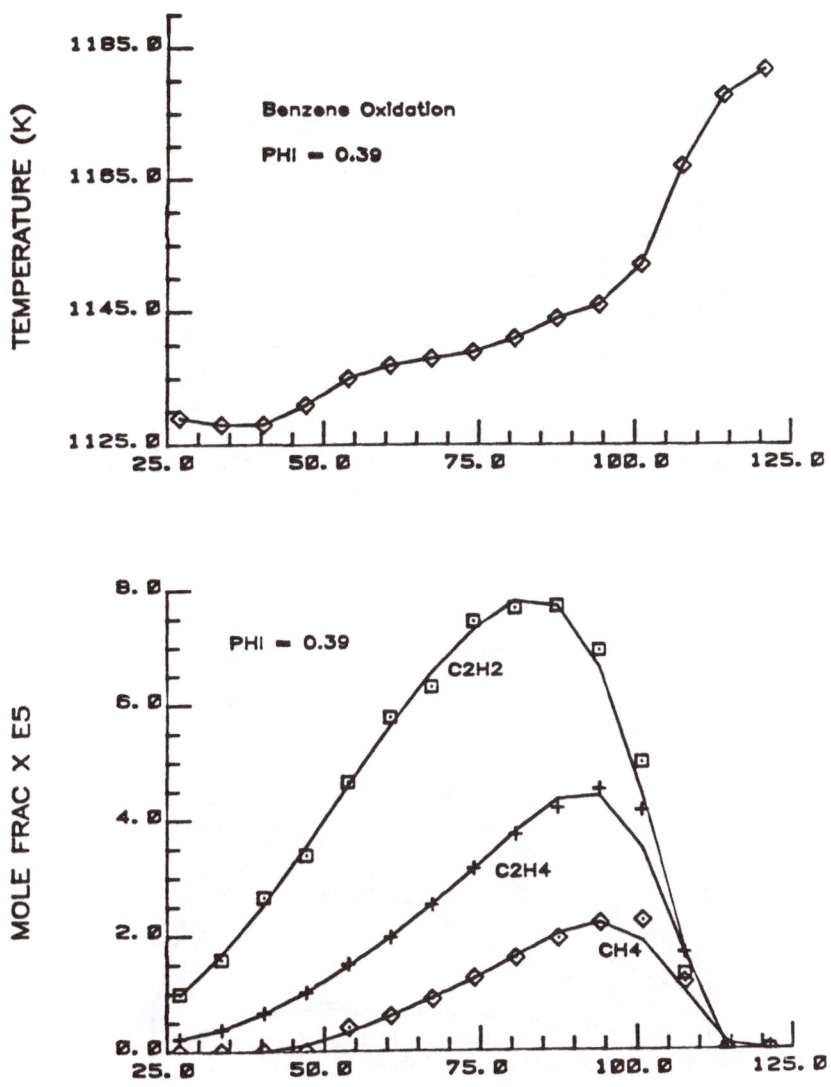

Fig. 6 (cont.).

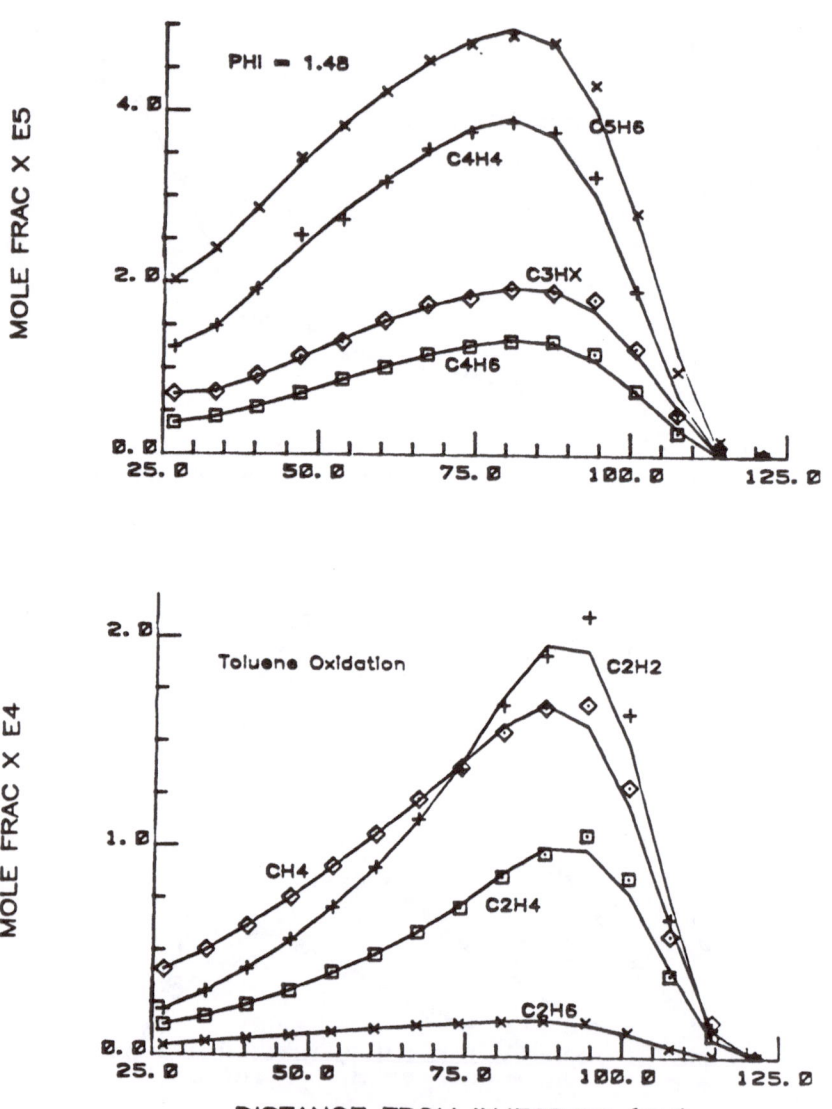

Fig. 7. Flow reactor specie concentration distribution for the oxidation of toluene.

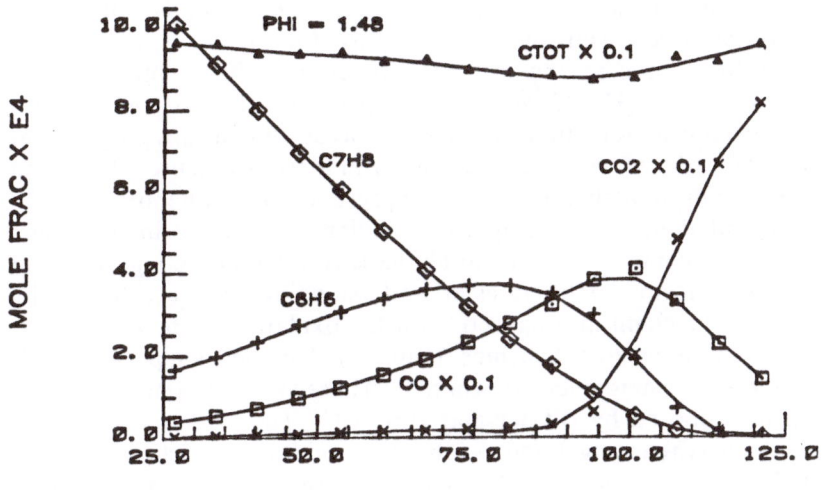

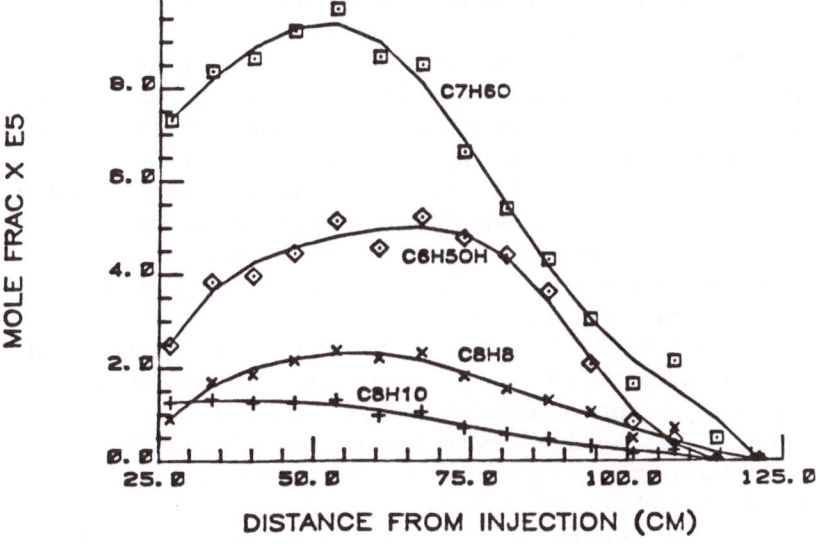

DISTANCE FROM INJECTION (CM)

Fig. 7 (cont.).

interesting and understandable steps, until the phenyl radical is formed. Then the mechanisms for benzene and its alkylated partners are practically the same. Analyses in the flow reactor are made by quenching samples and performing gas chromatography and GC/MS. Consequently only stable species are measured. Thus the species reported in Figures 6 and 7 can be the stable form of some radical present. It was the identification of cyclopentadiene by GC/MS analysis that gave the clue to postulating a general mechanistic trend in the aromatic process. For various purposes, at this point the order in which species appear in these two figures starts with phenol (which most likely arises from the presence of the phenoxy radical), then cyclopentadiene, vinyl acetylene, butadiene, acetylene, and ethylene. CO forms early compared to similar aliphatic runs (compare Figures 2 and 5) and CO_2 is formed late. A clear understanding can be achieved by discussing the mechanism proposed[39] for the oxidation of the phenyl part of aromatic systems (see Schemes 1 and 2). Detailed support for this proposed mechanism is given elsewhere.[39] There are many similarities to a mechanism proposed by Bittner and Howard.[5] The essential points are that the phenoxy rearranges and expels a CO and forms a cyclopentadienyl radical. This radical reacts with O_2 in a similar fashion to phenoxy to form an intermediate, which also expels CO and forms the butadienyl radical. The early formation of CO is thus explained, but most important for the discussion here is that the fate of the butadienyl radical is to form vinyl acetylene, butadiene, and acetylene in the shown relative proportions, which were found to be consistent with estimated rate constants.[39] Bittner and Howard[5] have postulated the necessity of the presence of vinyl acetylene for the soot-forming process from aromatics. Our[4] earlier emphasis also was that the C_4s vinyl acetylene and butadiene could be essential precursors in soot-nucleation process for aliphatics, as well as aromatics. Thus one can conclude that the great propensity of aromatic fuels to soot arises not so much from the initial fuel structure, but due to the fact that, at least in premixed flames, vinyl acetylene and butadiene intermediates form as the ring is broken.

Since the mechanism for aromatics to form the precursors is quite different than that of the aliphatics, it is not surprising that the aromatic sooting tendency is not consistent with the temperature trend discussed earlier. Indeed, the possibility exists that the rate of formation of the precursors from aromatics may rise faster with temperature than the rate of oxidative attack, and thus for aromatic premixed combustion systems the higher the flame temperature the greater the propensity to soot. The effect of temperature on the sooting tendency of aromatics is currently under experimental investigation at Princeton.*

---

* Recent results change some of the concepts in the preceding paragraphs, see *Combust. Sci. Technol.* **37**, 1 (1984).

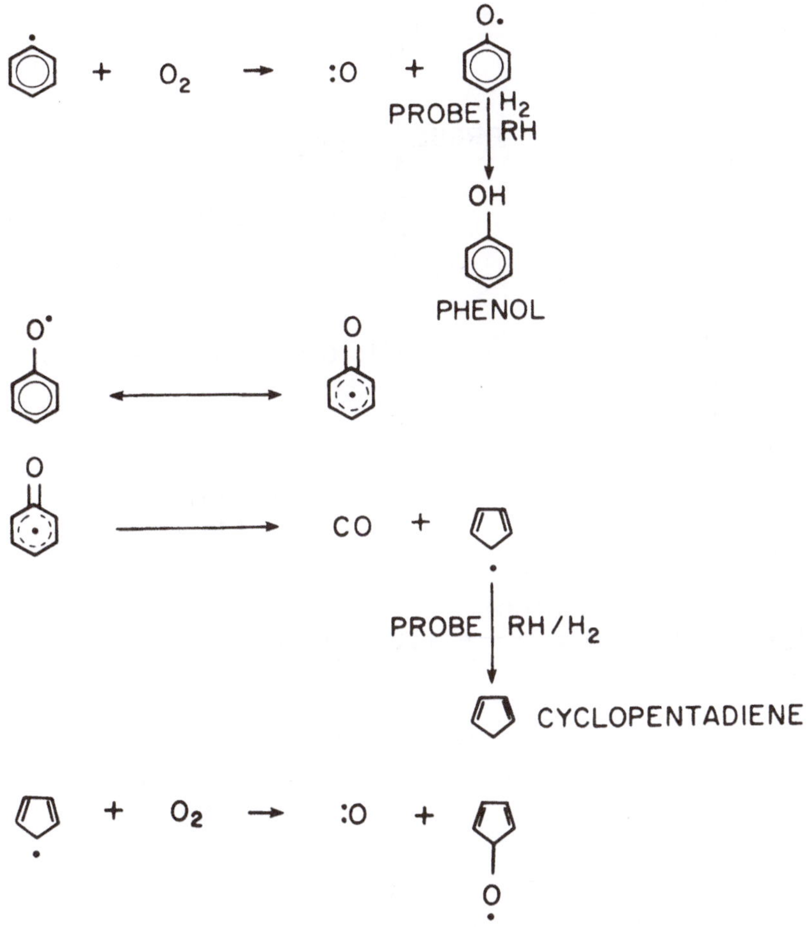

Scheme 1.

5. SOOTING TENDENCIES IN DIFFUSION FLAMES

In diffusion flames it is quite well established that soot formation takes place on the fuel side of the flame front.[40–42] Since there is no oxidative species at the position where the soot forms, then following the initial concept of Milliken for premixed flames, as the flame temperature increases the tendency of a fuel to soot also increases. Prior to reporting the new results on nonaliphatic compounds (the cyclohexadienes and aromatics), a brief review of our earlier work[18–20] on sooting diffusion flames will be given.

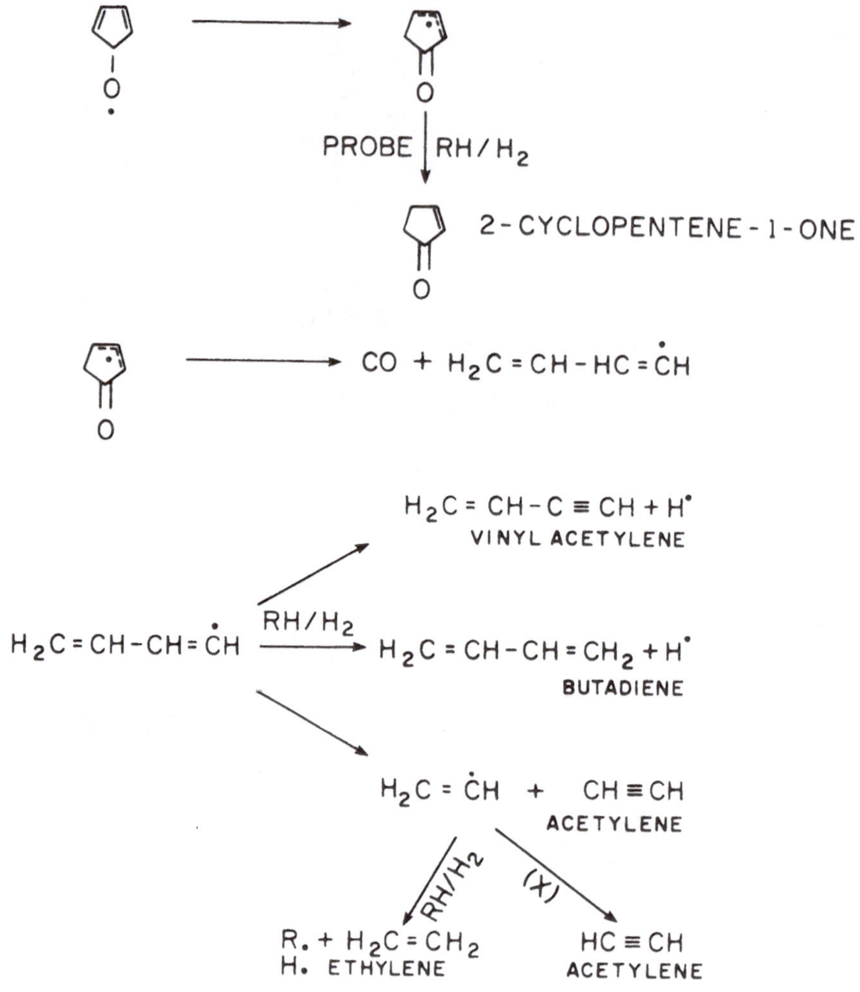
PROBE
RH/H2
2-CYCLOPENTENE-1-ONE
CO + H2C=CH-HC=ĊH
H2C=CH-C≡CH + H•
VINYL ACETYLENE
H2C=CH-CH=ĊH
RH/H2
H2C=CH-CH=CH2 + H•
BUTADIENE
H2C=ĊH + CH≡CH
ACETYLENE
RH/H2
(X)
R. + H2C=CH2
H. ETHYLENE
HC≡CH
ACETYLENE

Scheme 2.

The general procedure for measuring the tendency to soot under diffusion flame conditions is to surround a fuel jet with a concentric air stream. For consistent results it is necessary to operate this system in a highly overventilated hood.[16,19,43] The fuel flow rate is increased until soot particles break through the continuous luminous flame front. The results are reported either as the height at which breakthrough occurs (the smoke point) or the mass flow rate corresponding to this sooting height. In order to determine the effect of the flame temperature, inert diluents are added to the fuel jet. Once a smoke point is reached, inerts are added, and the mixture flow rate must be increased before a new smoke point is reached. An unresolved question is why the smoke point arises as the fuel mixture rate is increased. Since the fuel port size does not change, an increase in the mass flow rate is simply an increase in the velocity. If the velocity increases linearly with mass flow rate, then the average stay time of a fluid element in the jet from exit to the flame front must be the same regardless of height. Thus the fuel pyrolysis time remains the same. We have proposed[19] that the sooting increased as the fuel flow rate increased due to the induction of oxygen (air) into the fuel jet at the exit lip of the fuel tube. The greater the fuel velocity, the more oxygen (air) is induced. Smith and Gordon[44] had found oxygen on the fuel side of a laminar diffusion flame system. It is known[44,45] that small amounts of oxygen increase fuel pyrolysis rates. In fact Figure 8 taken from the paper by Schug et al.[18] shows the effect of adding small amounts of pyrolysis accelerators to an ethylene fuel jet. The ordinate is the ethylene flow rate at the smoke point. Figure 9, also taken from Schug et al.,[18] is a plot of smoke points and normal flame heights versus the volumetric flow rate of the ethylene fuel. The fact that all data points fall on a straight line confirms the simple result of diffusion-flame theory[46] that the flame height is strictly proportional to the volumetric flow rate of the fuel alone, even in fuel-inert mixtures. Further, these data show

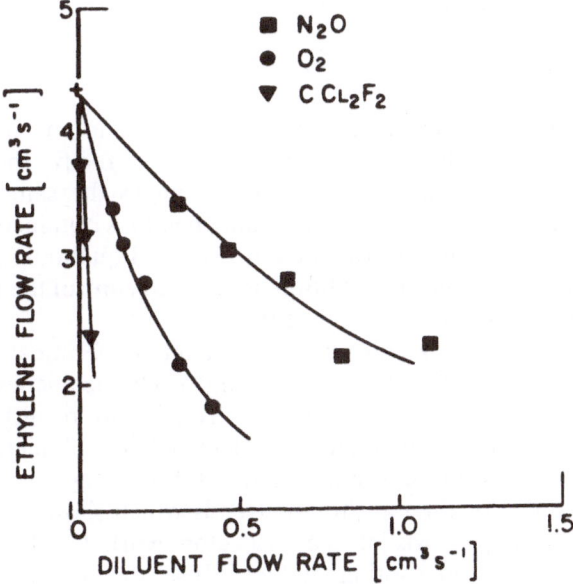

Fig. 8. The effect of reactive diluents on the sooting volumetric flow rate of ethylene in laminar diffusion flames.

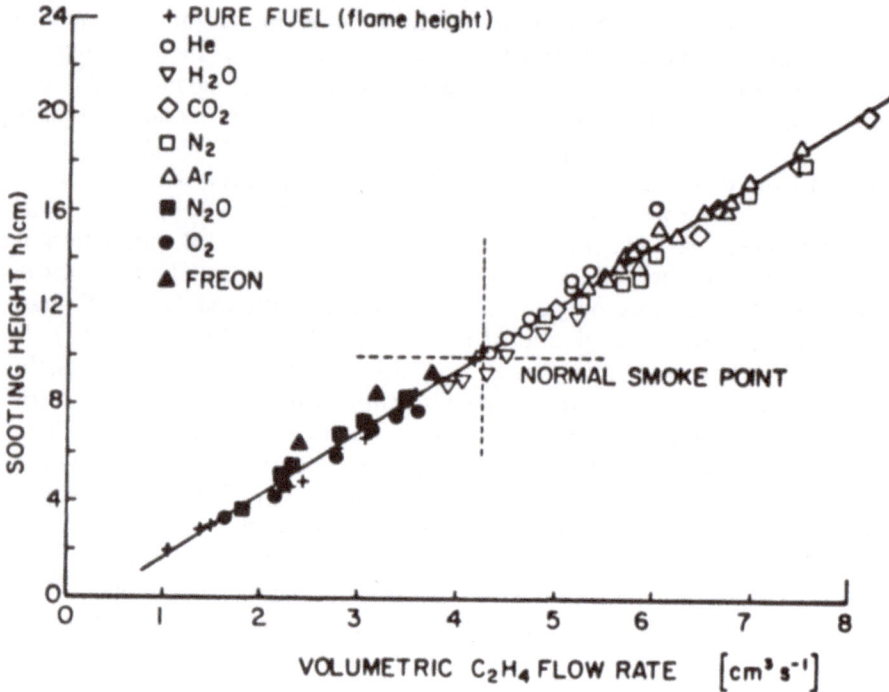

Fig. 9. The sooting height of ethylene diffusion flames with various additives as a function of volumetric flow rate.

that, even if the fuel molecule is broken down prior to entering the flame front, the flame height is not altered. Conceptually the fuel is a stoichiometric sink for the oxygen and as long as the same number of carbon and hydrogen atoms prevail the same height is obtained.[18] Thus we conclude that the so-called smoke height is a good qualitative measure of the sooting tendencies and the addition of inerts to control the temperature does not affect the basic flame structure.

Understanding this result, diffusion flame experiments were performed on aliphatic fuels,[19] aromatics, and cyclohexadienes. Shown in Figure 10 (taken from Glassman and Yaccarino[19]) are results of the mass flow rate at the smoke point versus the calculated stoichiometric temperature for the fuel-inert mixture in air. Without discussing any specifics, it is apparent that much of the data follow trends which show curvature. The important initial hypothesis of the Princeton work has been that fuel pyrolysis rates control the sooting tendency. If so, then the smoke point, or mass flow rate at the smoke point, is an indirect measure of the fuel pyrolysis rate. Since the larger the mass flow rate at the sooting point, the less the tendency to soot and consequently the less the pyrolysis rate, then it became apparent that the data of the type shown in Figure 9 should be plotted as ln(1/mass rate) vs. 1/T. If straight lines were obtained, then some confirmation of the hypothesis that fuel pyrolysis controls would be obtained. Figure 11 shows the results for 16 different fuels: all follow the proposed Arrhenius trends.

Some direct conclusions can be drawn from Figure 10. First, we must note that the highest temperature point for all fuels, except ethylene, is the case of no inert addition. For ethylene there was fuel preheating to increase the flame temperature. Acetylene is thus observed to be a prolific sooting fuel in diffusion flames only because its flame temperature is so high. At a

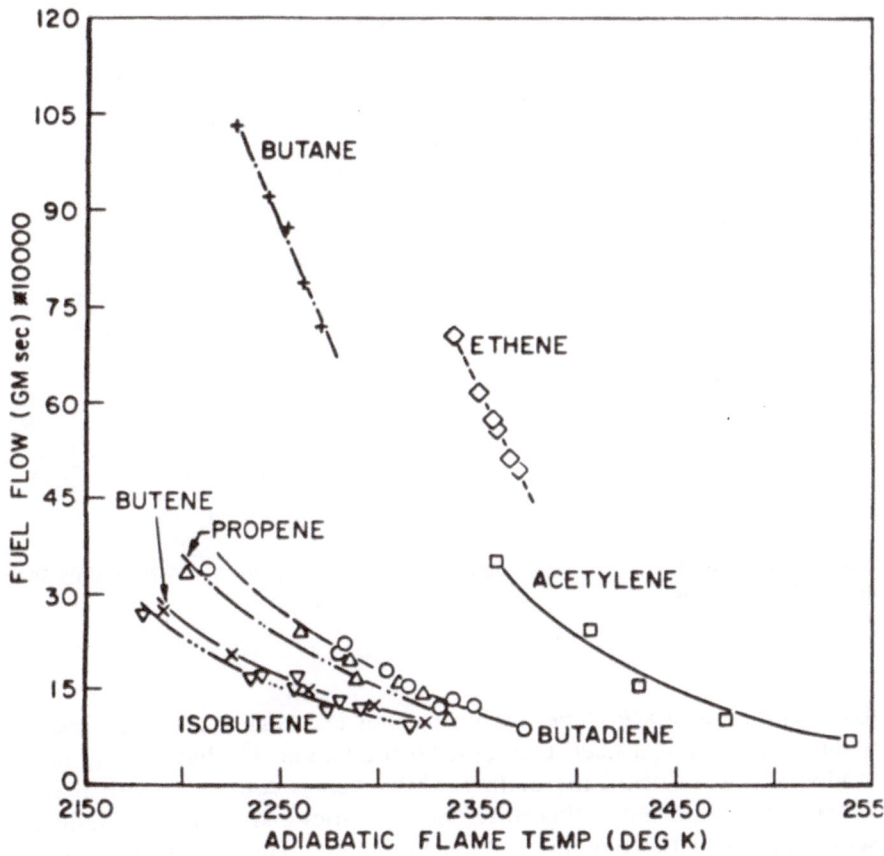

Fig. 10. The mass flow rate at the sooting height plotted as a function of temperature for various fuels.

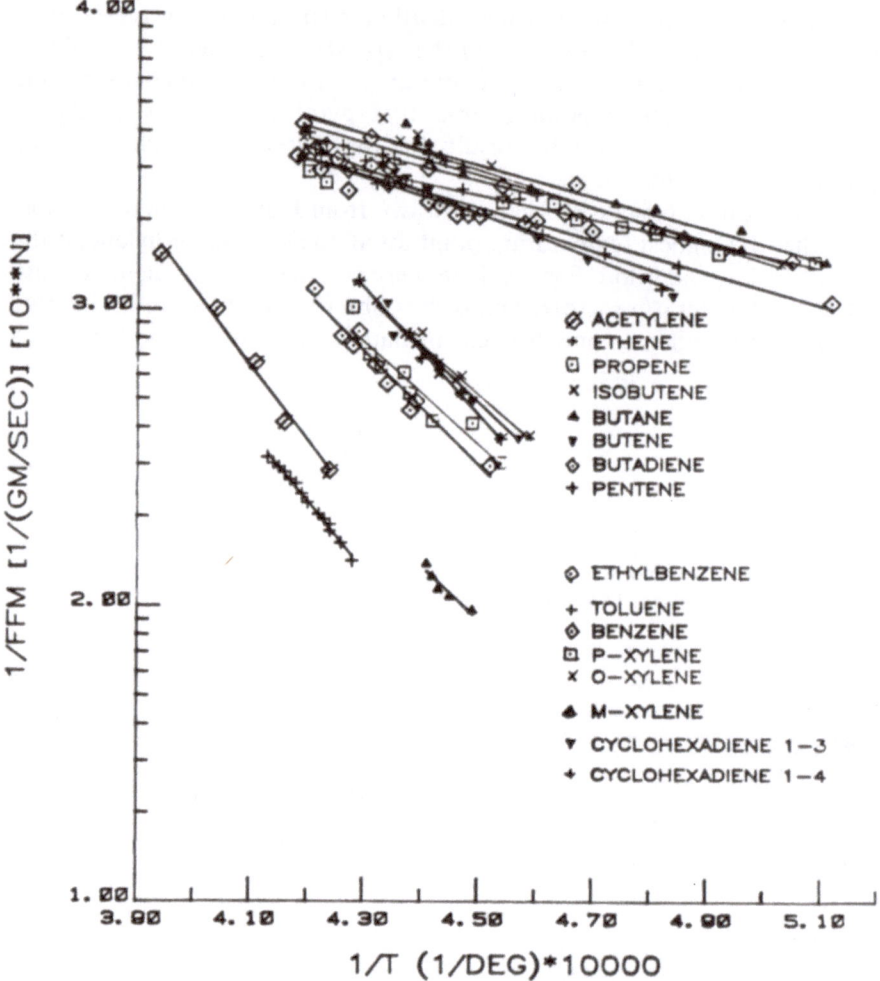

Fig. 11. The sooting mass flow rate–temperature dependency plotted as Arrhenius parameters for various fuels.

fixed temperature acetylene is seen to soot less than the C_4, C_5, and C_6 olefins. Further, at a fixed temperature the C_4 and C_5s have the greatest tendency to soot of all the olefins. This result is consistent with that of Schalla *et al.* reported in Figure 1. The aromatics and the cyclohexadienes soot more than the aliphatics, even butadiene. The butadiene result is contrary to what Schalla and co-workers[16,17] report and we have not been able to explain the difference.

Considering the experimental scatter it is difficult to separate the tendency among the aromatics. The trend does appear to follow ethyl benzene > xylenes > toluene > benzene, which would be the order expected for ease of initiation of pyrolysis. 1,3 and 1,4 cyclohexadiene were tested because one (1,3) is conjugated and the other is not and it was initially thought that the conjugated species would have the more noticeable tendency to soot. The experimental results do not indicate any difference. Review of the literature subsequently showed that the cylohexadienes immediately pyrolyze to benzene,[47] and it is rewarding to note that our experimental results show that all three compounds have the same tendency to soot.

It is debatable whether the slopes of the lines in Figure 11 represent any realistic activation energies. Nevertheless, the activation energy obtained[19] from the curves for acetylene was 118 kcal/mol, for ethene 114, and for all other aliphatics between 80 and 90. It is interesting to observe that the weakest bond in acetylene is the C—H with a value of approximately 110, the weakest bond in ethene is also a C—H with a value of about 108, the weakest bond in all others is a C—C bond which will always have a value around 85.[19,37]

The slopes for the aromatic fuels are all about 25–30 kcal/mol. There is no bond strength in any of the fuels that has a value so low. Another interesting observation from the literature is that the activation energy for the oxidative pyrolysis of benzene has been reported as approximately 35 kcal/mol.[48] Assuming that the type of agreement shown in these last two paragraphs is not fortuitous, it is feasible to speculate that, considering the temperatures which exist and the normal activation energies of the pure pyrolysis processes of the fuels examined, O_2 induction in the fuel stream does not play a role for the aliphatics, but may for the aromatics.

6. CONCLUSIONS

Pyrolysis kinetics plays a dominant role in the sooting tendencies of flames. Contrary to previous belief it has been shown by controlling the temperature of premixed flames that acetylene has a greater tendency to soot than most other aliphatic compounds. The oxidative kinetics of aromatics reveal the direct formation of intermediates such as vinyl acetylene and butadiene, which could be important precursors to soot formation and explain the great sooting propensity of aromatics. Sooting experiments with laminar diffusion flames show the unique importance of temperature of the inner stoichiometric flame. The aromatics and cyclohexadienes, although having a greater propensity to soot, show much less dependency on temperatures than do the aliphatics. The results appear to be consistent with the ease of onset of pyrolysis in the case of the aliphatics and the oxidative pyrolytic trends of the aromatics. Flow reactor[49] results have been particularly useful in analyzing the results of flame studies.

ACKNOWLEDGMENTS. The authors wish to gratefully acknowledge the support of the Air Force Office of Scientific Research under Contract F49620-82-K-0011. P. Yaccarino contributed to the early development of the diffusion flame apparatus and the associated techniques.

REFERENCES

1. H. B. Palmer and H. F. Cullis, The formation of carbon from gases, in: *The Chemistry and Physics of Carbon*, Vol. 1, pp. 265–325, Marcel Dekker, New York (1965).
2. K. H. Homann and H. Gg. Wagner, Some aspects of the mechanisms of carbon formation in pre-mixed flames, in: *Eleventh Symp. (Int.) on Combust.*, pp. 371–379, The Combustion Institute, Pittsburg, PA (1967).
3. H. Gg. Wagner, Soot formation in combustion, in: *Seventeenth Symp. (Int.) on Combust.*, pp. 3–19, The Combustion Institute, Pittsburgh, PA (1979).
4. I. Glassman, Phenomenological Models of Soot Processes in Combustion Systems, Princeton University Department of Mechanical and Aerospace Engineering, Report 1450 (1979); also Air Force Office of Scientific Research Report TR 79-1147 (1979).
5. J. P. Bittner and J. B. Howard, Role of aromatics in soot formation, in: *Alternative Hydrocarbon Fuels: Combustion and Chemical Kinetics* (C. T. Bowman and J. Birkeland, eds.), *Prog. Astronaut. Aeronaut.*, Vol. 62, pp. 335–358, American Institute of Aeronautics and Astronautics, New York (1978); also, Composition profiles and reaction mechanisms in near-sooting pre-mixed benzene/oxygen/argon flames, in: *Eighteenth Symp. (Int.) on Combust.*, pp. 1105–1116, The Combustion Institute, Pittsburgh, PA (1981).
6. B. S. Haynes and H. Gg. Wagner, Soot formation, *Prog. Energy Combust. Sci.* 7, 229–273 (1981).
7. O. I. Smith, Fundamentals of soot formation in flames with application to diesel engines particulate emissions, *Prog. Energy Combust. Sci.* 7, 275–292 (1981).
8. H. F. Calcote, Mechanisms of soot nucleation in flames — A critical review, *Combust. Flame* 42, 215–242 (1981).
9. C. F. Cullis, D. J. Huchnall, and J. V. Shepard, Studies of radical reactions leading to carbon formation, in: *Combustion Institute European Symposium*, pp. 111–118, Academic Press, New York (1973).
10. K. H. Homann, Carbon formation in pre-mixed flames, in: *The Mechanisms of Pyrolysis, Oxidation and Burning of Organic Materials* (L. A. Wall, ed.), U.S. National Bureau of Standards Publication 357, pp. 143–152 (1972).
11. T. Tanzawa and W. C. Gardiner, Jr., Mechanisms of acetylene thermal decomposition, in: *Seventeenth Symp. (Int.) on Combust.*, pp. 563–573, The Combustion Institute, Pittsburgh, PA (1979).
12. C. F. Cullis, Role of acetylenic species in carbon formation, *ACS Symp. Ser.* 21, 248–357 (1976).
13. L. E. Stein, On the high temperature chemical equilibria of polycyclic aromatic hydrocarbons, *J. Phys. Chem.* 82, 566–571 (1978).
14. K. H. Homann, Carbon formation in flames, *Combust. Flame* 11, 265–287 (1967).
15. R. Breslow, *Organic Reaction Mechanisms*, W. A. Benjamin, New York (1965).

16. R. L. Schalla, T. P. Clark, and G. E. McDonald, Formation and Combustion of Smoke in Laminar Flames, NACA Report 1186 (1954).

17. R. L. Schalla and R. R. Hibbard, Smoke and coke formation in combustion of hydrocarbon–air mixtures, in: Basic Considerations in the Combustion of Hydrocarbon Fuels in Air, Chapter 9, NACA Report 1300 (1957).

18. K. P. Schug, Y. Mannheimer-Timnat, P. Yaccarino, and I. Glassman, Sooting behavior of gaseous hydrocarbon diffusion flames and the influence of additives, *Combust. Sci. Technol.* **22**, 235–250 (1980).

19. I. Glassman and P. Yaccarino, The temperature effect in sooting diffusion flames, in: *Eighteenth Symp. (Int.) on Combust.*, pp. 1175–1183, The Combustion Institute, Pittsburg, PA (1981).

20. I. Glassman and P. Yaccarino, The effect of oxygen concentration on sooting diffusion flames, *Combust. Sci. Technol.* **24**, 107–114 (1980).

21. I. Glassman and P. Lara, The temperature effect in sooting pre-mixed flames, Paper presented at Eastern States Section/The Combustion Institute Meeting, Carnegie-Mellon University, Pittsburgh (October, 1981).

22. A. D'Alessio and A. Di Lorenzo, Optical and chemical investigation in fuel-rich methane-oxygen pre-mixed flames at atmospheric pressures, in: *Fourteenth Symp. (Int.) on Combust.*, pp. 941–953, The Combustion Institute, Pittsburgh, PA (1973).

23. A. D'Alessio, A. Di Lorenzo, A. Borghese, F. Beretta, and S. Masi, Study of the soot nucleation zone of rich methane-oxygen flames, in: *Sixteenth Symp. (Int.) on Combust.*, pp. 695–708, The Combustion Institute, Pittsburgh, PA (1977).

24. B. S. Haynes, H. Jander, and H. Gg. Wagner, The effect of metal additives on the formation of soot in pre-mixed flames, in: *Seventeenth Symp. (Int.) on Combustion*, pp. 1365–1374, The Combustion Institute, Pittsburgh, PA (1979).

25. J. C. Street and A. Thomas, Carbon formation in pre-mixed flames, *Fuel* **34**, 4–36 (1955).

26. S. T. Minchin, The chemical significance of tendency to soot, *World Pet. Congr. Proc.* **2**, 738–743 (1933).

27. A. E. Clarke, T. C. Hunter, and F. H. Garner, The tendency to smoke of organic substances on burning, *J. Inst. Petrol.* **32**, 627–642 (1946).

28. B. F. Magnussen and B. H. Hjerhager, An investigation into the behavior of soot in turbulent free jet C_2H_2 flame, in: *Fifteenth Symp. (Int.) on Combust.*, pp. 1415–1425, The Combustion Institute, Pittsburgh, PA (1975).

29. F. J. Wright, The formation of carbon under well-stirred conditions, in: *Twelfth Symp. (Int.) on Combust.*, pp. 867–874, The Combustion Institute, Pittsburgh, PA (1969).

30. W. S. Blazowski, Dependence of soot production on fuel structures on back-mixed combustion, *Combust. Sci. Technol.* **21**, 87–96 (1980).

31. W. J. Miller and H. F. Calcote, Ionic mechanisms of carbon formation in flames, Paper presented at Eastern States Section/The Combustion Institute Meeting, United Technologies Research Center, East Hartford, Conn. (November, 1977).

32. R. C. Milliken, Non-equilibrium soot formation in flames, *J. Phys. Chem.* **66**, 794–799 (1962).

33. F. L. Dryer and I. Glassman, Combustion chemistry of chain hydrocarbons, in: *Alternative Hydrocarbon Fuels: Combustion and Chemical Kinetics* (C. T. Bowman and J. Birkeland, eds.), *Prog. Astronaut. Aeronaut.*, Vol. 62, pp. 255–306, American Institute of Aeronautics and Astronautics, New York (1978).

34. D. J. Hautman, F. L. Dryer, K. P. Schug, and I. Glassman, A multiple-step overall kinetic mechanism for the oxidation of hydrocarbons, *Combust. Sci. Technol.* **25**, 219–235 (1981).

35. T. M. Dyer and W. F. Flower, A phenomenological description of particulate formation during constant volume combustion, in: *Particulate Carbon Formation During Combustion* (D. C. Siegla and G. W. Smith, eds.), pp. 363–390, Plenum Press, New York (1981).

36. J. Peeters and G. Mahner, Structure of ethylene-oxygen flames, reaction mechanisms and rate constants of elementary reactions, in: *Combustion Institute European Symposium*, pp. 53–59, Academic Press, New York (1973).

37. I. Glassman, The Homogeneous Oxidtion Kinetics of Hydrocarbons: Concisely and With Application, Princeton University Department of Mechanical and Aerospace Engineering Report 1446 (1980).

38. C. Venkat, High Temperature Oxidation of Aromatic Hydrocarbons, Princeton University Department of Chemical Engineering M.S.E. Thesis (1981).

39. C. Venkat, K. Brezinsky, and I. Glassman, High temperature oxidation of aromatic hydrocarbons, in: *Nineteenth Symp. (Int.) on Combust.*, The Combustion Institute, Pittsburgh, PA (to appear).

40. B. S. Haynes and H. Gg. Wagner, Sooting structure in a laminar diffusion flame, *Ber. Bunsenges. Phys. Chem.* **84**, 499–506 (1980).

41. J. H. Kent, J. Jander, and H. Gg. Wagner, Soot formation in laminar diffusion flames, in: *Eighteenth Symp. (Int.) on Combust.*, pp. 1117–1126, The Combustion Institute, Pittsburgh, PA (1981).

42. I. M. Kennedy, U. Vandsburger, F. L. Dryer, and I. Glassman, Soot Formation in the Forward Stagnation Region of a Porous Cylinder, Western States Section/The Combustion Institute Paper WSCI 82-39 (1982).

43. P. Yaccarino, Parametric Study of Sooting Diffusion Flames, Princeton University Department of Mechanical and Aerospace Engineering M.S.E. Thesis (1980).

44. S. R. Smith and A. S. Gordon, Studies of diffusion flames I. The methane diffusion flame, *J. Phys. Chem.* **60**, 759–763 (1956).

45. F. J. Wright, Effect of oxygen on the carbon-forming tendencies of diffusion flames, *Fuel* **53**, 232–235 (1974).

46. I. Glassman, *Combustion*, Academic Press, New York (1977).

47. D. G. L. James and R. D. Stuart, Kinetic study of the cyclohexadienyl radical, *Trans. Faraday Soc.* **64**, 2752–2769 (1968).

48. N. Fuji and T. Asaba, Shock-tube study of the reaction of rich mixtures of benzene and oxygen in: *Fourteenth Symp. (Int.) on Combust.*, pp. 433–442, The Combustion Institute, Pittsburgh, PA (1973).

49. L. Crocco, I. Glassman, and I. E. Smith, A flow reactor for high temperature reaction kinetics, *Jet Propulsion* **27**, 1266–1267 (1957).

18

Application of ESCIMO Theory to Turbulent Reacting Mixing Layers

A. S. C. MA, D. B. SPALDING, AND R. L. T. SUN

ABSTRACT. The ESCIMO theory of turbulent combustion is applied to turbulent plane reacting mixing layers. A set of transport equations describing population distributions of folds in such layers is solved, taking into account the property variation in the cross-stream direction.

Simplifications, made in the interest of reducing the computational task, include the following: (1) Fold properties depend on age alone, at a given point in the layer. (2) Fold properties at birth can be determined by tracking upstream along a trajectory of constant mean mixture fraction. (3) Fold size at birth is proportional to the local length scale of turbulence. (4) Fold-formation rate is proportional to the entrainment rate, with further assumptions for its distribution across the layer. (5) Fold-stretching rate is proportional to the time-mean velocity gradient.

The results are presented in the form of population-distribution functions for fold age, and of the mean properties and the fluctuation intensities of scalar quantities in the time-mean flow. The latter are compared with the experimental data of Batt (1977). The qualitative agreement between measured and predicted results is acceptable but quantitative discrepancies exist in the fluctuation intensities of temperature.

1. INTRODUCTION

The ESCIMO theory of turbulent combustion has been applied in the past to the well-stirred reactor[1] and to stabilized flames.[2] In both cases,

---

A. S. C. MA, D. B. SPALDING, AND R. L. T. SUN • Computational Fluid Dynamics Unit, Imperial College of Science and Technology, London SW7 2BX, U.K.

the distribution of folds was treated in a simplified way suitable for certain flow geometries to which a primitive demographic analysis could apply.

In the present paper, a set of transport equations describing the population distribution of folds in two-dimensional turbulent reacting mixing layers has been solved, taking into account the property variation in the cross-stream direction.

In order to assess the validity of the present model, the predictions have been compared with the experimental data of Batt.[3] The results to be presented are the population-distribution function of various folds, the mean turbulent properties and the fluctuation intensities of various scalar quantities.

2. MATHEMATICAL FORMULATION

2.1. Demographic Analysis

The governing differential equations solved in the current work are of the Patankar–Spalding[4] form for steady, two-dimensional, and parabolic flows:

$$\frac{\partial \phi}{\partial x} + (a + b\omega) \frac{\partial \phi}{\partial \omega} = \frac{\partial}{\partial \omega} \left(c_\phi \frac{\partial \phi}{\partial \omega} \right) + d_\phi \qquad (2.1)$$

The time-mean variables, ϕ, solved in the present work include the longitudinal velocity (u), the turbulent kinetic energy (k) and its dissipation rate (ε), the mixture fraction $(f \equiv$ total mass fraction of NO_2 and $N_2O_4)$, and the normalized probability-density function of the fold age $(\tilde{P})$. The independent variables, x and ω, represent the longitudinal distance and nondimensional stream function respectively, a and b describe rates of entry of fluid into the $x \sim \omega$ grid, c_ϕ represents the turbulent diffusion process in the cross-stream direction, and d_ϕ is the source of the relevant entity.

The calculations of a, b, c_ϕ, and d_ϕ have already been set up for u, k, ε, and f by Patankar and Spalding[4] and Launder and Spalding,[5] while those for $\tilde{P}$ have been described in the recent work of Ma et al.[6] Therefore only a brief description will be provided for the $\tilde{P}$-equations here.

The differential equation for $\tilde{P}$, written in terms of the normalized age of the folds, $\tilde{A}$, is given by

$$\frac{\partial \tilde{P}}{\partial x} + (a + b\omega) \frac{d\tilde{P}}{\partial \omega} = \frac{\partial}{\partial \omega} \left(c_P \frac{\partial \tilde{P}}{\partial \omega} \right) - \frac{\partial}{\partial \tilde{A}} \left[\left(\frac{F}{u} + \frac{\tilde{A}}{F} \frac{\partial F}{\partial x} \right) \tilde{P} \right]$$

$$+ \frac{1}{\rho u} [\rho \dot{R}_F \delta(\tilde{A}) - \rho \dot{R}_F \tilde{P}] \qquad (2.2)$$

where $F \equiv U_{ref}/x$ and $\tilde{A} \equiv AF$. The function F has been so chosen that $\tilde{P}$ is always close to zero for the higher range of $\tilde{A}$ values. The diffusion coefficient c_P is taken to be the same as that for the mass fraction, $\dot{R}_F$ is the fold-formation rate (to be explained in the next section), ρ is the density of the fluid, and $\delta(\tilde{A})$ is the Dirac delta function; U_{ref} is taken as $0.2\, U_{max}$.

It should be noted that the age coordinate $\tilde{A}$ has been discretized into a number of finite intervals (from 0 to 1) in the finite-difference formulation, which results in a set of coupled differential equations for $\tilde{P}_i$ (the $\tilde{P}$-value in the i-th $\tilde{A}$-interval). Also the upwind-difference scheme has been employed for the first term on the right-hand side of equation (2.2), as described in Ma *et al.*[6]

2.2. Biographic Analysis

The only governing equation that must be solved in the biographic analysis is the one for the variation of the mixture fraction f within a fold, given by

$$\frac{\partial f}{\partial \tilde{A}} = C \frac{\partial^2 f}{\partial \eta^2} \tag{2.3}$$

where C is the diffusion coefficient and η the nondimensional distance across the fold. The influence of stretching rate and fold size are contained in the coefficient C, while the fold compositions at birth serve as the initial conditions for equation (2.3).

The "profile" method[7] has been used to solve equation (2.3) in order to reduce the computation task, and the variation of f with position in the fold is assumed to be sinusoidal. Hence the differential equation is solved only in an integral manner with the plane-symmetric boundary conditions applied to both sides of a fold. The variation of temperature (T) and species concentration in the fold is uniquely related to the variation of f, according to the presumption of an adiabatic, equilibrium chemical-reaction system.

The fold average value of any property, ϕ, is calculated according to the formulas

$$\tilde{\phi} = \int_0^1 \phi\, d\eta$$

$$\approx \sum_{j=1}^{NP} \phi_j \Delta \eta_j \tag{2.4}$$

where $\tilde{\phi}$ is the fold-average value of ϕ, $\Delta \eta_j$ is the j-th interval of the η-domain, and NP is the number of subdivisions in the fold.

2.3. Assumptions

2.3.1. Fold Size at Birth

The fold thickness at formation time ("birth"), Z_0, has been assumed to be proportional to the local length scale of turbulence, given by the $k \sim \varepsilon$ model of turbulence. Thus

$$Z_0 = C_z k^{3/2}/\varepsilon \qquad (2.5)$$

where C_z is a constant (various values have been tried in the computations performed).

The thickness of the fold changes during the life-span of each fold under the influence of stretching, which is a consequence of turbulent shear strain.

2.3.2. Fold Composition at Birth

Within the framework of the ESCIMO concept, each fold is composed of two distinct parts at formation time, namely the fresh part and the re-engulfed part. The f-value of the fresh part (f_0) is equal to the free-stream value, while that of the latter (f_R) is determined by the local average value ($\bar{f}$) and its mean gradient $|\partial\bar{f}/\partial y|$. Hence

$$\left.\begin{aligned} f_0 &= 0 \\ f_R &= \bar{f} + C_F l \left|\frac{\partial\bar{f}}{\partial y}\right| \end{aligned}\right\} \qquad (2.6)$$

or

$$\left.\begin{aligned} f_0 &= 1 \\ f_R &= \bar{f} - C_F l \left|\frac{\partial\bar{f}}{\partial y}\right| \end{aligned}\right\} \qquad (2.7)$$

according to the position in the mixing layer; l is the local length scale of turbulence ($= 0.164\, k^{3/2}/\varepsilon$) and C_F is a constant.

Whether equation (2.6) or (2.7) is used depends on the "dividing" ω-value (ω^*) across the mixing layer, which is determined by the ratio of entrainment rates from both boundaries of the mixing layer as described in Sun.[8]

The mass ratio of the fresh material to the re-engulfed part, M_0, is calculated by the expression

$$M_0 = (f_R - \bar{f})/(f_R - f_0) \qquad (2.8)$$

indicating that the average value of f inside each fold is equal to $\bar{f}$.

2.3.3. Fold-Formation Rate and Destruction Rate

The total fold-formation rate at any cross-section of the mixing layer is directly related to the entrainment rate. However, the distribution of formation rate across the layer has to be postulated. Various assumptions have been made and will be described in the next section.

As a consequence of the ESCIMO concept, the fold-destruction rate is proportional to the fold-formation rate and the proportionality constant is deducible from M_0.

2.3.4. The Stretching Rate

We shall suppose that the stretching effect is dominated by the time-mean-velocity gradient $|\partial u/\partial y|$. The rate of change of fold thickness with time, R, is then given by

$$R = - C_s|\partial u/\partial y| \tag{2.9}$$

where C_s is a "stretching constant." The minus sign implies that stretching reduces the fold thickness.

For the evaluation of equation (2.3), the arithmetic mean of the R-values at the birth place and at the location in question is adopted.

The trajectory of the fold is taken as the constant mixture-fraction contour, since the elemental composition in each fold is assumed to be fixed during its lifetime from the coherence concept.

2.4. Coupling of Demographic and Biographic Analysis

The turbulent mean properties, $\bar{\phi}$, are obtained from the corresponding fold-average values multiplied by the population distribution function of various folds. Thus

$$\bar{\phi} = \int_0^1 \tilde{\phi}(A)\tilde{P}d\tilde{A}$$

$$\approx \sum_{j=1}^{NA} \tilde{\phi}_j\tilde{P}_j\Delta\tilde{A}_j \tag{2.10}$$

where $\Delta\tilde{A}_j$ is the jth interval in the $\tilde{A}$-domain and NA is the number of subdivisions.

The root-mean-square fluctuation intensities of ϕ, $\overline{\phi'^2}$, can be deduced as follows:

$$\overline{\phi'^2} = \int_0^1 \left\{ \int_0^1 [\phi(\hat{A}, \eta) - \bar{\phi}]^2 d\eta \right\} \hat{P} d\hat{A}$$

$$= \int_0^1 \bar{\phi}^2 \hat{P} d\hat{A} - 2\bar{\phi} \int_0^1 \bar{\phi} \hat{P} d\hat{A} + (\bar{\phi})^2 = \overline{\phi^2} - (\bar{\phi})^2 \qquad (2.11)$$

3. DESCRIPTION OF THE FLOW CONFIGURATION

The flow considered is a two-dimensional turbulent shear layer. The stream velocities are 23 ft/s and 2 ft/s, respectively. The fast stream is composed of cold air (252 K) seeded with dilute concentrations of N_2O_4, while the slow one consists of hot pure air (310 K). A schematic diagram of the shear layer, and a list of test conditions, are shown in Figure 1.

The chemical process is a first-order dissociation–recombination reaction represented by

$$N_2 + N_2O_4 \rightleftharpoons 2NO_2 + N_2 \qquad (3.1)$$

the constants of which have been measured by Wegener.[9] The heat released from the chemical reaction is negligible in the present flow process since the concentration of N_2O_4 is lower than 0.5% (by volume).

4. COMPUTATIONAL NOTES

4.1. The Grid Systems

The expanding grid of the GENMIX code[10] is employed in the present prediction, with 20 cross-stream grid nodes ($N = 20$). The intervals

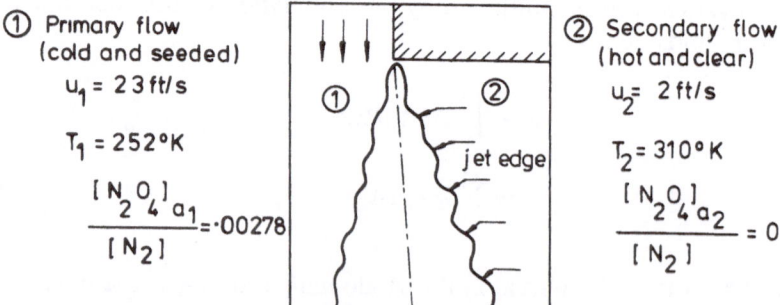

Fig. 1. Test configuration of reacting shear layer.

of ω (from 0 to 1) are distributed in accordance with the expression

$$\omega_i = \left(\frac{i-1}{N-1}\right)^{0.5}, \qquad i = 1, 2, 3, \ldots, N \qquad (4.1)$$

Therefore, the grids are more densely distributed near the external boundary where the velocity is lower.

The total number of marching steps to reach the downstream distance of $x = 1.52$ m (5 ft) is 300 in a typical computation. The initial width of the mixing layer is 0.001 m.

The total number of age intervals, NA, is equal to 10, uniformly distributed. The total number of subdivisions inside a fold for its biographic calculation, NP, is equal to 10.

4.2. The Chemical-Reaction Rate Constants

Local chemical equilibrium conditions are assumed to prevail within each fold, as the magnitude of the measured eddy-decay time of turbulence (30 ms) is large compared with the typical time for chemical changes (< 1 ms).

The chemical equilibrium constant is taken from the measured data of Wegener,[9] namely

$$k_D[N_2O_4][N_2] = k_R[NO_2]^2[N_2] \qquad (4.2)$$

and

$$K_C = k_D/k_R = \frac{1}{82T} \exp\left(20.72 - \frac{6747}{T}\right) \qquad (4.3)$$

where k_D, k_R, and K_C are the forward dissociation-rate constant, the recombination-rate constant, and the equilibrium constant, respectively; the square bracket indicates the mole fraction of the species enclosed therein.

The net-production rate for the total oxide mass fraction is zero, hence

$$\dot{m} = \dot{m}_{N_2O_4} + \dot{m}_{NO_2} = 0 \qquad (4.4)$$

where $\dot{m}$ stands for the production rate (mass fraction per unit time). The relation between the total mass fraction of oxides and the mole fractions of oxide is given by

$$m = m_{N_2O_4} + m_{NO_2}$$
$$= [N_2O_4]W_{N_2O_4} + [NO_2]W_{NO_2} \qquad (4.5)$$

where the W's are the molecular weights. Thus the mass fraction m obeys the species-conservation equation of boundary-layer form

$$\frac{\partial m}{\partial x} + (a + b\omega)\frac{\partial m}{\partial \omega} = \frac{\partial}{\partial \omega}\left(c\frac{\partial m}{\partial \omega}\right) \tag{4.6}$$

with the following boundary conditions:

$$m = m_1 \qquad \text{at } \omega = 0 \tag{4.7}$$

$$m = 0 \qquad \text{at } \omega = 1 \tag{4.8}$$

Hence, after dividing equation (4.5) by $W_{N_2O_4}$,

$$[N_2O_4]_a = [N_2O_4] + \tfrac{1}{2}[NO_2] \tag{4.9}$$

and

$$\frac{[N_2O_4]_a}{[N_2O_4]_{a_1}} = \frac{m}{m_1} \tag{4.10}$$

where $[N_2O_4]_a \equiv m/W_{N_2O_4}$ is defined as the "available" mole fraction of N_2O_4 and $[N_2O_4]_{a_1}$ is its value in the primary flow.

The local equilibrium mole fraction of NO_2 is obtained by inserting equation (4.9) into (4.3):

$$\frac{[NO_2]}{[N_2]} = \frac{K_C}{4}\left\{\left(1 + 16[N_2O_4]_{a_1}\frac{m}{m_1}\frac{1}{K_C}\right)^{1/2} - 1\right\} \tag{4.11}$$

The quantity m/m_1, varying from 0 to 1 across the mixing layer, can be treated as the mixture fraction. Its value will be solved at every ω-node with the turbulent-diffusion coefficient obtained from the turbulent viscosity and a uniform Schmidt number (the turbulent Schmidt number measured by Batt is equal to 0.5).

4.3. The Computations Performed

Several computations have been performed with combinations of input constants, as shown in Table 1. The column under "Mode" lists three presumptions for the distribution of fold-formation rate: (i) proportional to the local mean velocity gradient, (ii) proportional to the local mean velocity, and (iii) proportional to the normalized stream function.

Table 1. Characteristics of computer runs

Run No.	C_z	C_F	C_s	Mode
1	0.164	2	1	(i)
2	0.164	2	1	(ii)
3	0.164	2	1	(iii)
4	0.328	2	1	(i)
5	0.164	3	1	(i)
6	0.164	2	0.3	(i)
7	0.328	2	0.5	(i)

5. RESULTS OF DEMOGRAPHIC ANALYSIS

5.1. The Population Distribution vs. Age at a Fixed Position

The population distributions with respect to age are presented in Figures 2–7 showing the difference resulting from various assumptions on fold-formation rate. First, results obtained from Run No. 1 are shown in Figure 2. The ordinate represents the nondimensional population of folds having a particular age while the abscissa is the nondimensional age $\tilde{A}$. The three curves refer to the population distribution prevailing at a position near the high-speed edge ($\eta_T = -1.5$), the center ($\eta_T = -0.4$), and the low-speed edge ($\eta_T = 1.5$).

The similarity parameter η_T is defined by

$$\eta_T \equiv \frac{12(y - y_{0.5})}{x - x_0} \tag{5.1}$$

where $y_{0.5}$ is the value of y at which $\bar{T}$ is equal to $0.5\,(T_1 + T_2)$ and x_0 is the

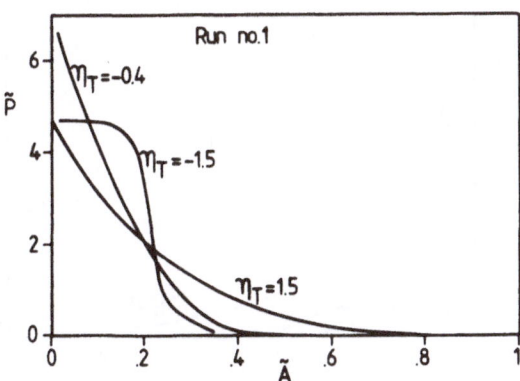

Fig. 2. Population distribution function with respect to age.

effective origin of the mixing layer (taken to be 3″ from Batt's measurements).

It can be observed that the $\tilde{P}$ vs. $\tilde{A}$ curve has the steepest slope in the center of the mixing layer, corresponding to the maximum fold-formation rate assumed. Near the low-speed edge, the curve has a similar exponential decay but a less steep gradient. Near the high-speed boundary, the curve is almost flat for $0 \leqslant \tilde{A} \leqslant 0.15$ followed by a sudden drop in the range $0.2 \leqslant \tilde{A} \leqslant 0.35$. The flat part of the curve shows the relatively low fold-formation rate near the high-speed boundary, according to the distribution hypothesis.

The overall feature is that the youngest folds are popular and the population of the old ones (say $\tilde{A} < 0.6$) is negligible, signifying that most folds are born nearby upstream.

Second, the results from Run No. 2 are demonstrated in Figure 3 in which the distribution of fold-formation rate is assumed to be proportional to the local mean velocity. The figure reveals that all curves belong to the type of exponential decay with the steepest one near the high-speed edge where the fold-formation rate is highest. The curve near the low-speed stream is more uniform, with a significant amount of old folds (e.g., $\tilde{A} > 0.6$) present.

Finally, the results from Run No. 3 are provided in Figure 4, where the distribution of fold-formation rate is supposed to be proportional to the "local-entrainment rate" $(r\dot{m}'')_l$, defined by

$$(r\dot{m}'')_l \equiv (1 - \omega)r_l\dot{m}_l'' + \omega r_E\dot{m}_E'' \qquad (5.2)$$

It can be seen that the fastest diminution of population with respect to age is again found near the high-speed boundary where the entrainment rate is

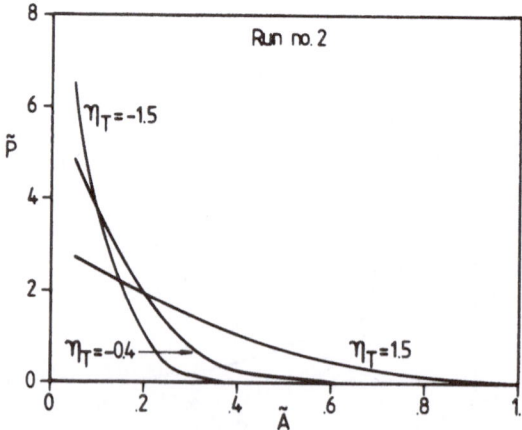

Fig. 3. Population distribution function with respect to age.

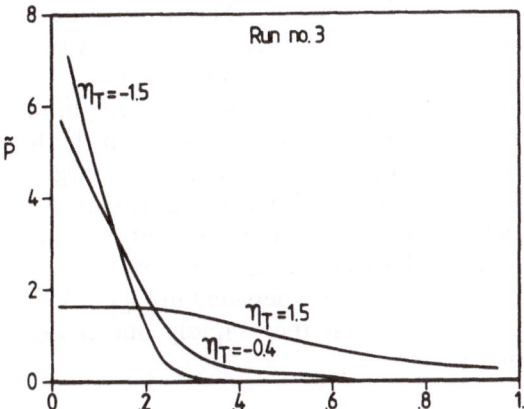

Fig. 4. Population distribution function with respect to age.

at its peak. The distribution at the low-speed side is now even more uniform, while a finite amount of the oldest folds born in the far-upstream region have survived.

5.2. The Population Distribution of Age Groups Across the Mixing Layer

The spatial variation of the population distribution is illustrated by the plot of $\tilde{P}_j$ (for fixed $\tilde{A}_j$) vs. η_T in Figures 5–7, obtained in the self-similar region ($x = 0.47$ m) of the layer.

Figure 5 presents the results obtained from Run No. 1. Each curve stands for the population of folds belonging to a particular age interval. Only four age groups are shown here, for clarity. For the youngest folds, $\tilde{P}_1$ has its peak value at $\eta_T \approx -0.4$ where the velocity gradient and the

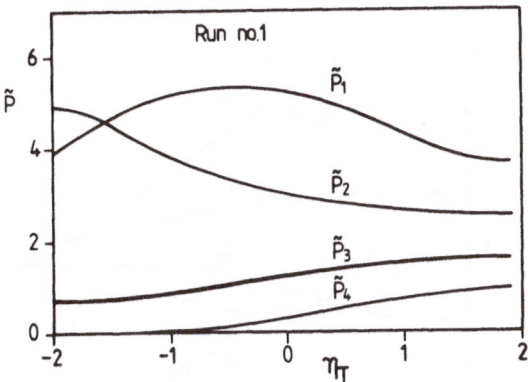

Fig. 5. Cross-layer distribution of fold population in age groups.

fold-formation rate attain maxima. The $\tilde{P}_2$-distribution behaves as a monotonic decreasing function with respect to η_T, while the population of older folds P_3 and P_4 reflects the opposite tendency.

The corresponding results from Run No. 2 are plotted in Figure 6. The curve representing $\tilde{P}_1$ now has its highest value at the high-speed edge of the mixing layer, where the fold-formation rate is largest according to the presumption. The $\tilde{P}_2$-distribution is fairly uniform for $-2 \leqslant \eta_T \leqslant -0.4$ and then decreases smoothly toward the low-speed boundary. The shapes of the $\tilde{P}_3$- and $\tilde{P}_4$-distributions are similar to those in Figure 5.

The curves for Run No. 3 are depicted in Figure 7. They are similar to those in Figure 6, except that they rapidly bunch together toward the low-speed boundary.

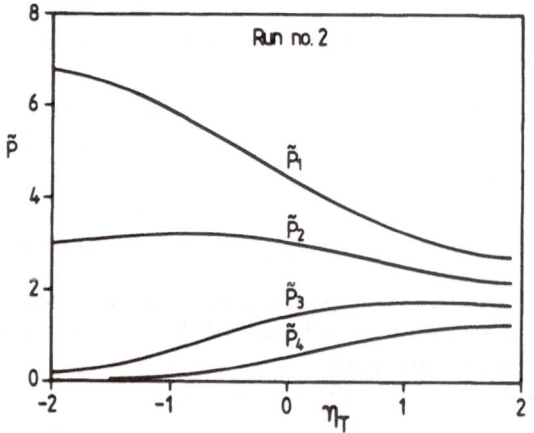

Fig. 6. Cross-layer distribution of fold population in age groups.

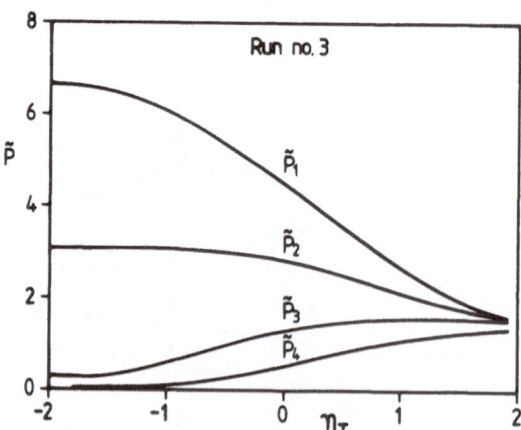

Fig. 7. Cross-layer distribution of fold population in age groups.

5.3. The Average-Age Distribution Across the Mixing Layer

Another interesting quantity in the demographic analysis is the average age of the folds defined by

$$\tilde{A}_{av} \equiv \int_0^1 \tilde{A} \tilde{P} d\tilde{A} \qquad (5.3)$$

Its variation across the mixing layer is plotted in Figure 8 for Run Nos. 1 to 3. In the case of Run No. 1, the minimum value occurs at $\eta_T \approx -0.8$, instead of $\eta_T \approx -2.0$ for other cases. The average age in the center of the layer ($-0.2 \leq \eta_T \leq 0.2$) is not larger than 0.2 for all cases, signifying that most folds are created within a distance of $0.2X_D$ from the point in question.

However, the curves differ significantly near the slow boundary, e.g., they vary from 0.23 to 0.4 at $\eta_T = 1.6$. (See Section 7.1 for a discussion.)

6. RESULTS OF THE COMBINED ANALYSIS

6.1. The Mean-Temperature Profile

The normalized mean-temperature profile across the mixing layer for Run No. 1 (at $x = 0.4$ m) is plotted in Figure 9 together with the measurements from Batt.[3] It should be noted that the mean temperature is calculated from the solution of an enthalpy equation that does not

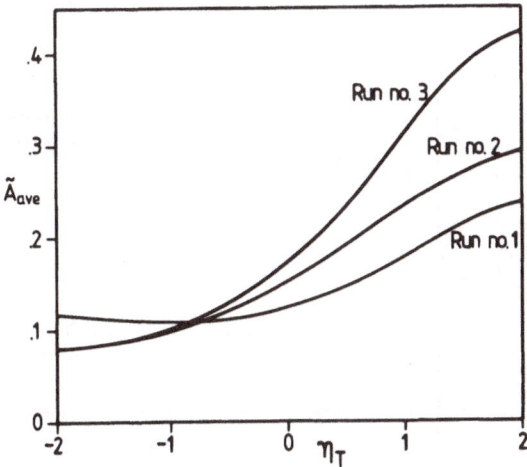

Fig. 8. Average age across the mixing layer.

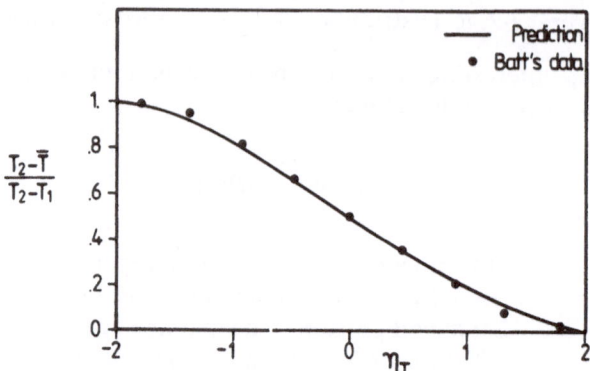

Fig. 9. Cross-layer profile of mean normalized temperature at $x = 0.47$; Run No. 1.

contain the heat release of the reaction. Therefore, the contribution of the ESCIMO theory is not fully appreciated here.

The quantitative agreement between the predictive results and the measured data is satisfactory, implying that the turbulent-diffusion coefficient is adequately determined. The mean temperature is important in the calculation of NO_2 concentration since the chemical-equilibrium constant is temperature-dependent.

6.2. The Mean Concentration Profile of Nitrogen Dioxide

The normalized mean NO_2 concentration profile across the mixing layer for Run No. 1 is plotted in Figure 10 against the experimental results from Batt.[3] The calculation is in line with the full ESCIMO approach described earlier.

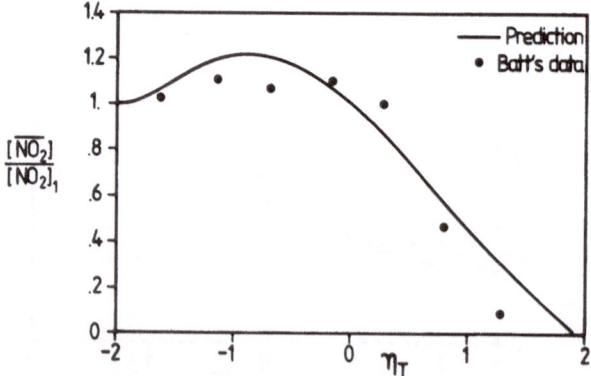

Fig. 10. Cross-layer profile of mean normalized concentration of NO_2 at $x = 0.47$; Run No. 1.

The prediction exhibits a maximum value of 1.2 near $\eta_T = -1$, while there appears a slight double hump when the measured data are connected with a smooth curve, the existence of which has not been explained in Batt's paper. But the author did mention that the accuracy of his measurements is roughly $\pm 10\%$ of the core-flow concentration level, and that may account for the discrepancy.

The level of NO_2 concentration decays slightly faster according to Batt's data than the present prediction, in the region where $\eta_T > 0.5$.

6.3. The RMS Fluctuation of Temperature

The fluctuation intensity of temperature for Run No. 1 across the mixing layer is presented in Figure 11. The predictions show a maximum intensity of 0.23 at $\eta_T \approx 0.5$, followed by a sharp decrease until the magnitude is smaller than 0.1. The experimental data, however, reveal a slight double hump with a rather flat plateau in the central region. The maximum measured value is around 0.14, considerably lower than the present prediction. Fiedler[11] has found the maximum fluctuating temperature intensities as large as 0.20, also very much larger than the corresponding results for the Batt[3] study.

6.4. The RMS Fluctuation of Species Concentration

The fluctuation intensities of NO_2 across the mixing layer, obtained from Run No. 1, are plotted in Figure 12. The predicted results exhibit a small hump (≈ 0.12) near $\eta_T = -1.0$, and the maximum value of 0.36 located at $\eta_T = 0.7$. The experimental data share similar characteristics with the predictions, though some quantitative discrepancies still exist.

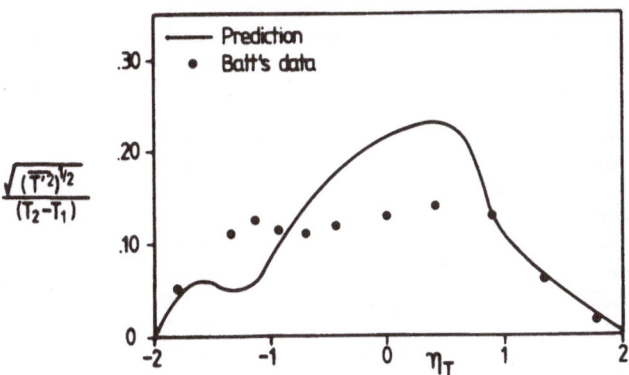

Fig. 11. Cross-layer profile of temperature fluctuation intensities; Run No. 1.

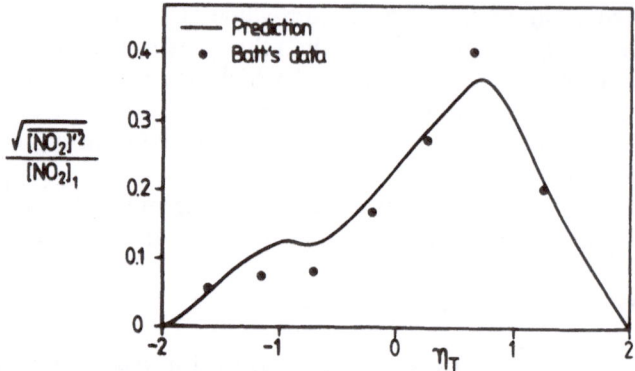

Fig. 12. Cross-layer profile of concentration fluctuation intensities; Run No. 1.

The prediction of concentration fluctuation is one of the main contributions made by ESCIMO theory, since the effect of local unmixedness has been taken into account (folds arrive from various places of birth).

6.5. Consequences of Assumptions

6.5.1. Fold-Formation Rate

The influence of the hypothetical distributions about the fold-formation rates on the mean NO_2-concentration profile is presented in Table 2. It is tabulated because the difference between the hypotheses would be indistinguishable on a graphic presentation.

Table 2. The influence of distribution of fold-formation rate on NO_2 concentration

Location across the mixing layer, η_T	Normalized mean NO_2 concentration obtained from each mode number		
	Mode (i)	Mode (ii)	Mode (iii)
−2.00	1.00	1.00	1.00
−1.50	1.10	1.10	1.10
−1.07	1.21	1.20	1.20
−0.674	1.21	1.22	1.21
−0.300	1.13	1.15	1.15
0.104	0.947	0.989	0.988
0.508	0.712	0.756	0.764
0.889	0.575	0.582	0.586
1.17	0.398	0.400	0.403
1.57	0.178	0.179	0.179

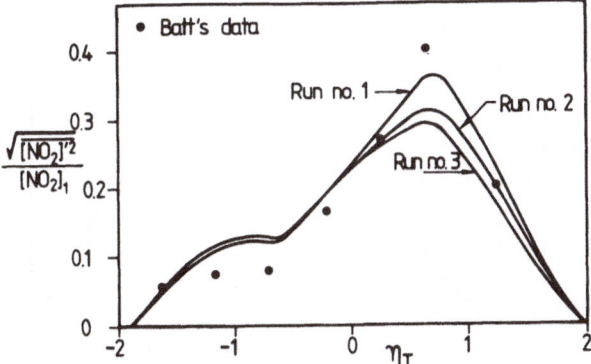

Fig. 13. Influence of distribution of fold-formation rate on the concentration fluctuation intensity.

The relative difference reaches a maximum of 7% around $\eta_T = 0.508$ but is generally lower than 2%. On the other hand, the influence on the concentration fluctuation intensities of NO_2 is more significant, as shown in Figure 13. The results are almost identical in the region $-2 \leqslant \eta_T \leqslant 0$, but the maximum fluctuation intensity varies from 0.36 for Run No. 1 to 0.29 for Run No. 3.

6.5.2. Fold Size at Birth

The results produced by Run No. 1 and Run No. 4 are compared in Table 3 and Figure 14. It should be noted that the fold size at birth is taken

Table 3. Influence of fold size on NO_2 concentration

Location across the mixing layer, η_T	Mean NO_2 concentration	
	Run No. 1 $C_Z = 0.164$	Run No. 4 $C_Z = 0.328$
-2.00	1.00	1.00
-1.50	1.10	1.09
-1.07	1.21	1.20
-0.674	1.21	1.18
-0.300	1.13	1.10
0.104	0.947	0.907
0.508	0.712	0.669
0.889	0.575	0.565
1.17	0.398	0.397
1.57	0.178	0.180

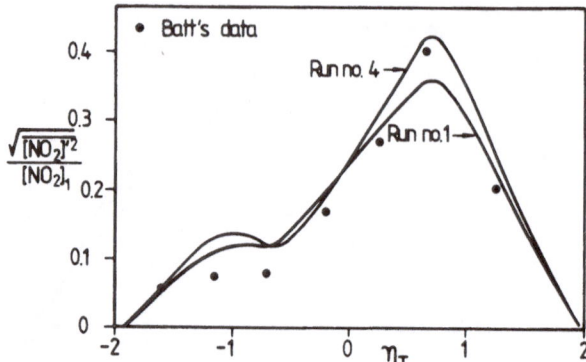

Fig. 14. Influence of fold size on the concentration fluctuation intensity.

as the length scale of turbulence in Run No. 1, while it is doubled in Run No. 4. Table 3 shows that the influence of fold size on mean NO_2 concentration is visible only in the central part of the mixing layer, up to 6% at $\eta_T = 0.508$, revealing that the assumption regarding fold size within the present range is not crucial.

The influence is more significant in Figure 14 for the fluctuation intensities of NO_2. The predicted peak is 0.42 in Run No. 4 compared with 0.36 in Run No. 1, indicating that larger fold size will yield higher fluctuation intensities.

6.5.3. Fold Composition at Birth

The results of mean NO_2 concentration from Run No. 1 and Run No. 5 are given in Table 4, in which C_F is a measure for the fold composition at birth [equations (2.6) and (2.7)]. Again, the discrepancies are apparent only in the midregion of the mixing layer, where the relative difference is around 7% (e.g., at $\eta_T = 0.104$). The peak value is almost identical in both cases.

The fluctuation intensities are plotted in Figure 15. Considerable difference is observed in the range $-0.5 < \eta_T < 0.6$ and the percentage difference can reach 40% (at $\eta_T = 0$). The peak value computed from Run No. 5 is 0.43 (the measured value is 0.40) at $\eta_T = 0.4$. But the peak value is shifted to 0.36 at $\eta_T = 0.8$ when C_F changes from 3 to 2. All these data indicate that C_F is an important parameter in determining the fluctuation level.

Table 4. Influence of fold composition on NO_2 concentration

Location across the mixing layer, η_T	Mean NO_2 concentration	
	Run No. 1 $C_F = 2.0$	Run No. 5 $C_F = 3.0$
−2.00	1.00	1.00
−1.50	1.10	1.10
−1.07	1.21	1.21
−0.674	1.21	1.20
−0.300	1.13	1.07
0.104	0.947	0.884
0.508	0.712	0.663
0.889	0.575	0.574
1.17	0.398	0.398
1.57	0.178	0.179

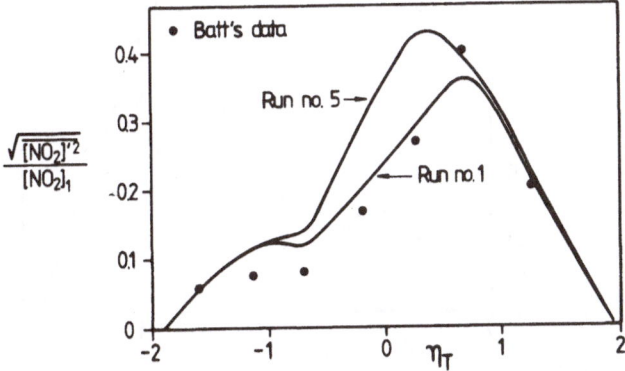

Fig. 15. Influence of fold composition on the concentration fluctuation intensity.

6.5.4. The Influence of the Stretching Rate

The mean NO_2 concentrations computed from Run No. 1 and Run No. 6 are provided in Table 5, in which C_S indicates the levels of stretching [equation (2.9)]. Inspection of this table reveals that the maximum difference between the results computed from two runs is now around 11% at $\eta_T \approx 0.508$. The peak value is lower in Run No. 6 and equal to 1.19 at $\eta_T \approx -1.07$.

More evident influence is apparent in the concentration fluctuation intensities than in the other cases, presented in Figure 16. The secondary hump (at $\eta_T \approx -1.1$) is magnified in the case of low stretching rate, for

Table 5. Influence of stretching rate on NO_2 concentration

Location across the mixing layer, η_T	Mean NO_2 concentration	
	Run No. 1 $C_s = 1.0$	Run No. 6 $C_s = 0.3$
-2.00	1.00	1.00
-1.50	1.10	1.09
-1.07	1.21	1.19
-0.674	1.21	1.14
-0.300	1.13	1.03
0.104	0.947	0.840
0.508	0.712	0.621
0.889	0.575	0.558
1.17	0.398	0.391
1.57	0.178	0.177

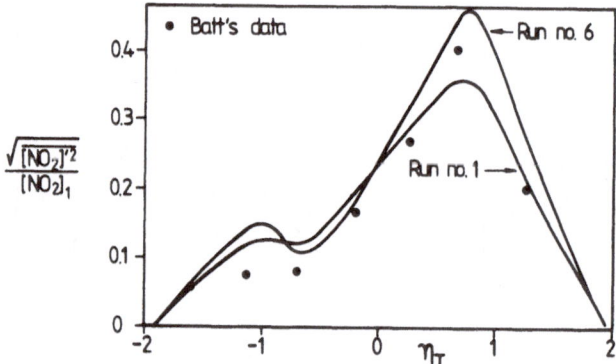

Fig. 16. Influence of stretching rate on the concentration fluctuation intensity.

which the maximum fluctuation increases up to 0.46. The predictions in Run No. 6 are in good quantitative agreement with the measurements over the range $-0.5 \leqslant \eta_T \leqslant 0.6$. However, the fluctuation level is overpredicted in the region near the fast stream boundary for both runs.

7. DISCUSSION

7.1. The Population Distribution of Folds

The major significance of the results obtained from the demographic analysis is that the cross-stream variation in the population distribution of folds has been calculated from the transport equations, as shown in Figures

2–4. The basic factors that contribute to the variation are the turbulent-diffusion effect and the distribution of fold-formation rate.

The turbulent-diffusion effect is determined by the turbulent viscosity and trubulent Schmidt number, which are taken from the existing turbulence model as the necessary input to ESCIMO theory.

On the other hand, the hypothesis about the distribution of fold-formation rate is a new and unique feature of the present approach. As far as the authors know, no direct experimental measurements are available for verification. The hypothesis can only be tested by its influence on the mean properties and their RMS quantities.

Nevertheless, there is a common trend in Figure 8, where $\hat{A}_{av}$ is always larger on the slower-moving side of the layer for all three runs. The explanation lies in the velocity distribution across the mixing layer, since higher velocity results in shorter time (and hence smaller age) for a fold born in the upstream position to travel a certain distance.

7.2. The Mean NO$_2$ Concentration Profile

It has been pointed out that the quantitative agreement between the present predictions and measurements are reasonably good in Figure 10. However, some degree of discrepancy does exist near the hump of the profile and may be explained as follows.

Apart from the uncertainties in experimental measurements noted by Batt,[3] the effects of nonequilibrium chemistry and the turbulence intermittency factor may cause the deviation of predictions from measured values.

Alber and Batt[12] have estimated the time scales involved in the flow process. The chemical reaction is so rapid that only the turbulent dissipation time scale associated with the smallest scale eddies is of the same order of magnitude. Hence, the nonequilibrium phenomena (i.e., finite-rate chemistry) may only appear in the finest scale of turbulence structure, and Batt[3] has evaluated that it would reduce the reaction rate by approximately 5%.

The turbulence intermittency factor, defined as the time portion when the flow exhibits turbulent characteristics, has not been accounted for in the present model. According to the experimental data of Sunyach and Mathieu[13] and Batt,[3] it varies from zero near both edges of the mixing layer to unity in the midregion. Strictly speaking, the results calculated from ESCIMO theory refer to the fully turbulent part only and should therefore be weighted by it. Consequently, the contribution of the irrotational potential flow will be more dominant near the boundaries of the mixing layer and the results will be given by

$$\overline{[NO_2]} = (1 . I)[NO_2]_p + I \cdot \overline{[NO_2]} \tag{7.1}$$

where $\overline{[NO_2]}$ is the mean concentration of NO_2 including the intermittency factor I, $[NO_2]_p$ is the concentration of NO_2 in the potential flow, and $[\overline{NO_2}]$ is the mean concentration obtained from the ESCIMO theory excluding intermittency. Therefore, equation (7.1) will be expected to produce different results from those formerly reported.

7.3. The Fluctuation Intensities of NO_2 Concentration

There are more influential factors in the determination of fluctuation intensities of NO_2 concentration than the mean quantities described above. All parameters employed in ESCIMO theory are significant in this respect.

The following remarks can be made from the sensitivity analysis performed in Section 6.5:

1. The fluctuation intensities at a fixed position increase when the population of young folds becomes more prominent. This is reflected in the results of Figure 13, where the population of the youngest folds is highest for Run No. 1 near $\eta_T \approx 0.8$. The profile within a fold will be flattened by molecular diffusion and stretching as the fold ages.
2. The fluctuation intensities magnify as the fold size at birth increases, because the diffusion distance is longer.
3. The fluctuation level heightens as the properties of the re-engulfed part in the newly formed fold further deviate from the local mean value (e.g., with a larger C_F), since the profile in the fold is initially steeper and it remains so if other parameters are the same.
4. Slower stretching rate yields larger fluctuations over most parts of the mixing layer, except in the region $-0.8 \leqslant \eta_T \leqslant 0$ where the mean concentration is high. The reason is that the thickness of the fold does not decrease so rapidly and the nonuniformity of the properties within each fold can last longer. Since the relation between the temperature and NO_2 concentration is neither linear nor monotonic, it is possible that the fluctuation intensity of concentration becomes lower even when that of the temperature is higher under the low-stretching condition.

Therefore satisfactory agreement with experimental data can be realized if a set of optimized parameters is employed in the computation.

7.4. The Temperature Fluctuation Intensities

The maximum fluctuation intensities may have been overpredicted by about 60%, as shown in Figure 11. This is in remarkable contrast with the fluctuation intensities of NO_2 concentration (from Run No. 1) that are underpredicted by only 10%.

Batt[3] suggested that the relative low fluctuation level was caused by random motion or three-dimensional effects in the experiment and less influenced by large-scale two-dimensional coherent structures that have been observed in other shear-layer investigations.

8. CONCLUSIONS

The application of ESCIMO theory to the turbulent reacting plane mixing layer has been described in this chapter. The population distribution of folds has been presented in two-dimensional flows for the first time and the influence of various hypotheses about the distribution of fold-formation rate tested.

The mean turbulent properties and their fluctuation intensities have been calculated and compared with the experimental data of Batt.[3] The sensitivity analysis about various presumptions made in the present theory has been carried out. The grid independence test has already been performed as a prerequisite.

The quantitative agreement between measured and predicted results is acceptable although some discrepancies exist in the fluctuation intensities of temperature.

ACKNOWLEDGMENTS. The authors thank Dr. Andrzej Przekwas for his contribution to the early stages of the present work, the Safety in Mines Research Establishment, Health and Safety Executive of Great Britain for financial support of the project, and Mrs. F. Oliver for typing the manuscript.

REFERENCES

1. D. B. Spalding, The influences of laminar transport and chemical kinetics on the time-mean reaction rate in a turbulent flame, in: *Seventeenth (Int.) Symp. on Combust.*, pp. 431–440, The Combustion Institute, Pittsburgh, PA (1979).
2. A. S. C. Ma, M. A. Noseir, and D. B. Spalding, An application of the ESCIMO theory of turbulent combustion. Paper presented at the AIAA Aerospace Science Meeting, January 1980, Pasadena, California (1980).
3. R. G. Batt, Turbulent mixing of passive and chemically reacting species in a low-speed shear layer, *J. Fluid Mech.* **82**, Part 1, 53–95 (1977).
4. S. V. Patankar and D. B. Spalding, *Heat and Mass Transfer in Boundary Layers*, 2nd edn., Intertext Books, London (1970).
5. B. E. Launder and D. B. Spalding, The numerical computation of turbulent flows, *Comput. Methods Appl. Mech. Eng.* **3**, 269–289 (1974).

6. A. S. C. Ma, D. B. Spalding, and R. L. T. Sun, Application of "ESCIMO" to the turbulent hydrogen–air diffusion flame, in: *Nineteenth (Int.) Symp. on Combust.*, The Combustion Institute, Pittsburgh, PA (1982) (to appear).

7. D. B. Spalding, Approximate solutions of transient and two-dimensional flame phenomena: Constant-enthalpy flames, *Proc. R. Soc. London, Ser. A* **245**, 352–372 (1958).

8. R. L. T. Sun, Application of "ESCIMO" Theory to Turbulent Diffusion Flames, Ph.D thesis, University of London (1982).

9. P. P. Wegener, Supersonic nozzle flow with a reacting gas mixture, *Phys. Fluids* **2**, 264–275 (1959).

10. D. B. Spalding, *GENMIX. A General Computer Program for Two-Dimensional Parabolic Phenomena*, Pergamon Press, Oxford (1978).

11. H. E. Fiedler, On turbulent structure and mixing mechanism in free turbulent·shear flows, in: *Turbulent Mixing in Nonreactive and Reactive Flows*, A Project SQUID Workshop, pp. 381–410, Plenum Press, New York (1975).

12. I. E. Alber and R. G. Batt, Diffusion-limited chemical reactions in a turbulent shear layer, *AIAA J.* **14**, 70–76 (1975).

13. M. Sunyach and J. Mathieu, Mixing zone of a two-dimensional jet, *Int. J. Heat Mass Transfer* **12**, 1679–1697 (1969).

19

Growth of a Diffusion Flame in the Field of a Vortex

F. E. Marble

ABSTRACT. A simple diffusion flame with fast chemical kinetics is initiated along the horizontal axis between a fuel occupying the upper half-plane and an oxidizer below. Simultaneously a vortex of circulation Γ is established at the origin. As time progresses the flame is extended and "wound up" by the vortex flow field and the viscous core of the vortex spreads, converting the motion in the core to a solid-body rotation.

The kinematics of the flame extension and distortion is described and the effect of the local-flow field upon local-flame structure is analyzed in detail. It is shown that the combustion field consists of a totally reacted core region, whose radius is time dependent, and an external flame region consisting of a pair of spiral arms extending off at large radii toward their original positions on the horizontal axis.

The growth of the reacted core, and the reactant-consumption rate augmentation by the vortex field in both core and outer-flame regions were determined for values of the Reynolds number ($\Gamma/2\pi\nu$) between 1 and 10^3 and for a wide range of Schmidt numbers (ν/D) covering both gas and liquid reactions.

For large values of Reynolds number the radius r_* of the reactant grows much more rapidly than the viscous core so that only the nearly inviscid portion of the flow is involved. The more accurate condition for this behavior is that $R(\mathrm{Sc})^{1/2} > 50$ and, under this restriction, the similarity rule for the core radius growth is shown to be

$$\frac{r_*}{(\Gamma^{2/3} D^{1/3} t)^{1/2}} \equiv 0.5092 + O(D/\Gamma)^{1/2}$$

In this case also the reactant consumption rate becomes independent of time and, for the complete diffusion flame in the vortex field, the augmentation of reactant-consumption rate due to the vortex field

F. E. MARBLE • California Institute of Technology, 205–45, Pasadena, CA 91125, U.S.A.

satisfies

$$\frac{\text{Augmented consumption rate}}{\Gamma^{2/3} D^{1/3}} \equiv 1.2327 - 1.4527(D/\Gamma)^{1/6} + O(D/\Gamma)^{1/2}$$

Both of these similarity rules are, as is appropriate for high Reynolds number, independent of kinematic viscosity.

1. INTRODUCTION

The development of the theory of flames in nonuniform flow fields has been limited by the complications introduced into the flame structure itself as well as the rather strong effect of sensible heat release on the flow field. The diffusion flame has a relative simplicity that allows more latitude in such an analysis particularly when the chemistry is fast and the diffusion is the only rate-controlling process. In this way it has been possible to examine diffusion flames under conditions of straining in the plane of the flame,[1] a condition that occurs in the immediate vicinity of a stagnation point,[2] and describes locally some aspects of turbulent diffusion flames.[3]

The analysis of combustion-flow fields in the large is a much more challenging problem and the difficulty is amplified when the flow field and flame structure are nonstationary. The distortion of a diffusion flame by a viscous vortex flow, the problem considered here, has features that make it particularly significant. First, the vortex field provides a strong, variable, time-dependent strain rate that can still be handled analytically. Second, the example provides an excellent opportunity to study the behavior of diffusion flames in a viscous-dominated region of the flow, a situation that may be of significance in turbulent diffusion flames. Finally, this particular chemically reacting flow field occurs naturally in the large-scale structures of turbulent diffusion flames and in combustion-instability problems.

2. KINEMATICS OF THE DIFFUSION FLAME–VORTEX INTERACTION

Suppose that initially the upper and lower half-planes consist of a fuel and oxidizer respectively, that the chemical kinetics are sufficiently fast, that the reaction is diffusion controlled, and that the stoichiometry of the reaction is unity, i.e., the mass consumption of the two reactants is equal. In the absence of any fluid motion, this situation would lead to a diffusion flame located on the horizontal axis. The effective thickness of the diffusion zones would behave as $(Dt)^{1/2}$ while the reactant consumption rate would vary as approximately $(D/t)^{1/2}$, where D is the common binary diffusion coefficient of either reactant in the product. In the following analysis we shall neglect the variation of density resulting from the release of chemical energy by the reaction; it will be indicated later that this has little effect upon the results.

Now if, at $t = 0$, we situate a point vortex at the origin, the induced-flow field is given by the tangential velocity[4]

$$v_\theta \equiv r \frac{\partial \theta}{\partial t} = -\frac{\Gamma}{2\pi r} (1 - e^{-\eta^2}) \tag{1}$$

where r and θ are the polar coordinates of an arbitrary point in the plane, Γ is the circulation of the vortex for large radii, ν is the kinematic viscosity, and

$$\eta = r/(4\nu t)^{1/2}$$

is the dimensionless radius. The distortion of the flame front by this flow follows by noting that any element, originally on the horizontal axis, is transported at constant radius through an angle θ obtained by integrating equation (1) at constant radius. Then if the initial position of the flame element lies to the right of the vortex position, its angle at time t is

$$\theta(r, t) = -\frac{\Gamma}{2\pi r^2} \int_0^t [1 - \exp(-r^2/4\nu t_1)] dt_1 \tag{2}$$

which has a logarithmic singularity at $r = 0$, $t > 0$. A second branch of the curve arises from points initially to the left of the vortex and moves to an angle

$$\theta(r, t) = \pi - \frac{\Gamma}{2\pi r^2} \int_0^t [1 - \exp(-\eta_1^2)] dt_1 \tag{3}$$

The dimensionless group $\Gamma/2\pi\nu$ will be denoted the Reynolds number, R, of the vortex. A flame contour is shown in Figure 1 for a Reynolds number equal to 40 and at a time after starting the motion such that $(4\nu t)^{1/2} = 0.1$ cm. For larger radii the displacement is essentially that corresponding to an inviscid vortex flow while for $r < 1$ mm the viscous effects are dominant and the displacements, at this particular time, correspond to those of a solid-body rotation. At earlier time, before the viscous core has spread to

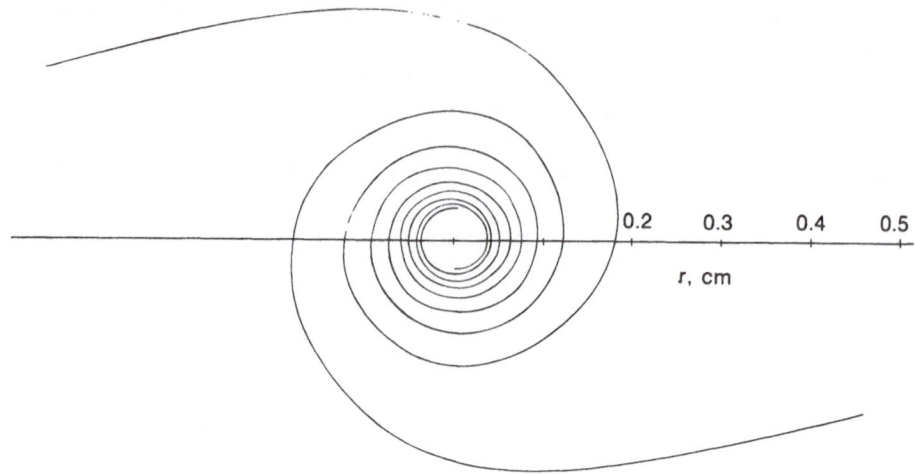

Fig. 1. Core region of distorted flame sheet, $\Gamma/2\pi\nu = 40$, $(4\nu t)^{1/2} = 0.1$ cm.

$r \sim 0.1$ cm, a considerable degree of winding has taken place and, as the growing viscous core dominates the flow to that radius, the spiral-like distortion is "frozen" into the fluid. The contours for $r < 0.1$ cm are essentially frozen into the flow at this time. Note that the distances between neighboring flame surfaces are of the order of tenths of a millimeter. The Reynolds number determines the degree of winding that has taken place before the viscous core has spread and stopped further distortion. For values of R equal to approximately 1000, the structure is extremely fine; for R about 5, the structure is rather coarse.

To analyze the effect of the flow field on the local-flame structure it is necessary to determine the history of the flow that is observed by a flame element in its immediate neighborhood. Figure 2 shows a flame element at three successive times; the termini of the element lie on fixed circles and the increase in element length and its rotation with respect to a local tangent are established by the velocity difference between the two radii.

If this flame element lies outside the region of viscous influence, then at a radius r the gas will move distance $\Gamma t/2\pi r$ in time t, and the length ds of a flame element initially of length dr and situated on the horizontal axis is given by

$$ds = \left[1 + \left(\frac{\Gamma t}{\pi r^2}\right)^2\right]^{1/2} dr \equiv (1 + 4\zeta^2)^{1/2} dr$$

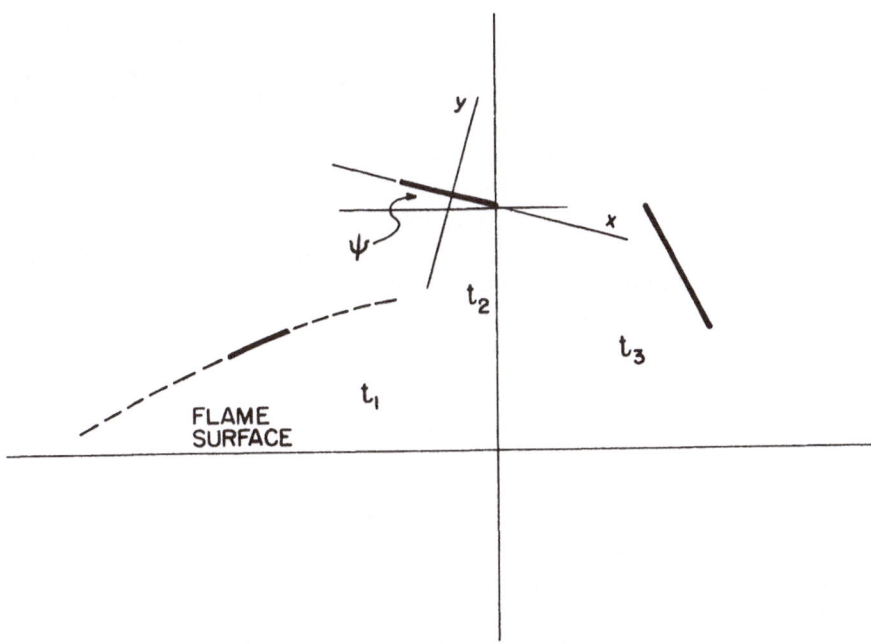

Fig. 2. Schematic drawing of distortion and rotation of a flame element at successive times.

for $\zeta \gg 1$, the ratio of flame-element length to its initial length is

$$ds/dr \cong 2\zeta \equiv \Gamma t/\pi r^2 \tag{4}$$

This quantity will be convenient in our calculations because it clarifies the effect of Reynolds number upon an otherwise ideal-fluid result. For example, the angle θ though which a point has moved becomes, from equation (1),

$$\theta = -\int_0^\zeta (1 - e^{-R/4\zeta})d\zeta \tag{1'}$$

The angle ψ that the flame element makes with respect to the local tangential direction follows directly by differentiating equation (2) or (3) with respect to r at constant t. Thus $r\partial\theta/\partial r = \cot \psi$ and hence

$$\tan \psi = 1/[2\zeta(1 - e^{-R/4\zeta})] \tag{5}$$

At a fixed radius, as time increases, the flame angle ψ decreases toward $\tan^{-1}(2/R)$, demonstrating the very tight winding that occurs for large

Reynolds numbers before viscosity prevents further changes. For large radii at fixed time, ζ becomes small and $\psi \cong \pi/2$, indicating that the interface is only slightly displaced from its initial position on the horizontal axis.

To examine the flow field in the neighborhood of the flame element, construct a coordinate system from the midpoint of the element with x-axis in the plane of the element and y-axis normal to it. To determine the u and v velocity components in this reference frame we must use the velocity field given by equation (1) and account for the orientation and the angular velocity $\partial \psi/\partial t$ of the flame element. After some calculation it follows that these velocity components are given by

$$u = r \frac{\partial}{\partial r} (v_\theta/r)[(\tfrac{1}{2}\sin 2\psi)x + (\cos 2\psi)y] \tag{6}$$

$$v = -r \frac{\partial}{\partial r} (v_\theta/r)(\tfrac{1}{2}\sin 2\psi)y \tag{7}$$

where x and y represent distances in the x–y coordinate system appropriately erected at the point r, θ in the polar-coordinate system. Equations (6) and (7) represent, in fact, the first terms in the Taylor expansion of the velocity field about r, θ and hence x and y must be small in comparison with distances over which significant changes take place in the velocity field. Simultaneously we may deduce that the angular velocity of the element is

$$\partial \psi/\partial t = -r \frac{\partial}{\partial r} (v_\theta/r)\sin^2 \psi \tag{8}$$

which follows from the condition that the velocity component v, normal to the element, must vanish on the element.

Now the local-flow field, given by equations (6) and (7), may be decomposed into two components: a normal straining motion

$$u_1 = r \frac{\partial}{\partial r} (v_\theta/r)(\tfrac{1}{2}\sin 2\psi)x \tag{9}$$

$$v_1 = -r \frac{\partial}{\partial r} (v_\theta/r)(\tfrac{1}{2}\sin 2\psi)y \tag{10}$$

and a shearing motion

$$u_2 = r \frac{\partial}{\partial r} (v_\theta/r)(\cos 2\psi)y \tag{11}$$

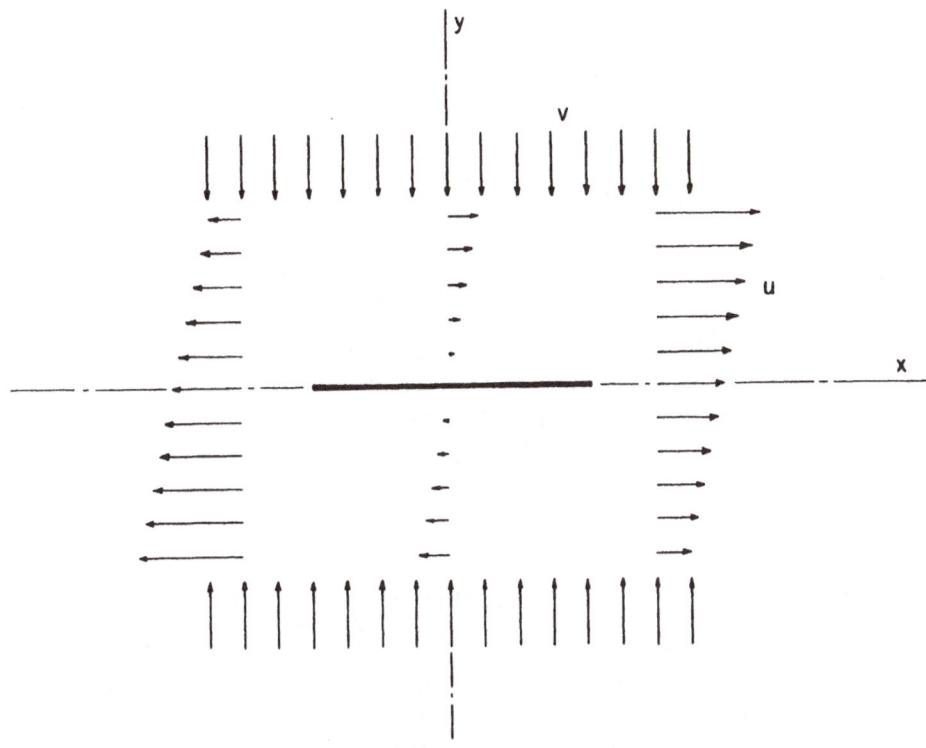

Fig. 3. Velocity field observed from a flame element in its immediate neighborhood, showing combined normal strain rate and shear rate fields.

This field is indicated schematically in Figure 3. It is particularly to be noted that when the motion corresponds to a solid-body rotation, namely $v_\theta \sim r$, both the straining and shearing motions vanish. This situation corresponds to our earlier observation that the pattern of the interface becomes fixed in the fluid when the viscous core has grown to the radius at which the observation is made. As the motion progresses the flame element tends to align itself with the circular arc passing through its midpoint, $\psi \to 0$, and the straining motion becomes small. The shearing motion will prove to be unimportant in the flame analysis.

The actual normal strain rate components $\varepsilon_{xx} = -\varepsilon_{yy} \equiv \varepsilon$ follow from equation (9) or (10) as

$$\varepsilon = \frac{\partial u}{\partial x} = -\frac{\partial v}{\partial y} = r \frac{\partial}{\partial r} (\sin \psi \cos \psi) \qquad (12)$$

Then recalling the expression for $\tan \psi$ in equation (5), it follows that

$$\sin \psi \cos \psi = 1/\{2\zeta(1 - e^{-R/4\zeta}) - [2\zeta(1 - e^{-R/4\zeta})]^{-1}\} \qquad (13)$$

Likewise from equation (1) we find that

$$r\frac{\partial}{\partial r}(v_\theta/r) = 2\zeta/t[1-(1+R/4\zeta)e^{-R/4\zeta}] \tag{14}$$

and hence the strain rate may be written explicitly as

$$\varepsilon t = \frac{4\zeta^2(1-e^{-R/4\zeta})[1-(1+R/4\zeta)e^{-R/4\zeta}]}{1+4\zeta^2(1-e^{-R/4\zeta})^2} \tag{15}$$

For values of radius and time such that both $\zeta \ll 1$ and $\zeta/R \ll 1$ we find that $\varepsilon t \cong (\Gamma t/\pi r^2)^2$ is independent of viscosity and hence correponds to the strain rate encountered in a potential vortex. For long values of time or smaller values of r, i.e., $\zeta \gg 1$ and $\zeta/R \gg 1$, the viscous core dominates the field,

$$\varepsilon t \cong (1/2\zeta)(R/2)^3/[1+(R/2)^2]$$

and is strongly Reynolds-number dependent. In particular, for long time values, the straining motion in the core vanishes. In the range of time and radius values such that $\zeta \gg 1$ but $\zeta/R \ll 1$ we find that $\varepsilon t \cong 1$. The range of validity is larger for larger Reynolds number, as indicated in Figure 4.

The fact that, even for very large Reynolds numbers, $\varepsilon t \leqslant 1$ has considerable significance in the time-dependent diffusion-flame theory, which will be developed subsequently.

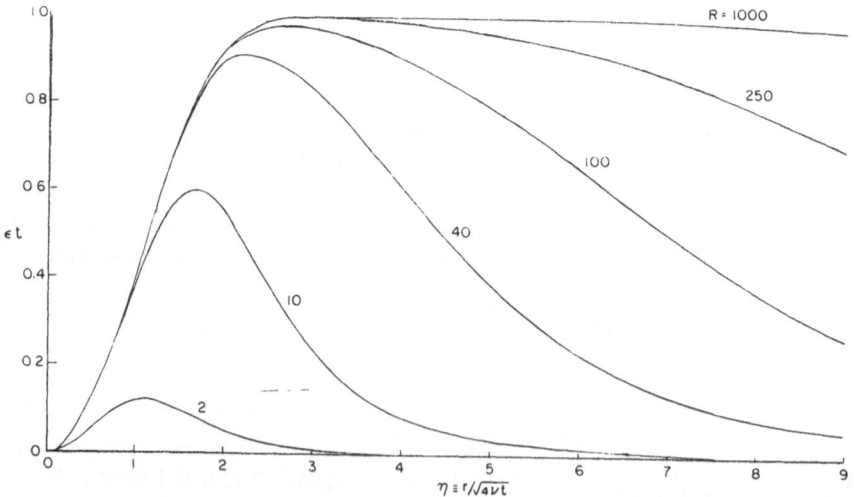

Fig. 4. Dependence of normal strain rate–time product on dimensionless radius for various values of the Reynolds number $R \equiv \Gamma/2\pi\nu$.

3. FLAME STRUCTURE IN THE VORTEX FIELD

We assume, principally to assure clarity in the final results, that the changes in density of the reactants do not seriously alter the results that we are seeking. Consider first the time-dependent diffusion flame, with infinitely fast kinetics, which would occur if the vortex were absent. Then at $t = 0$ the mass fraction of fuel in the upper half-plane is $K_1(y, 0) = 1$ and similarly the mass fraction of oxidizer in the lower half-plane is $K_2(y, 0) = 1$. When the stoichiometry of the reaction is such that equal masses of each reactant are consumed at the infinitely thin reaction zone located on the horizontal axis, the reactant mass fractions satisfy identical diffusion equations which, together with the requirement that both reactants vanish on the axis, determine the reactant distributions as

$$K_1 = \mathrm{erf}(y/\sqrt{4Dt}) \qquad \text{and} \qquad K_2 = -\mathrm{erf}(y/\sqrt{4\nu t})$$

The volume consumption rate of fuel K_1 is then, in units of volume per unit time and per unit area of the interface,

$$D(\partial K_1/\partial y)(0, t) = \sqrt{D/\pi t} \tag{16}$$

and the consumption rate of oxidizer is identical. The reactant consumption decreases with time as $(D/t)^{1/2}$ while the thickness of the diffusion zone is growing at a rate of approximately $(Dt)^{1/2}$.

Now we employ the same ideas in performing a local analysis of the flame structure when the vortex field is present. The only difference is that the local-flame structure in the presence of the vortex must account for the velocity field in the neighborhood of the flame. These velocities are given by equations (6) and (7) in the x–y coordinate systems situated at the center of the element (Figures 2 and 3). This notation is consistent with that used for the plane diffusion flame because, in that instance, all flame elements were permanently located on the horizontal axis. We make the approximation here that the normal strain rate ε and the shearing strain rate $\partial u/\partial y$ are nearly independent of x, essentially a boundary-layer approximation, and thus the flow is locally a one-dimensional, time-dependent field. Proceeding in a manner analogous to that employed above, the conservation equation for the fuel mass fraction is

$$\frac{\partial K_1}{\partial t} - \varepsilon y \frac{\partial K_1}{\partial y} = D \frac{\partial^2 K_1}{\partial y^2} \tag{17}$$

where the relation $v = -\varepsilon y$ has been used in accordance with equations (7) and (12). If we recall that we are analyzing the flame structure in the neighborhood of a point r, θ on the flame interface and that x and y represent infinitesimal displacements from this point, then it is clear that, so far as equation (17) is concerned, ε is a function of time only, $\varepsilon = \varepsilon(t)$. Furthermore it appears that, because of the independence of x locally, the shearing-strain-rate term plays no part in the local flame structure.

The statement of the problem is simplified if we note that a material point, originally at $y = \bar{y}$, is transported with a velocity $v = -\varepsilon t$ and hence arrives at the location $y = \bar{y} \exp(-\int_0^t \varepsilon dt_1)$ after a time t has elapsed. Writing equation (17) in terms of this "material" coordinate $\bar{y}$ then yields

$$\frac{\partial K_1}{\partial t} = \exp\left(2 \int_0^t \varepsilon dt_1\right)\left(D \frac{\partial^2 K_1}{\partial \bar{y}^2}\right) \tag{18}$$

which may be viewed as a diffusion equation for a medium with time-dependent properties. Then by transforming our time scale according to

$$d\tau = \exp\left(2 \int_0^t \varepsilon dt_1\right) dt \tag{19}$$

equation (18) takes on the familiar form

$$\frac{\partial K_1}{\partial \tau} = D \frac{\partial^2 K_1}{\partial \bar{y}^2} \tag{20}$$

which may be treated in the manner of the plane flame analyzed previously. An identical argument holds for the mass fraction K_2 at negative values of y. The distribution of fuel mass fraction is thus

$$K_1 = \mathrm{erf}[\bar{y}/(4D\tau)^{1/2}] \tag{21}$$

and the fuel consumption rate is

$$D \frac{\partial K_1}{\partial y}(0, t) = (D/\pi t)^{1/2} \exp\left(\int_0^t \varepsilon dt_1\right) \tag{22}$$

The strain rate ε, in equation (15), depends upon the variables r and t but the integral in equation (22) is carried out at a fixed radius in the vortex structure. Thus we may write, using equation (15),

$$\int_0^t \varepsilon dt_1 = \int_0^\zeta f(\zeta; R) d\zeta \tag{23}$$

where

$$f(\zeta; R) = \varepsilon t/\zeta \tag{24}$$

Now the transformed time scale τ, which also appears in the expression for the fuel-consumption rate, is obtained by integrating equation (19). If we call

$$\exp\left\{\int_0^t \varepsilon dt_1\right\} = \exp\left\{\int_0^\zeta f(\zeta_1; R)d\zeta_1\right\} \equiv F(\zeta; R) \qquad (25)$$

then the transformed time may be written in the form

$$\Gamma\tau/2\pi r^2 = \int_0^\zeta F^2(\zeta_1; R)d\zeta_1 \equiv G^2(\zeta; R) \qquad (26)$$

and, from equation (22), the volume rate of fuel consumption per unit area of flame surface is

$$[(D\Gamma)^{1/2}/(\sqrt{2}\,\pi r)]F(\zeta; R)/G(\zeta; R) \qquad (27)$$

This expression is valid at all points on the flame in the vortex field under the assumptions that the variations of the strain components along the flame are small over distances comparable with the flame-zone thickness and that adjacent flame surfaces do not interfere.

Eventually, however, adjacent flame sheets will interact in the sense that the diffusion-zone thickness about the flames becomes comparable with the spacing of the wound-up flame surfaces. As the overlap of the diffusion zones depletes the concentration of reactant between adjacent flame surfaces, the reactant consumption rate will decrease and the two flame sheets eventually annihilate each other. The nature of the end process may be analyzed by considering a strip of reactant (say, the fuel component), which is being strained along its length at a rate ε. If the two edges of the fuel strip are at the points $y = \pm y_0$ at $t = 0$, the time the diffusion flames start, then the straining motion itself moves these flame fronts to the instantaneous position $\pm y_0 \exp\{-\int_0^t \varepsilon dt_1\}$, hence decreasing the width of the fuel strip. Then the transformations used for the single strained flame are appropriate; the mass-fixed coordinate

$$\bar{y} = y \exp\left(-\int_0^t \varepsilon dt_1\right) \qquad (28)$$

and the effective time

$$\tau = \int_0^t \left[\exp\left(2\int_0^{t_2} \varepsilon dt_1\right)\right] dt_2 \qquad (29)$$

reduce the problem to a conventional diffusion equation. It is not difficult to show that, within the fuel strip, the mass fraction of fuel is

$$K_1 = -1 - \text{erf}[(\bar{y} - y_0)/(4D\tau)^{1/2}] + \text{erf}[(\bar{y} + y_0)/(4D\tau)^{1/2}] \tag{30}$$

At either flame sheet $K_1 = 0$, a condition which determines the positions $\bar{y}_f$ and $-\bar{y}_f$ of the flame sheets. The volume consumption rate of fuel by each of the flame sheets is

$$\sqrt{\frac{D}{\pi\tau}} \left\{ \exp\left[-\left(\frac{\bar{y}_f - y_0}{(4D\tau)^{1/2}}\right)^2\right] - \exp\left[-\left(\frac{\bar{y}_f + y_0}{(4D\tau)^{1/2}}\right)^2\right] \right\} \tag{31}$$

which vanishes as the flames collapse, i.e., as $\bar{y}_f \to 0$. It is a straightforward calculation to show that the two flames behave independently, except for a brief final period $\cong 1/\varepsilon$ when the combustion of the fuel strip is abruptly completed.

For our purpose here, this abrupt termination of the burning process may be used to advantage. We shall adopt the approximation that the reactant consumption of the flames proceeds as if they were independent until the fuel supply is exhausted, at which time the flames bounding the fuel (or oxidizer) strip terminate instantaneously.

4. INCREASE OF REACTANT CONSUMPTION CAUSED BY THE VORTEX FIELD

The length ds of the flame element that extends from a radius r to $r + dr$ is

$$ds = \left[1 + \left(r\frac{\partial\theta}{\partial r}\right)^2\right]^{1/2} dr$$

$$= \{1 + [2\zeta(1 - e^{-R/4\zeta})]^2\}^{1/2} \tag{32}$$

where equation (2) has been used to obtain the final expression. The rate at which fuel is being consumed by the two flames that bound this annulus is then

$$(1/\pi)(2D\Gamma)^{1/2}(F/G)\{1 + [2\zeta(1 - e^{-R/4\zeta})]^2\}^{1/2} dr/r \tag{33}$$

Corresponding to our earlier discussion we shall make the quite accurate approximation that the reactant consumption rate is described by equation

(33) until the fuel in the annulus (r, dr) has been consumed and, at that point, the local consumption rate vanishes. In other words, the condition for burnout of the annulus (r, dr) is that the integral of the volume rate of fuel consumption, equation (33), over time shall equal the volume of fuel within the annulus, namely $\pi r dr$. Thus the time t_* at which the fuel and oxidizer in the annulus (r, dr) are consumed is determined by the integral equation

$$(1/\pi)(2D\Gamma)^{1/2} dr/r \int_0^{t_*} (F/G)\{1 + [2\zeta(1 - e^{-R/4\zeta})]^2\}^{1/2} dt = \pi r dr$$

and because r is constant during this integration we may write

$$(2/\pi)(2D/\Gamma)^{1/2} \int_0^{\zeta_*} (F/G)\{1 + [2\zeta(1 - e^{-R/4\zeta})]^2\}^{1/2} d\zeta = 1 \tag{34}$$

The value of ζ_* determined by this integral may be interpreted as giving the radius r_* at which the reactants have been consumed at any time t. In fact it is clear now that the structure of the vortex combustion is divided into two distinct parts: (i) an inner core in which the combustion has been completed and (ii) an outer flame structure in which the two arms of the flame sheet may be described locally by the nonsteady diffusion-flame theory of Section 3. The circle which separates these two zones has a radius $r_* = (\Gamma/2\pi\zeta_*)^{1/2} t^{1/2}$.

It is a consequence of the fact that the radius of the core increases as $t^{1/2}$ that the rate at which combustion products are accumulated within the core is a constant, independent of time. This follows because the volume of fuel consumed is just half of the volume swept out by the increasing core radius and consequently the volume rate of fuel consumption in the core is $\frac{1}{2}d(\pi r_*^2)/dt = 4\Gamma/\zeta_*$. Referring to equation (34) it is clear that ζ_* depends upon the Reynolds number $\Gamma/2\pi\nu$ and the ratio D/Γ; thus both the radius of the reacted core and the reactant consumption rate will vary with these quantities. It is clear that this dependence can equally well be expressed in terms of the Reynolds number and ν/D, the Schmidt number.

It is a matter of some calculation to determine ζ_* as a function of Reynolds number and Schmidt number, but another representation proves to be more convenient and physically more penetrating. When the Reynolds number is large and $\zeta_* = \Gamma t/2\pi r_*^2$ is large, it transpires that (i) the winding of the flame sheet is quite tight and (ii) the core of combustion products grows more rapidly than the viscous core. In this limit the behavior of the field is essentially inviscid and equation (34) may be examined by elementary asymptotic methods. Consider $R/4\zeta_* \gg 1$ and $\zeta_* \gg 1$; then it is easily shown that $F(\zeta; R) \rightarrow (1 + 4\zeta^2)^{1/2}$ and

$G^2(\zeta; R) \to \zeta(1 + 4\zeta^2/3)$, with the consequence that the integral relation (34) becomes

$$(2/\pi)(2D/\Gamma)^{1/2} \int_0^{\zeta_*} \frac{1 + 4\zeta^2}{(1 + 4\zeta^2/3)^{1/2}} \zeta^{-1/2} d\zeta \cong 1 \qquad (35)$$

The integral behaves as $\zeta^{-1/2}$ for small ζ and as $2\sqrt{3}\,\zeta^{1/2}$ for large ζ and it is convenient to rewrite equation (35) as

$$(2/\pi)(2D/\Gamma)^{1/2} \left[\frac{4}{\sqrt{3}} \zeta_*^{3/2} + I(\infty) - 2\sqrt{3} \int_{\zeta_*}^{\infty} \left(\frac{1 + 1/4\zeta^2}{(1 + 3/4\zeta^2)^{1/2}} - 1 \right) \zeta^{-1/2} d\zeta \right] \qquad (36)$$

where

$$I(\infty) = 2\sqrt{3} \int_0^{\infty} \left[\frac{1 + 4\zeta^2}{(1 + 4\zeta^2/3)^{1/2}} - 1 \right] \zeta^{1/2} d\zeta = 0.1994 \qquad (37)$$

The dominant term $\zeta_*^{3/2}$ yields the lowest-order approximation,

$$\zeta_*(D/\Gamma)^{1/3} \cong (\sqrt{3/2}\,\pi/e)^{2/3}$$

The dimensionless group

$$\zeta_*(D/\Gamma)^{1/3} = \Gamma^{2/3} D^{1/3} t/2\pi r_*^2 \qquad (38)$$

is of a form common to diffusion problems except that it contains an unusual composite transport property $\Gamma^{2/3} D^{1/3}$. If the core were growing by a molecular-diffusion process we should expect its area to increase as Dt but the growth rate for our problem, $\sim (\Gamma/D)^{2/3} Dt$, is enormously faster. Physically this result arises because the full diffusion potential occurs across each narrow strip of the strongly wound-up pattern.

Based upon this observation it is appropriate to use the dependent variable $\Gamma^{2/3} D^{1/3} t/2\pi r_*^2$ in preference to ζ_* because it is, in the lowest approximation, independent of both Reynolds and Schmidt numbers. Then a conventional asymptotic evaluation of equation (35) for large ζ_* may be used to express

$$2\pi r_*^2/\Gamma^{2/3} D^{1/3} t \cong (8\sqrt{2/3}/\pi)^{2/3} + O(D/\Gamma)^{1/2} \qquad (39)$$

Returning now to the complete formulation given by equation (34), it is, not difficult to evaluate ζ_* numerically and thereby determine $2\pi r_*^2/\Gamma^{2/3} D^{1/3} t$ as a function of Reynolds number and Schmidt number. Figure 5 shows such a result for $\nu/D = 1.0$ and shows the essential

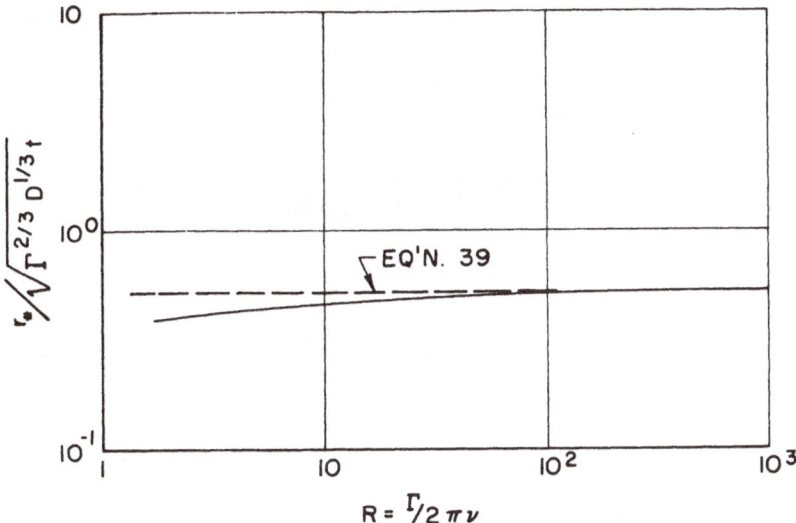

Fig. 5. Dimensionless radius of reacted core $r_*/(\Gamma^{2/3}D^{1/3}t)^{1/2}$ as a function of Reynolds number $R \equiv \Gamma/2\pi\nu$ for Schmidt number $(\nu/D) = 1.0$.

independence of Reynolds number, certainly for $R > 50$. The asymptotic value calculated from equation (39) is also shown and clearly provides a very satisfactory representation over most of the range.

To account for the reactant-consumption rate we again proceed from equation (32), which gives the fuel-consumption rate for an isolated flame element situated in an annulus of radius r and thickness dr. It is convenient to compute the augmentation of reactant-consumption rate caused by the vortex-flow field and consequently we subtract the consumption rate that would be experienced by an element of length dr if it remained on the horizontal axis in its original position, in the absence of the vortex. This reactant consumption rate for the undistorted flame element is given by equation (16).

The net increase in fuel-consumption rate within dr, accounting for both branches of the flame, is

$$(1/\pi)(2D\Gamma)^{1/2}(F/G)\{1 + [2\zeta(1 - e^{-R/4\zeta})]^2\}^{1/2}\frac{dr}{r} - 2(D/\pi t)^{1/2}dr \quad (40)$$

The net increase in fuel-consumption rate, associated with the vortex-induced motion, over the entire field follows by integrating this expression from r_* to ∞, i.e., from $\zeta = 0$ to $\zeta = \zeta_*$. The result is

$$(1/\pi)(D\Gamma/2)^{1/2}\int_0^{\zeta_*} [(F/G)\{1 + [2\zeta(1 - e^{-R/4\zeta})]^2\}^{1/2} - \zeta^{-1/2}]d\zeta/\zeta \quad (41)$$

Again, as in our evaluation of the core radius r_*, it is instructive to find the ideal limit of large Reynolds number where the conditions $\zeta_* \gg 1$ and $R/4\zeta_* \gg 1$ are satisfied. Utilizing some previous results, the expression for the augmented consumption rate is

$$\cong (1/\pi)(D\Gamma/2)^{1/2} \int_0^{\zeta_*} \left(\frac{1+4\zeta^2}{(1+4\zeta^2/3)^{1/2}} - 1 \right) \zeta^{-3/2} d\zeta \qquad (42)$$

For asymptotic evaluation rewrite this as

$$(1/\pi)(D\Gamma/2)^{1/2} \left\{ 4\sqrt{3}\zeta_*^{1/2} + J(\infty) - \int_{\zeta_*}^{\infty} \left[\frac{1+4\zeta^2}{(1+4\zeta^2/3)^{1/2}} - 2\sqrt{3}\zeta - 1 \right] \zeta^{-3/2} d\zeta \right\}$$

$$(43)$$

where

$$J(\infty) = \int_0^{\infty} \left[\frac{1+4\zeta^2}{(1+4\zeta^2/3)^{1/2}} - 2\sqrt{3}\zeta - 1 \right] \zeta^{-3/2} d\zeta$$

$$= -6.4543 \qquad (44)$$

The integral in equation (43) may be evaluated asymptotically for large ζ_* and, regrouping the variables in the manner found convenient for core-radius evaluation, the augmented consumption rate is

$$\Gamma^{2/3} D^{1/3} \left\{ \frac{4}{\pi} \sqrt{\frac{3}{2}} \left(\frac{\Gamma^{2/3} D^{1/3} t}{2\pi r_*^2} \right)^{1/2} + \frac{1}{\sqrt{2\pi}} J(\infty) \left(\frac{D}{\Gamma} \right)^{1/6} + O\left(\frac{D}{\Gamma} \right)^{1/2} \right\} \qquad (45)$$

Because the second term in this expression is decreased only by the factor $(D/\Gamma)^{1/6} = (2\pi)^{-1/6}(Sc \cdot R)^{-1/6}$, it shall be included although higher-order terms may be neglected.

We conclude from equation (45) that the augmented reactant consumption rate scales as $\Gamma^{2/3} D^{1/3}$, the composite transport property we found earlier. Inserting known values of $\Gamma^{2/3} D^{1/3} t/2\pi r_*^2$ and $J(\infty)$, equation (45) for the augmented consumption rate becomes

$$\cong \Gamma^{2/3} D^{1/3} [1.2327 - 1.0694(Sc \cdot R)^{-1/6} + O(D/\Gamma)^{1/2}] \qquad (46)$$

Note particularly that writing $D/\Gamma \sim (Sc \cdot R)^{-1}$ has a certain physical convenience but does not imply the involvement of viscosity.

Now it is a simple matter to compute from the complete representation, equation (41), the quantity $\Gamma^{-2/3} D^{-1/3}$ (augmented consumption rate) as suggested by our considerations for high Reynolds numbers. Such

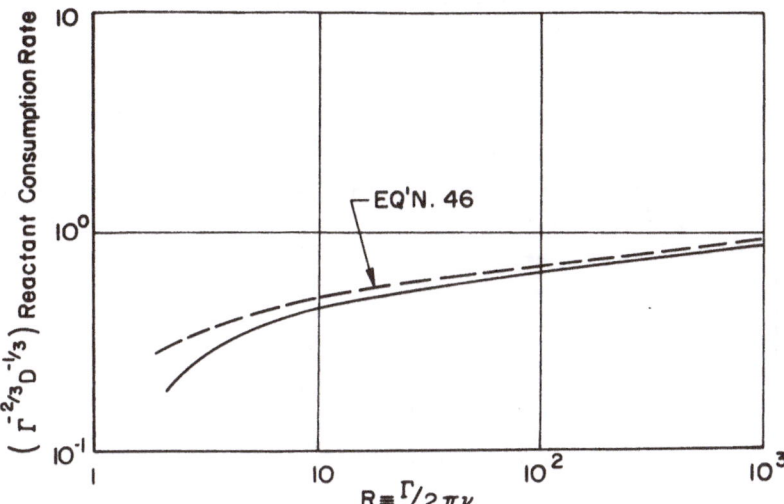

Fig. 6. Reactant consumption rate as a function of Reynolds number $R \equiv \Gamma/2\pi\nu$ for Schmidt number $(\nu/D) = 1.0$.

calculations are given in Figure 6 as a function of Reynolds number for $\nu/D = 1.0$. The scaling with the composite transport coefficient $\Gamma^{2/3}D^{1/3}$ is generally quite adequate. The asymptotic formula given by equation (46) is surprisingly accurate and the significance of the $R^{-1/6}$ term is particularly to be noted. Again, it is to be emphasized that the augmentation in combustion rate caused by the vortex field is independent of time, in spite of the fact that the velocity field, flame shape, and core size are time-dependent.

Because these results are significant not only in the gas-phase combustion process but also for reactions in liquids, it is of interest to explore the influence of the ratio of kinematic viscosity to diffusion coefficients, or Schmidt number. This ratio is the significant quantity that changes drastically between the two cases. The Schmidt number is also of interest because it is frequently convenient to explore combustion processes in gases by examining their analogues in liquids. Figure 7 shows the effect of ν/D upon the reactant consumption rates; the Reynolds number is constant at a value $R = 1000$ and the ν/D ratio ranges from 10^{-2} to 10^{3}. The computed values of the dimensionless core size are essentially independent of the Schmidt number and are indistinguishable from the asymptotic formula, equation (39), over the same range of Schmidt numbers.

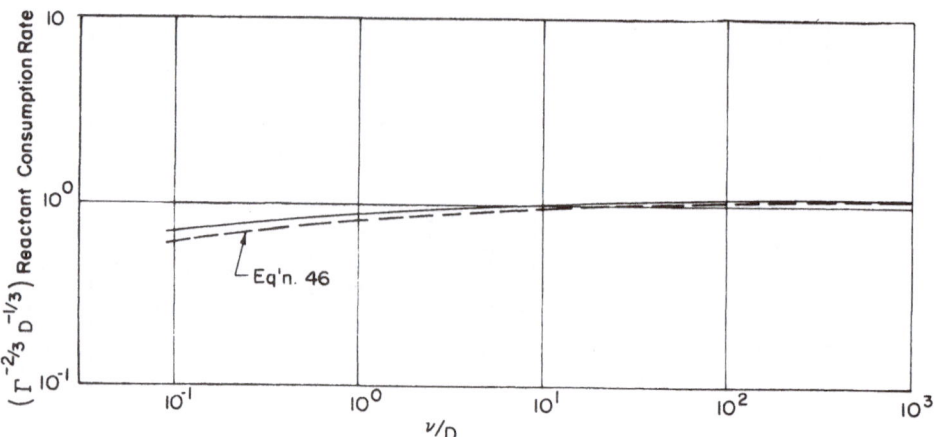

Fig. 7. Reactant consumption rates as function of $\nu/D = Sc$, Reynolds number $\Gamma/2\pi\nu = 10^3$.

5. CONCLUDING REMARKS

The foregoing results have a relatively simple asymptotic behavior because most of the diffusion and combustion process lies sufficiently outside the viscous core that the velocity field may be considered a potential vortex. Because of this it is possible to arrive at these asymptotic forms in a manner physically more instructive than formal analysis.

Recall that for a potential vortex the length of flame ds that spans a radial element dr is given by equation (4), so that when the flame makes a complete circuit of the origin, $ds = 2\pi r$ and the distance between these neighboring flame sheets is $dr = (2\pi r^2/\Gamma t)(2\pi r)$. Now the thickness of the diffusion layer accompanying this flame sheet grows as $(D/\varepsilon)^{1/2} \cong (Dt)^{1/2}$, since in this range of Reynolds number (cf Figure 4) $\varepsilon t \cong 1$. The flame sheets interact and consume the intervening reactant when $(Dt)^{1/2} \cong dr$. From the above relations $(Dt)^{1/2} \sim r_*^3/\Gamma t$, from which it follows that

$$r_*/(\Gamma^{2/3} D^{1/3} t)^{1/2} = \text{const}$$

Therefore the proposed similarity-law radius of the reaction product core, equation (39), follows immediately from physical reasoning.

Likewise the reactant consumption rate may be deduced by noting that the rate of accumulation of products in the core is proportional to

$$\frac{d}{dt}(\pi r_*^2) \sim \frac{d}{dt}(\Gamma^{2/3} D^{1/3} t) \sim \Gamma^{2/3} D^{1/3}$$

Thus the similarity law for the time-invariant reactant-consumption rate, equation (46), follows from the same approximate physical picture. The lower limit of validity for these relations must depend upon $\Gamma^{2/3} D^{1/3}$ and hence it is appropriate to state it in terms of $R \cdot Sc^{1/2}$. From the calculations it appears that the combustion similarity laws are accurately valid for $R \cdot Sc^{-1/2} \geqslant 50$.

A final word may be added regarding the effects of a change in fluid species, especially between gases and liquids. If we have species designated 1 and 2, the similarity law suggests that

$$\frac{r_{*2}}{r_{*1}} = \frac{\Gamma_2^{1/3} D_2^{1/6}}{\Gamma_1^{1/3} D_1^{1/3}} = \left(\frac{\nu_2}{\nu_1}\right)^{1/2} \left(\frac{R_2}{R_1}\right)^{1/3} \left(\frac{Sc_1}{Sc_2}\right)^{1/6}$$

and that the reactant consumption rates scale as

$$\frac{\Gamma_2^{2/3} D_2^{1/3}}{\Gamma_1^{2/3} D_1^{1/3}} = \left(\frac{\nu_2}{\nu_1}\right) \left(\frac{R_2}{R_1}\right)^{2/3} \left(\frac{Sc_1}{Sc_2}\right)^{1/3}$$

Comparing, for example, a liquid (2) with a gas (1) for the same Reynolds number, the reacted-core radii would be in the ratio of about 1/10 while the production of reactant products is in the ratio of about 1/100.

ACKNOWLEDGMENT. This work was supported, in part, by grant AFOSR-80-0265 and by NASA grant NAG 3-70.

REFERENCES

1. G. F. Carrier, F. E. Fendell, and F. E. Marble, Effect of strain rate on diffusion flames, *SIAM J. Appl. Math.* **28**, 453 (1975).
2. D. B. Spalding, Theory of mixing and chemical reaction in the opposed-jet diffusion flame, *ARS J.* **31**, 763 (1961).
3. F. E. Marble and J. E. Broadwell, The Coherent Flame Model for Turbulent Chemical Reactions, Project Squid Technical Report TRW-9-PU (January, 1977).
4. Sir H. Lamb, *Hydrodynamics*, sixth edn., p. 592, Cambridge University Press (1932).
5. H. C. Hottel, Burning in laminar and turbulent fuel jets, in: *Fourth Symp. (Int.) on Combust.*, p. 97, William & Wilkins Co., Baltimore (1952).

20

Crocco Variables for Diffusion Flames

F. A. Williams

ABSTRACT. The use of variables, such as mixture fractions, in the theory of diffusion-flame structure is examined in relationship to early work of Crocco and from the viewpoint of providing simplifications in the analysis of problems of current interest. It is found that these variables offer advantages in a number of respects.

1. INTRODUCTION

The equations of chemically reacting flows that describe diffusion-flame structure are complicated enough to motivate searches for simplifications. In general, restrictive assumptions of one kind or another must be introduced to obtain useful simplifications. Assumptions that aid in the analysis of diffusion flames are considered here, and variables that facilitate flame studies are explored. It is indicated that by the introduction of suitable variables many flame properties can be clarified without reference to specific flow configurations. This is helpful in particular for investigations of turbulent diffusion flames.

2. ASSUMPTIONS AND CONSERVATION EQUATIONS

Equations for conservation of mass, momentum, energy, and chemical species describe diffusion flames. Here attention is focused mainly on the last two of these. Much of the complexity of diffusion-flame analysis arises from the large number of species-conservation equations involved and from the fact that in general the dependent variables in these equations are not related simply either to one another or to an energy variable. Thus, a

---

F. A. WILLIAMS • Department of Mechanical and Aerospace Engineering, Princeton University, Princeton, NJ 08544, U.S.A.

large number of partial differential equations must be considered. A major reason for this is the differing diffusion coefficients for different species. However, it is often found experimentally that, with good accuracy, relationships exist among the species concentrations of greatest interest that would be anticipated if the effects of different diffusion were negligible.[1] This suggests that diffusional separation can be neglected or treated as a perturbation. Therefore in a first approximation, binary diffusion coefficients of all pairs of species may be set equal, and the Soret and Dufour effects, pressure-gradient diffusion, and body-force diffusion may be neglected. The fundamental reason for this simplification is that, in gases, the diffusion coefficients of different species seldom differ greatly. The conservation equation for chemical species i becomes

$$\rho \frac{\partial Y_i}{\partial t} + \rho \mathbf{v} . \nabla Y_i = \nabla . (\rho D \nabla Y_i) + w_i \tag{1}$$

where ρ is density, $\mathbf{v}$ velocity, Y_i mass fraction, D diffusion coefficient, and w_i the mass rate of production of species i by chemical reactions.

Most diffusion flames of interest involve flows at low Mach number. Although radiant energy emission is often important in applications, the amount of radiant energy emitted typically is a small fraction of the thermal enthalpy locally and may therefore be treated as a perturbation. With these approximations and with the work done by body forces neglected, it may be shown that if the Lewis number is unity then the total enthalpy per unit mass h obeys the equation

$$\rho \frac{\partial h}{\partial t} + \rho \mathbf{v} . \nabla h = \nabla . (\rho D \nabla h) + \frac{\partial p}{\partial t} \tag{2}$$

where p denotes pressure. Typically the most severe of these assumptions is that the Lewis number is unity (species and heat diffusivities are equal), and the accuracy of this approximation is seldom worse than that of equal diffusivities for the various species. A chemical source term does not appear in equation (2) because the reactions do not change the sum of the thermal and chemical enthalpy. For boundary-layer flows the restriction to low Mach number may be replaced by an assumption that the Prandtl number is unity when the kinetic energy is included in h.

3. THE MIXTURE FRACTION AND RELATED VARIABLES

Diffusion flames are characterized by separate fuel and oxidizer streams that are brought together in the course of the combustion process. It is possible to define a mixture fraction Z in terms of local, instantaneous element concentrations that is zero in the oxidizer stream and unity in the fuel stream. Under the assumptions that have been introduced, Z obeys the equation

$$\rho \frac{\partial Z}{\partial t} + \rho \mathbf{v} \cdot \nabla Z = \nabla \cdot (\rho D \nabla Z) \tag{3}$$

which lacks a chemical source term because chemical elements are neither created nor destroyed by chemical reactions. There is a long history of use of a variable like Z for diffusion flames.[2] At least implicit in the work of Burke and Schumann[3] and related to the formulation of Shvab and Zel'dovich,[4] a variable of this type was introduced for turbulent diffusion flames by Hawthorne, Weddell, and Hottel.[5] Worthwhile reviews are given by Bilger.[2,6]

In the flame-sheet approximation, or more generally under conditions of chemical equilibrium, h and all Y_i are expressible in terms of Z, provided the assumptions underlying equations (1) and (2) are satisfied, that $\partial p / \partial t$ is negligible in equation (2) (e.g., for steady flows), and that the resulting expressions are consistent with initial and boundary conditions.[6] Problems for which this is true are significantly simpler than others in that only one partial differential equation occurs for describing the enthalpy and all species concentrations. Simplifications of this type have been developed on the basis of the flame-sheet approximation, for example by Shvab and Zel'dovich,[4] and on the basis of element conservation equations, for example by Libby.[7]

From equations (2) and (3) it is seen that, when it exists, the relationship obtained between h and Z must be linear. It may also be shown that a linear relationship between h and the streamwise component of velocity u is consistent with the differential equation for momentum conservation in a restricted class of problems that involves boundary-layer flows with a Prandtl number of unity (steady, $\nabla p = 0$, and constant wall enthalpy). This last result was derived by Crocco in an early paper[8] and is called Crocco's energy integral.[9] It has been used in combustion problems, e.g., by Emmons[10] in studying the burning of a semi-infinite flat plate of a liquid or solid fuel in an oxidizing gas stream. By analogy, linear relationships among u and other flow variables, e.g., between u and Z, have been called Crocco relations.

4. APPROACH TO ANALYSIS OF STRUCTURES OF THIN FLAME SHEETS

Recently there has been interest in ascertaining structures of diffusion flames that do not maintain chemical equilibrium. Without equilibrium, algebraic relationships among the Y_i and Z no longer exist in general, and complications associated with an increased number of dependent variables arise again. Often the most important departures from equilibrium occur only over a small range of values of the variable Z, usually near the hottest part of the flame. In this range the time-derivative and convective terms in equations (1), (2), and (3) are sometimes small, and equation (1) then expresses a diffusive–reactive balance in a first approximation, thereby enabling stretched coordinates to be employed to simplify the structure analysis. However, this is not always true; the analysis of reaction zones in which convective or time-derivative terms are important is more difficult because essentially the full form of equation (1) must then be retained.

To circumvent this difficulty in a counterflow diffusion-flame problem Liñán[11] introduced a convection-free form of the conservation equations for energy and species. Development of the convection-free form involves a transformation of the space variable x to, say, $z = f(x)$, such that

$$d(\rho D d Y_i / dx) / dx - \rho u d Y_i / dx = d^2 Y_i / dz^2$$

Thus, the new variable z combines effects of both convection and diffusion into an apparently purely diffusive term. This transformation facilitates the reaction-zone analysis; in particular, it enabled Liñán to conveniently employ asymptotic methods for analyzing influences of finite-rate chemistry on the flame structure, ignition, and extinction for a one-step, irreversible reaction with a large ratio of the overall activation temperature to the adiabatic flame temperature.

To find a convection-free form of the conservation equations has often been a central step in diffusion-flame analysis. For example, a form of this type was identified in a study of the diffusion flame in a stagnation-point flow of oxidizer directed onto the flat surface of a burning liquid or solid fuel[12]; the identification of this form enabled Liñán's results[11] to be applied directly to the new problem. In the past, convection-free forms had to be developed separately for each problem and each geometrical configuration investigated, and there was no guarantee in advance that such a form could be found. Recently Peters[13] discovered that by transforming the equations by introducing the mixture fraction Z as one of the independent variables, a convection-free form was automatically obtained upon introduction of a stretching in Z. This provides a systematic approach to the identification of a convection-free form and demonstrates the existence of such a form in a general class of problems. It also enables various aspects of flame-structure analyses to be performed without reference to specific flow configurations.

To illustrate the transformation to the mixture-fraction variable, consider an orthogonal coordinate system with Z being one of the coordinates and with x and y, distances along surfaces of constant Z, being the other two. In this system let u and v be velocity components in the x and y directions. It may then be shown from equations (1) and (3) that

$$\rho \frac{\partial Y_i}{\partial t} + \rho \mathbf{v}_\perp . \nabla_\perp Y_i = w_i + \rho D |\nabla Z|^2 \frac{\partial^2 Y_i}{\partial Z^2}$$

$$+ \nabla_\perp . (\rho D \nabla_\perp Y_i) - \rho D \nabla_\perp (\ln |\nabla Z|) . \nabla_\perp Y_i \qquad (4)$$

where the x and y components of the transverse velocity vector $\mathbf{v}_\perp$ are u and v, and where $\nabla_\perp$ is the two-dimensional gradient involving x and y.

In equation (4), the convective term that remains involves velocities parallel to surfaces of constant Z, i.e., transverse to flame sheets. Convection in the normal direction is now implicit in the term involving $\partial^2 Y_i / \partial Z^2$, which superficially exhibits a purely diffusive character. In analyzing structures of thin flame sheets a stretching of Z about a fixed value Z_0, usually the stoichiometric value, is often useful. Unless the time-derivative or transverse-derivative terms in equation (4) become large, the stretching in Z approximately reduces equation (4) to the simplified equation

$$\partial^2 Y_i / \partial Z^2 = w_i / (\rho D |\nabla Z|^2) \qquad (5)$$

which exhibits an apparent diffusive–reactive balance in the new coordinate. Equation (5) has proven to be quite useful for analyses of flame structure, e.g., by providing a criterion for local extinction of laminar diffusion flamelets in turbulent diffusion flames.[14]

5. ASPECTS OF THE USE OF THE MIXTURE FRACTION AS AN INDEPENDENT VARIABLE

The introduction of Z as an independent variable can be found in earlier literature. Notably, Bilger[2] writes ordinary differential equations for Y_i as a function of Z. Peters[13] refers to the use of Z as a Crocco transformation. Crocco[15] introduced the streamwise velocity component u as an independent variable in boundary-layer theory and wrote a differential equation for the viscous stress τ as a function of u and of the streamwise coordinate x. This change of variables is called the Crocco transformation (see, for example, Lagerstrom's work,[9] pp. 215–216). Since u was originally a dependent variable, the use of the dependent variable Z as an independent variable may by analogy be called a type of Crocco transformation.

There are differences between the use of u in boundary layers and of

Z in diffusion flames. For example, the reduction in the order of the system of equations helped to make the Crocco transformation useful in boundary-layer calculations.[16] This reduction does not occur in equation (4).

The introduction of u or Z as an independent variable introduces difficulties in seeking complete solutions for the flow field if the variable does not vary monotonically with the space coordinate that it replaces; for most of the simpler problems in laminar diffusion flames Z possesses the requisite monotonicity. The appearance of ∇Z in equation (5) implies that the equation can be employed to find complete solutions only if this gradient can be expressed in terms of Z and other variables retained; this can be done in the simpler applications.[11-13]

In more complicated situations, for example in turbulent flows, ∇Z cannot be calculated directly; nevertheless, equation (5) remains useful if only partial information is sought. After the stretching is performed in producing equation (5) for analyzing thin flame sheets, $|\nabla Z|$ remains constant across the sheet in the first approximation, and equation (5) may be studied with suitable matching conditions by treating this factor as constant. The flame structure is thereby obtained as a function of the local, instantaneous value of $|\nabla Z|$. In turbulent flows[2,6,14] this quantity is related to the so-called scalar dissipation, mean values of which have been subjected to turbulence modeling. Thus, the results of the partial analysis provide an understanding of flamelet structure and of its relationship to turbulence properties that can be employed in investigating the behavior of turbulent diffusion flames. The monotonicity restrictions are absent when only partial information is sought; for example, at zero-gradient points (where $\nabla Z = 0$), Damköhler's second similarity group (the ratio of a diffusion time to a reaction time) is infinite, and equation (5) predicts the occurrence of chemical equilibrium. Study of phenomena that may cause equation (5) to cease to be a good approximation to equation (4) could help to define limitations to the use of the transformation and to indicate new influences on flame-sheet structure.

6. CONCLUSIONS

Studies of diffusion flames through the introduction of formulations equivalent to that indicated by equation (5) have been helpful in revealing important aspects of flame structure. Under flow conditions that do not permit the attainment of complete chemical equilibrium, the large number of species concentration fields described by differential equations requires the introduction of simplifying approximations in investigations designed to improve understanding. These approximations concern chemical mechanisms, diffusion, radiation, etc. Equation (5) can be used as a starting point to begin to investigate influences of removing these approximations (e.g., accounting for radiant heat loss,[17] for effects of dissociation,[18] or for preferential diffusion of different species), either by solving various forms of the equation itself or by first augmenting the equation through perturbation methods. Results of the investigations should further clarify diffusion-flame structure. The mixture fraction as an independent variable may be expected to play an important role in these continuing diffusion-flame studies.

REFERENCES

1. R. W. Bilger, *Combust. Flame* **30**, 277 (1977).
2. R. W. Bilger, *Combust. Sci. Tech.* **13**, 155 (1976).
3. S. P. Burke and T. E. W. Schumann, *Ind. Eng. Chem.* **20**, 998 (1928).
4. Y. B. Zel'dovich, *Zh. Tekh. Fiz.* **19**, 1199 (1949).
5. W. R. Hawthorne, D. S. Weddell, and H. C. Hottel, in: *3rd Symposium on Combustion, Flame and Explosion Phenomena*, pp. 266–288, Williams and Wilkins, New York (1949).
6. R. W. Bilger, Turbulent flows with nonpremixed reactants, in: *Turbulent Reacting Flows* (P. A. Libby and F. A. Williams, eds.), pp. 65–113, Springer-Verlag, Berlin (1980).
7. P. A. Libby, *ARS J.* **32**, 388 (1962).
8. L. Crocco, *Aerotecnica* **12**, 181 (1932).
9. P. A. Lagerstrom, Laminar flow theory, in: *Theory of Laminar Flows* (F. K. Moore, ed.), Vol. IV of *High Speed Aerodynamics and Jet Propulsion*, pp. 230–232, Princeton University Press, Princeton, New Jersey (1964).
10. H. W. Emmons, *Z. Angew. Math. Mech.* **36**, 60 (1956).
11. A. Liñán, *Acta Astronautica* **1**, 1007 (1979).
12. L. Krishnamurthy, F. A. Williams, and K. Seshadri, *Combust. Flame* **26**, 363 (1976).
13. N. Peters, *Combust. Sci. Tech.* **30**, 1 (1983).
14. N. Peters and F. A. Williams, *AIAA J.* **21**, 423 (1983).
15. L. Crocco, *Atti di Guidonia* **17**, 118 (1939).
16. S.-H. Lam, personal communication.
17. S. H. Sohrab, A. Liñán, and F. A. Williams, *Combust. Sci. Tech.* **27**, 143 (1982).
18. N. Peters and F. A. Williams, Effects of chemical equilibrium on the structure and extinction of laminar diffusion flames, to appear.

Index